ELEMENTS OF PHOTOGRAMMETRY
With Applications in GIS

THIRD EDITION

Paul R. Wolf, Ph.D.

Professor Emeritus of Civil and Environmental Engineering
The University of Wisconsin, Madison

Bon A. Dewitt, Ph.D.

Associate Professor of Geomatics
The University of Florida, Gainesville

Boston   Burr Ridge, IL   Dubuque, IA   Madison, WI   New York
San Francisco   St. Louis   Bangkok   Bogotá   Caracas
Lisbon   London   Madrid   Mexico City   Milan   New Delhi
Seoul   Singapore   Sydney   Taipei   Toronto

McGraw-Hill

A Division of The McGraw·Hill Companies

ELEMENTS OF PHOTOGRAMMETRY
WITH APPLICATIONS IN GIS

This book is printed on acid-free paper.

1 2 3 4 5 6 7 8 9 0 DOC/DOC 0 9 8 7 6 5 4 3 2 1 0

ISBN 0-07-292454-3

Publisher: *Thomas Casson*
Executive editor: *Eric M. Munson*
Editorial coordinator: *Michael Jones*
Senior marketing manager: *John T. Wannemacher*
Project manager: *Laura Ward Majersky*
Production supervisor: *Debra R. Sylvester*
Freelance design coordinator: *Pam Verros*
Cover illustration: *Paul D. Turnbaugh*
Photo research coordinator: *Sharon Miller*
Senior supplement coordinator: *Carol Loreth*
Compositor: *Interactive Composition Corporation*
Typeface: *10/12 Times Roman*
Printer: *R. R. Donnelley & Sons Company*

Library of Congress Cataloging-in-Publication Data

Wolf, Paul R.
    Elements of photogrammetry : with applications in GIS. — 3rd ed. / Paul R. Wolf, Bon A. Dewitt.
        p.   cm.
    Includes index.
    ISBN 0-07-292454-3
    1. Photogrammetry.   2. Aerial photogrammetry.   3. Image processing—Digital techniques.   4. Remote sensing.   5. Geographic information systems.   I. Dewitt, Bon A.   II. Title.
    TR693.W63   2000
    526.9'82—dc21   99-057138

http://www.mhhe.com

CONTENTS

Differences / 8-9 Measurement of Parallax Differences /
8-10 Computing Flying Height and Air Base / 8-11 Error
Evaluation / References / Problems

PREFACE

Since the first edition of this book was published more than 25 years ago, many new technological developments have occurred in the areas of optics, electronics, computers, and satellite and space technology. Advancements in these and other areas have brought about significant changes in the instruments and procedures now used in the practice of photogrammetry. The changes have been particularly prevalent since the second edition of this book was published. Therefore, the foremost objective in preparing this third edition has been to update the book and incorporate these new developments. Another important objective, however, has been to include the many valuable improvements that have been suggested to the authors by professors, students, and practicing photogrammetrists who have used the first and second editions.

During the past decade, geographic information systems (GISs) have come of age, a development that has placed great new demands upon photogrammetry. GISs are computer-based systems that enable the storing, integrating, manipulating, analyzing, and displaying of virtually any type of spatially related information about our environment. These systems are now being used globally at all levels of government, and by businesses, private industry, and public utilities to assist in planning, design, maintenance, management, and decision making. Photogrammetry has arguably emerged as the most important discipline employed in the collection of spatially related information for use in GIS databases. A variety of photogrammetric products are generated for use in GISs; they include unrectified and rectified aerial photos, satellite images, maps of various types; digital orthophotos, digital elevation models, and other types of digital files containing topographic information. Because of the importance of photogrammetry in geographic information systems, a substantial amount of new GIS coverage is included in this edition. In fact, two new chapters have been added that are completely devoted to this topic: Chap. 20, Introduction to Geographic Information Systems, and Chap. 21, Photogrammetric Applications in GIS. Other coverage describing the role of photogrammetry in GISs is also given throughout the book. Even the title of this third edition has been expanded to identify the important relationship between photogrammetry and geographic information systems.

Coverage in this third edition has been updated to include many other major advances in modern photogrammetry. Substantial new material has been added, for example, on the subject of digital photogrammetry, also called *softcopy photogramme-try*. To support this area of study, digital images are discussed, digital cameras are described, and instruments and techniques for scanning hard-copy photos to create digital images are covered. Furthermore, a separate section on softcopy plotters, Part IV, has been added to Chap. 12, Stereoscopic Plotting Instruments. Finally, two completely new chapters on digital photogrammetry have been added; Chap. 14, Fundamental Principles of Digital Image Processing, and Chap. 15, Principles of Softcopy Photogrammetry.

Because photogrammetry plays such a vital role in generating spatially related information, it is imperative that students of this discipline have a firm grasp of the fundamentals of object space coordinate systems. To meet this goal, a new chapter has been added on this topic. It includes some elementary fundamentals of geodesy, describes various object space reference coordinate systems that are used in photogrammetry, and discusses map projections and datums. Also a new appendix has been added that describes and illustrates procedures for converting coordinates of points from one object space coordinate system to another.

Many other important additions have been made to this third edition. The chapter on control surveys for photogrammetry has been expanded to include new sections that describe the global positioning system (GPS) and the use of this equipment in performing ground control surveys. New coverage has also been added on procedures for incorporating a GPS receiver with an aerial camera during photographic missions. The subject of analytical photogrammetry has been significantly updated. This includes expanded coverage on the simultaneous bundle adjustment method, the addition of a new section on computing the bundle adjustment with airborne GPS control, and a new section that describes procedures for performing aerotriangulation by using satellite images. Other new material presented in this third edition includes discussions of instruments and computational techniques for producing digital orthophotos and digital elevation models, and a description of the new airborne laser mapping system.

The authors have endeavored to maintain the level and scope of coverage, and style of presentation, in this third edition consistent with those of the first two editions. The intent continues to be the production of a book that will be suitable for an introductory course in photogrammetry at the college level. This includes courses given at universities, junior colleges, technology colleges, schools of applied arts, and military schools. The book has also been written with the intent of making it suitable for self-study and reference. Thus it should be a valuable addition to the libraries of practicing photogrammetrists, geographic information specialists, cartographers, engineers, foresters, geologists, geographers, landscape architects, and others who use maps and photographs in their work.

As in the first two editions, the material in this third edition has been written using elementary terms as much as possible, and extensive use has been made of illustrations and diagrams. Many example problems have been included to clarify computational procedures, and the International System (*Système International,* or SI) units is used predominately throughout the book. In order that the book be suitable for students of varying levels of mathematical competence, the body of the text has been largely presented so that knowledge of only algebra, geometry, and trigonometry is necessary. More challenging mathematical developments have been placed in the appendixes, where they are available either for reference or for students having the necessary mathematics background.

The order of presentation has been arranged so that early chapters establish fundamental principles, while later chapters discuss more specialized aspects of photogrammetry. In general, each chapter is arranged so that the more important material is presented first, a feature that makes it convenient to cover only the first parts of certain chapters if course-time limitations will not allow covering the entire book. While the material has been kept as elementary as possible, the book's depth and breadth of coverage are sufficient to make it suitable for a second, more advanced course in photogrammetry. The coverage of the later chapters and the appendixes would be particularly useful in an advanced course.

A selected list of references is given at the end of each chapter, and from the materials cited, students can expand their knowledge on subjects of particular interest. Although only a limited number of key references are given, many also have extensive bibliographies which will lead students to numerous additional articles. Computer programs for solving selected photogrammetry problems are available to students and can be downloaded from the following Web site address: http://www.surv.ufl.edu/wolfdewitt. A solutions manual which contains answers to all after-chapter problems is also available to instructors from the publisher.

We wish to again acknowledge our sincere appreciation to the many individuals who contributed to the first and second editions of this book. And we gratefully acknowledge the people, agencies, and firms who contributed to this third edition. In particular, we express our thanks to Professors Paul Hopkins of the State University of New York at Syracuse, Robert Schultz of Oregon State University, and Steven Johnson of Purdue University for reviewing substantial portions of the manuscript; to Gary Johanning of Geodetic Services, Inc., who provided materials for Chap. 19 on close-range photogrammetry; to Professors Alan Vonderohe and Bernard Nieman of the University of Wisconsin–Madison for supplying materials used in the chapters on geographic information systems; to Diann Danielsen and Tim Confare of the Dane County Land Information Office, to Michelle Richardson of the Dane County Land Conservation Department, to Michael Bohn and Tom Simmons of the Wisconsin Department of Natural Resources, to Jay Arnold and Jeremy Conner of 3001, Inc., to Cyril Fernandez, Kay Brothers and Jason Mace of the Southern Nevada Water Authority, and to Tim Wolf (the author's son) of the Las Vegas Valley Water District, for materials used to illustrate applications of photogrammetry in geographic information systems; to Ken Worden and Steve Root of the Wisconsin Department of Transportation; to Fred Halfen of Ayres Associates; and to the numerous undergraduate and graduate students at both the University of Wisconsin–Madison and the University of Florida who provided suggestions, feedback, and assistance.

We also express our appreciation to the instrument manufacturers, government agencies, and private photogrammetric firms who supplied figures and other information for this book. We especially wish to thank Aerial Cartographics of America, Inc., and U.S. Imaging, Inc., for providing special aerial photography for the book.

Finally we wish to thank our wives, Lynn Wolf and Monica Dewitt, and our children for their patience, understanding, endurance, support, and love.

Paul R. Wolf
Bon A. Dewitt

CHAPTER
1

INTRODUCTION

1-1 DEFINITION OF PHOTOGRAMMETRY

Photogrammetry has been defined by the American Society for Photogrammetry and Remote Sensing as the art, science, and technology of obtaining reliable information about physical objects and the environment through processes of recording, measuring, and interpreting photographic images and patterns of recorded radiant electromagnetic energy and other phenomena. As implied by its name, the science originally consisted of analyzing photographs. Although photogrammetry has expanded to include analysis of other records, such as digital imagery, radiated acoustical energy patterns, laser ranging measurements, and magnetic phenomena, photographs are still the principal source of information. In this text *photographic* and *digital* photogrammetry are emphasized, but other sources of information are also discussed.

Included within the definition of photogrammetry are two distinct areas: (1) *metric* photogrammetry and (2) *interpretative* photogrammetry. Metric photogrammetry consists of making precise measurements from photos and other information sources to determine, in general, the relative locations of points. This enables finding distances, angles, areas, volumes, elevations, and sizes and shapes of objects. The most common applications of metric photogrammetry are the preparation of planimetric and topographic maps from photographs (see Secs. 13-2 through 13-6), and the production of digital orthophotos from scanned photography (see Sec. 13-7). The photographs are most often *aerial* (taken from an airborne vehicle), but *terrestrial* photos (taken from earth-based cameras) and satellite imagery are also used.

Interpretative photogrammetry deals principally in recognizing and identifying objects and judging their significance through careful and systematic analysis. It includes branches of *photographic interpretation* and *remote sensing*. Photographic interpretation includes the study of photographic images, while remote sensing includes not only the analysis of photography but also the use of data gathered from a wide variety of sensing instruments, including multispectral cameras, infrared sensors, thermal scanners, and side-looking airborne radar. Remote sensing instruments, which are often carried in vehicles as remote as orbiting satellites, are capable of providing quantitative as well as qualitative information about objects. At present, with our recognition of the importance of preserving our environment and natural resources, photographic interpretation and remote sensing are both being employed extensively as tools in management and planning.

Of the two distinct areas of photogrammetry, concentration in this book is on metric photogrammetry. Interpretative photogrammetry is discussed only briefly, and those students interested in further study in this area should consult the references cited at the end of this chapter.

1-2 HISTORY OF PHOTOGRAMMETRY

Developments leading to the present-day science of photogrammetry occurred long before the invention of photography. As early as 350 B.C. Aristotle had referred to the process of projecting images optically. In the early 18th century Dr. Brook Taylor published his treatise on linear perspective, and soon afterward, J. H. Lambert suggested that the principles of perspective could be used in preparing maps.

The actual practice of photogrammetry could not occur, of course, until a practical photographic process was developed. This occurred in 1839, when Louis Daguerre of Paris announced his direct photographic process. In his process the exposure was made on metal plates that had been light-sensitized with a coating of silver iodide. This is essentially the photographic process in use today.

A year after Daguerre's invention, Francois Arago, a geodesist with the French Academy of Science, demonstrated the use of photographs in topographic surveying. The first actual experiments in using photogrammetry for topographic mapping occurred in 1849 under the direction of Colonel Aimé Laussedat of the French Army Corps of Engineers. In Colonel Laussedat's experiments kites and balloons were used for taking aerial photographs. Due to difficulties encountered in obtaining aerial photographs, he curtailed this area of research and concentrated his efforts on mapping with terrestrial photographs. In 1859 Colonel Laussedat presented an accounting of his successes in mapping using photographs. His pioneering work and dedication to this subject earned him the title "father of photogrammetry."

Topographic mapping using photogrammetry was introduced to North America in 1886 by Captain Eduard Deville, the Surveyor General of Canada. He found Laussedat's principles extremely convenient for mapping the rugged mountains of western Canada. The U.S. Coast and Geodetic Survey (now the National Geodetic Survey) adopted photogrammetry in 1894 for mapping along the border between Canada and the Alaska Territory.

Meanwhile new developments in instrumentation, including improvements in cameras and films, continued to nurture the growth of photogrammetry. In 1861 a three-color photographic process was developed, and roll film was perfected in 1891. In 1909 Dr. Carl Pulfrich of Germany began to experiment with overlapping pairs of photographs. His work formed much of the foundation for the development of many instrumental photogrammetric mapping techniques in use today.

The invention of the airplane by the Wright brothers in 1902 provided the great impetus for the emergence of modern aerial photogrammetry. Until that time, almost all photogrammetric work was, for the lack of a practical means of obtaining aerial photos, limited to terrestrial photography. The airplane was first used in 1913 for obtaining photographs for mapping purposes. Aerial photos were used extensively during World War I, primarily in reconnaissance. In the period between World War I and World War II, aerial photogrammetry for topographic mapping progressed to the point of mass production of maps. Within this period many private firms and government agencies in North America and in Europe became engaged in photogrammetric work. During World War II, photogrammetric techniques were used extensively to meet the great new demand for maps. Air photo interpretation was also employed more widely than ever before in reconnaissance and intelligence. Out of this war-accelerated mapping program came many new developments in instruments and techniques.

Advancements in instrumentation and techniques in photogrammetry have continued at a rapid pace during the past 50 years. The many advancements are too numerous to itemize here, but collectively they have enabled photogrammetry to become the most accurate and efficient method available for compiling maps and generating topographic information. The improvements have affected all aspects of the science, and they incorporate many new developments such as those in optics, electronics, computers and satellite technology. While this text does include some historical background, its major thrust is to discuss and describe the current state of the art in photogrammetric instruments and techniques.

1-3 TYPES OF PHOTOGRAPHS

Two fundamental classifications of photography used in the science of photogrammetry are *terrestrial* and *aerial*. Terrestrial photographs (see Chap. 19) are taken with ground-based cameras, the position and orientation of which are often measured directly at the time of exposure. A great variety of cameras are used for taking terrestrial photographs, and these may include anything from simple hobby cameras, which are handheld, to precise specially designed cameras mounted on tripods. A *phototheodolite,* as shown in Fig. 1-1, is a combination camera and theodolite mounted on a tripod used for taking terrestrial photographs. The theodolite, a surveying instrument which is used to measure angles, facilitates aligning the camera in a desired or known azimuth and measuring its position and elevation. Figure 1-2 shows a terrestrial photograph taken with a camera of the type shown in Fig. 1-1.

Another special type of terrestrial camera, which was employed for important work in the past, is the *ballistic camera,* shown in Fig. 1-3. These large cameras were mounted at selected ground stations and used to obtain photographs of orbiting artificial satellites against a star background. The photographs were analyzed to calculate satellite trajectories; the size, shape, and gravity of the earth; and the precise positions of the camera stations. This procedure utilized precisely known camera constants, together with the

FIGURE 1-1
Phototheodolite used for taking terrestrial photographs.
(*Courtesy LH Systems, LLC.*)

FIGURE 1-2
Terrestrial photograph. (*Courtesy LH Systems, LLC.*)

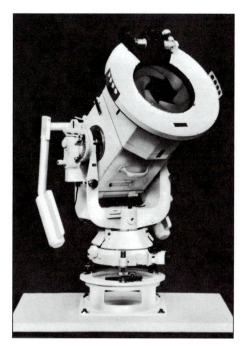

FIGURE 1-3
BC-4 ballistic camera. (*Courtesy LH Systems, LLC.*)

known positions of the background stars at the instants of exposure. Ballistic cameras played an essential role in establishing a worldwide network of control points and in accurately determining the relative positions of the continents, remote ocean islands, etc. Use of ballistic cameras for this purpose has been made obsolete by the Global Positioning System (GPS), a network of transmitting satellites and ground-based receivers which enables extremely accurate positions to be determined anywhere on or near the earth.

Aerial photography is commonly classified as either vertical or oblique. *Vertical photos* are taken with the camera axis directed as nearly vertically as possible. If the camera axis were perfectly vertical when an exposure was made, the photographic plane would be parallel to the datum plane and the resulting photograph would be termed *truly vertical*. In practice, the camera axis is rarely held perfectly vertical due to unavoidable aircraft tilts. When the camera axis is unintentionally tilted slightly from vertical, the resulting photograph is called a *tilted photograph*. These unintentional tilts are usually less than 1° and seldom more than 3°. For many practical applications, simplified procedures suitable for analyzing truly vertical photos may also be used for tilted photos without serious consequence. Precise photogrammetric instruments and procedures have been developed, however, that make it possible to rigorously account for tilt with no loss of accuracy. Figure 1-4 shows an aerial camera with its electric control mechanism and the mounting framework for placing it in an aircraft. The vertical photograph illustrated in Fig. 1-5 was taken with a camera of the type illustrated in Fig. 1-4 from an altitude of 470 meters (m) above the terrain.

FIGURE 1-4
Zeiss RMK TOP 15, aerial camera, with electronic controls and aircraft mountings.
(*Courtesy Carl Zeiss, Inc.*)

Oblique aerial photographs are exposed with the camera axis intentionally tilted away from vertical. A *high oblique* photograph includes the horizon; a *low oblique* does not. Figure 1-6 illustrates the orientation of the camera for vertical, low oblique, and high oblique photography and also shows how a square grid of ground lines would appear in each of these types of photographs. Figures 1-7 and 1-8 are examples of low oblique and high oblique photographs, respectively.

1-4 TAKING VERTICAL AERIAL PHOTOGRAPHS

When an area is covered by vertical aerial photography, the photographs are usually taken along a series of parallel passes, called *flight strips*. As illustrated in Fig. 1-9, the photographs are normally exposed in such a way that the area covered by each successive photograph along a flight strip duplicates or overlaps part of the coverage of the previous photo. This lapping along the flight strip is called *end lap,* and the area of coverage common to an adjacent pair of photographs in a flight strip is called the *stereoscopic overlap area*. The overlapping pair of photos is called a *stereopair*. For reasons which will be given in subsequent chapters, the amount of end lap is normally between 55 and 65 percent. The positions of the camera at each exposure, e.g., positions 1, 2, 3 of Fig. 1-9, are called the *exposure stations,* and the altitude of the camera at exposure time is called the *flying height*.

Adjacent flight strips are photographed so that there is also a lateral overlapping of adjacent strips. This condition, as illustrated in Fig. 1-10, is called *side lap,* and it is normally held at approximately 30 percent. The photographs of two or more side-lapping strips used to cover an area is referred to as a *block* of photos.

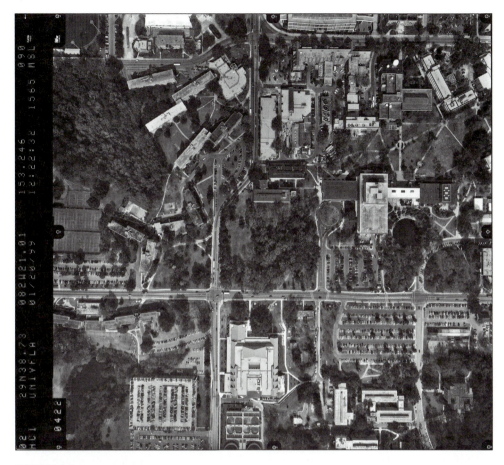

FIGURE 1-5
Vertical aerial photograph. (*Courtesy Hoffman and Company, Inc.*)

1-5 EXISTING AERIAL PHOTOGRAPHY

Photogrammetrists and photo interpreters can obtain aerial photography in one of two ways: (1) They can purchase photographs from existing coverage, or (2) they can obtain new coverage. It is seldom economical to use existing coverage for mapping, because it rarely meets the needs of the user; but existing coverage may prove suitable for other uses such as reconnaissance, project planning, historical records, or photo interpretation. If existing photography is not satisfactory because of age, scale, camera, etc., it will be necessary to obtain new coverage. Of course, before the decision can be made whether to use existing photography or obtain new, it is necessary to ascertain exactly what coverage exists in a particular area.

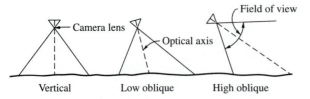

Camera orientation for various types of aerial photographs

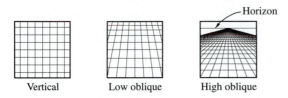

How a grid of section lines appears on various types of photos

FIGURE 1-6
Camera orientation for various types of aerial photographs.

Existing aerial photography is available for nearly all the United States and Canada. Some areas have been covered several times, so that various scales and qualities of photography are available. Most of the coverage is vertical photography.

The U.S. Geological Survey, at its *Earth Resources Observation System* (EROS) Data Center in Sioux Falls, South Dakota, has archived, for distribution upon request,[1] millions of aerial photos and satellite images. These include air photo coverage of virtually all areas of the United States as well as images from several series of satellites which provide global coverage. Their archived air photo coverage includes photos taken through the *National Aerial Photography Program* (NAPP); they were taken from a flying height of 20,000 feet (ft) above ground and are in black and white and color-infrared. It also includes photos from the *National High Altitude Photography* (NHAP) Program, also in black and white and color-infrared and taken from 40,000 ft above ground. The EROS Data Center also archives photos that were taken by the U.S. Geological Survey for its topographic mapping projects as well as photos taken by other federal agencies including the *National Aeronautics and Space Administration* (NASA), the *Bureau of Reclamation*, the *Environmental Protection Agency* (EPA), and the *U.S. Army Corps of Engineers*.

---

[1] Requests for information about aerial photographic coverage should be directed to Customer Services, U.S. Geological Survey, EROS Data Center, 47914 252nd Street, Sioux Falls, SD 57198-0001. Information can also be obtained at the following Web site: <http://nsdi.usgs.gov/products/aerial.html>; or contact can be made through e-mail at: custserv@edcmail.cr.usgs.gov.

FIGURE 1-7
Low oblique photograph of Madison, Wisconsin (note that the horizon is not shown). (*Courtesy State of Wisconsin, Department of Transportation.*)

The U.S. Department of Agriculture[2] is another useful resource for obtaining existing aerial photography. Their archives contain extensive coverage for the United States. Available products include black and white, color, and color infrared prints at negative scales of 1 : 20,000 and smaller.

Existing aerial photography can also be obtained from the department of transportation of most states. These photos have usually been taken for use in highway plan-

---

[2] Requests for information about aerial photographic coverage should be directed to Sales Branch, USDA FSA Aerial Photography Field Office, 2222 West 2300 South, Salt Lake City, UT 84119-2020. Information can also be obtained at the following Web site: <http://www.fsa.usda.gov/dam>; or contact can be made through e-mail at: sales@apfo.usda.gov.

FIGURE 1-8
High oblique photograph of Tampa, Florida (note that the horizon shows on the photograph). (*Courtesy US Imaging, Inc.*)

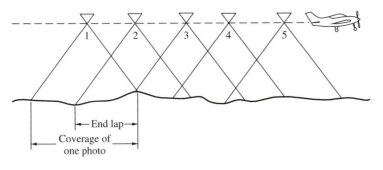

FIGURE 1-9
End lap of photographs in a flight strip.

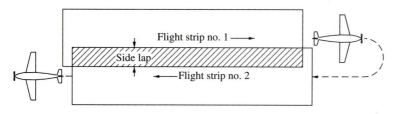

FIGURE 1-10
Side lap of adjacent flight strips.

ning and design; thus the scales are generally relatively large, and coverage typically follows state and federal highways. In addition, many counties have periodic coverage.

1-6 USES OF PHOTOGRAMMETRY

The earliest applications of photogrammetry were in topographic mapping, and today that use is still the most common of photogrammetric activities. At present, the U.S. Geological Survey (USGS), the federal agency charged with mapping the United States, performs nearly 100 percent of its map compilation photogrammetrically. State departments of transportation also use photogrammetry almost exclusively in preparing their topographic maps. In addition, private engineering and surveying firms prepare many special-purpose topographic maps photogrammetrically. These maps vary in scale from large to small and are used in planning and designing highways, railroads, rapid transit systems, bridges, pipelines, aqueducts, transmission lines, hydroelectric dams, flood control structures, river and harbor improvements, urban renewal projects, etc. A huge quantity of topographic maps are prepared for use in providing spatial data for geographic information systems (see Sec. 1-7).

Two newer photogrammetric products, *orthophotos,* and *digital elevation models* (DEMs), are now often used in combination to replace traditional topographic maps. As described in Sec. 13-7, an orthophoto is an aerial photograph that has been modified so that its scale is uniform throughout. Thus orthophotos are equivalent to planimetric maps, but unlike planimetric maps which show features by means of lines and symbols, orthophotos show the actual images of features. For this reason they are more easily interpreted than planimetric maps, and hence are preferred by many users. A DEM, as discussed in Sec. 13-6, consists of an array of points in an area that have had their X, Y, and Z coordinates determined. Thus they provide a numerical representation of the topography in the area, and contours, cross sections, profiles, etc. can be computed from them. Orthophotos and DEMs are both widely applied in all fields where maps are used, but because they are both in digital form, one of their most common applications, as discussed in Sec. 1-7 and Chap. 21, is their use in connection with geographic information systems.

Photogrammetry has become an exceptionally valuable tool in land surveying. To mention just a few uses in the field, aerial photos can be used as rough base maps for relocating existing property boundaries. If the point of beginning or any corners can be located with respect to ground features that can be identified on the photo, the entire parcel can be plotted on the photo from the property description. All corners can then be located on the photo in relation to identifiable ground features which, when located in the field, greatly assist in finding the actual property corners. Aerial photos can also be used in planning ground surveys. Through stereoscopic viewing, the area can be studied in three dimensions. Access routes to remote areas can be identified, and surveying lines of least resistance through difficult terrain or forests can be found. The photogrammetrist can prepare a map of an area without actually setting foot on the ground—an advantage which circumvents problems of gaining access to private land for ground surveys.

The field of highway planning and design provides an excellent example of how important photogrammetry has become in engineering. In this field, high-altitude photos or satellite images are used to assist in area and corridor studies and to select the best route; small-scale topographic maps are prepared for use in preliminary planning; large-scale topographic maps and *digital elevation models* (DEMs) are compiled for use in final design; and earthwork cross sections are taken to obtain contract quantities. In many cases the plan portions of the "plan profile" sheets of highway plans are prepared from aerial photographs. Partial payments and even final pay quantities are often calculated from photogrammetric measurements. Map information collected from modern photogrammetric instruments is directly compatible with *computer-aided drafting* (CAD) systems commonly used in highway design. The use of photogrammetry in highway engineering not only has reduced costs but also has enabled better overall highway designs to be created.

Many areas outside of engineering have also benefitted from photogrammetry. Nonengineering applications include the preparation of tax maps, soil maps, forest maps, geologic maps, and maps for city and regional planning and zoning. Photogrammetry is used in the fields of astronomy, architecture, archaeology, geomorphology, oceanography, hydrology and water resources, conservation, ecology, and mineralogy. Stereoscopic photography literally enables the outdoors to be brought into the comfortable confines of the laboratory or office for viewing and study in three dimensions.

Photogrammetry has been used successfully in traffic management and in traffic accident investigations. One advantage of its use in the latter area is that photographs overlook nothing that may be needed later to reconstruct the accident, and it is possible to restore normal traffic flow quickly. Even in the fields of medicine and dentistry, measurements from X-ray and other photographs and images have been useful in diagnosis and treatment. Of course, one of the oldest and still most important uses of aerial photography is in military intelligence. Space exploration is one of the new and exciting areas where photogrammetry is being utilized.

Photogrammetry has become a powerful research tool because it affords the unique advantage of permitting instantaneous recordings of dynamic occurrences to be captured in images. Measurements from photographs of quantities such as beam or pavement deflections under impact loads may easily be obtained photographically where such measurements would otherwise be nearly impossible.

As noted earlier, photogrammetry has become extremely important in the developing field of *geographic information systems* (GISs). This area of application is described in the following section.

It would be difficult to cover all the many situations in which photogrammetric principles and methods could be or have been used to solve measurement problems. Photogrammetry, although still a relatively new science, has already contributed substantially to engineering and nonengineering fields alike. New applications appear to be bounded only by our imagination, and the science should continue to grow in the future.

1-7 PHOTOGRAMMETRY AND GEOGRAPHIC INFORMATION SYSTEMS

Geographic information systems, although relatively new, have rapidly gained a position of prominence in many fields. These computer-based systems enable storing, integrating, manipulating, analyzing, and displaying virtually any type of spatially related information about the environment. They are being used at all levels of government, and by businesses, private industry, and public utilities to assist in planning, design, management, and decision making. Later chapters in this text describe GISs in detail and illustrate some of their applications. It is important at this juncture, however, to mention the major role that photogrammetry plays in these systems.

An essential element of any GIS is a complex relational database. The information that comprises the database usually includes both natural and cultural features. Specific types of information, or *layers,* within the database may include political boundaries, individual property ownership, transportation networks, utilities, topography, hydrography, soil types, land use, vegetation types, wetlands, etc. To be of use in a GIS, however, all data must be spatially related; i.e., all the data must be in a common geographic frame of reference. Photogrammetry is ideal for deriving much of this layered information. As noted in the preceding section, topographic maps, digital elevation models, and digital orthophotos are examples of photogrammetric products which are now commonly employed in developing these spatially related layers of information. By employing photogrammetry the data can be compiled more economically than through ground surveying methods, and this can be achieved with comparable or even greater spatial accuracy. Furthermore, the data are compiled directly in digital format, and thus are compatible for direct entry into GIS databases.

The photogrammetric procedures involved in developing topographic maps, digital elevation models, digital orthophotos, and other products used in GISs are described in later chapters of this text. Chapters 20 and 21 discuss and demonstrate the application of these products within GISs for specific purposes.

1-8 PROFESSIONAL PHOTOGRAMMETRY ORGANIZATIONS

There are several professional organizations in the United States serving the interests of photogrammetry. Generally these organizations have as their objectives the advancement of knowledge in the field, encouragement of communication among photogrammetrists, and upgrading of standards and ethics in the practice of photogrammetry.

The American Society for Photogrammetry and Remote Sensing (ASPRS), formerly known as the American Society of Photogrammetry, founded in 1934, is the foremost professional photogrammetric organization in the United States. One of this society's most valuable contributions has been its publication of various manuals, such as the *Manual of Photogrammetry,* the *Manual of Remote Sensing,* the *Manual of Photographic Interpretation, Digital Photogrammetry: An Addendum to the Manual of Photogrammetry,* and the *Manual of Color Aerial Photography.* In preparing these volumes, leading photogrammetrists from government agencies as well as private and commercial firms and educational institutions have authored and coauthored chapters in their various special areas of expertise. The American Society for Photogrammetry and Remote Sensing also publishes *Photogrammetric Engineering and Remote Sensing,*[3] a monthly journal which brings new developments and applications to the attention of its readers. The society regularly sponsors technical meetings, at various locations throughout the United States, which bring together large numbers of photogrammetrists for the presentation of papers, discussion of new ideas and problems, and firsthand viewing of the latest photogrammetric equipment.

The fields of photogrammetry and surveying are so closely knit that it is difficult to separate them. Both are measurement sciences dealing with the production of maps. The American Congress on Surveying and Mapping (ACSM), although primarily concerned with more traditional types of ground surveys, is also vitally interested in photogrammetry. The quarterly journal of ACSM, *Surveying and Land Information Systems,* frequently carries photogrammetry-related articles.

The Geomatics Division of the American Society of Civil Engineers (ASCE) is also dedicated to surveying and photogrammetry. Articles on photogrammetry are frequently published in its *Journal of Surveying Engineering.*

The Canadian Institute of Geomatics (CIG) is the foremost professional organization of Canada concerned with photogrammetry. The CIG regularly sponsors technical meetings, and its journal, *Geomatica,* carries photogrammetry articles. The *Australian Surveyor* and *Photogrammetric Record* are similar journals with wide circulation, published in English, by professional organizations in Australia and Great Britain, respectively.

The International Society for Photogrammetry and Remote Sensing (ISPRS), founded in 1910, fosters the exchange of ideas and information among photogrammetrists all over the world. Approximately a hundred foreign countries having professional organizations similar to the American Society for Photogrammetry and Remote Sensing form the membership of ISPRS. This society fosters research, promotes education, and sponsors international conferences at four-year intervals. Its organization consists of seven technical commissions, each concerned with a specialized area in photogrammetry and remote sensing. Each commission holds periodic symposia where photogrammetrists gather to hear presented papers on subjects of international interest.

---

[3] The title of this journal was changed from *Photogrammetric Engineering* to *Photogrammetric Engineering and Remote Sensing* in 1975.

The society's official journal is the *ISPRS Journal of Photogrammetry and Remote Sensing,* which is published in English.

1-9 THE WORLD WIDE WEB

A recent innovation of the information age is the *World Wide Web,* which is a product of the *Internet,* a network which connects computers across the globe. The World Wide Web is a source of vast amounts of information, and it includes much about photogrammetry, remote sensing, and mapping, as well as nearly any other topic imaginable. Access to the World Wide Web requires a computer equipped with a dial-up modem or a hardwired or wireless Internet connection. Computer users with modems require an Internet service provider which they access via telephone lines in order to make the network connection. Web browsing software is also required on the user's computer to access Web pages available over the Internet.

Network links known as *universal resource locators* (URLs) are essentially addresses to the various sources of information on the Internet. Users enter the URL for the information resource they wish to access; e.g., the URL for the American Society for Photogrammetry and Remote Sensing is *http://www.asprs.org.* A wealth of information on the theory and applications of photogrammetry is available on the World Wide Web. Discussion of specific information that is available on the World Wide Web is presented at various locations throughout this text, and important Web site addresses are frequently given for convenience in accessing the information.

REFERENCES

American Society for Photogrammetry and Remote Sensing: *Manual of Remote Sensing,* 3d ed., Bethesda, MD, 1998.

————: *Manual of Photographic Interpretation,* 2d ed., Bethesda, MD, 1997.

————: *Digital Photogrammetry: An Addendum to the Manual of Photogrammetry,* Bethesda, MD, 1996.

————: *Close-Range Photogrammetry and Machine Vision,* Bethesda, MD, 1996.

American Society of Photogrammetry: *Manual of Photogrammetry,* 4th ed., Bethesda, MD, 1980.

————: *Manual of Color Aerial Photography,* Bethesda, MD, 1968.

Doyle, F. J.: "Photogrammetry: The Next Two Hundred Years," *Photogrammetric Engineering and Remote Sensing,* vol. 43, no. 5, 1977, p. 575.

Gruner, H.: "Photogrammetry 1776–1976," *Photogrammetric Engineering and Remote Sensing,* vol. 43, no. 5, 1977, p. 569.

Gutelius, B.: "Engineering Applications of Airborne Scanning Lasers: Reports from the Field," *Photogrammetric Engineering and Remote Sensing,* vol. 64, no. 4, 1998, p. 246.

Konecny, G.: "Paradigm Changes in ISPRS from the First to the Eighteenth Congress in Vienna," *Photogrammetric Engineering and Remote Sensing,* vol. 62, no. 10, 1996, p. 1117.

Kraus, K.: *Photogrammetry, Fundamentals and Standard Process,* vol. 1, Dummler Publishers, Bonn, Germany, 1993.

Lillesand, T. M., and R. W. Kiefer: *Remote Sensing and Image Interpretation,* 4th ed., Wiley, New York, 1999.

Merchant, J. W., et al: "Special Issue: Geographic Information Systems," *Photogrammetric Engineering and Remote Sensing,* vol. 62, no. 11, 1996, p. 1243.

Mikhail, E.: "Is Photogrammetry Still Relevant?" *Photogrammetric Engineering and Remote Sensing,* vol. 65, no. 7, 1999, p. 740.

Poore, B. S., and M. DeMulder: "Image Data and the National Spatial Data Infrastructure," *Photogrammetric Engineering and Remote Sensing,* vol. 63, no. 1, 1997, p. 7.

Ridley, H. M., P. M. Atkinson, P. Aplin, J. P. Muller, and I. Dowman: "Evaluating the Potential of the Forth-coming Commercial U.S. High-Resolution Satellite Sensor Imagery at the Ordnance Survey," *Photogrammetric Engineering and Remote Sensing,* vol. 63, no. 8, 1997, p. 997.

Terry, N. G., Jr.: "Field Validation of the UTM Gridded Map," *Photogrammetric Engineering and Remote Sensing,* vol. 63, no. 4, 1997, p. 381.

PROBLEMS

1-1. Explain the differences between metric and interpretative photogrammetry.

1-2. Describe the different classifications of aerial photographs.

1-3. What is the primary difference between high and low oblique aerial photographs?

1-4. What is a phototheodolite?

1-5. Define the following photogrammetric terms: end lap, side lap, stereopair, exposure station, and flying height.

1-6. Discuss some of the principal uses of aerial photogrammetry.

1-7. Discuss some of the principal uses of terrestrial photogrammetry.

1-8. Describe how you would go about obtaining existing aerial photographic coverage of an area.

1-9. To what extent is photogrammetry being used in highway planning in your state?

1-10. Discuss the importance of photogrammetry in geographic information systems.

1-11. Access the World Wide Web, visit the following Web sites, and briefly discuss the information they provide regarding photogrammetry and mapping.

(*a*) http://www.asprs.org/

(*b*) http://www.p.igp.ethz.ch/isprs/isprs.html

(*c*) http://www.cig-acsg.ca/

(*d*) http://www.isaust.org.au/

(*e*) http://cesgi1.city.ac.uk/photsoc/

(*f*) http://www.nima.mil/

(*g*) http://www.fgdc.gov/

(*h*) http://mapping.usgs.gov/

CHAPTER
2

PRINCIPLES OF PHOTOGRAPHY AND IMAGING

2-1 INTRODUCTION

Photography, which means "drawing with light," originated long before cameras and light-sensitive photographic films came into use. Ancient Arabs discovered that when inside a dark tent, they could observe inverted images of illuminated outside objects. The images were formed by light rays which passed through tiny holes in the tent. The principle involved was actually that of the pinhole camera of the type shown in Fig. 2-1. In the 1700s French artists used the pinhole principle as an aid in drawing perspective views of illuminated objects. While inside a dark box, they traced the outlines of objects projected onto the wall opposite a pinhole. In 1839 Louis Daguerre of France developed a photographic film which could capture a permanent record of images that illuminated it. By placing this film inside a dark "pinhole box," a picture or *photograph* could be obtained without the help of an artist. This box used in conjunction with photographic film became known as a *camera*. Tremendous improvements have been made in photographic films and film cameras over the years; however, their basic principle has remained essentially unchanged.

A recent innovation in imaging technology is the digital camera which relies on electronic sensing devices rather than conventional film. The resulting image, called a *digital image,* is stored in computer memory which enables direct computer manipulation.

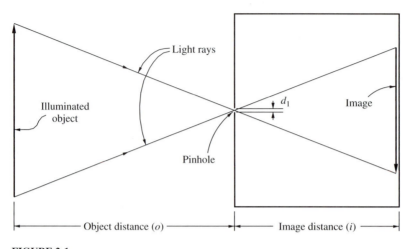

FIGURE 2-1
Principle of the pinhole camera.

This chapter describes basic principles that are fundamental to understanding imaging processes. Concepts germane to both photography and digital imaging are included.

2-2   FUNDAMENTAL OPTICS

Both film and digital cameras depend upon optical elements, especially lenses, to function. Thus an understanding of some of the fundamental principles of optics is essential to the study of photography and imaging.

The science of optics consists of two principal branches: *physical optics* and *geometric optics*. In physical optics, light is considered to travel through a transmitting medium such as air in a series of electromagnetic waves emanating from a point source. Conceptually this can be visualized as a group of concentric circles expanding or *radiating* away from a light source, as illustrated in Fig. 2-2. In nature, a good resemblance of this manner in which light waves propagate can be created by dropping a small pebble into a pool of still water, to create waves radiating from the point where the pebble was dropped. As with water, each light wave has its own *frequency, amplitude,* and *wavelength.* Frequency is the number of waves that pass a given point in a unit of time; amplitude is the measure of the height of the crest or depth of the trough; and wavelength is the distance between any wave and the next succeeding one. The speed with which a wave moves from a light source is called its *velocity.* Velocity is related to frequency and wavelength according to the equation

$$V = f\lambda \tag{2-1}$$

In Eq. (2-1), V is velocity, usually expressed in units of meters per second; f is frequency, generally given in cycles per second, or *hertz;* and λ is wavelength, usually expressed in meters. Light has an extremely high velocity, moving at the rate of 2.9979246×10^8 meters per second (m/s) in a vacuum.

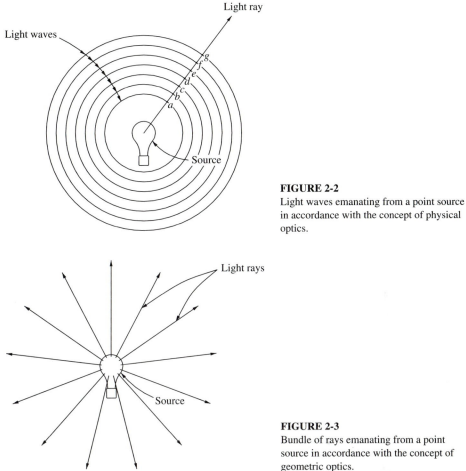

FIGURE 2-2
Light waves emanating from a point source in accordance with the concept of physical optics.

FIGURE 2-3
Bundle of rays emanating from a point source in accordance with the concept of geometric optics.

In geometric optics, light is considered to travel from a point source through a transmitting medium in straight lines called *light rays*. As illustrated in Fig. 2-3, an infinite number of light rays radiate in all directions from any point source. The entire group of radiating lines is called a *bundle of rays*. This concept of radiating light rays develops logically from physical optics if one considers the travel path of any specific point on a light wave as it radiates away from the source. In Fig. 2-2, for example, point *a* radiates to *b*, *c*, *d*, *e*, *f*, etc. as it travels from the source, thus creating a light ray.

In analyzing and solving photogrammetric problems, rudimentary line diagrams are often necessary. Their preparation generally requires tracing the paths of light rays through air and various optical elements. These same kinds of diagrams are often used as a basis for deriving fundamental photogrammetric equations. For these reasons, a basic knowledge of the behavior of light, and especially of geometric optics, is prerequisite to a thorough understanding of the science of photogrammetry.

When light passes from one transmitting material to another, it undergoes a change in velocity in accordance with the composition of the substances through which it travels. Light achieves its maximum velocity traveling through a vacuum, it moves more slowly through air, and travels still more slowly through water and glass.

The rate at which light travels through any substance is represented by the *refractive index* of the material. Refractive index is simply the ratio of the speed of light in a vacuum to its speed through a substance, or

$$n = \frac{c}{V} \tag{2-2}$$

In Eq. (2-2), n is the refractive index of a material, c is the velocity of light in a vacuum, and V is its velocity in the substance. The refractive index for any material, which depends upon the wavelength of the light, is determined through experimental measurement. Typical values for indexes of refraction of common media are vacuum, 1.0000; air, 1.0003; water, 1.33; and glass, 1.5 to 2.0.

When light rays pass from one homogeneous, transparent medium to a second such medium having a different refractive index, the path of the light ray is bent or *refracted,* unless it intersects the second medium normal to the interface. If the intersection occurs obliquely, as shown in Fig. 2-4, then the *angle of incidence, ϕ,* is related to the *angle of refraction, ϕ',* by the law of refraction, frequently called *Snell's law.* This law is stated as follows:

$$n \sin \phi = n' \sin \phi' \tag{2-3}$$

where n is the refractive index of the first medium and n' is the refractive index of the second medium. The angles ϕ and ϕ' are measured from the normal to the incident and refracted rays, respectively.

Light rays can also be made to change directions by reflection. When a light ray strikes a smooth surface such as a highly polished metal mirror, it is reflected so that the *angle of reflection ϕ''* is equal to the incidence angle ϕ, as shown in Fig. 2-5a. Both angles lie in a common plane and are measured from NN', the normal to the reflecting surface.

Plane mirrors used for nonscientific purposes generally consist of a plane sheet of glass with a thin reflective coating of silver on the back. This type of "back-surfaced" mirror is optically undesirable, however, because it creates multiple reflections that interfere with the primary reflected light ray, as shown in Fig. 2-5b. These undesirable

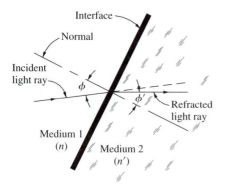

FIGURE 2-4
Refraction of light rays.

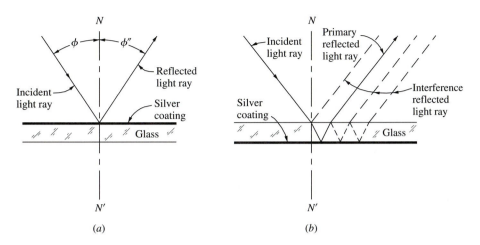

FIGURE 2-5
(*a*) First-surface mirror demonstrating the angle of incidence ϕ and angle of reflectance ϕ'';
(*b*) back-surfaced mirror.

reflections may be avoided by using *first-surface* mirrors, which have their silver coating on the front of the glass, as shown in Fig. 2-5*a*.

2-3 LENSES

A simple lens consists of a piece of optical glass that has been ground so that it has either two spherical surfaces or one spherical surface and one flat surface. Its primary function is to gather light rays from object points and bring them to focus at some distance on the opposite side of the lens. A lens accomplishes this function through the principles of refraction. The simplest and most primitive device that performs the functions of a lens is a tiny pinhole which theoretically allows a single light ray from each object point to pass. The tiny hole of diameter d_1 of the pinhole camera illustrated in Fig. 2-1 produces an inverted image of the object. The image is theoretically in focus regardless of the distance from the pinhole to the camera's image plane. Pinholes allow so little light to pass, however, that they are unsuitable for photogrammetric work. For practical purposes they are replaced by glass lenses.

The advantage of a lens over a pinhole is the increased amount of light that is allowed to pass. A lens gathers an entire *pencil of rays* from each object point instead of only a single ray. As discussed earlier and illustrated in Fig. 2-3, when an object is illuminated, each point in the object reflects a bundle of light rays. This condition is also illustrated in Fig. 2-6. A lens placed in front of the object gathers a pencil of light rays from each point's bundle of rays and brings these rays to focus at a point in a plane on the other side of the lens, called the *image plane*. An infinite number of image points, focused in the image plane, form the image of the entire object. Note from Fig. 2-6 that the image is inverted by the lens.

The *optical axis* of a lens is defined as the line joining the centers of curvature of the spherical surfaces of the lens (points O_1 and O_2 of Fig. 2-7). In this figure R_1 and R_2

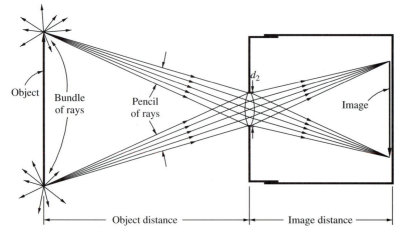

FIGURE 2-6
Pencils of light and image formation in a single-lens camera.

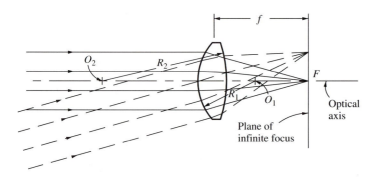

FIGURE 2-7
Optical axis, focal length, and plane of infinite focus of a lens.

are the radii of the lens surfaces, and the optical axis is the line $O_1 O_2$. Light rays that are parallel to the optical axis as they enter a lens come to focus at F, the *focal point* of the lens. The distance from the focal point to the center of a lens is f, the *focal length* of the lens. A plane perpendicular to the optical axis passing through the focal point is called the *plane of infinite focus,* or simply the *focal plane*. Parallel rays entering a *converging lens* (one with two convex exterior surfaces, as shown in Fig. 2-7), regardless of the angle they make with the optical axis, are ideally brought to focus in the plane of infinite focus (see the dashed rays of the figure).

> **Example 2-1.** A single ray of light traveling through air ($n = 1.0003$) enters a convex glass lens ($n' = 1.52$) having a radius of 5.00 centimeters (cm), as shown in Fig. 2-8. If the light ray is parallel to and 1.00 cm above the optical axis of the lens, what are the angles of incidence ϕ and refraction ϕ' for the air-to-glass interface?

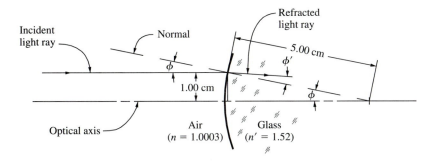

FIGURE 2-8
Refraction of an incident light ray parallel to the optical axis of a lens.

Solution. From the figure,

$$\sin \phi = \frac{1.00 \text{ cm}}{5.00 \text{ cm}} \qquad \text{from which} \qquad \phi = 11.5°$$

Applying Snell's law [Eq. (2-3)] gives

$$n \sin \phi = n' \sin \phi'$$

$$1.0003 \left(\frac{1.00 \text{ cm}}{5.00 \text{ cm}} \right) = 1.52 \sin \phi' \qquad \text{from which} \qquad \phi' = 7.56°$$

A pencil of incident light rays coming from an object located an infinite distance away from the lens will be parallel, as illustrated in Fig. 2-7, and the image will come to focus in the plane of infinite focus. For objects located some finite distance from the lens, the *image distance* (distance from lens center to image plane) is greater than the focal length. The following equation, called the *lens formula,* expresses the relationship of object distance o and image distance i to the focal length f of a converging lens:

$$\frac{1}{o} + \frac{1}{i} = \frac{1}{f} \tag{2-4}$$

If the focal length of a lens and the distance to an object are known, the resulting distance to the image plane can be calculated by using the lens formula.

Example 2-2. Find the image distance for an object distance of 50.0 m and a focal length of 50.0 cm.

Solution. By Eq. (2-4),

$$\frac{1}{50.0} + \frac{1}{i} = \frac{1}{0.500}$$

$$\frac{1}{i} = \frac{1}{0.500} - \frac{1}{50.0} = 2.00 - 0.0200 = 1.98$$

$$i = \frac{1}{1.98} = 0.505 \text{ m} = 50.5 \text{ cm}$$

The preceding analysis of lenses was simplified by assuming that their thicknesses were negligible. With *thick lenses,* this assumption is no longer valid. Thick lenses may consist of a single thick element or a combination of two or more elements which are either cemented together in contact or otherwise rigidly held in place with airspaces between the elements. A thick "combination" lens used in an aerial camera is illustrated in Fig. 2-9. Note that it consists of 15 individual elements.

Two points called *nodal points* must be defined for thick lenses. These points, termed the *incident* nodal point and the *emergent* nodal point, lie on the optical axis. They have the property that conceptually, any light ray directed toward the incident nodal point passes through the lens and emerges on the other side in a direction parallel to the original incident ray and directly away from the emergent nodal point. In Fig. 2-10, for example, rays AN and $N'a$ are parallel, as are rays BN and $N'b$. Points N and N'

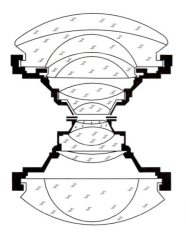

FIGURE 2-9
Cross section of SAGA-F lens. (*Drawing from brochure courtesy of LH Systems, LLC.*)

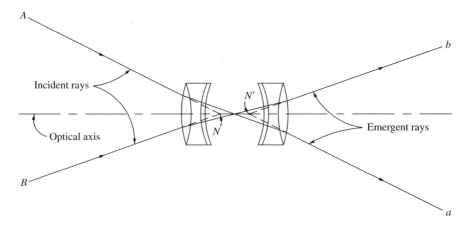

FIGURE 2-10
Nodal points of a thick lens.

are the incident and emergent nodal points, respectively, of the thick lens. Such light rays do not necessarily pass through the nodal points, as illustrated by the figure.

If parallel incident light rays (rays from an object at an infinite distance) pass through a thick lens, they will come to focus at the plane of infinite focus. The focal length of a thick lens is the distance from the emergent nodal point N' to this plane of infinite focus.

It is impossible for a single lens to produce a perfect image; it will, instead, always be somewhat blurred and geometrically distorted. The imperfections that cause blurring, or degrade the sharpness of the image, are termed *aberrations*. Through the use of additional lens elements, lens designers are able to correct for aberrations and bring them within tolerable limits. *Lens distortions*, on the other hand, do not degrade image quality but deteriorate the geometric quality (or positional accuracy) of the image. Lens distortions are classified as either *symmetric radial* or *decentering*. Both occur if light rays are bent, or change directions, so that after they pass through the lens, they do not emerge parallel to their incoming directions. Symmetric radial distortion, as its name implies, causes imaged points to be distorted along radial lines from the optical axis. Outward radial distortion is considered positive, and inward radial distortion is considered negative. Decentering distortion, which has both tangential and asymmetric radial components, causes an off-center distortion pattern. These lens distortions are discussed in greater detail in Secs. 3-9 and 4-13.

Resolution or *resolving power* of a lens is the ability of the lens to show detail. One common method of measuring lens resolution is to count the number of *line pairs* (black lines separated by white spaces of equal thickness) that can be clearly distinguished within a width of 1 millimeter (mm) in an image produced by the lens. The *modulation transfer function* (MTF) is another way of specifying the resolution characteristics of a lens. Both methods are discussed in Sec. 3-13, and a line-pair test pattern is shown in Fig. 3-16. Good resolution is important in photogrammetry because photo images must be sharp and clearly defined for precise measurements and accurate interpretative work. Photographic resolution is not just a function of the camera lens, however, but also depends on other factors, as described in later sections of this book.

The *depth of field* of a lens is the range in object distance that can be accommodated by a lens without introducing significant image deterioration. For a given lens, depth of field can be increased by reducing the size of the lens opening (*aperture*). This limits the usable area of the lens to the central portion. For aerial photography, depth of field is seldom of consequence, because variations in the object distance are generally a very small percentage of the total object distance. For close-range photography, however (see Chap. 19), depth of field is often extremely critical. The shorter the focal length of a lens, the greater its depth of field, and vice versa. Thus, if depth of field is critical, it can be somewhat accommodated either through the selection of an appropriate lens or by reducing aperture size.

Vignetting and *falloff* are lens characteristics which cause resultant images to appear brighter in the center than around the edges. Compensation can be provided for these effects in the lens design itself, by use of an antivignetting filter in the camera, or through lighting adjustments in the printing process (see Sec. 2-12).

2-4   SINGLE-LENS CAMERA

One of the most fundamental instruments used in photogrammetry is the single-lens camera. The geometry of this device, depicted in Fig. 2-6, is similar to that of the pinhole camera, shown in Fig. 2-1. In the single-lens camera, the size of the aperture (d_2 in Fig. 2-6) is much larger than a pinhole, requiring a lens in order to maintain focus.

Instead of object distances and image distances being unrestricted as with the pinhole camera, with the lens camera these distances are governed by the lens formula, Eq. (2-4). To satisfy this equation, the lens camera must be focused for each different object distance by adjusting the image distance. When object distances approach infinity, such as for photographing objects at great distances, the term $1/o$ in Eq. (2-4) approaches *zero* and image distance i is then equal to f, the lens focal length. With aerial photography, object distances are very great with respect to image distances; therefore aerial cameras are manufactured with their focus fixed for infinity. This is accomplished by fixing image distance equal to the focal length of the camera lens.

2-5   ILLUMINANCE

Illuminance of any photographic exposure is the brightness or amount of light received per unit area on the image plane surface during exposure. A common unit of illuminance is the *meter-candle*. One meter-candle (1 m · cd) is the illuminance produced by a standard candle at a distance of 1 m.

Illuminance is proportional to the amount of light passing through the lens opening during exposure, and this, of course, is proportional to the area of the opening. Since the area of the lens opening is $\pi d^2/4$, illuminance is proportional to the variable d^2, the square of the diameter of the lens opening.

Image distance i is another factor which affects illuminance. Illuminance is an effect that adheres to the *inverse square law,* which means that the amount of illuminance is inversely proportional to the square of distance from the aperture. According to this law, at the center of the photograph, illuminance is proportional to $1/i^2$. As distances increase away from the center of the photograph, distances from the aperture likewise increase. This causes decreased illuminance, an effect which can be quite severe for wide-angle lenses. This is one aspect of the physical basis for lens falloff, mentioned in Sec. 2-3. Normally in photography, object distances are sufficiently long that the term $1/o$ in Eq. (2-4) is nearly zero, in which case i is equal to f. Thus, at the center of a photograph, illuminance is proportional to the quantity $1/f^2$, and the two quantities may be combined so that illuminance is proportional to d^2/f^2. The square root of this term is called the *brightness factor,* or

$$\sqrt{\frac{d^2}{f^2}} = \frac{d}{f} = \text{brightness factor} \tag{2-5}$$

The inverse of Eq. (2-5) is also an inverse expression of illuminance and is the very common term *f-stop,* also called *f-number*. In equation form,

$$f\text{-stop} = \frac{f}{d} \tag{2-6}$$

According to Eq. (2-6), *f*-stop is the ratio of focal length to the diameter of the lens opening, or *aperture*. As the aperture increases, *f*-stop numbers decrease and illuminance increases, thus requiring less exposure time, i.e., faster shutter speeds. Because of this correlation between *f*-stop and shutter speed, *f*-stop is the term used for expressing *lens speed* or the "light-gathering" power of a lens. Illuminance produced by a particular lens is correctly expressed by Eq. (2-6), whether the lens has a very small diameter with short focal length or a very large diameter with a long focal length. If *f*-stop is the same for two different lenses, the illuminance at the center of each of their images will be the same.

2-6   RELATIONSHIP OF APERTURE AND SHUTTER SPEED

A sunbather gets a suntan or sunburn when exposed to sunshine. The darkness of tan or severity of burn is a function of the sun's brightness and the time of exposure to the sun. *Total exposure* of photographic film is likewise the product of illuminance and time of exposure. Its unit is *meter-candle-seconds,* although in certain applications a different unit, the *microlux-second* is used.

In making photographic exposures, the correct amounts of illuminance and time may be correlated using a *light meter*. Illuminance is regulated by varying *f*-stop settings on the camera, while time of exposure is set by varying the shutter speed. Variations in *f*-stop settings are actually variations in the diameter of the aperture, which can be controlled with a *diaphragm*—a circular shield that enlarges or contracts, changing the diameter of the opening of the lens and thus regulating the amount of light that is allowed to pass through the lens.

With a lens camera, as the diameter of the aperture increases, enabling faster exposures, the depth of field becomes less and lens distortions become more severe. At times a small diaphragm opening is desirable, and there are times when the reverse is true. To photograph a scene with great variations in object distances and yet retain sharp focus of all images, a large depth of field is required. In this case, to maximize depth of field, the picture would be taken at a slow shutter speed and large *f*-stop setting, corresponding to a small-diameter lens opening. On the other hand, in photographing rapidly moving objects or in making exposures from a moving vehicle such as an airplane, a fast shutter speed is essential, to reduce image motion. In this situation a small *f*-stop setting corresponding to a large-diameter lens opening would be necessary for sufficient exposure.

From the previous discussion it is apparent that there is an important relationship between *f*-stop and shutter speed. If exposure time is cut in half, total exposure is also halved. Conversely, if aperture area is doubled, total exposure is doubled. If shutter time is halved and aperture area is doubled, total exposure remains unchanged.

Except for inexpensive models, cameras are manufactured with the capability of varying both shutter speed and *f*-stop setting, and many modern cameras do this function automatically. The nominal *f*-stop settings are 1, 1.4, 2.0, 2.8, 4.0, 5.6, 8.0, 11, 16, 22, and 32. Not all cameras have all these, but the more expensive cameras have many of them. The camera pictured in Fig. 2-11, for example, has *f*-stops ranging from *f*-1.4 to *f*-16. This camera is also equipped for varying shutter speeds down to $\frac{1}{1000}$ second (s).

Shutter speed settings

f-stop settings ranging from f-1.4 – f-1.6

Lens focal length 50 mm

Maximum aperture (f-stop = 1.4)

FIGURE 2-11

Single-lens reflex camera having f-stop settings ranging from f-1.4 to f-16 and variable shutter speeds ranging down to $\frac{1}{1000}$ s. (*Courtesy Paillard Inc.*)

An f-stop number 1, or f-1, occurs, according to Eq. (2-6), when the aperture diameter equals the lens focal length. A setting at f-1.4 halves the aperture area from that of f-1. In fact, each succeeding number of the nominal f-stops listed previously halves the aperture area of the preceding one, and it is seen that each succeeding number is obtained by multiplying the preceding one by $\sqrt{2}$. This is illustrated as follows:

Let $d_1 = f$, where d_1 is the aperture diameter. Then

$$\frac{f}{d_1} = 1 = f\text{-stop}$$

At f-stop = 1,

$$\text{Aperture area} = A_1 = \frac{\pi(d_1)^2}{4}$$

If the aperture diameter is reduced to d_2, giving a lens opening area of one-half of A_1, then

$$A_2 = \frac{A_1}{2} = \frac{\pi(d_2)^2}{4} = \frac{\pi(d_1)^2}{2(4)}$$

From the above, $d_2 = d_1/\sqrt{2}$, and the corresponding f-stop number is

$$f\text{-stop} = \frac{f\sqrt{2}}{d_1} = 1\sqrt{2} = 1.4$$

The relationship between f-stop and shutter speed leads to many interesting variations in obtaining correct exposures.

Example 2-3. Suppose that a photographic film is optimally exposed with an f-stop setting of f-4 and a shutter speed of $\frac{1}{500}$ s. What is the correct f-stop setting if shutter speed is changed to $\frac{1}{1000}$ s?

Solution. Total exposure is the product of diaphragm area and shutter speed. This product must remain the same for the $\frac{1}{1000}$-s shutter speed as it was for the $\frac{1}{500}$-s shutter speed, or

$$\text{Area}_1 \times \text{time}_1 = \text{area}_2 \times \text{time}_2$$

Rearranging, we have

$$\text{Area}_2 = \text{area}_1 \times \frac{\text{time}_1}{\text{time}_2} \qquad (a)$$

Let d_1 and d_2 be diaphragm diameters for $\frac{1}{500}$- and $\frac{1}{1000}$-s shutter times, respectively. Then the respective diaphragm areas are

$$\text{Area}_1 = \frac{\pi(d_1)^2}{4} \qquad \text{and} \qquad \text{area}_2 = \frac{\pi(d_2)^2}{4} \qquad (b)$$

By Eq. (2-6),

$$d_1 = \frac{f}{f\text{-stop}_1} \qquad \text{and since } f\text{-stop}_1 = 4 \qquad d_1 = \frac{f}{4} \qquad (c)$$

Substituting (b) and (c) into (a) gives

$$\frac{\pi(d_2)^2}{4} = \frac{\pi(f)^2}{4(4)^2} \times \frac{1/500}{1/1000}$$

Reducing gives

$$\frac{f}{d_2} = \sqrt{\frac{500 \times 16}{1000}} = 2.8$$

Hence f-2.8 is the required f-stop. The above is simply computational proof of an earlier statement that each successive nominal f-stop setting halves the aperture area of the previous one; or in this case f-2.8 doubles the aperture area of f-4, which is necessary to retain the same exposure if shutter time is halved.

2-7 CHARACTERISTICS OF PHOTOGRAPHIC EMULSIONS

Photographic films consist of two parts: *emulsion* and *backing* or *support*. The emulsion contains light-sensitive silver halide crystals. These are placed on the backing or support in a thin coat, as shown in Fig. 2-12. The support material is usually paper, plastic film, or glass.

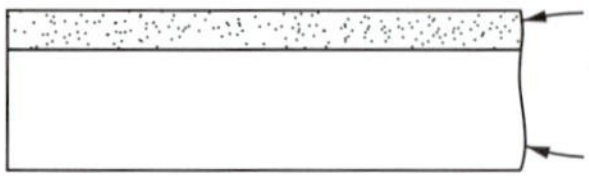

← Emulsion of silver halide crystals

← Support material

FIGURE 2-12
Cross section of a photographic film.

When silver halide crystals are exposed to light, the bond between the silver and the halide is weakened. An emulsion that has been exposed to light contains an invisible image of the object, called the *latent image*. When the latent image is developed, areas of the emulsion that were exposed to intense light turn to free silver and become black. Areas that received no light become white if the support is white paper. (They become clear if the support is glass or transparent plastic film.) The degree of darkness of developed images is a function of the total exposure (product of illuminance and time) that originally sensitized the emulsion to form the latent image. In any photographic exposure, there will be variations in illuminance received from different objects in the photographed scene, and therefore between black and white there will exist various tones of gray which result from these variations in illuminance. Actually the crystals turn black, not gray, when exposed to sufficient light. However, if the light received in a particular area is sufficient to sensitize only a portion of the crystals, then a gray tone results from a mixture of the resulting black and white. The greater the exposure, the greater the percentage of black in the mixture and hence the darker the shade of gray.

The degree of darkness of a developed emulsion is called its *density*. The greater the density, the darker the emulsion. Density of a developed emulsion on a transparent film can be determined by subjecting the film to a light source, and then comparing the intensity of incident light upon the film to that which passes through (transmitted light). The relationship is expressed in Eq. (2-7), where D is the density. Since the intensity response of a human eye is nonlinear, the base-ten logarithm (log) is used so that density will be nearly proportional to perceived brightness. A density value of zero corresponds to a completely transparent film, whereas a film that allows 1 percent of the incident light to pass through has a density of 2. The amount of light incident to an emulsion and the amount transmitted can be measured with an instrument called a *densitometer*.

$$D = \log\left(\frac{\text{incident intensity}}{\text{transmitted intensity}}\right) \tag{2-7}$$

If exposure is varied for a particular emulsion, corresponding variations in densities will be obtained. A plot of density on the ordinate versus logarithm of exposure on the abscissa for a given emulsion produces a curve called the *characteristic curve,* also known as the *D–log E curve,* or the *H-and-D curve.* A typical characteristic curve is shown in Fig. 2-13. Characteristic curves for different emulsions vary somewhat, but they all have the same general shape. The lower part of the curve, which is concave upward, is known as the *toe* region. The upper portion, which is concave downward, is the *shoulder* region. A *straight-line* portion occurs between the toe and shoulder regions.

Characteristic curves are useful in describing the characteristics of photographic emulsions. The slope of the straight-line portion of the curve for example, is a measure of the contrast of the film. The steeper the slope, the greater the *contrast* (change in density for a given range of exposure). Contrast of a given film is expressed as *gamma,* the slope of the straight-line portion of the curve, as shown on Fig. 2-13. From the figure it is

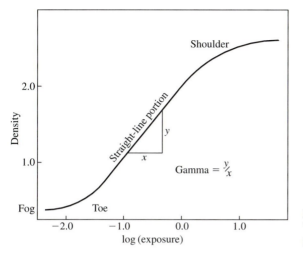

FIGURE 2-13
Typical "characteristic curve" of a photographic emulsion.

evident that for an exposure of zero the film has some density. The density of an unexposed emulsion is called *fog,* and on the curve it is the density corresponding to the low portion of the toe region. It is also apparent from Fig. 2-13 that exposure must exceed a certain minimum before density greater than fog occurs. Also, exposures within the shoulder region affect the density very little, if any. Thus, a properly exposed photograph is one in which the entire range of exposure occurs within the straight-line portion of the curve.

Just as the skin of different people varies in sensitivity to sunlight, so does the sensitivity of emulsions vary. Light sensitivity of photographic emulsions is a function of the size and number of silver halide crystals or *grains* in the emulsion. When the required amount of light exposes a grain in the emulsion, the entire grain becomes exposed regardless of its size. If a certain emulsion is composed of grains smaller than those in another emulsion, such that approximately twice as many grains are required to cover the film, then this emulsion will also require about twice as much light to properly expose *the emulsion.* Conversely, as grain size increases, the total number of grains in the emulsion decreases and the amount of light required to properly expose the emulsion decreases. Film is said to be more sensitive and faster when it requires less light for proper exposure. Faster films can be used advantageously in photographing rapidly moving objects.

As sensitivity and grain size increase, the resulting image becomes coarse and *resolution* (sharpness or crispness of the picture) is reduced. Thus, for highest pictorial quality, such as portrait work, slow, fine-grained emulsions are preferable. Film resolution can be tested by photographing a standard test pattern, as previously discussed in Sec. 2-3 and covered in greater detail in Sec. 3-13.

Photographers have developed exposure guides for films of various sensitivities. For films not used in aerial photography, the *International Standards Organization* (ISO) number is used to indicate film sensitivity or speed. The ISO number assigned to a film is roughly equal to the inverse of shutter speed (in seconds) required for proper exposure in pure sunlight for a lens opening of *f*-16. According to this rule of thumb, if a film is properly exposed in pure sunlight at *f*-16 and $\frac{1}{200}$ s, it is classified ISO 200. This rule of thumb is seldom needed today because of the availability of microprocessor-controlled cameras which, given the ISO rating of the film being used, automatically yield proper exposures (*f*-stops and shutter speeds) for particular lighting conditions.

The foregoing discussion of film speed applies primarily to ordinary ground-based photography. In aerial photography, the range of illuminance at the focal plane is significantly lower due to the narrower range of ground illuminance and atmospheric haze. For this reason, the sensitivity of films used in aerial photography is expressed as *aerial film speed* (AFS), which is different from, and should not be confused with, the ISO number. Aerial film speed is determined by the point on the characteristic curve where density is 0.3 unit above the fog density.

2-8 PROCESSING BLACK-AND-WHITE EMULSIONS

The five-step darkroom procedure for processing an exposed black-and-white emulsion is as follows:

1. *Developing*. The exposed emulsion is placed in a chemical solution called *developer*. The action of the developer causes grains of silver halide that were exposed to light to be reduced to free black silver. The free silver produces the blacks and shades of gray of which the image is composed. Developers vary in strength and other characteristics and must therefore be carefully chosen to produce the desired results. Generally the time of immersion of the film in the developer is between one and 15 minutes, depending upon the particular film and developer used. The contrast of the final image can be changed somewhat by changing the development time and the temperature of the developer.

2. *Stop bath*. When proper darkness and contrast of the image have been attained in the developing stage, it is necessary to stop the developing action. This is done with a stop bath—an *acidic* solution which neutralizes the *basic* developer solution. The emulsion is immersed in the stop bath for just a few seconds.

3. *Fixing*. Not all the silver halide grains are turned to free black silver as a result of developing. Instead, there remain many undeveloped grains which would also turn black upon exposure to light if they were not removed. To prevent further developing which would ruin the image, the undeveloped silver halide grains are dissolved out in the fixing solution. The fixing bath also hardens the emulsion. Normal immersion time in the fixing bath is from 10 to 20 min.

4. *Washing*. The emulsion is washed in clean running water to remove any remaining chemicals. If not removed, these chemicals could cause spotting or haziness of the image. Normal washing time is from 10 to 20 min. A detergent may be added to decrease the washing time.

5. *Drying*. The emulsion is dried to remove the water from the emulsion and backing material. This can be done in a variety of ways, from simply air drying to drying in elaborate heated dryers.

Modern equipment is capable of automatically performing the entire five-step darkroom procedure nonstop. The result obtained from developing black-and-white film is a *negative*. It derives its name from the fact that it is reversed in tone and geometry from the original scene that was photographed; i.e., black objects appear white and vice versa, and images are inverted.

A *positive* is obtained from the negative by repeating the photographic process. This reverses tone and geometry again, thereby producing an image in which those two characteristics are true. In producing a paper print positive from a negative, printing paper which is covered with a layer of emulsion is exposed by passing light through the negative onto the emulsion. Light is transmitted through the various areas of the negative in proportion to the lightness of the negative; e.g., black areas will not transmit light at all, and therefore they will not expose their corresponding areas of the printing paper. As a result of exposure through the negative, a latent image is formed on the printing paper. This latent image is processed using the same five-step darkroom procedure outlined above.

Besides using printing paper, positives may also be prepared on plastic film or glass plates. In photogrammetric terminology, positives prepared on glass plates or transparent plastic materials are called *diapositives*.

2-9 SPECTRAL SENSITIVITY OF EMULSIONS

The sun and various artificial sources such as lightbulbs emit a wide range of *electromagnetic energy*. The entire range of this electromagnetic energy is called the *electromagnetic spectrum*. X-rays, visible light rays, and radio waves are some familiar examples of energy variations within the electromagnetic spectrum. Electromagnetic energy travels in sinusoidal oscillations called *waves*. Variations in electromagnetic energy are classified according to variations in their wavelengths or frequencies of propagation. The velocity of electromagnetic energy in a vacuum is constant and is related to frequency and wavelength through the following expression (see also Sec. 2-2):

$$c = f\lambda \tag{2-8}$$

In Eq. (2-8), c is the velocity of electromagnetic energy in a vacuum, f is frequency, and λ is wavelength. Figure 2-14 illustrates the wavelength classification of the electromagnetic spectrum. Visible light (that electromagnetic energy to which our eyes are sensitive) is composed of only a very small portion of the spectrum (see the figure). It consists of energy with wavelengths in the range of from about 0.4 to 0.7 micrometer (μm). Energy having wavelengths slightly shorter than 0.4 μm is called *ultraviolet*, and

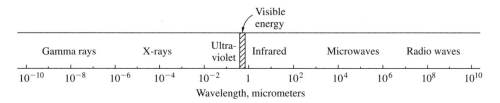

FIGURE 2-14
Classification of the electromagnetic spectrum by wavelength.

energy with wavelengths slightly longer than 0.7 μm is called *near-infrared*. Ultraviolet and near-infrared cannot be detected by the human eye.

Within the wavelengths of visible light, the human eye is able to distinguish different colors. The primary colors—blue, green, and red—are composed of slightly different wavelengths: Blue is composed of energy having wavelengths of about 0.4 to 0.5 μm, green from 0.5 to 0.6 μm, and red from 0.6 to 0.7 μm. To the human eye, other hues can be represented by combinations of the primary colors; e.g., yellow is perceived when red and green light are combined. There are multitudes of these combination colors. White light is the combination of all the visible colors. It can be broken down into its component colors by passing it through a prism, as shown in Fig. 2-15. Color separation occurs because of different refractions that occur with energy of different wavelengths.

To the human eye, an object appears a certain color because the object reflects energy of the wavelengths producing that color. If an object reflects all the visible energy that strikes it, that object will appear white. But if an object absorbs all light and reflects none, that object will appear black. If an object absorbs all green and red energy but reflects blue, that object will appear blue.

Just as the retina of the human eye is sensitive to variations in wavelength, photographic emulsions can also be manufactured with variations in wavelength sensitivity.

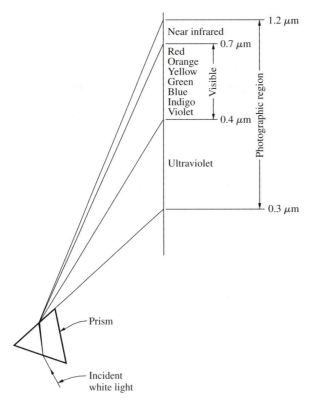

FIGURE 2-15

White light broken into the individual colors of the visible and near-visible spectrum by means of a prism. (Note that the range of wavelengths of transmitted light is non-linear.)

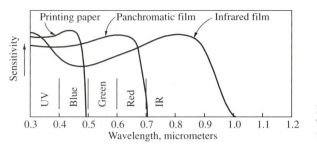

FIGURE 2-16

Typical sensitivities of various black-and-white emulsions.

Black-and-white emulsions composed of untreated silver halides are sensitive only to blue and ultraviolet energy. A red object, for example, will not produce an image on such an emulsion. These untreated emulsions are usually used on printing papers for making positives from negatives. When these printing papers are used, red or yellow lights called *safe lights* can conveniently be used to illuminate the darkroom because these colors cannot expose a paper that is sensitive only to blue light.

Black-and-white silver halide emulsions can be treated by use of fluorescent dyes so that they are sensitive to other wavelengths of the spectrum besides blue. Emulsions sensitive to blue, green, and red are called *panchromatic*. Emulsions can also be made to respond to energy in the near-infrared range. These emulsions are called *infrared,* or IR. Infrared films make it possible to obtain photographs of energy that is invisible to the human eye. An early application of this type of emulsion was in camouflage detection, where it was found that dead foliage or green netting, which had the same green color as live foliage to the human eye, reflected infrared energy differently. This difference could be detected through infrared photography. Infrared film is now widely used for a variety of applications such as detection of crop stress, tree species mapping, etc. Fig. 2-16 illustrates sensitivity differences of various emulsions.

2-10   FILTERS

The red or yellow safe light described in the previous section usually is simply an ordinary white light covered with a red or yellow filter. If the filter is red, it blocks passage of blue and green wavelengths and allows only red to pass. Filters placed in front of camera lenses also allow only certain wavelengths of energy to pass through the lens and expose the film. The use of filters on cameras can be very advantageous for certain types of photography.

Atmospheric haze is largely caused by the scattering of ultraviolet and short blue wavelengths. Pictures which are clear in spite of atmospheric haze can be taken through haze filters. These filters block passage of objectionable scattered short wavelengths (which produce haze) and prevent them from entering the camera and exposing the film. Because of this advantage, haze filters are almost always used on aerial cameras.

Filters for aerial mapping cameras are manufactured from high-quality optical glass. This is the case because light rays that form the image must pass through the filter before entering the camera. In passing through the filter, light rays are subjected to distortions caused by the filter. The camera should therefore be calibrated (see Secs. 3-9 through 3-13), with the filter locked firmly in place; after calibration, the filter should not be removed, for this would upset the calibration.

2-11 COLOR FILM

Normal color and *color infrared* emulsions are relatively recent advancements which are now finding widespread use in photogrammetry. Color emulsions consist of three layers of silver halides, as shown in Fig. 2-17. The top layer is sensitive to blue light, the second layer is sensitive to green and blue light, and the bottom layer is sensitive to red and blue light. A blue-blocking filter is built into the emulsion between the top two layers, thus preventing blue light from exposing the bottom two layers. The result is three layers sensitive to blue, green, and red light, respectively, from top to bottom. The sensitivity of each layer is indicated in Fig. 2-18.

In making a color exposure, light entering the camera sensitizes the layer(s) of the emulsion that correspond(s) to the color or combination of colors of the original scene.

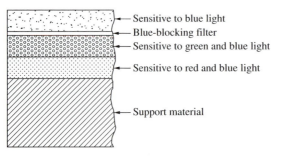

— Sensitive to blue light
— Blue-blocking filter
— Sensitive to green and blue light

— Sensitive to red and blue light

— Support material

FIGURE 2-17
Cross section of normal color film.

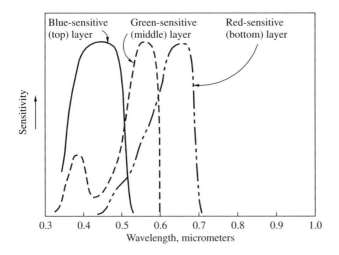

FIGURE 2-18
Typical color sensitivity of three layers of normal color film.

There are a variety of color films available, each requiring a slightly different developing process. The first step of color developing accomplishes essentially the same result as the first step of black-and-white developing. The exposed halides in each layer are turned into black crystals of silver. The remainder of the process depends on whether the film is *color negative* or *color reversal* film. With color negative film, a negative is produced and color prints are made from the negative. Color reversal film produces a true color transparency directly on the film. It is used for making color slides.

During World War II there was great interest in increasing the effectiveness of films in the infrared region of the spectrum. This interest led to the development of *color infrared* or *false-color* film. The military called it *camouflage detection* film because it allowed photo interpreters to easily differentiate between camouflage and natural foliage. Like normal color film, color IR film also has three emulsion layers, each sensitive to a different part of the spectrum. Figure 2-19 illustrates the sensitivity curves for each layer of color IR film. The top layer is sensitive to ultraviolet, blue, and green energy. The middle layer has its sensitivity peak in the red portion of the spectrum, but it, too, is sensitive to ultraviolet and blue light. The bottom layer is sensitive to ultraviolet, blue, and infrared. Color IR film is commonly used with a yellow filter, which blocks wavelengths shorter than about 0.5 μm. The shaded area of Fig. 2-19 illustrates the blocking effect of a yellow filter.

To view the exposure resulting from invisible IR energy, the IR sensitive layer must be represented by one of the three primary (visible) colors. With color IR film and a yellow filter, any objects that reflect infrared energy appear red on the final processed picture. Objects that reflect red energy appear green, and objects reflecting green energy appear blue. It is this misrepresentation of color which accounts for the name *false color*. Although color IR film was developed by the military, it has found a multitude of uses in civilian applications.

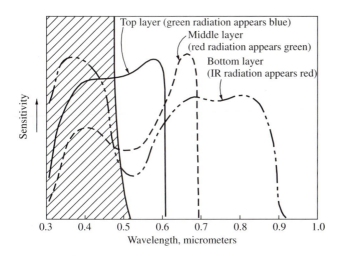

FIGURE 2-19
Typical sensitivity of color infrared (false-color) film.

2-12   CONTACT AND PROJECTION PRINTING

Contact printing is the direct process of making a photo positive from a negative. The emulsion side of a negative is placed in direct contact with the unexposed emulsion contained on printing paper, plastic material, or glass. Together these are placed in a contact printer and exposed with the emulsion of the positive facing the light source. Figure 2-20 is a schematic of a single-frame contact printer. In contact printing, the positive that is obtained is the same size as the negative from which it was made.

A processed negative will generally have a nonuniform darkness due to lens falloff, vignetting, and other effects. Consequently, uniform lighting across the negative during exposure of a positive print will underexpose the emulsion in darker areas of the negative and overexpose it in lighter areas. Compensation for this can be made in a process called *dodging*. It consists of adjusting the amount of light passing through different parts of the negative so that optimum exposure over the entire print is obtained in spite of darkness variations. Some contact printers have a bank of lights arranged in a rectangular pattern of rows beneath the exposure stage. Dodging is done manually by turning off lamps in various positions until optimum overall lighting is achieved.

Contact printers are available which automatically perform dodging. With these instruments a spotlight source makes the exposure by scanning systematically back and forth across the negative. The intensity of light transmitted through the negative is monitored, and the scanning speed is automatically increased or decreased to achieve optimum exposure in spite of varying negative density. Figure 2-21 shows the EPC model UCP-2 automatic dodging contact printer.

If positives are desired at a scale either enlarged or reduced from the original negative size, a *projection printing* process can be used. The geometry of projection printing is illustrated in Fig. 2-22. In this process, the negative is placed in the projector of the printer and illuminated from above. Light rays carry images *c* and *d*, for example, from the negative, through the projector lens, and finally to their locations *C* and *D* on the positive, which is situated on the easel plane beneath the projector. The emulsion of the positive, having been exposed, is then processed in the manner previously described.

Distances *A* and *B* of Fig. 2-22 can be varied so that positives can be printed at varying scales, and at the same time the lens formula, Eq. (2-4), can be satisfied for the projector's lens. The enlargement or reduction ratio from negative to positive size is equal to the ratio *B/A*.

The easel of a projection printer often has many small holes in it which are connected to a vacuum system. When the exposure is made, this vacuum system holds the

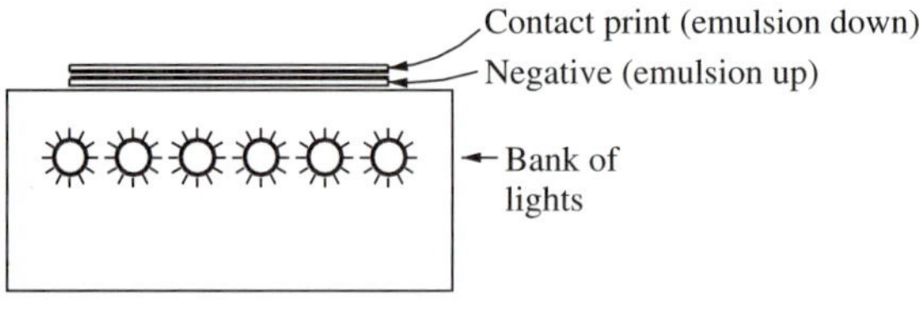

FIGURE 2-20
A contact printer.

FIGURE 2-21
EPC automatic dodging contact printer.
(*Courtesy Electronic Photo Controls, Inc.*)

positive flat, to prevent distortions due to buckling. The easel and lens of some en-largers are capable of being tilted, which makes it possible to remove the distortions inherent in tilted photographs. A print made from a tilted photograph in which these distortions have been removed has the same geometry as a vertical photograph and is called a *rectified photograph*. Rectification is discussed in greater detail in Secs. 10-8 through 10-13. A projection printer capable of enlarging and rectifying is shown in Fig. 10-10.

2-13 DIGITAL IMAGES

A digital image is a computer-compatible pictorial rendition in which the image is divided into a fine grid of "picture elements," or *pixels*. The image in fact consists of an array of integers, often referred to as *digital numbers,* each quantifying the *gray level,* or degree of darkness, at a particular element. When an output image consisting of many thousands or millions of these pixels is viewed, the appearance is that of a continuous-tone picture. The image of the famous Statue of Liberty shown in Fig. 2-23*a* illustrates

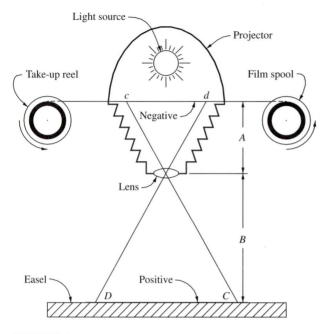

FIGURE 2-22
Geometry of enlargement with a projection printer.

the principle. This image has been divided into a pixel grid of 72 rows and 72 columns, with each pixel being represented by a value from 0 (dark black) to 255 (bright white). A portion of the image near the mouth of the statue is shown enlarged in Figure 2-23*b*, and the digital numbers associated with this portion are listed in Fig. 2-23*c*. Note the correspondence of low numbers to dark areas and high numbers to light areas.

The seemingly unusual range of values (0 to 255) can be explained by examining how computers deal with numbers. Since computers operate directly in the binary number system, it is most efficient and convenient to use ranges corresponding to powers of 2. Numbers in the range 0 to 255 can be accommodated by 1 *byte,* which consists of 8 binary digits, or *bits*. An 8-bit value can store 2^8, or 256, values, which exactly matches the range of 0 to 255 (remember to count the 0 as a value). The entire image of Fig. 2-23*a* would require a total of $72 \times 72 = 5184$ bytes of computer storage.

Digital images are produced through a process referred to as *discrete sampling*. In this process, a small image area (a pixel) is "sensed" to determine the amount of electromagnetic energy given off by the corresponding patch of surface on the object. Discrete sampling of an image has two fundamental characteristics, *geometric resolution* and *radiometric resolution*. Geometric (or spatial) resolution refers to the physical size of an individual pixel, with smaller pixel sizes corresponding to higher geometric resolution. The four illustrations of Fig. 2-24 show the entire image of the Statue of Liberty and demonstrate the effect of different geometric resolutions on image clarity. The original 72×72 pixel image of Fig. 2-24*a* and the half-resolution 36×36 image

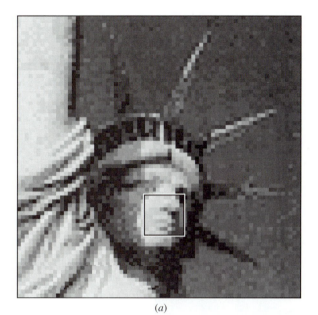

(a)

(b)

190	237	234	223	227	220	219	231	115	2
237	227	223	228	229	237	229	219	190	1
231	227	223	227	229	219	196	216	217	96
229	218	220	219	160	120	164	183	127	136
219	218	219	213	214	210	113	2	54	127
217	213	223	227	223	222	199	54	70	128
219	217	207	196	183	187	207	149	74	126
217	216	210	218	217	203	145	70	73	127
207	223	227	203	145	127	200	136	75	80
227	219	218	223	219	190	115	70	71	74

(c)

FIGURE 2-23
Digital image of Statue of Liberty.

of Fig. 2-24*b* are readily discernible. The 18 × 18 image of Fig. 2-24*c* is barely recognizable, and then only when the identity of the actual feature is known. At the resolution of the 9 × 9 image of Fig. 2-24*d*, one sees a semiorganized collection of blocks bearing little resemblance to the original image, although the rough position of the face and the arm can be detected. Obviously, geometric resolution is important for feature recognition in digital photographs.

Another fundamental characteristic of digital imagery is radiometric resolution, which can be further broken down into level of *quantization* and *spectral resolution*. Quantization refers to the conversion of the amplitude of the original electromagnetic

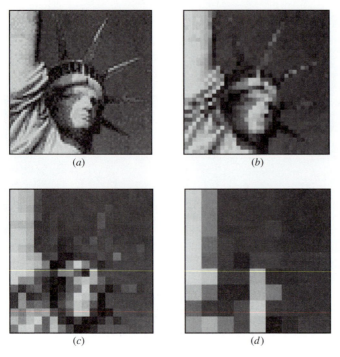

(a)   (b)

(c)   (d)

FIGURE 2-24
Digital image of Statue of Liberty at various spatial resolutions.

energy (analog signal) into a number of discrete levels (digital signal). Greater levels of quantization result in more accurate digital representations of the analog signal. While the image of Fig. 2-23 has 8-bit quantization, some digital images use 10-bit (1024 gray levels) or even 12-bit (4096 gray levels) quantization.

Figure 2-25 illustrates the effect of various levels of quantization for the statue image. In this figure, part (a) shows the original image with 256 discrete quantization levels, while parts (b), (c), and (d) show 8, 4, and 2 levels, respectively. Notice that in lower-level quantizations, large areas appear homogeneous and subtle tonal variations can no longer be detected. At the extreme is the 2-level quantization, which is also referred to as a *binary* image. Quantization level has a direct impact on the amount of computer memory required to store the image. The binary image requires only 1 bit of computer memory for each pixel ($2^1 = 2$), the 4-level image requires 2 bits ($2^2 = 4$), and the 8-level image requires 3 bits ($2^3 = 8$).

Spectral resolution is another aspect of radiometric resolution. Recall from Sec. 2-9 that the electromagnetic spectrum covers a continuous range of wavelengths. In imaging, samples of electromagnetic energy are normally sensed only within narrow bands in the spectrum. For example, the color film discussed in Sec. 2-9 samples three bands within the visible portion of the spectrum—blue, green, and red. The spectral resolution of color film corresponds to three spectral bands or channels. The black-and-white statue image of Fig. 2-23 has only one spectral band which covers the entire

(a)   (b)

(c)   (d)

FIGURE 2-25
Digital image of Statue of Liberty at various quantization levels.

visible range from violet through red. Other imaging devices may take samples in several, or even hundreds, of bands. With a higher level of spectral resolution, a more accurate representation of an object's *spectral response pattern* can be made. Since each spectral band must be stored individually in computer memory, the number of bands has a direct impact on the amount of image storage required.

> **Example 2-4.** A 3000-row by 3000-column satellite image has three spectral channels. If each pixel is represented by 8 bits (1 byte) per channel, how many bytes of computer memory are required to store the image? If this image is transferred over a computer network line having a data transfer speed of 100,000 bytes per second (bytes/s), how long will it take to transfer the image?
>
> *Solution*
>
> Number of pixels $= 3000 \times 3000 = 9{,}000{,}000$ pixels
>
> Bytes per pixel $= (3 \text{ channels}) \times (1 \text{ byte}) = 3$ bytes
>
> Image size $= (9{,}000{,}000 \text{ pixels}) \times (3 \text{ bytes/pixel}) = 27{,}000{,}000$ bytes
>
> Transfer time $= \dfrac{27{,}000{,}000 \text{ bytes}}{100{,}000 \text{ bytes/s}} = 270$ s, or $4\frac{1}{2}$ min

2-14 COLOR IMAGE REPRESENTATION

A human's perception of color is a complex interaction involving an analog spectral signal. Highly accurate and precise portrayal of the analog signal in the digital realm is prohibitively expensive in terms of memory storage. Color digital images generally utilize a simplification based on the same principle as a television screen. With the set turned on, hold a magnifying lens up to a white area on your TV (or computer) screen. You should see that the region is covered by thousands of tiny red, green, and blue dots or rectangles. If you look at an area of the screen that is a bluish color, you will see that the blue dots remain brightly lit, whereas the red and green dots are dim. At normal viewing distance, however, your eyes perceive seemingly continuous shades of color. In this same fashion, digital images represent shades of color by varying the levels of brightness of the three primary colors or channels, individually. The samples from these three channels form the digital representation of the original analog signal. This scheme is appropriately based on the theory that three types of cones (color-sensing components of the retina) in the human eye are likewise most sensitive to blue, green, and red energy.

The color of any pixel can be represented by three-dimensional coordinates in what is known as BGR (or RGB) color space. Assume for example that the digital number in each channel is quantified as an 8-bit value, which can range from 0 to 255. The color of a pixel is thus represented by an ordered triplet consisting of a blue value, a green value, and a red value, each of which can range from 0 to 255. This can be represented by the three-dimensional axis system shown in Fig. 2-26. This figure shows the *color cube,* where a particular color has a unique set of B, G, R coordinates.

Even though a B,G,R position within the color cube is adequate to specify any given color, this scheme does not lend itself to convenient human interpretation. The *intensity-hue-saturation* (IHS) system, on the other hand, is more readily understood by humans. This system can be defined as a set of cylindrical coordinates in which the height, angle, and radius represent intensity, hue, and saturation, respectively. The axis of the cylinder is the *gray line* which extends from the origin of the color cube to the opposite corner where the levels of blue, green, and red are maximum, as shown in Fig. 2-27. Thus, for any particular color, intensity represents overall brightness irrespective of color, hue represents the specific mixture of wavelengths that define the color, and saturation represents the boldness of the color. The representations of hue and saturation are illustrated in Fig. 2-28, which is a two-dimensional view showing the projection of the color cube and base of the cylinder as viewed along the gray line toward the origin. Here, hue and saturation appear as a set of polar coordinates with the direction of the 0° hue axis being arbitrarily chosen as halfway between blue and magenta. As an example, in Fig. 2-28 the hue value of 240° ($-120°$) corresponds to the color orange, and the large saturation value (radius) represents a very bold or vivid color.

Conversion from the system of B, G, R coordinates to I, H, S is accomplished by starting with the intensity (cylinder) axis lined up with the red axis and then rotating so the intensity axis lines up with the gray line. Then the standard cartesian-to-cylindrical coordinate conversion is applied. To convert from I, H, S to B, G, R the process is reversed. The derivation is based on a rigorous three-dimensional coordinate conversion,

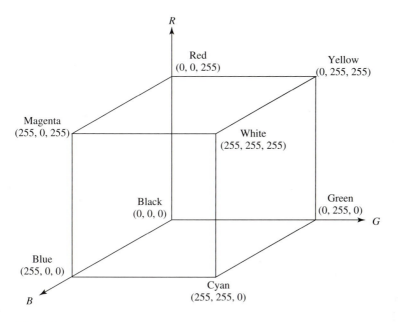

FIGURE 2-26
Blue-green-red color cube.

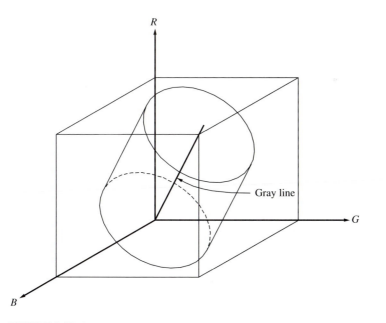

FIGURE 2-27
Position of gray line as axis of cylindrical intensity-hue-saturation system.

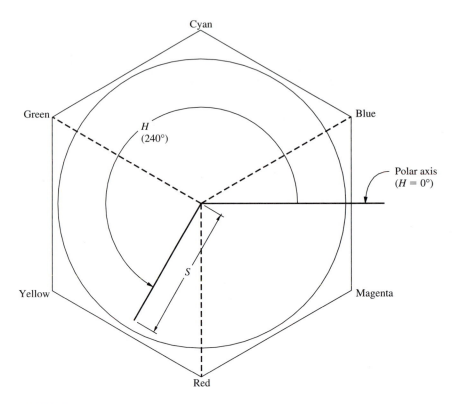

FIGURE 2-28
Representation of hue and saturation with respect to color values.

the details of which are not presented in this text. To convert from B, G, R to I, H, S, first compute two intermediate variables X and Y by Eqs. (2-8) and (2-9). Then intensity, hue, and saturation can be computed by Eqs. (2-10), (2-11), and (2-12), respectively. Example 2-5 demonstrates the procedure.

$$X = \frac{B - G}{\sqrt{2}} \tag{2-8}$$

$$Y = \frac{B + G - 2R}{\sqrt{6}} \tag{2-9}$$

$$I = \frac{B + G + R}{\sqrt{3}} \tag{2-10}$$

$$H = \tan^{-1}\left(\frac{Y}{X}\right) \tag{2-11}$$

$$S = \sqrt{X^2 + Y^2} \tag{2-12}$$

In order for the full range of hues to be accommodated, the full-circle inverse tangent function (e.g. ATAN2 in the FORTRAN language) must be used in Eq. (2-11). To convert back from I, H, S to B, G, R intermediate values of X and Y are computed by Eqs. (2-13) and (2-14). Then blue, green, and red can be computed by Eqs. (2-15), (2-16), and (2-17), respectively.

$$X = S\cos H \tag{2-13}$$

$$Y = S\sin H \tag{2-14}$$

$$B = \frac{X}{\sqrt{2}} + \frac{Y}{\sqrt{6}} + \frac{I}{\sqrt{3}} \tag{2-15}$$

$$G = -\frac{X}{\sqrt{2}} + \frac{Y}{\sqrt{6}} + \frac{I}{\sqrt{3}} \tag{2-16}$$

$$R = -\frac{2Y}{\sqrt{6}} + \frac{I}{\sqrt{3}} \tag{2-17}$$

When conversion is done from B, G, R to I, H, S the resulting value of intensity can range from 0 to $255\sqrt{3}$, hue can range from $-\pi$ to $+\pi$, and saturation can range from 0 to $255\sqrt{\frac{2}{3}}$ (assuming the ranges for the B, G, R coordinates were each 0 to 255). Since it is often desirable to store the I, H, S values as 1-byte integers, they may have to be rescaled to the 0-to-255 range. When these values are rescaled, the conversion process may no longer be perfectly reversible due to loss of precision. In other words, given a particular set of B, G, R coordinates, if they are converted to I, H, S scaled to a 1-byte integer from 0 to 255, rescaled to the original ranges, and converted back to B, G, R then the final values may be slightly different from the original ones.

Example 2-5. Convert the B, G, R coordinates (165, 57,105) to I, H, S values. Then convert the resultant values back to B, G, R.

Solution. By Eqs. (2-8) through (2-12):

$$X = \frac{165 - 57}{\sqrt{2}} = 76.37$$

$$Y = \frac{165 + 57 - 2 \times 105}{\sqrt{6}} = 4.899$$

$$I = \frac{165 + 57 + 105}{\sqrt{3}} = 188.8$$

$$H = \tan^{-1}\left(\frac{Y}{X}\right) = 0.06406 \text{ rad} = 3.670°$$

$$S = \sqrt{X^2 + Y^2} = 76.52$$

Reverse by using Eqs. (2-13) through (2-17):

$$X = 76.52 \cos{(0.06406 \text{ rad})} = 76.36$$

$$Y = 76.52 \sin{(0.06406 \text{ rad})} = 4.899$$

$$B = \frac{76.36}{\sqrt{2}} + \frac{4.899}{\sqrt{6}} + \frac{188.8}{\sqrt{3}} = 165$$

$$G = -\frac{X}{\sqrt{2}} + \frac{Y}{\sqrt{6}} + \frac{I}{\sqrt{3}} = 57$$

$$R = -\frac{2Y}{\sqrt{6}} + \frac{I}{\sqrt{3}} = 105$$

2-15   DIGITAL IMAGE PRINTING

The *B, G, R* system discussed in the preceding section is an example of a *color-additive* process. In that process, different colors of actively produced light are combined to yield other colors. When a color digital image is printed on paper, the situation is different. Instead of each pixel being an active light source, ambient white light reflects off colored inks, which absorb selected colors or channels of energy. Digital printing is a *color-subtractive* process because certain colors are absorbed (subtracted) from the ambient white light.

Controlling the "intensity" of ink in a digital printing environment would be a difficult task. With current printer technology, an ink dot is either placed on a page or it is not, there is no in-between. However, tonal variations can be simulated by varying the spatial density of the various ink dots. This concept is straightforward to explain for a black-and-white image. When a black-and-white (panchromatic) digital image is printed, tonal variations can be simulated in a variety of ways. One method, often referred to as *patterning,* redefines each pixel so it corresponds to a small two-

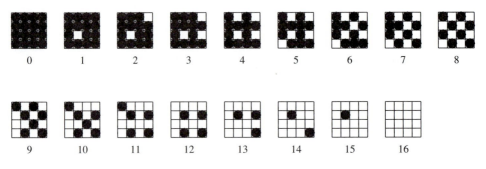

FIGURE 2-29
Ink dot patterns for 17 gray levels using a 4 × 4 output array.

dimensional array of ink dots on the output device. Figure 2-29 illustrates the idea using a 4 × 4 output array corresponding to a single pixel. In this figure, 17 different gray levels are represented, ranging from black (all 16 dots printed) to white (no dots printed). Another way of looking at it is that each dot subtracts a bit of reflected white light in the output print. If all dots are filled, all light is subtracted.

A fundamental disadvantage of this approach is the loss of spatial resolution. There is, in fact, a direct tradeoff between radiometric and geometric resolution. For example, with the 4 × 4 output array, 17 distinct radiometric values can be accommodated, but the spatial resolution will be one-fourth that of the output device. If the array is increased to 6 × 6, radiometric resolution increases to 37 distinct levels, but the spatial resolution drops to one-sixth that of the output device. The following example illustrates this tradeoff.

Example 2-6. A particular laser printer can print 600 dots per inch (dpi) over an area 8 inches (in) wide by 10 in high. If a 4 × 4 output array is used for each pixel, how many rows and columns of an image can be printed? How many if a 6 × 6 array is used?

Solution

$$4 \times 4: \quad \text{Number of columns} = (600 \text{ dpi}) (8 \text{ in}) \div 4 = 1200 \text{ columns}$$
$$\text{Number of rows} = (600 \text{ dpi}) (10 \text{ in}) \div 4 = 1500 \text{ rows}$$
$$6 \times 6: \quad \text{Number of columns} = (600 \text{ dpi}) (8 \text{ in}) \div 6 = 800 \text{ columns}$$
$$\text{Number of rows} = (600 \text{ dpi}) (10 \text{ in}) \div 6 = 1000 \text{ rows}$$

Two alternate approaches to digital printing are the related processes known as *dithering* and *error diffusion*. In these approaches, the output is produced in a manner such that an image pixel corresponds directly to a single ink dot. This results in an image which is printed at the same spatial resolution as the output device. Different radiometric levels are approximated by causing the average densities of dots in small localized areas of the image to correspond to the local brightness value. For example, a dark gray region of the image may have 95 percent of the spots covered with ink dots, with the remaining 5 percent left white. This technique results in a subtle loss in both radiometric resolution and crispness of the picture, and the printed image will appear to be slightly blurred. An advantage of these methods over the patterning approach described earlier is that a larger portion of the image, in terms of number of pixels, can be printed on the same size paper area.

Printing of color images can be accomplished by methods which are fundamentally the same as those used for black-and-white images. The process is complicated by the inclusion of spectral information; however, the principle of color subtraction still applies. The *complementary* colors of yellow, magenta, and cyan are used instead of the primary colors of blue, green, and red. For example, yellow ink appears yellow because the blue portion of the incident white light is absorbed by the ink, and only the green and red energy are reflected. Similarly, magenta absorbs green energy, reflecting only blue and red, and cyan absorbs red energy, reflecting only blue and green. To produce the color red, yellow ink and magenta ink are used, which will absorb blue and green, respectively, leaving a red hue. By varying the spatial density of the ink dots, the amounts of blue, green, and red energy being absorbed will likewise vary, allowing the full range of colors to be produced. Generally, black ink is incorporated into the scheme, which allows darker grays and black to be produced.

REFERENCES

American Society of Photogrammetry: *Manual of Color Photography,* Bethesda, MD, 1968.
_____: *Manual of Photogrammetry,* 4th ed., Bethesda, MD, 1980, chaps. 3 and 6.
Baines, H., and E. S. Bomback: *The Science of Photography,* Halsted Press, New York, 1974.
Carlson, F. P.: *Introduction to Applied Optics for Engineers,* Academic Press, New York, 1972.
Fent, L., R. J. Hall, and R. K. Nesby: "Aerial Films for Forest Inventory: Optimizing Film Parameters," *Photogrammetric Engineering and Remote Sensing,* vol. 61, no. 3, 1995, p. 281.
Ghatak, A.: *An Introduction to Modern Optics,* McGraw-Hill, New York, 1972.
Handbook of Chemistry and Physics, 56th edition, CRC Press, Cleveland, OH, 1975.
Harris, J. R., R. Murray, and T. Hirose: "IHS Transform for the Integration of Radar Imagery with Other Remotely Sensed Data," *Photogrammetric Engineering and Remote Sensing,* vol. 56, no. 12, 1990, p. 1631.
Klimes, D., and D. I. Ross: "A Continuous Process for the Development of Kodak Aerochrome Infrared Film 2433 as a Negative," *Photogrammetric Engineering and Remote Sensing,* vol. 59, no. 2. 1993, p. 209.
Knuth, D. E.: "Digital Halftones by Dot Diffusion," *ACM Transactions on Graphics,* vol. 6, no. 4, 1987, p. 245.
Laikin, M.: *Lens Design,* 2d ed., Marcel Dekker, New York, 1995.
Lillesand, T. M., and R. W. Kiefer: *Remote Sensing and Image Interpretation,* 4th ed., Wiley, New York, 1999.
Malacara, D., and Z. Malacara: *Handbook of Lens Design,* Marcel Dekker, New York, 1994.
Smith, W. J., and Genesee Optics Software, Inc.: *Modern Lens Design—A Resource Manual,* McGraw-Hill, New York, 1992.
Tani, T.: *Photographic Sensitivity—Theory and Mechanisms,* Oxford University Press, New York, 1995.
Walther, A.: *The Ray and Wave Theory of Lenses,* Cambridge University Press, New York, 1995.

PROBLEMS

2-1. Briefly explain the difference between physical and geometric optics.

2-2. Under certain conditions, the speed of light through air is 2.99688×10^8 m/s. What is the index of refraction of air under these conditions?

2-3. A certain electromagnetic energy is propagated in a vacuum at a frequency of 2,361,000 cycles per second. What is the wavelength (to the nearest meter) of this energy?

2-4. The wavelength of visible light ranges from 0.4 to 0.7 μm. Express this range in terms of frequency (to the nearest cycle per second), based on the speed of light in a vacuum.

2-5. If a certain type of glass has an index of refraction of 1.550, what is the speed of light through this glass?

2-6. A ray of light enters glass (index 1.570) from air (index 1.0003) at an incident angle of 36.9°. Find the angle of refraction.

2-7. A ray of light enters glass (index 1.515) from air (index 1.0003) at an incident angle of 67.8°. Find the angle of refraction.

2-8. A light ray emanating from under water (index 1.333) makes an angle of 37.5° with the normal to the surface. What is the angle that the refracted ray makes with the normal as it emerges into air (index 1.0003)?

2-9. What is the incident angle for a light ray emanating from glass (index 1.53) as it passes into air (index 1.0003) so that the angle of refraction is 90.0°? (This angle is known as the *critical angle,* above which total internal reflection takes place.)

2-10. Repeat Prob. 2-10, except the light ray emanates from water (index 1.333) into air.

2-11. A single ray of light traveling through air (index 1.0003) enters a convex glass lens (index 1.575) having a radius of 47.5 mm. If the light ray is parallel to and 9.5 mm above the optical axis of the lens, what are the angles of incidence and refraction?

2-12. An object located 1.8 m in front of a thin lens has its image in focus 72.5 mm from the lens on the other side. What is the focal length of the lens?

2-13. An object is located 12 m in front of a thin lens having a focal length of 50.0 mm. At what image distance will the object's image be in perfect focus?

2-14. A camera lens can accommodate object distances ranging from 1.2 m to infinity. If the focal length of the lens is 38 mm, what is the corresponding range of image distances?

2-15. A lens has a focal length of 70.0 mm. What is the object distance for an image that is perfectly focused at an image distance of 125.0 mm?

2-16. Prepare a table of image distances (in millimeters) versus object distances of exactly 1, 2, 5, 10, 100, 1000, and 5000 m for a lens having a 152.416-mm focal length, such that the images are in perfect focus.

2-17. Explain why the lens camera replaced the early pinhole camera.

2-18. Define the photographic terms *illuminance, aperture, emulsion, latent image,* and *fog.*

2-19. A camera lens has a focal length of 35.0 mm. Its *f*-stop settings range from *f*-1.4 to *f*-22. What is the maximum diameter of the aperture? Minimum diameter?

2-20. A camera with a 50.0-mm focal length lens has the *f*-stop set at 4. A 50.0-mm cylindrical extension is inserted between the lens and the camera body, increasing the nominal image distance from 50 to 100-mm. What true *f*-stop corresponds to the original setting of 4?

2-21. Prepare a table of lens aperture diameters versus nominal *f*-stop settings ranging from *f*-1 to *f*-32 for a 70.0-mm focal length lens.

2-22. An exposure is optimum at a shutter speed of $\frac{1}{250}$ s and *f*-8. If it is necessary to change the shutter speed to $\frac{1}{500}$ s, what should be the corresponding *f*-stop, to retain optimum exposure?

2-23. Repeat Prob. 2-22, except change the shutter speed to $\frac{1}{125}$ s.

2-24. An exposure is optimum at a shutter speed of $\frac{1}{1000}$ s and *f*-5.6. To increase the depth of field, it is necessary to expose at *f*-22. What is the required shutter speed to retain optimum exposure?

2-25. What is the relationship between film speed and emulsion grain size?

2-26. What is the relationship between resolution and emulsion grain size?

2-27. Explain how the slope of the straight-line portion of the *D*–log *E* curve relates to contrast.

2-28. Explain when and why a safe light can be used in a darkroom.

2-29. Explain why a haze filter is used on aerial cameras.

2-30. A panchromatic (one-band) digital image has 16,000 rows and 16,000 columns. Assuming each pixel takes 1 byte, how many bytes are required to store the image?

2-31. A 6200-row by 6200-column digital image has six spectral channels. Assuming each pixel requires 1 byte per channel, how many bytes are required to store the image?

2-32. Convert the *B, G, R* color of (115, 55, 135) to the *I, H, S* representation.

2-33. Convert the *I, H, S* color of (221, 225°, 65) to the *B, G, R* representation.

2-34. A particular ink jet printer can print 600 dpi over a 22-in-wide by 34-in-high area. If a 4 × 4 output array is used for each pixel, how many rows and columns of an image can be printed? How many if an 8 × 8 array is used?

CHAPTER

3

CAMERAS AND OTHER IMAGING DEVICES

3-1 INTRODUCTION

Perhaps the most fundamental device in the field of photogrammetry is the camera. It is the basic instrument which acquires images, from which photogrammetric products are produced.

The fourth edition of the *Manual of Photogrammetry* defines a *camera* as "a light-proof chamber or box in which the image of an exterior object is projected upon a sensitized plate or film, through an opening usually equipped with a lens or lenses, shutter, and variable aperture." That definition has been broadened in recent years with the advent of the *digital camera* which senses light energy through the use of semiconductor electronics instead of film. In many cases a more general term such as *imaging device* may be more appropriate to describe the instrument used for primary photogrammetric data acquisition.

Whatever the characteristics of the imaging device may be, an understanding of the underlying geometry is essential for precise and accurate applications of photogrammetry. The remarkable success of photogrammetry in recent years is due in large part to the progress that has been made in developing precision cameras. Perhaps the most noteworthy among recent camera developments has been the perfection of lenses of extremely high resolving power and almost negligible distortion. This has greatly increased the accuracy of photogrammetry. There have also been many significant improvements in general camera construction and operation.

Imaging devices can be categorized according to how the image is formed. Devices that acquire the image simultaneously over the entire format are *frame cameras* (or *frame sensors*). Frame cameras generally employ shutters which open and allow light from the field of view to illuminate a two-dimensional (usually rectangular) image plane before closing. Other imaging devices sense only a linear projection (strip) of the field of view at a given time and require that the device move or sweep across the area being photographed in order to acquire a two-dimensional image. Devices of this second type are referred to as *strip cameras, linear array sensors, or pushbroom scanners.* A third type of device builds an image by detecting only a small spot at a time, requiring movements in two directions (sweep and scan) in order for the two-dimensional image to be formed. These devices are often referred to as *flying spot scanners* or *mechanical scanners.*

The traditional imaging device used in photogrammetry is the aerial mapping camera, and its use is widespread in the photogrammetric industry. The requirements of aerial mapping cameras are quite different from those of ordinary handheld cameras, such as that shown in Fig. 2-11. The primary requirement of any photogrammetric aerial camera is a lens of high geometric quality. Aerial cameras must be capable of exposing in rapid succession a great number of photographs to exacting specifications. Since these cameras must perform this function while moving in an aircraft at high speed, they must have short cycling times, fast lenses, and efficient shutters. They must be capable of faithful functioning under the most extreme weather conditions and in spite of aircraft vibrations. Aerial cameras generally use roll film and have magazine capacities of several hundred exposures. Because the aerial photographic flight mission is fairly expensive and since weather and other conditions may prevent aerial photography for long periods of time, it is imperative that every precaution be taken in the manufacture of aerial cameras to guarantee the quality and reliability of the photography on each mission.

This chapter discusses various types of imaging devices, but the standard aerial mapping camera is presented in the greatest detail. This is primarily due to its wide usage; however, this is also a practical approach because other imaging devices can then conveniently be described as variations of this basic instrument.

3-2 METRIC CAMERAS FOR AERIAL MAPPING

Single-lens frame cameras are by far the most common cameras in use today. They are used almost exclusively in obtaining photographs for mapping purposes because they provide the highest geometric picture quality. With a single-lens frame camera, the lens is held fixed relative to the focal plane. The film is generally fixed in position during exposure, although it may be advanced slightly during exposure to compensate for image motion. The entire format is exposed simultaneously with a single click of the shutter.

Single-lens frame cameras are often classified according to their *angular field of view*. Angular field of view, as illustrated in Fig. 3-1, is the angle α subtended at the rear nodal point of the camera lens by the diagonal d of the picture format. [The most common frame or format size of aerial mapping cameras is 230 mm (9 in) square.]

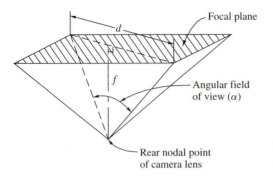

FIGURE 3-1
Angular field of view of a camera.

Classifications according to angular field of view are

1. Normal angle (up to 75°)
2. Wide angle (75° to 100°)
3. Superwide angle (greater than 100°)

Angular field of view may be calculated as follows (see Fig. 3-1):

$$\alpha = 2 \tan^{-1}\left(\frac{d}{2f}\right) \tag{3-1}$$

For a nominal 152-mm-focal-length camera with a 230-mm-square format, the angular field of view is

$$\alpha = 2 \tan^{-1}\left[\frac{\sqrt{230^2 + 230^2}}{2(152)}\right] = 94° \qquad \text{wide angle}$$

Single-lens frame cameras are available in a variety of lens *focal lengths,* and the choice depends on the purpose of the photography. The most common one in use today for mapping photography has a 152-mm (6-in) focal length and 230-mm- (9-in-) square format, although 89-mm ($3\frac{1}{2}$-in), 210-mm ($8\frac{1}{4}$-in), and 305-mm (12-in) focal lengths with 230-mm formats are also used. The 152-mm focal length with a 230-mm format provides the best combination of geometric strength and photographic scale for mapping. Longer focal lengths such as 305 mm are used primarily for obtaining photographs for aerial mosaics and for reconnaissance and interpretation purposes. They enable reasonably large photographic scales to be obtained in spite of high flying heights, and they reduce image displacements due to relief variations (see Sec. 6-8).

From Eq. (3-1), it is seen that for a particular format size, the angular field of view increases as focal length decreases. Short focal lengths, therefore, yield wider ground coverage at a given flying height than longer focal lengths.

In Fig. 1-4 the Zeiss RMK TOP 15 aerial mapping camera was shown, and Fig. 3-2 illustrates the Leica RC30 aerial mapping camera. These cameras and their predecessors, together with a few others, are being used today to take the bulk of aerial photos for mapping purposes. Both are precision single-lens frame cameras having 230-mm-square formats and film capacities of approximately 500 exposures. The TOP 15 has a nominal

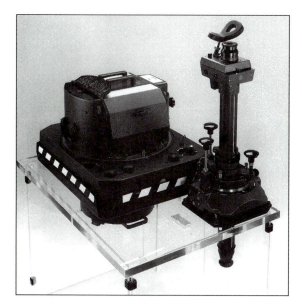

FIGURE 3-2
Leica RC30 aerial mapping camera.
(*Courtesy LH Systems, LLC.*)

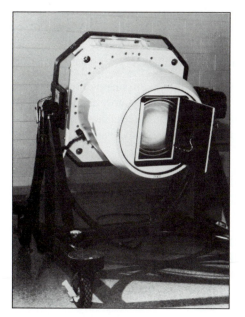

FIGURE 3-3
LFC large-format single-lens frame
camera. (*Courtesy ITEK Optical Systems.*)

152-mm-focal-length lens. The RC30 is capable of accepting interchangeable cones with lenses having nominal focal lengths of 89, 152, 210, or 305 mm.

Figure 3-3 shows the ITEK LFC large-format single-lens frame camera. This camera's lens has a 305-mm focal length, its format is 230 by 460 mm, and its film capacity is 1220 m. Designed principally for the space program, it was carried in orbit by the Space Shuttle, although it can also be used in standard aircraft. The principal uses of

FIGURE 3-4
Hasselblad 500 ELX camera. (*Courtesy University of Florida.*)

the LFC camera, in addition to topographic mapping, are in environmental monitoring and geologic exploration.

The Hasselblad camera of Fig. 3-4 is a small-format single-lens frame camera which has been employed extensively for space photography. It uses 70-mm film and can be obtained with various focal-length lenses.

3-3   MAIN PARTS OF FRAME AERIAL CAMERAS

Although all frame aerial cameras are somewhat different in construction, they are enough alike that a general description can be given which adequately encompasses all of them. The three basic components or assemblies of a frame aerial camera, as shown in the generalized cross section of Fig. 3-5, are the *magazine,* the *camera body,* and the *lens cone assembly*.

3-3.1   Camera Magazine

The camera magazine houses the reels which hold exposed and unexposed film, and it also contains the *film-advancing* and *film-flattening* mechanisms. Film flattening is very important in aerial cameras, for if the film buckled during exposure, the image positions on the resulting photographs would be incorrect. Film flattening may be accomplished in any of the following four ways: (1) by applying tension to the film during exposure; (2) by pressing the film firmly against a flat focal-plane glass that lies in front of the film; (3) by applying air pressure into the airtight camera cone, thereby forcing the film against a flat plate lying behind the focal plane; or (4) by drawing the film tightly up against a vacuum plate or platen whose surface lies in the focal plane. The vacuum system has proved most satisfactory and is the most widely used method of film flattening in aerial cameras. A focal-plane glass in front of the film is objectionable because image positions are distorted due to refraction of light rays passing through the glass (see Sec. 2-2). These distortions can be determined through calibration, however, and their effect eliminated in subsequent photogrammetric operations.

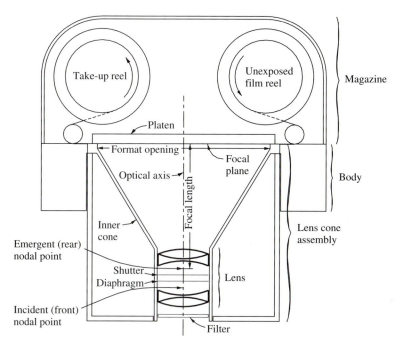

FIGURE 3-5
Generalized cross section of a frame aerial mapping camera.

3-3.2 Camera Body

The camera body is a one-piece casting which usually houses the drive mechanism. The drive mechanism operates the camera through its cycle; the cycle consists of (1) advancing the film, (2) flattening the film, (3) cocking the shutter, and (4) tripping the shutter. Power for the drive mechanism is most commonly provided by an electric motor. The camera body also contains carrying handles, mounting brackets, and electrical connections.

3-3.3 Lens Cone Assembly

The lens cone assembly contains a number of parts and serves several functions. Contained within this assembly are the *lens, shutter,* and *diaphragm* (see Fig. 3-5). With most mapping cameras, the lens cone assembly also contains an *inner cone* or *spider.* The inner cone rigidly supports the lens assembly and focal plane in a fixed relative position. This fixes the so-called elements of *interior orientation* of the camera. These elements are carefully determined through camera calibration (see Sec. 3-9) so that they are available for photogrammetric calculations. The inner cone is made of metal having a low coefficient of thermal expansion, so that changes in operating temperatures do not upset the calibration. In some aerial cameras which do not have inner cones, the body and outer lens cone act together to hold the lens with respect to the focal plane.

The *camera lens* is the most important (and most expensive) part of an aerial camera. It gathers light rays from the object space and brings them to focus in the focal plane behind the lens. Lenses used in aerial cameras are highly corrected compound lenses consisting of several elements. The lens of Fig. 2-9, for example, is the 88-mm-focal-length *SAGA-F* used in the superwide-angle lens cone of the Leica RC30 aerial mapping camera.

The *filter* serves three purposes: (1) It reduces the effect of atmospheric haze, (2) it helps provide uniform light distribution over the entire format, and (3) it protects the lens from damage and dust.

The *shutter* and *diaphragm* together regulate the amount of light which will expose the photograph. The shutter controls the length of time that light is permitted to pass through the lens. Shutters are discussed in detail in Sec. 3-5. As discussed in Sec. 2-5, the diaphragm regulates the *f*-stops of the camera by varying the size of the aperture to control the amount of light passing through the lens. Typically *f*-stops of aerial cameras range from about *f*-4 down to *f*-22. Thus, for a nominal 152-mm-focal-length lens, the diameter of the aperture ranges from about 38 mm at *f*-4 to about 7 mm at *f*-22. The diaphragm is normally located in the airspace between the lens elements of an aerial camera and consists of a series of leaves which can be rotated to vary the size of the opening.

3-4 FOCAL PLANE AND FIDUCIAL MARKS

The *focal plane* of an aerial camera is the plane in which all incident light rays are brought to focus. In aerial photography, object distances are great with respect to image distances. Aerial cameras therefore have their focus fixed for infinite object distances. This is done by setting the focal plane as exactly as possible at a distance equal to the focal length behind the rear nodal point of the camera lens. The focal plane is defined by the upper surface of the focal-plane frame. This is the surface upon which the film emulsion rests when an exposure is made.

Camera *fiducial marks* are usually four or eight in number, and they are situated in the middle of the sides of the focal plane opening, in its corners, or in both locations. Blinking lamps cause these marks to be exposed onto the negative when the picture is taken. Modern mapping cameras generally expose the fiducials at the midpoint of the duration that the shutter is open. This defines the *instant of exposure,* which is critical when incorporating airborne GPS control. This topic is discussed in Chap. 17. The aerial photographs of Figs. 1-5, 1-7, and 1-8 have four corner and four side fiducial marks.

Fiducial marks (or fiducials) serve to establish a reference *xy* photo coordinate system for image locations on the photograph. In essence, fiducials are two-dimensional control points whose *xy* coordinates are precisely and accurately determined as a part of camera calibration (see Sec. 3-9). Lines joining opposite fiducials intersect at a point called the *indicated principal point,* and aerial cameras are carefully manufactured so that this occurs very close to the true *principal point,* which is defined as the point in the focal plane where a line from the rear nodal point of the camera lens, perpendicular to the focal plane, intersects the focal plane. As will be demonstrated in subsequent chapters, it is an exceedingly important reference point in photogrammetric work. Besides providing a coordinate reference for the principal point and image points, fiducials allow for correction of film distortion (shrinkage and expansion) since each photograph contains the images of these stable control points (see Sec. 4-11).

Cameras have been perfected to compensate for image motion that occurs because of the forward movement of the aircraft during the time that the shutter is open. *Forward-motion compensation* (FMC) is usually accomplished by moving the film slightly across the focal plane during exposure, in the direction of, and at a rate just equal to, the rate of image movement. In cameras equipped with FMC, since the film moves with respect to the focal plane, it is important for fiducial lamps to blink for only an instant so that the images of the fiducial marks are sharp and distinct.

Example 3-1. An aerial camera with forward-motion compensation and a 152.4-mm focal length is carried in an airplane traveling at 200 kilometers per hour (km/h). If the flying height above the terrain is 3500 m and if the exposure time is $\frac{1}{500}$ s, what distance (in millimeters) must the film be moved across the focal plane during exposure in order to obtain a clear image?

Solution. The distance D traveled by the airplane during exposure is

$$D = \left(200\ \frac{\text{km}}{\text{h}}\right)\left(\frac{1}{500}\ \text{s}\right)\left(\frac{1\ \text{h}}{3600\ \text{s}}\right)\left(\frac{1000\ \text{m}}{1\ \text{km}}\right) = 0.11\ \text{m}$$

The distance d that the image moves during exposure [based on scale Eq. (6-1)] is

$$d = 0.11\ \text{m}\left(\frac{152.4\ \text{mm}}{3500\ \text{m}}\right) = 0.005\ \text{mm}$$

3-5 SHUTTERS

As noted in the previous section, because the aircraft carrying a camera is typically moving at a rapid speed, images will move across the focal plane during exposure. If exposure times are long or flying heights low, blurred images may result. It is important, therefore, that the shutter be open for a very short duration when aerial photographs are taken. Short exposure times also reduce the detrimental effects of aircraft vibrations on image quality. The shutter speeds of aerial cameras typically range from about $\frac{1}{100}$ to $\frac{1}{1000}$ s. Shutters are designed to operate efficiently so that they open instantaneously, remain open the required time, and then instantaneously close, thus enabling the most uniform exposure possible over the format. (Other effects such as vignetting and lens falloff, as discussed in Sec. 2-3, cause nonuniform exposure, which is unavoidable.)

There are a number of different types of camera shutters. Those used in aerial cameras are generally classified as either *between-the-lens* shutters or *focal-plane* shutters. Between-the-lens shutters are most commonly used in mapping cameras. These shutters are placed in the airspace between the elements of the camera lens, as illustrated in Fig. 3-5. Common types of between-the-lens shutters are the *leaf* type, *blade*

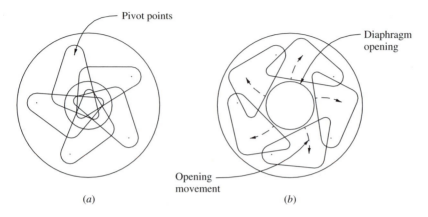

FIGURE 3-6
Schematic diagrams of a leaf-type shutter. (*a*) Shutter closed; (*b*) shutter open.

type, and *rotating-disk* type. A schematic diagram of the leaf type is shown in Fig. 3-6. It consists usually of five or more leaves mounted on pivots and spaced around the periphery of the diaphragm. When the shutter is tripped, the leaves rotate about their pivots to the open position of Fig. 3-6*b*, remain open the desired time, and then snap back to the closed position of Fig. 3-6*a*. Some camera shutters use two sets of leaves, one for opening and the other for closing. This increases shutter efficiency, shutter speed, and shutter life.

The blade-type shutter consists of four blades, two for opening and two for closing. Its operation is similar to that of a guillotine. When the shutter is triggered, the two thin *opening* plates or blades move across the diaphragm to open the shutter. When the desired exposure time has elapsed, two *closing* blades close it.

The rotating-disk type of shutter consists of a series of continuously rotating disks. Each disk has a cutaway section, and when these cutaways mesh, the exposure is made. The speed of rotation of the disks can be varied so that the desired exposure times are obtained. This type of shutter is very efficient because no starting or stopping of parts is required, as is with other types.

Focal-plane shutters are so named because they are located directly in front of the focal plane. The most common type of focal-plane shutter, the *curtain* type, consists of a curtain containing a slit. The curtain width equals the width of the focal plane. When the shutter is tripped, the slit moves across the focal plane. Exposure time is varied by varying either the speed at which the curtain moves or the width of the slit. These shutters expose different areas of the focal plane at slightly different times, and this causes relative image position errors in the resulting pictures. They are therefore not suitable for mapping cameras, but are used in reconnaissance cameras.

Another type of focal-plane shutter is the *louver* shutter. It consists of a number of louvers which are operated simultaneously in a manner similar to the operation of venetian blinds. These shutters are not as efficient as other types, and shadows created by the open louvers cause uneven lighting over the focal plane.

3-6 CAMERA MOUNTS

The camera mount is the mechanism used to attach the camera to the aircraft. Its purpose is to constrain the angular alignment of the camera so that the optical axis is vertical and the format is squarely aligned with the direction of travel. A minimal mount is equipped with dampener devices which prevent (or at least reduce) aircraft vibrations from being transmitted to the camera, and a mechanism that allows rotation in azimuth to correct for *crab*. Crab is a disparity in the orientation of the camera in the aircraft with respect to the aircraft's actual travel direction. It is usually the result of side winds which cause the aircraft's direction of heading to deviate from its actual travel direction, as shown in Fig. 3-7a. Crab can be of variable amounts, depending on the wind velocity and direction. It has the undesirable effect of reducing the stereoscopic ground coverage (see Sec. 1-4) of aerial photos, as shown in Fig. 3-7b. Figure 3-7c shows the ground coverage when the camera has been rotated within the mount in the aircraft to make two sides of the format parallel to the actual direction of travel.

More elaborate mounts like the Leica PAV 30, shown in Fig. 3-8, provide gyro stabilization of the camera. Gyroscopic devices in the housing sense the rotational movements of the aircraft, which in turn are counteracted by microprocessor-controlled

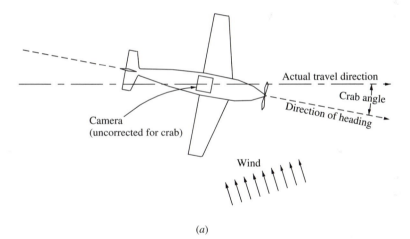

(a)

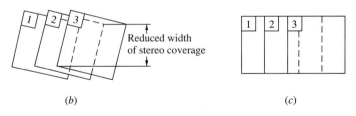

(b)   (c)

FIGURE 3-7
(a) Camera exposing aerial photography with crab present. (b) Crabbed overlapping aerial photographs.
(c) Overlapping aerial photographs with no crab.

FIGURE 3-8
Leica PAV 30 gyro-stabilized aerial camera mount.
(*Courtesy LH Systems, LLC.*)

motors that keep the camera properly oriented. Control is provided in three directions: rotation about the longitudinal axis (roll), rotation about the transverse axis (pitch), and rotation about the optical axis (yaw or drift). In addition to simply counteracting the aircraft's rotational movements, the three rotations are measured and can be recorded at the instant of exposure. These three rotational quantities are essential for proper data reduction when using airborne GPS control. When combined with forward-motion compensation, a gyro-stabilized mount results in the sharpest images by minimizing image movement during exposure.

3-7 CAMERA CONTROLS

Camera controls are those devices necessary for operating the camera and varying settings according to conditions at the time of photography. The *intervalometer* is a device which automatically trips the shutter and actuates the camera cycle as required by the flight plan. Early intervalometers could be set to automatically make exposures at fixed intervals of time. The time interval depended upon the camera focal length and format size, desired end lap (see Sec. 1-4), flying height above ground, and aircraft velocity. The disadvantage of this type of intervalometer is that with fixed time intervals, variations in end lap occur with variations in terrain elevation, flying height, or aircraft velocities. Subsequent intervalometer designs, such as that shown in Fig. 3-9, maintain

FIGURE 3-9
NS 1 Navigation Sensor. (*Courtesy Carl Zeiss, Inc.*)

exposures at the desired percentage of end lap in spite of such variations. This is done by means of a rotating chain, shown in the *viewfinder*. (The viewfinder enables the operator to continually view the terrain beneath the aircraft and to see the ground coverage of each photo.) The chain moves in the viewfinder in the same direction as the passing images. By means of a rheostat, the operator can vary the rate of movement of the chain and can make it travel at the same rate as the passing images. The desired end lap setting is entered on the intervalometer, and when the chain has moved the corresponding amount, the intervalometer automatically actuates the camera cycle.

Modern intervalometers are part of an integrated automatic control unit which incorporates a GPS receiver, enabling the exposures to be made at preprogrammed locations as dictated by the flight plan. Examples of this type of control unit are the Zeiss T-CU, shown in Fig. 3-10, and the Leica ACU30, shown in Fig. 3-11. The Zeiss unit interfaces with the camera as well as an external GPS receiver and a navigation computer. The Leica unit interfaces with up to two cameras and mounts, and it includes an integrated GPS receiver and allows for connection to an external GPS receiver. Special flight planning software allows the operator, in conjunction with GPS navigation, to establish coordinates for the planned exposure positions that will occur during the flight. The system will then trigger the camera shutter at the proper locations.

FIGURE 3-10
Zeiss T-CU control unit (upper left) with camera, GPS receiver, and computer terminal.
(*Courtesy Carl Zeiss, Inc.*)

FIGURE 3-11
Leica ACU30 control unit (upper left) with pilot display, GPS antenna, and operator terminal.
(*Courtesy LH Systems, LLC.*)

Another aerial camera control device is the *exposure control*. This consists of an exposure meter which measures terrain brightness and correlates it with the optimum combination of aperture size and shutter speed, given a particular film speed and filter factor. Exposure control units are available which operate automatically, and they constantly vary camera settings to provide optimum exposures.

3-8 AUTOMATIC DATA RECORDING

Most modern aerial mapping cameras are equipped with a data-recording system which automatically produces pertinent data on the pictures. The usual data consist of date, flying altitude, calibrated focal length of the camera, photograph number, job identification, etc. This information is entered on a *data block,* which is exposed onto the film when the photograph is taken. The images of an automatic data-recording system are shown along the left border of the aerial photograph in Fig. 1-5. Automatic data recording is a convenience which saves time and prevents mistakes in later use of the photographs.

3-9 CAMERA CALIBRATION

After manufacture and prior to use, aerial cameras are carefully calibrated to determine precise and accurate values for a number of constants. These constants, generally referred to as the *elements of interior orientation,* are needed so that accurate spatial information can be determined from photographs.

In general, camera calibration methods may be classified into one of three basic categories: *laboratory* methods, *field* methods, and *stellar* methods. Of these, laboratory methods are most frequently utilized and are normally performed by either camera manufacturers or agencies of the federal government. In one particular method of laboratory calibration, which uses a *multicollimator,* as well as in the field and stellar procedures, the general approach consists of photographing an array of targets whose relative positions are accurately known. Elements of interior orientation are then determined by making precise measurements of the target images and comparing their actual image locations with the positions they should have occupied had the camera produced a perfect perspective view. In another laboratory method, which employs a *goniometer,* direct measurements are made of projections through the camera lens of precisely positioned grid points located in the camera focal plane. Comparisons are then made with what the true projections should have been.

The elements of interior orientation which can be determined through camera calibration are as follows:

1. *Calibrated focal length* (CFL). This is the focal length that produces an overall mean distribution of lens distortion. Actually this parameter would be better termed *calibrated principal distance* since it represents the distance from the rear nodal point of the lens to the principal point of the photograph. When aerial mapping cameras are manufactured, this distance is set to correspond to the optical focal length of the lens as nearly as possible, hence the more common, though somewhat misleading, term *calibrated focal length.*

2. *Symmetric radial lens distortion.* This is the symmetric component of distortion that occurs along radial lines from the principal point. Although the amount may be negligible, this type of distortion is theoretically always present even if the lens system is perfectly manufactured to design specifications. Figure 3-12*a* shows a typical symmetric radial lens distortion pattern with magnitudes of distortion greatly exaggerated. Notice that distortion occurs in a direction inward toward, or outward from, the center of the image.

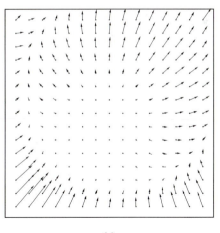

(a)

(b)

(c)

FIGURE 3-12
Lens distortion patterns: (*a*) symmetric radial, (*b*) decentering, and (*c*) combined symmetric radial and decentering.

3. *Decentering lens distortion.* This is the lens distortion that remains after compensation for symmetric radial lens distortion. Decentering distortion can be further broken down into *asymmetric radial* and *tangential* lens distortion components. These distortions are caused by imperfections in the manufacture and alignment of the lens system. Figure 3-12*b* shows a typical decentering distortion pattern, again with the magnitudes greatly exaggerated. Figure 3-12*c* shows a typical pattern of combined symmetric radial and decentering distortion.

4. *Principal point location.* This is specified by coordinates of the principal point given with respect to the *x* and *y* coordinates of the fiducial marks. (Although it is the intent in camera manufacture to place the fiducial marks so that lines between opposite pairs intersect at the principal point, there is always some small deviation from this ideal condition.)

5. *Fiducial mark coordinates.* These are the *x* and *y* coordinates of the fiducial marks which provide the two-dimensional positional reference for the principal point as well as images on the photograph.

In addition to the determination of the above elements of interior orientation, several other characteristics of the camera are often measured. *Resolution* (the sharpness or crispness with which a camera can produce an image) is determined for various distances from the principal point. Due to lens characteristics, highest resolution is achieved near the center, and lowest is at the corners of the photograph. *Focal-plane flatness* (deviation of the platen from a true plane) is measured by a special gauge. Since photogrammetric relationships assume a flat image, the platen should be nearly a true plane, generally not deviating by more than 0.01 mm. Often the *shutter efficiency*— the ability of the shutter to open instantaneously, remain open for the specified exposure duration, and close instantaneously—is also quantified.

3-10   LABORATORY METHODS OF CAMERA CALIBRATION

As noted in the previous section, the *multicollimator* method and the *goniometer* method are two types of laboratory procedures of camera calibration. The multicollimator method consists of photographing, onto a glass plate, images projected through a number of individual collimators mounted in a precisely measured angular array. A single collimator consists of a lens with a cross mounted in its plane of infinite focus. Therefore, light rays carrying the image of the cross are projected through the collimator lens and emerge parallel. When these light rays are directed toward the lens of an aerial camera, the cross will be perfectly imaged on the camera's focal plane because aerial cameras are focused for parallel light rays (those having infinite object distances).

A multicollimator for camera calibration consists of several individual collimators mounted in two perpendicular vertical planes (often, more than two planes are used). One plane of collimators is illustrated in Fig. 3-13. The individual collimators are rigidly mounted so that the optical axes of adjacent collimators intersect at known

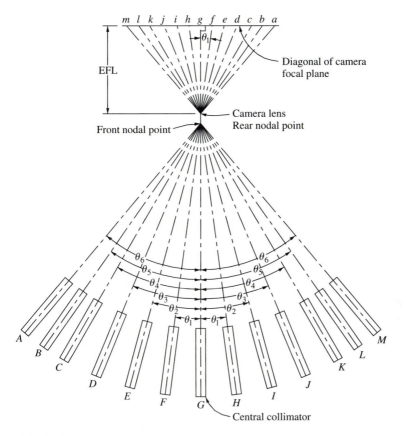

FIGURE 3-13
Bank of 13 collimators for camera calibration.

(measured) angles, such as θ_1 of Fig. 3-13. The camera to be calibrated is placed so that its focal plane is perpendicular to the central collimator axis and so that the front nodal point of its lens is at the intersection of all collimator axes. In this orientation, image g of the central collimator, which is called the *principal point of autocollimation,* occurs very near the principal point, and also very near the intersection of lines joining opposite fiducials. The camera is further oriented so that when the calibration exposure is made, the collimator crosses will be imaged along the diagonals of the camera format, as shown in Fig. 3-14.

Figure 3-14 also contains a magnified view of the very center, which illustrates several key features. In the close-up, the fiducial lines are indicated which are simply lines joining opposite pairs of fiducials, and their intersection defines the *indicated principal point.* The position of the center collimator cross (principal point of autocollimation) also serves as the origin of the photo coordinate system. The *calibrated principal point* (also known as the *point of best symmetry*) is the point whose position is determined as a result of the camera calibration. This point is the principal

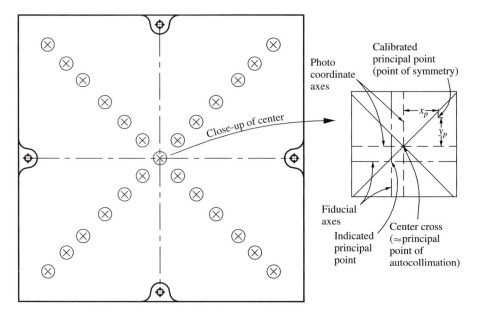

FIGURE 3-14
Images of photographed collimator targets and principal point definitions.

point that should be used to make the most precise and accurate photogrammetric calculations.

In determining the calibration parameters, a complex mathematical model is used which includes terms for the calibrated focal length and calibrated principal point coordinates as well as coefficients of symmetric radial lens distortion and decentering distortion. A least squares solution is performed which computes the most probable values for the above-mentioned terms and coefficients. The details of the calculations are beyond the scope of this text, but the fundamental concepts are presented in Chap. 11.

The goniometer laboratory procedure of camera calibration is very similar to the multicollimator method, but consists of centering a precision grid plate in the camera focal plane. The grid is illuminated from the rear and projected through the camera lens in the reverse direction. The angles at which the projected grid rays emerge are measured with a goniometer, a device similar to a surveyor's theodolite. CFL and lens distortion parameters are then computed with a mathematical model similar to that used in the multicollimator approach.

3-11   STELLAR AND FIELD METHODS OF CAMERA CALIBRATION

Both the multicollimator and goniometer methods of laboratory camera calibration require expensive and precise special equipment. An advantage of stellar and field methods is that this special equipment is not necessary. Several different stellar and field methods of camera calibration have been developed. In the stellar method, a target array consisting of identifiable stars is photographed, and the instant of exposure is recorded. Right ascensions and declinations of the stars can be obtained from an ephemeris for the precise instant of exposure so that the angles subtended by the stars at the camera station become known. Then these are compared to the angles obtained from precise measurements of the imaged stars. A drawback of this method is that since the rays of light from the stars pass through the atmosphere, compensation must be made for atmospheric refraction. On the other hand, there will be a large number of stars distributed throughout the camera format, enabling a more precise determination of lens distortion parameters.

Field procedures require that an array of targets be established and that their positions with respect to the camera station be measured precisely and accurately in three dimensions. This can be achieved conveniently using GPS methods. The targets are placed far enough from the camera station so that there is no noticeable image degradation. (Recall that an aerial camera is fixed for infinite focus.) In this configuration, the camera must be placed in a special apparatus such as a fixed tower, so that camera station coordinates are correctly related to target coordinates. This enables the CFL and principal point location to be determined as well as lens distortion parameters, even if the target configuration is essentially a two-dimensional plane. If the targets are well distributed in depth as well as laterally, accurate location of the camera is less important.

A variation of the field method described above, termed *in-flight* camera calibration, can also be employed. In this approach, the aircraft carrying the camera makes multiple passes in different directions over a target range. Based on a high number of redundant measurements of target images, additional parameters (i.e., calibration parameters) can be calculated. This method has become more practical due to advancements in airborne GPS techniques which enable accurate camera station coordinates for each exposure (see Sec. 17-11). The in-flight method can also be generalized to the point where calibration parameters are determined in conjunction with the photographs taken during the actual job. This approach, known as *analytical self-calibration,* is further described in Sec. 11-12.

3-12 CALIBRATION OF NONMETRIC CAMERAS

In certain situations where accuracy requirements and budgets are low, photogrammetrists may employ nonmetric cameras for acquisition of imagery. Nonmetric cameras are characterized by an adjustable principal distance, no film flattening or fiducial marks, and lenses with relatively large distortions. Calibration of a nonmetric camera allows at least some compensation to be made for these systematic errors.

Perhaps the most difficult aspect of calibrating a nonmetric camera is dealing with film distortion. First, by definition, a nonmetric camera has no fiducials which can be used as control points to correct for film distortion. Second, since there is no film-flattening mechanism, the film distortion can be highly variable. One approach for dealing with

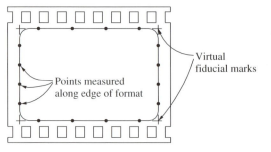

FIGURE 3-15
Point measurements to establish virtual
fiducials on a 35-mm film negative.

these problems is to establish *virtual* fiducial marks, as illustrated in Fig. 3-15. This figure depicts a 35-mm format film negative. The opening in the focal plane of the camera is nominally rectangular, although the corners are generally rounded. Using a comparator (see Sec. 4-7), *xy* coordinates of a series of points along the edges of the format can be measured. The four format border lines can be fit to these sets of points by linear regression, which is then followed by line-line intersection to determine the coordinates of the virtual fiducial marks. Once coordinates have been determined in this way, distances between fiducials can be calculated and used as a basis for computing calibrated fiducial coordinates. This operation should be repeated for several different negatives, and average values of the fiducial distances used to establish final values for calibrated fiducial coordinates.

Once calibrated fiducial coordinates have been established, the same measurement procedure can be performed on project photographs, and a two-dimensional coordinate transformation (see App. C) can be used to transform to these established fiducials.

A further complication in the calibration of nonmetric cameras arises when one is dealing with different focusing distances. Lens distortion values vary with different focus settings (i.e., different principal distances), thereby requiring a more general lens distortion model. This problem can be avoided by setting the focus of the camera to infinity during calibration as well as during normal use. That way, one set of lens distortion parameters can be determined to account for this consistent focus setting. If the camera is used for close-range work, the aperture size should be minimized in order to produce the sharpest image.

3-13   CALIBRATING THE RESOLUTION OF A CAMERA

In addition to determining interior orientation elements, laboratory methods of camera calibration provide an evaluation of the camera's resolving power. As noted in Sec. 2-3, there are two common methods of specifying lens resolving power. One is a direct count of the maximum number of lines per millimeter that can be clearly reproduced by a lens; the other is the *modulation transfer function* (MTF) of the lens. The method of calibration employed to determine the line count consists of photographing resolution test

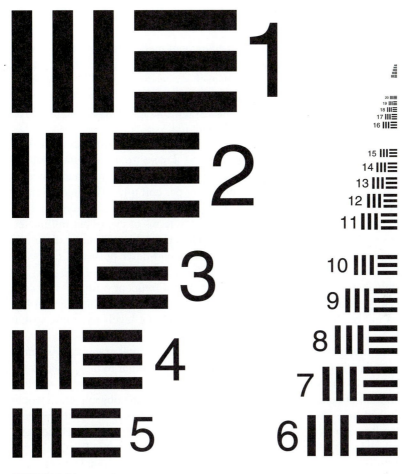

FIGURE 3-16
Resolution test pattern for camera calibration.

patterns using a very high-resolution emulsion. The test patterns (an example is shown in Fig. 3-16) are comprised of numerous sets of *line pairs* (parallel black lines of varying thickness separated by white spaces of the same thickness). The measure of line thickness for each set is its number of line pairs per millimeter. Line thickness variations in a typical test pattern may range from 10 to 80 or more line pairs per millimeter. If the multicollimator method is used to calibrate a camera, the test patterns may be projected by the collimators simultaneously with the collimator crosses and imaged on the diagonals of the camera format. After the photograph is made, the resulting images are examined under magnification to determine the finest set of parallel lines that can be clearly resolved. The average of the four resolutions at each angular increment from the central collimator is reported in the calibration certificate. Another parameter generally reported is the *area-weighted average resolution* (AWAR), which is an indication of resolution over the entire format.

Whereas the above-described maximum-line-count method appears to be a relatively simple and effective way of quantifying resolving power, it is not without its shortcomings. In the line count procedure, with each succeedingly smaller test pattern, the sharpness of distinction between lines and spaces steadily diminishes, and the smallest pattern that can clearly be discerned becomes somewhat subjective. The preferred measure of resolution is the modulation transfer function.

A fundamental concept involved in quantifying the modulation transfer function is the notion of *spatial frequency*. Spatial frequency is a measure of the number of cycles of a sinusoidal wave per unit distance. An analogy can be drawn from audio signals where frequency concerns the number of sound waves per unit time. The units of audio frequency are generally specified in cycles per second, or hertz (abbreviated Hz), whereas units of spatial frequency are typically given in terms of cycles per millimeter. Spatial frequency is directly related to the count of line pairs per millimeter discussed above. A black-and-white line pair corresponds to the up-and-down pulse of a sine wave and thus can be defined as one cycle of a wave. Therefore the number of line pairs per millimeter is equivalent to cycles per millimeter, or spatial frequency. Images that contain areas of rapidly changing levels of brightness and darkness have high spatial frequency, whereas images that contain areas of gently changing levels have low spatial frequency.

To determine modulation transfer, density scans using a photogrammetric scanner (see Sec. 4-8) are taken in a single trace across test patterns similar to those used in the line count procedure, as shown in Fig. 3-17a and c. For heavy lines with wide spacing, the actual distribution of density (brightness variations) across the object pattern would appear as the dashed lines shown in Fig. 3-17b, whereas brightness distributions measured with a densitometer across the image of this pattern would appear as the solid lines. Note that the edges of the image patterns are rounded somewhat in Fig. 3-17b, but the amplitude of brightness differences is the same as that for the original object. Thus at this spatial frequency of the pattern, modulation transfer is said to be 100 percent. Figure 3-17c shows an object pattern at a frequency 4 times that of the pattern shown in Fig. 3-17a. The density distributions of the object and resulting image of this higher-frequency pattern are shown in Fig. 3-17d. Note that in this figure, not only are the edges rounded, but also the amplitude of brightness differences is about one-half that of the original object. This indicates a modulation transfer of 50 percent from object to image. Actually, Fig. 3-17 is a somewhat simplified illustration of the quantification of modulation transfer. In the rigorous determination of modulation transfer, exposure values (rather than densities), which have a logarithmic relationship to density, are employed. The reader may consult references at the end of this chapter for more details on the modulation transfer function.

By measuring densities across many patterns of varying spatial frequencies, and plotting the resulting modulation transfer percentages on the ordinate versus corresponding spatial frequencies on the abscissa, a curve such as that illustrated in Fig. 3-18 is obtained. This curve is the modulation transfer function. The MTF has a number of advantages over the simple line count method. It is a very sensitive indicator of *edge*

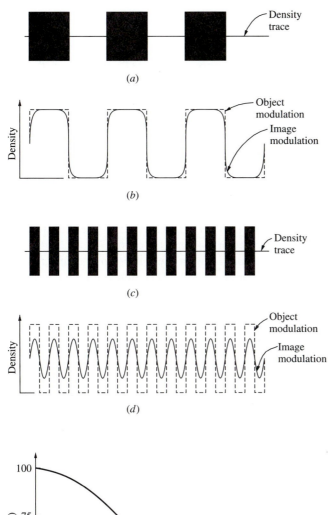

FIGURE 3-17
(*a*) Test object at low spatial frequency with density trace. (*b*) Density modulation of object (dashed) and image (solid). (*c*) Test object at high spatial frequency with density trace. (*d*) Density modulation of object (dashed) and image (solid). [Note that in part (*b*), the amplitude of the image modulation is the same as that of the object, corresponding to 100 percent modulation transfer. In (*d*) however, amplitude of the image modulation is one-half that of the object, corresponding to reduced modulation transfer.]

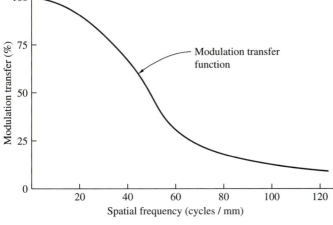

FIGURE 3-18
Curve of modulation transfer function (MTF).

effects, and it also affords the capability of predicting the resolution that may be expected at any given degree of detail. Furthermore, MTF curves can be combined for different lenses, films, and film processes; thus, it is possible to estimate the combined effects of any given imaging system. For these reasons, the MTF has become the preferred method of expressing resolution.

3-14 CHARACTERISTICS OF DIGITAL IMAGING DEVICES

The majority of the material of this chapter to this point has pertained to traditional film cameras. As noted earlier, an alternative to film cameras is the digital imaging device. A digital image, as described in Sec. 2-13, is a rectangular array of pixels in which the brightness of a scene at each discrete location has been quantified. Rather than record the reflected electromagnetic energy through silver halide crystals as in a film emulsion, digital imaging devices use solid-state detectors to sense the energy.

A common type of solid-state detector in current use is the *charge-coupled device* (CCD). Although there are a number of variations of CCD configurations, the basic operating principle is the same. At a specific pixel location, the CCD element is exposed to incident light energy, and it builds up an electric charge proportional to the intensity of the incident light. The electric charge is subsequently amplified and converted from analog to digital form. A large number of CCDs can be combined on a silicon chip in a one-dimensional or two-dimensional array.

3-14.1 Digital Frame Cameras

A digital frame camera has similar geometric characteristics to a single-lens frame camera which employs film as its recording medium. It consists of a two-dimensional array of CCD elements, called a *full-frame sensor*. The sensor is mounted in the focal plane of a single-lens camera. Acquisition of an image exposes all CCD elements simultaneously, thus producing the digital image. Figure 3-19 shows a full-frame sensor. It consists of an array of 4096 rows by 4096 columns of CCD elements, and thus it produces an image having nearly 16,800,000 pixels. Figure 3-20 shows a schematic illustration of a digital frame camera capturing an image of the ground. Light rays from all points in the scene pass through the center of the lens before reaching the CCD

FIGURE 3-19
Kodak Digital Science KAF-16800 Image Sensor.
(*Courtesy © Eastman Kodak Company.*)

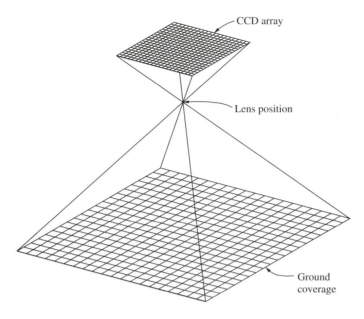

FIGURE 3-20
Geometry of a digital frame camera.

elements, thus producing the same type of point-perspective image as would have occurred if film were used.

Digital frame cameras can be classified in terms of the number of pixels in the digital image. Inexpensive digital cameras may have arrays of roughly 500 rows and 500 columns for a total of $500 \times 500 = 250{,}000$ pixels. *Megapixel* arrays have at least 1 million pixels (for example, 1024×1024) over the format. Large arrays such as that of Fig. 3-19 may have 16 million or more. Current technology can produce chips with individual CCD elements approximately 5 to 15 μm in size; thus a 4096×4096 array format will be from 20 to 60 mm square. The array of Fig. 3-19 has a 9-μm pixel size and thus can capture a 37-mm-square image in the focal plane of a camera.

Calibration of a digital frame camera is in many ways more straightforward than calibrating a film camera. Since the CCD elements are embedded in silicon, they are essentially fixed in position, which practically eliminates film distortion considerations. In a manner of speaking, each CCD element constitutes a fiducial mark. Each pixel in the resulting image, being represented by a digital number in a rectangular array, is initially assumed to be in a perfect grid pattern. In reality, however, the arrangement of CCD elements on the chip may have departed from this assumed arrangement. This departure from a perfect grid pattern can be calibrated by incorporating a two-dimensional transformation such as an affine transformation (see Sec. C-6) in the calibration model. This topic is discussed in greater detail under analytical self-calibration in Sec. 11-12.

The upper limit of resolution for a digital frame camera is absolutely fixed because of sampling into discrete elements. Since a full cycle of a wave in terms of spatial frequency must consist of a dark-to-light transition (line pair), two CCD elements are the minimum number that can capture information at the highest frequency. Thus the maximum spatial frequency (best resolution) at image scale that can be detected is

$$f_{max} = \frac{1}{2w} \tag{3-2}$$

where w = width between centers of adjacent CCD elements
        f_{max} = maximum detectable frequency

Example 3-2. A digital frame camera consists of a 5120 × 5120 array of CCD elements at a pixel size of 12 μm square. The nominal focal length of the camera is 40 mm. What is the maximum spatial frequency that can be detected (at image scale)? What is the angular field of view for this camera?

Solution. By Eq. (3-2),

$$f_{max} = \frac{1}{2w} = \frac{1}{2(0.012)} = 42 \text{ cycles/mm}$$

Format diagonal d can be calculated by

$$d = \sqrt{(5120 \times 0.012)^2 + (5120 \times 0.012)^2} = 87 \text{ mm}$$

By Eq. (3-1),

$$\alpha = 2 \tan^{-1}\left(\frac{d}{2f}\right) = 2 \tan^{-1}\left(\frac{87}{2 \times 40}\right) = 95°$$

3-14.2   Linear Array Sensors

The geometric characteristics of a linear array sensor are different from those of a single-lens frame camera. At first glance, an image obtained from a linear array sensor may appear to be the same as a digital frame camera image, but there are subtle geometric differences. A linear array sensor acquires an image by sweeping a line of detectors across the terrain and building up the image. A linear array sensor consists of a one-dimensional array or strip of CCD elements mounted in the focal plane of a single-lens camera. Since the two-dimensional image is acquired in a sweeping fashion, the image is not exposed simultaneously. Figure 3-21 shows a schematic illustration of a linear array sensor capturing an image of the ground. At a particular instant, light rays from all points along a perpendicular to the vehicle trajectory pass through the center of the lens before reaching the CCD elements, thus producing a single row of the two-dimensional image. An instant later, the vehicle has advanced to its position for the next contiguous row, and the pixels of this row are imaged. The sensor proceeds in this fashion until the entire image is acquired. Since the image was acquired

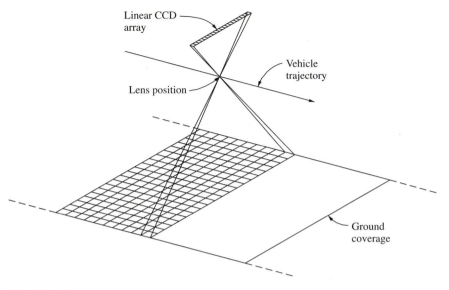

FIGURE 3-21
Geometry of a linear array sensor.

at a multitude of points along the line of the vehicle's trajectory, the resulting geometry is called *line perspective*.

In order for the image to have stable geometry suitable for photogrammetric
analysis, it is essential that the vehicle carrying the linear array sensor travel smoothly
as the image is acquired. If the sensor tips and tilts erratically as the image is acquired,
the resulting image will be correspondingly distorted. Since this smooth-flight condition
is difficult to achieve with aircraft flying in the atmosphere, linear array sensors applied
in photogrammetric mapping are more commonly used in satellite vehicles where, because of the lack of atmosphere and air currents, the vehicle trajectory is smoother.

Example 3-3. A satellite containing a linear array sensor (pointed vertically)
travels at a speed of 7300 m/s with respect to the earth's surface. If the size of the
individual pixels of the image is 10.0 m at ground scale, at what rate (in hertz)
must the detectors be read in order to produce the image?

Solution. The time interval Δt required to travel the width of 1 pixel is

$$\Delta t = \frac{10.0 \text{ m}}{7300 \text{ m/s}} = 0.0014 \text{ s}$$

Sample rate r is the reciprocal of Δt:

$$r = \frac{1}{0.0014 \text{ s}} = 730 \text{ s}^{-1} = 730 \text{ Hz}$$

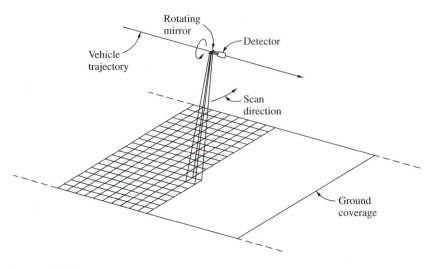

FIGURE 3-22
Geometry of a flying spot scanner.

3-14.3   Flying Spot Scanners

The third method of obtaining digital imagery is with a flying spot scanner. This device commonly uses a mirror which rotates or oscillates in a direction transverse to the vehicle trajectory. The method is similar to that of the linear array sensor except that the image is formed one pixel at a time instead of one row at a time. As the vehicle advances, the mirror scans from side to side to acquire a single row of pixels, although some designs acquire a number of rows in a single scan. The timing is such that after a row is finished being imaged, the vehicle has advanced to the beginning of the next row. A schematic illustration of the geometry of the flying spot scanner is given in Fig. 3-22. Although the figure shows the individual rows perpendicular to the vehicle trajectory, these scan lines (rows) are skewed somewhat due to the forward motion of the vehicle during scanning. This effect as well as other distortion characteristics of flying spot scanners requires that extensive geometric corrections be applied to the raw image to make it suitable for photogrammetric use.

The geometric stability of flying spot scanner images is even more susceptible to distortion effects due to erratic vehicular movement. As such, they are seldom used for photogrammetric mapping, and then only for low-accuracy work.

REFERENCES

Abdel-Aziz, Y.: "Asymmetrical Lens Distortion," *Photogrammetric Engineering and Remote Sensing,* vol. 41, no. 3, 1975, p. 337.

American Society of Photogrammetry: *Manual of Photogrammetry,* 4th ed., Bethesda, MD, 1980, chaps. 4 and 6.

———: *Manual of Remote Sensing,* 2d ed., Bethesda, MD, 1983, chaps. 8 and 21.

————: *Digital Photogrammetry: An Addendum to the Manual of Photogrammetry,* Bethesda, MD, 1996, chap. 2.

Anderson, J. M., and C. Lee: "Analytical In-Flight Calibration," *Photogrammetric Engineering and Remote Sensing,* vol. 41, no. 11, 1975, p. 1337.

Brock, G. C.: "The Possibilities for Higher Resolution in Air Survey Photography," *Photogrammetric Record,* vol. 8, no. 47, 1976, p. 589.

Brown, D. C.: "Close-Range Camera Calibration," *Photogrammetric Engineering,* vol. 37, no. 8, 1971, p. 855.

Carman, P. D.: "Camera Vibration Measurements," *Canadian Surveyor,* vol. 27, no. 3, 1973, p. 208.

Doyle, F. J.: "A Large Format Camera for Shuttle," *Photogrammetric Engineering and Remote Sensing,* vol. 45, no. 1, 1979, p. 737.

Fraser, C. S., and M. R. Shortis: "Variation of Distortion within the Photographic Field," *Photogrammetric Engineering and Remote Sensing,* vol. 58, no. 6, 1992, p. 851.

Hakkarainen, J.: "Image Evaluation of Reseau Cameras," *Photogrammetria,* vol. 33, no. 4, 1977, p. 115.

Karren, R. J.: "Camera Calibration by the Multicollimator Method," *Photogrammetric Engineering,* vol. 34, no. 7, 1968, p. 706.

Kenefick, J. F., M. S. Gyer, and B. F. Harp: "Analytical Self-Calibration," *Photogrammetric Engineering,* vol. 38, no. 11, 1972, p. 1117.

King, D., P. Walsh, and F. Ciuffreda: "Airborne Digital Frame Camera Imaging for Elevation Determination," *Photogrammetric Engineering and Remote Sensing,* vol. 60, no. 11, 1994, p. 1321.

Lei, F., and H. J. Tiziani: "A Comparison of Methods to Measure the Modulation Transfer Function of Aerial Survey Lens Systems from the Image Structures," *Photogrammetric Engineering and Remote Sensing,* vol. 54, no. 1, 1988, p. 41.

Light, D. L.: "The New Camera Calibration System at the U.S. Geological Survey," *Photogrammetric Engineering and Remote Sensing,* vol. 58, no. 2, 1992, p. 185.

————: "Film Cameras or Digital Sensors? The Challenge ahead for Aerial Imaging," *Photogrammetric Engineering and Remote Sensing,* vol. 62, no. 3, 1996, p. 285.

Merchant, D. C.: "Calibration of the Air Photo System," *Photogrammetric Engineering,* vol. 40, no. 5, 1974, p. 605.

Nielsen, V.: "More on Distortions by Focal Plane Shutters," *Photogrammetric Engineering and Remote Sensing,* vol. 41, no. 2, 1975, p. 199.

Rampal, K. K.: "System Calibration of Metric Cameras," *ASCE Journal of the Surveying and Mapping Division,* vol. 41, no. SU1, 1978, p. 51.

Rhody, B.: "A New Versatile Stereo-Camera System for Large-Scale Helicopter Photography of Forest Resources," *Photogrammetria,* vol. 32, no. 5, 1977, p. 183.

Scholer, H.: "On Photogrammetric Distortion," *Photogrammetric Engineering and Remote Sensing,* vol. 41, no. 6, 1975, p. 761.

Tayman, W. P.: "Calibration of Lenses and Cameras at the USGS," *Photogrammetric Engineering,* vol. 40, no. 11, 1974, p. 1331.

————: "User Guide for the USGS Aerial Camera Report of Calibration," *Photogrammetric Engineering and Remote Sensing,* vol. 50, no. 5, 1984, p. 577.

Welch, R., and J. Halliday: "Imaging Characteristics of Photogrammetric Camera Systems," *Photogrammetria,* vol. 29, no. 1, 1973, p. 1.

PROBLEMS

3-1. List and briefly describe the three geometric categories of imaging devices.

3-2. List the requirements of a precision mapping camera.

3-3. What is the angular field of view of a camera having a 230-mm-square format and a 305-mm focal length?

3-4. Repeat Prob. 3-3 except that the format is 55-mm-square and the focal length is 70 mm.

3-5. For a camera having a 230-mm-square format, what range of focal lengths could it have to be classified as wide angle?

3-6. An aerial camera makes an exposure at a shutter speed of $\frac{1}{500}$ s. If aircraft speed is 490 km/h, how far does the aircraft travel during the exposure?

3-7. Repeat Prob. 3-6, except that the shutter speed is $\frac{1}{1000}$ s and aircraft speed is 550 mi/h.

3-8. An aerial camera with forward-motion compensation and a 152.4-mm focal length is carried in an airplane traveling at 350 km/h. If flying height above terrain is 6100 m and if the exposure time is $\frac{1}{250}$ s, what distance (in millimeters) must the film be moved across the focal plane during exposure in order to obtain a clear image?

3-9. Repeat Prob. 3-8, except that the focal length is 305 mm and the flying height above terrain is 3200 m.

3-10. Name and briefly describe the main parts of a frame aerial mapping camera.

3-11. Discuss briefly the different types of camera shutters.

3-12. What is the purpose of the camera mount?

3-13. What is the primary benefit of gyro-stabilized camera mounts?

3-14. What is crab, and how may it be caused?

3-15. Why is camera calibration important?

3-16. What are the elements of interior orientation that can be determined in camera calibration?

3-17. List and briefly describe the various definitions of *principal point*.

3-18. Briefly describe the advantages of using the modulation transfer function to quantify the resolution of a lens over the simple line pairs per millimeter threshold.

3-19. Illustrate and briefly describe the concept of spatial frequency.

3-20. A digital frame camera consists of a 4096 × 4096 array of CCD elements at a pixel size of 9 μm square. The nominal focal length of the camera is 50 mm. What is the maximum spatial frequency that can be detected (at image scale)? What is the angular field of view for this camera?

CHAPTER
4

IMAGE
MEASUREMENTS
AND REFINEMENTS

4-1 INTRODUCTION

The solution of most photogrammetric problems generally requires some type of photographic measurement. For certain problems the measurements may simply be the lengths of lines between imaged points. However, rectangular coordinates of imaged points are the most common type of photographic measurement, and they are used directly in many photogrammetric equations. Photographic measurements are usually made on positives printed on paper, film, or glass, or in digital images manipulated on a computer. They could also be made directly on the negatives; however, this is seldom done because it can deface the imagery, and it is important to preserve the negatives for making additional prints. It is common, however, to make digital scans directly from the negatives, thus avoiding additional expense associated with making positive prints.

Equipment used for making photographic measurements varies from inexpensive, simple scales to very precise and complex machines that provide computer-compatible digital output. These various types of instruments and the manner in which they are used are described in this chapter. Because of several effects, there will be systematic errors associated with practically all photographic measurements. The sources of these errors and the manners by which they are eliminated are also discussed in this chapter.

4-2 COORDINATE SYSTEMS FOR IMAGE MEASUREMENTS

For metric cameras with side fiducial marks, the commonly adopted reference system for photographic coordinates, is the rectangular axis system formed by joining opposite fiducial marks with straight lines, as shown in Fig. 4-1. The x axis is usually arbitrarily designated as the fiducial line most nearly parallel with the direction of flight, positive in the direction of flight. The positive y axis is 90° counterclockwise from positive x. The origin of the coordinate system is the intersection of fiducial lines. This point is often called the indicated principal point, as discussed in Sec. 3-4; for a precise mapping camera it is very near the true principal point.

The position of any image on a photograph, such as point a of Fig. 4-1, is given by its rectangular coordinates x_a and y_a, where x_a is the perpendicular distance from the y axis to a and y_a is the perpendicular distance from the x axis to a. Similarly, the photographic position of image point b is given by its rectangular coordinates x_b and y_b.

It is very common for aerial cameras to have eight fiducials installed, in both side and corner locations. Figures 1-5, 1-7, and 1-8 show this fiducial mark configuration. The photographic coordinate system in this case is still defined as in Fig. 4-1. Eight fiducials enable somewhat more accurate corrections to be made for systematic errors in measured image coordinates.

When measurements are made of images taken with nonmetric cameras, the virtual fiducials (described in Sec. 3-12 and shown in Fig. 3-15) can be used to define the reference axis system. In this situation, the x axis is generally taken as the line parallel to the long sides of the focal-plane opening, and which passes through the midpoints of the two short sides. The y axis passes through the midpoint of the x axis and is perpendicular to x.

Rectangular coordinates are a very basic and useful type of photographic measurement. They are used in many different types of computations. As an example, they can be used to calculate the photo distances between points by using simple analytic geometry.

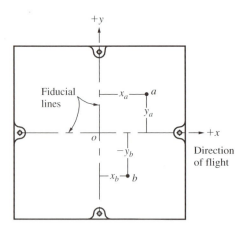

FIGURE 4-1
Photographic coordinate system based on side fiducials.

Photographic distance *ab* of Fig. 4-1, for example, may be calculated from rectangular coordinates as follows:

$$ab = \sqrt{(x_a - x_b)^2 + (y_a - y_b)^2}\tag{4-1}$$

4-3   SIMPLE SCALES FOR PHOTOGRAPHIC MEASUREMENTS

There are a variety of simple scales available for photographic measurements. The particular choice of measuring instrument will depend upon the accuracy required for the photogrammetric problem at hand. If a low order of accuracy is acceptable, an ordinary *engineer's scale* may prove satisfactory. Engineer's scales, as shown in Fig. 4-2, are available in both metric and English units and have several different graduation intervals so as to accommodate various nominal scales. Precision and accuracy can be enhanced by using a magnifying glass and the finest set of graduations, as long as suitable care is taken in aligning and reading the scale.

When greater accuracy is desired, a device such as the glass scale of Fig. 4-3 may be used. Glass scales have fine graduations etched on the bottom surface and are equipped with magnifying eyepieces which can slide along the scale. Glass scales of the

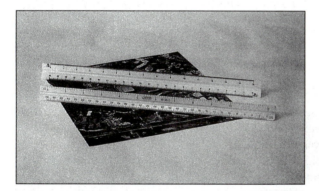

FIGURE 4-2
Metric (top) and English (bottom) engineer's scales.

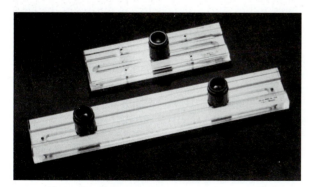

FIGURE 4-3
Glass scales for photographic measurements. (*Courtesy Teledyne-Gurley Co.*)

type shown in Fig. 4-3 can be obtained in either 6- or 12-in length with either millimeter graduations (least graduations of 0.1 mm) or inch graduations (least graduations of 0.005 in). With a glass scale, readings may be estimated quite readily to one-tenth of the smallest division, but these scales cannot be used to lay off distances.

4-4   MEASURING PHOTO COORDINATES WITH SIMPLE SCALES

The conventional procedure for measuring photo coordinates when using an engineer's scale generally consists of first marking the photo coordinate axis system. This may be done by carefully aligning a straightedge across the fiducial marks and lightly making a line with a razor blade, pin, or very sharp 4H or 5H pencil. Rectangular coordinates are then obtained by direct measurement of the perpendicular distances from these axes. Use of glass scales is generally not warranted when this method of marking axes is used, since the enhanced precision and accuracy of the scale are overshadowed by the errors associated with marking the axes.

   If the points whose coordinates are to be measured are sharp, distinct points, they may need no further identification. If not, they may be identified with a small pinprick. This should be carefully done under magnification, however, because systematic error will be introduced into measured photo coordinates if points are erroneously marked.

   It is important to affix the proper algebraic sign to measured rectangular coordinates; failure to do so will result in frustrating mistakes in solving photogrammetry problems. Points situated to the right of the y axis have positive x coordinates, and points to the left have negative x coordinates. Points above the x axis have positive y coordinates, and those below the x axis have negative y coordinates.

4-5   TRILATERATIVE METHOD OF PHOTO COORDINATE MEASUREMENT

It is possible to obtain photo coordinates using the simple scales described in Sec. 4-3 but without cutting or scratching fiducial lines. In this procedure, called the *trilaterative method,* distances such as D_1, D_2, D_3, and D_4 may be measured from fiducial marks to an image point, as illustrated in Fig. 4-4. From photo coordinates of the fiducial marks obtained in camera calibration, coordinates of image points may then be calculated using trigonometry. The procedure is applicable with corner as well as side fiducials.

   Example 4-1. Suppose that calibrated coordinates of fiducials 1 and 2 of Fig. 4-4 are $x_1 = -113.00$ mm, $y_1 = 0.00$ mm, $x_2 = 0.00$ mm, and $y_2 = -113.00$ mm. Calculate x_a and y_a if D_1 and D_2 are measured as 189.89 and 100.47 mm, respectively.

   Solution. By Eq. (4-1),

$$D_{12} = \sqrt{(-113.00 - 0.00)^2 + (0.00 + 113.00)^2} = 159.81 \text{ mm}$$

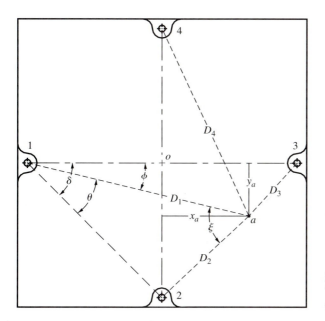

FIGURE 4-4
Trilaterative method of photo
coordinate measurement.

From the law of cosines,

$$\cos \theta = \frac{D_1^2 + D_{12}^2 - D_2^2}{2D_1 D_{12}}$$

$$= \frac{(189.89)^2 + (159.81)^2 - (100.47)^2}{2(189.89)(159.81)} = 0.84859$$

$$\theta = 31.941°$$

Also

$$\delta = \tan^{-1}\left(\frac{113.00}{113.00}\right) = 45.000°$$

Then

$$\phi = \delta - \theta = 45.000° - 31.941° = 13.059°$$

$$x_a = x_1 + D_1 \cos \phi = -113.00 + 189.89 \cos 13.059° = 71.98 \text{ mm}$$

$$y_a = y_1 - D_1 \sin \phi = 0.00 - 189.89 \sin 13.059° = -42.91 \text{ mm}$$

The trilaterative solution becomes weak as the ξ angle of Fig. 4-4 approaches either 180° or 0°. Strongest solutions are obtained for ξ angles near 90°. Since any two of four possible measured D distances yield a unique solution for the photo coordinates of an image point, the choice should be those two distances yielding an ξ angle nearest 90°. Also, accuracy may be improved by computing photo coordinates in more than one independent solution and then taking the average; e.g., one solution for image coordinates of point a of Fig. 4-4 can be made using D_1 and D_2, another can be made using D_1 and D_4, etc. A sketch should be used along with the calculations to reduce the likelihood of mistakes.

A further extension of the trilateration problem is to use all available distance measurements and to compute the coordinates of the image point by least squares. In a least squares solution, a mathematical equation is written for each distance measurement, expressing the distance as a function of the unknown coordinates of the point. For example, in Fig. 4-4, four distances are shown, therefore four independent equations can be written. There are, however, only two unknowns, x and y, associated with the point. This results in an overdetermined system of four equations with two unknowns which has, in general, no unique solution. By using the concepts of least squares, however, this system of equations can be solved to determine the most probable values for the x and y coordinates. Details of the method of least squares can be found in App. B.

A drawback of the foregoing trilaterative approach is that film shrinkage or expansion will cause a systematic error in the computed coordinates. An approach that avoids this source of systematic error is to measure distances from fiducials to image points as well as fiducial-to-fiducial distances. To add strength to the network, additional distances between image points may also be measured. However, instead of using coordinates of the fiducial marks to control the solution, the resulting trilateration network can be adjusted by least squares using *minimal constraint*. A minimal constraint is one that provides a nonredundant definition for the position and orientation of the fiducials and image points. Figure 4-5 shows how the trilateration network might be created for three image points a, b, and c. This figure shows the situation in which all possible distances are measured between each pair of points, including fiducials. When these

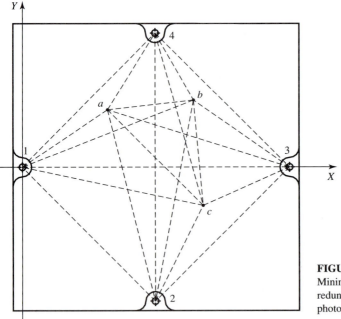

FIGURE 4-5
Minimally constrained, highly redundant, trilateration network for photo coordinate measurement.

measurements are adjusted by least squares, minimal constraint can be applied by defining the x and y coordinates for fiducial number 1 as (0, 0) and making the y coordinate of fiducial number 3 equal to 0. In this way, differences in scale between the position of fiducial images on the film and the measuring device will not stretch or compress the measured distances. The trilateration network is then adjusted by least squares, which results in most probable values for the x and y coordinates of fiducials 2 and 4, and points a, b, and c, as well as the x coordinate of fiducial 3. The x and y coordinates of fiducial 1 and the y coordinate of fiducial 3 will remain as defined by the minimal constraint.

After the minimally constrained trilateration network has been adjusted, the resulting coordinates can be used in a two-dimensional coordinate transformation (such as the affine transformation discussed in Sec. C-6) in which the calibrated fiducial coordinates are used to control the transformation. A coordinate transformation uses a mathematical model to relate the arbitrary coordinate system defined by the minimal constraint to the calibrated fiducial coordinates. Parameters of the mathematical model are computed by least squares and subsequently applied to the image points to obtain their coordinates relative to the fiducial axis system.

If the distance measurements are carefully made and sufficient redundancy exists, coordinate accuracies on the order of one-tenth to one-half of the least count of the distance-measuring device can be achieved by this method. The details of the least squares adjustments involved are beyond the scope of this text, although they may be found in references cited at the end of this chapter. Computer programs are available to perform the necessary calculations.[1]

Advantages of the trilaterative method over direct measurement from marked fiducial lines are that (1) accuracy is increased, (2) systematic errors of marking fiducial lines are eliminated, and (3) fiducial lines which deface the imagery are not necessary. If the highly redundant network approach is used in conjunction with a glass scale, coordinate measurements can be nearly as precise and accurate as those made with an expensive *comparator* (see Sec. 4-7). The drawback to this method is that substantial human effort is required, and therefore this approach is only practical for limited applications.

4-6 MEASUREMENT OF PHOTO COORDINATES WITH TABLET DIGITIZERS

Another device which is useful for making photo coordinate measurements is a *tablet digitizer*. A tablet digitizer, as shown in Fig. 4-6, consists of a large plastic base which contains a grid of wires precisely spaced and oriented in the x and y directions. A movable cursor containing a wire coil and a fine crosshair is used to point at images in the photograph. Electric current in the cursor coil generates a magnetic field which is detected by the wire grid in the tablet. Based upon the proximity of the coil to the nearest x and y wires within the grid, the magnetic field generates a corresponding electric current which is then detected and converted to x and y coordinates. These coordinates are then transferred to a computer.

---

[1]Web site address is http://www.surv.ufl.edu/wolfdewitt

FIGURE 4-6
Tablet digitizer.

Tablet digitizers generally yield a precision (repeatability) of about 0.0025 to 0.025 mm, and their accuracy varies a great deal from manufacturer to manufacturer and from model to model. The level of accuracy is affected by systematic errors in the wire grid alignment as well as other factors. Accuracies range from approximately 0.05 mm for high-quality digitizers to as course as 1 mm for inexpensive models.

To use a tablet digitizer for measuring photo coordinates, the photograph is securely mounted to the digitizer surface. Tape can be used to mount the photo although this can result in an accumulation of sticky residue on the surface. To avoid this accumulation, it is advisable to place a clear plastic sheet over the photo and fasten it with clamps at the edges. Care must be exercised so as not to shift the photograph during the measurement process. Once the photograph is placed, x and y coordinates of the fiducial marks and the image points are then measured with the digitizer. Since the xy digitizer axes are not aligned with the fiducial axis system, a two-dimensional coordinate transformation, such as the affine transformation discussed in Sec. C-6, must be applied to obtain coordinates related to the fiducial axes.

4-7 MONOCOMPARATOR MEASUREMENT OF PHOTO COORDINATES

For the ultimate in photo coordinate measurement accuracy, precise instruments called *comparators* have been used traditionally. These instruments are so named because they compare the photographic positions of imaged points with respect to the measurement scales of the devices. There are two basic types of comparators, *monocomparators* and *stereocomparators*. Monocomparators, which are discussed in this section, make measurements on one photograph at a time. With stereocomparators, image positions are measured by simultaneously viewing an overlapping stereo pair of photographs. Stereocomparators are briefly described in Sec. 17-2. Comparators are used primarily

Binocular
microscope   Stage

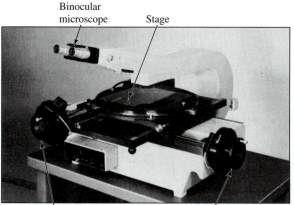

"Y" leadscrew   "X" leadscrew
drive wheel   drive wheel

FIGURE 4-7
Mann type 422-F monocomparator.
(*Courtesy David W. Mann Co.*)

to obtain precise photo coordinates necessary for camera calibration and for analytical photogrammetry. Their accuracy capability is typically in the range of from 2 to 3 micrometers (μm).

One type of monocomparator is shown in Fig. 4-7. This instrument is classified as a *leadscrew* monocomparator. It is capable of making both angle and coordinate measurements on photos having formats as large as 25 cm square. The film or diapositive to be measured is first mounted on the stage of the comparator. The stage may be rotated about a vertical axis and is equipped with a slow-motion screw for fine settings to the nearest twenty seconds of arc. The stage may be moved longitudinally along the X axis of the instrument by means of a leadscrew drive mechanism actuated by the handwheel on the right side. A similar leadscrew drive and handwheel on the left side moves the stage transversely along the Y axis of the instrument, the Y axis being perpendicular to the X axis. The instrument can be obtained with small motors for driving the stage along the leadscrews. The pitch of the leadscrew is 1 mm. A micrometer which records the nearest 0.001 turn of the leadscrew makes it possible to read X and Y coordinates to the nearest 0.001 mm (nearest micrometer). Usually a pair of digital encoders is attached to the X and Y axes so that the coordinates can be automatically transferred to a computer. The instrument is equipped with a binocular microscope, having variable magnification of from 10X to 40X, which facilitates setting the reference mark on points at which measurements are to be taken.

Another similar type of monocomparator, the Kern MK2, is shown in Fig. 4-8. This instrument has two fixed glass scales mounted orthogonally to each other from which X and Y coordinates are obtained. The photo is mounted on a stage which cannot be rotated but which can be translated in the X and Y directions. Movements of the stage are accomplished freehand to bring the desired image point to the approximate location of the reference-measuring reticle. The stage is then clamped, and using slow-motion screws, a precise setting can be made while observing under magnification through the eyepiece. By means of a photoelectric cell which monitors the positions of light beams that move along the glass scales when the stage is translated,

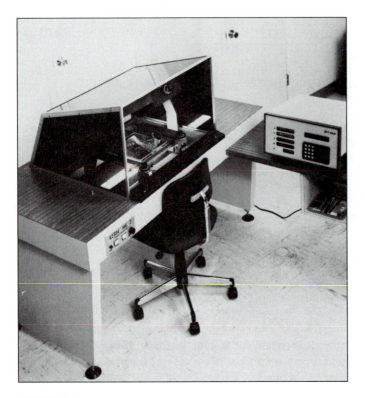

FIGURE 4-8
Kern MK2 monocomparator. (*Courtesy LH Systems, LLC.*)

measurements to the nearest micrometer are obtained and transferred automatically to a computer.

The usual approach for measuring photographic coordinates with a comparator is to measure the coordinates of all image points as well as all fiducial marks. Then an affine or other two-dimensional coordinate transformation is performed in order to relate the arbitrary comparator coordinates to the axis system defined by the fiducials. Example C-3 (of App. C) illustrates the procedure using a two-dimensional affine transformation.

Traditional monocomparators are no longer being manufactured today, although specialized devices are still being produced for limited markets. A more versatile device, the *analytical stereoplotter,* is commonly used to perform the function of both monocomparators and stereocomparators. Analytical stereoplotters are described in greater detail in Chap. 12.

Although monocomparators are generally very precise, small systematic errors do occur as a result of imperfections in their measurement systems. The magnitudes of these errors can be determined by measuring coordinates of a precise grid plate and then comparing the results with known coordinate values of the grid plate. The overall pattern of differences (errors) can be modeled with polynomials in a manner similar to that described in Sec. B-11. Measured photo coordinates can then be processed through the polynomial to effectively eliminate the systematic errors of the comparator.

4-8 PHOTOGRAMMETRIC SCANNERS

Photogrammetric scanners are devices used to convert the content of photographs from *analog form* (a continuous-tone image) to *digital form* (an array of pixels with their gray levels quantified by numerical values). These concepts were introduced in Sec. 2-13. Once the image is in digital form, coordinate measurement can take place in a computer environment, either through a manual process involving a human operator who points at features displayed on a computer screen, or through automated image-processing algorithms. A number of photogrammetric quality scanners are commercially available. They vary in approaches taken in the digital conversion (or *quantization*); however, their fundamental concepts are the same. It is essential that a photogrammetric scanner have sufficient geometric and radiometric resolution as well as high geometric accuracy.

The notions of geometric and radiometric resolution were previously discussed in Sec. 2-13. Geometric or spatial resolution of a scanner is an indication of the pixel size of the resultant image. The smaller the pixel size, the greater the detail that can be detected in the image. High-quality photogrammetric scanners should be capable of producing digital images with minimum pixel sizes on the order of 5 to 15 μm. This roughly corresponds to the resolution threshold of typical aerial photographs under actual flight conditions. Radiometric resolution of a scanner is an indication of the number of quantization levels (corresponding to image density differences) associated with a single pixel. Minimum radiometric resolution should be 256 levels (8-bit) with most scanners being capable of 1024 levels (10-bit) or higher.

The geometric quality of a scanner can be expressed by the positional accuracy of the pixels in the resultant image. If a digital image is to produce the same level of accuracy as is attainable by using film images and a comparator, the positions of the pixels in the digital image need to be at the same spatial accuracy. Hence, the geometric positional accuracy of a high-quality photogrammetric scanner should be at the 2- to 3-μm level.

Figure 4-9 shows the Zeiss PHODIS SC automatic scanner. This device is capable of scanning continuous roll or cut film at a minimum pixel size of 7 μm at 8-bit radiometric resolution and a geometric accuracy of better than 2 μm root mean square (rms). It scans by use of a linear CCD array module which makes several sweeps across the film in order to form the image. It can scan both color (red-green-blue) and panchromatic film. A one-band (black-and-white) 23 cm square image requires about 5 min of scan time at a 7-μm pixel size.

Figure 4-10 shows the DSW300 photogrammetric scanner by LH Systems. This device is also capable of scanning continuous roll or cut film with a minimum pixel size of 4 μm at 10-bit radiometric resolution and a geometric accuracy of better than 2 μm rms. It scans by use of a 2000 × 2000 CCD frame camera in a "step and stare" mode, which scans the entire format in a "tiled" fashion. It is capable of scanning both color and panchromatic film, with a one-band, 23 cm square image requiring about 4 min of scan time at 15-μm pixel size.

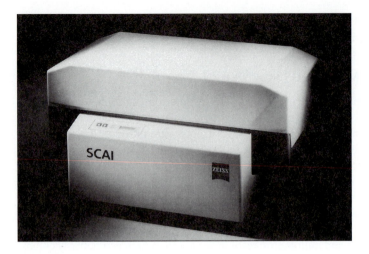

FIGURE 4-9
Zeiss PHODIS SC Scanning station with SCAI Precision Scanner. (*Courtesy Carl Zeiss, Inc.*)

FIGURE 4-10
Leica DSW300 high-performance photogrammetric scanner for roll film and cut film. (*Courtesy LH Systems, LLC.*)

4-9 REFINEMENT OF MEASURED IMAGE COORDINATES

The preceding sections of this chapter have discussed instruments and techniques for measuring photo coordinates. Procedures have also been described for eliminating systematic errors in the measurements and for reducing the coordinates to the fiducial axis system. These photo coordinates will still contain systematic errors from various other sources, however. The major sources of these errors are

1. Film distortions due to shrinkage, expansion, and lack of flatness
2. Failure of fiducial axes to intersect at the principal point
3. Lens distortions
4. Atmospheric refraction distortions
5. Earth curvature distortion

Corrections may be applied to eliminate the effects of these systematic errors. However, not all corrections need to be made for all photogrammetric problems; in fact, for work of coarse accuracy they may all be ignored. If, for example, an engineer's scale has been used to make the measurements, uncertainty in the photo coordinates may be so great that the small magnitudes of these systematic errors become insignificant. On the other hand, if precise measurements for an analytical photogrammetry problem have been made with a comparator, all the corrections may be significant. The decision as to which corrections are necessary for a particular photogrammetric problem can be made after considering required accuracy versus magnitude of error caused by neglecting the correction.

4-10 DISTORTIONS OF PHOTOGRAPHIC FILMS AND PAPERS

In photogrammetric work, true positions of images in the picture are required. Photo coordinates measured by any of the previously discussed methods will unavoidably contain small errors due to shrinkage or expansion of the photographic materials that support the emulsion of the negative and positive. In addition, since photogrammetric equations derived for applications involving frame cameras assume a flat image plane, any lack of flatness will likewise cause errors. These errors can also be categorized as film distortions, and they are generally the most difficult to compensate for, due to their nonhomogeneous nature. Photo coordinates must be corrected for these errors before they are used in photogrammetric calculations; otherwise, errors will be present in the computed results. The magnitude of error in computed values will depend upon the severity of the film distortions, which depends upon the type of emulsion support materials used and the flatness of the camera platen.

Most photographic films used to produce negatives for photogrammetric work have excellent dimensional stability, but some small changes in size do occur during processing and storage. Dimensional change during storage may be held to a minimum by maintaining constant temperature and humidity in the storage room. The actual amount of distortion present in a film is a function of several variables, including the type of film and its thickness. Typical values may vary from almost negligible amounts up to approximately 0.2 percent.

A variety of materials are available upon which positive photographic prints are made. Glass is unsurpassed in dimensional stability; when it is used, shrinkage or expansion may be considered nonexistent. If polyester material is used, a high degree of dimensional stability exists, but small distortions are usually present. If the material is paper, however, a much lower degree of dimensional stability exists. For this reason, paper prints are not used for precise photogrammetric work. The amount of paper shrinkage or expansion is a function of temperature, humidity, and paper type and thickness, but to a large degree it is a function of the method of drying the prints. If a hot-drum dryer is used or if the prints are hung to dry, greater distortions can be expected than if the prints are air-dried lying flat at room temperature. Paper shrinkages or expansions of up to 1 percent are not uncommon, and it is sometimes as large as 2 or 3 percent for single-weight papers that are hung to dry. Distortion in the x direction often is markedly different from distortion in y, and this type of *differential* distortion can lead to rather serious consequences if neglected.

4-11   SHRINKAGE CORRECTION

The nominal amount of shrinkage or expansion present in a photograph can be determined by comparing measured photographic distances between opposite fiducial marks with their corresponding values determined in camera calibration. Photo coordinates can be corrected if discrepancies exist, and the approach differs depending on the necessary level of accuracy. For lower levels of accuracy (corresponding to measurements with an engineer's scale on paper prints) the following approach may be used. If x_m and y_m are measured fiducial distances on the positive, and x_c and y_c are corresponding calibrated fiducial distances, then the corrected photo coordinates of any point a may be calculated as

$$x'_a = \left(\frac{x_c}{x_m}\right) x_a \tag{4-2}$$

$$y'_a = \left(\frac{y_c}{y_m}\right) y_a \tag{4-3}$$

In Eqs. (4-2) and (4-3), x'_a and y'_a are corrected photo coordinates and x_a and y_a are measured coordinates. The ratios x_c / x_m and y_c / y_m are simply scale factors in the x and y directions, respectively.

> **Example 4-2.** For a particular photograph, the measured x and y fiducial distances were 233.8 and 233.5 mm, respectively. The corresponding x and y calibrated fiducial distances were 232.604 and 232.621 mm, respectively. Compute

the corrected values for the measured photo coordinates which are listed in columns (*b*) and (*c*) in the table below.

Solution. From Eq. (4-2),

$$x' = \frac{232.604}{233.8}x = 0.9949x$$

$$y' = \frac{232.621}{233.5}y = 0.9962y$$

Each of the measured values is multiplied by the appropriate constant above, and the corrected coordinates are entered in columns (*d*) and (*e*) of the table below.

(*a*)	Measured coordinates		Corrected coordinates	
Point no.	(*b*) *x*, mm	(*c*) *y*, mm	(*d*) *x'*, mm	(*e*) *y'*, mm
1	−102.6	95.2	−102.1	94.8
2	−98.4	−87.8	−97.9	−87.5
3	16.3	−36.1	16.2	−36.0
4	65.7	61.8	65.4	61.6
5	104.9	−73.5	104.4	−73.2

For high-accuracy applications, shrinkage or expansion corrections may be applied through the *x* and *y* scale factors of a two-dimensional affine coordinate transformation. This method is particularly well-suited for analytical photogrammetric calculations. This procedure is described in Sec. C-6, and a numerical example is presented.

In addition to fiducial marks, or instead of them, some cameras are equipped with a *reseau*. As illustrated in Fig. 4-11, a reseau consists of a regular series of precise index marks which appear in the image, and whose coordinates are accurately known from calibration. A reseau may be superimposed on the image from finely etched marks on a piece of glass which is mounted in front of the film. When exposures are made, the grid is imprinted on the negatives, and of course, the grid then appears on all positives made from the negative. The measured positions of the grid marks on the positives can be related to their precisely known locations in the camera through local two-dimensional coordinate transformations. The usual manner of making these corrections is to individually compute an affine coordinate transformation (see Sec. C-6) for each image point, using only the nearest reseau crosses to control the transformation. Alternatively, a weighted least squares transformation (see Secs. B-6 through B-9) can be computed for each image point where the weights are inversely proportional to the distance from the image point to the reseau mark.

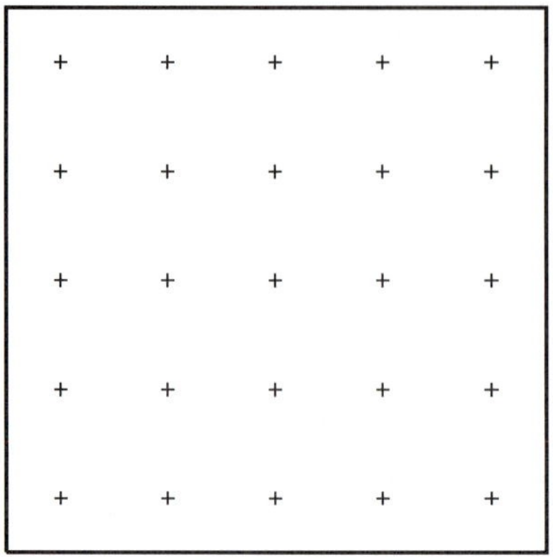

FIGURE 4-11
Reseau grid.

The advantage of using a reseau is that the grid pattern is distributed uniformly throughout the entire picture format, and thus corrections can be made for nonuniform shrinkage or expansion as well as lack of flatness of the image plane. This is not possible if only side and/or corner fiducials are available, and therefore, for the most precise and accurate analytical photogrammetric work, a reseau is preferred. A drawback associated with a glass plate reseau is that it can cause additional distortions and aberrations in the resulting image. As an alternative to using a glass plate reseau, in some cameras a series of tiny holes in a grid pattern can be drilled through the platen which holds the film firmly against the image plane during exposure. Fiber-optic filaments carry light from a lamp to each of the holes, resulting in tiny dots appearing in the image by virtue of exposure of the lights through the back of the film. With this system, distortions and aberrations are avoided since there is no glass in front of the film.

4-12 REDUCTION OF COORDINATES TO AN ORIGIN AT THE PRINCIPAL POINT

It has been previously stated that the principal point of a photograph rarely occurs precisely at the intersection of fiducial lines. The actual coordinates of the principal point with respect to the calibrated xy coordinate system of the camera are x_p and y_p, shown in Fig. 3-14. These coordinates specify the principal point location about which the lens distortions are most symmetric.

Photogrammetric equations that utilize photo coordinates are based on projective geometry and assume an origin of photo coordinates at the principal point. Therefore it is theoretically correct to reduce photo coordinates from the measurement or fiducial axis system to the axis system whose origin is at the principal point. Manufacturers of precision mapping cameras attempt to mount the fiducial marks and camera lens so that the principal point and intersection of fiducial lines coincide. Normally they accomplish this to within a few micrometers, and therefore in work of coarse accuracy using engineering scales and paper prints, this correction can usually be ignored. For precise analytical photogrammetric work, it is necessary to make the correction for the coordinates of the principal point. The correction is applied after a two-dimensional coordinate transformation (e.g., affine) is made to the coordinates measured by comparator. The principal point coordinates x_p and y_p from the camera calibration report are subtracted from the transformed x and y coordinates, respectively. Most appropriately, the correction for the principal point offset is applied in conjunction with lens distortion corrections.

4-13 CORRECTION FOR LENS DISTORTIONS

As described in Chap. 2, lens distortion causes imaged positions to be displaced from their ideal locations. The mathematical equations that are used to model lens distortions are typically comprised of two components: symmetric radial distortion and decentering distortion. In modern precision aerial mapping cameras, lens distortions are typically less than 5 μm and are only applied when precise analytical photogrammetry is being performed.

Symmetric radial lens distortion is an unavoidable product of lens manufacture, although with careful design its effects can be reduced to a very small amount. Decentering distortion, on the other hand, is primarily a function of the imperfect assembly of lens elements, not the actual design. Historically, metric aerial mapping cameras had significantly larger amounts of symmetric radial lens distortion than decentering distortion. Traditional camera calibration procedures provided information regarding only the symmetric radial component. For instance, an early approach used by the U.S. Geological Survey was to indicate radial distortion values for each of the angles of the multicollimator (see Sec. 3-10). The radial distortion value was the radial displacement from the ideal location to the actual image of the collimator cross, with positive values indicating outward displacements. The approach used for determining radial lens distortion values for these older calibration reports was to fit a polynomial curve to a plot of the displacements (on the ordinate) versus radial distances (on the abscissa). The form of the polynomial, based on lens design theory, is

$$\Delta r = k_1 r^1 + k_2 r^3 + k_3 r^5 + k_4 r^7 \tag{4-4}$$

In Eq. (4-4), Δr is the amount of radial lens distortion, r is the radial distance from the principal point, and k_1, k_2, k_3, and k_4 are coefficients of the polynomial. The coefficients of the polynomial are solved by least squares using the distortion values from the calibration report. To correct the x, y position of an image point, the distance r from the image point to the principal point is computed and used to compute the value of Δr from Eq. (4-4). This is done by first converting the fiducial coordinates x and y, to coordinates $\bar{x}$ and $\bar{y}$, relative to the principal point, by Eqs. (4-5) and (4-6). Then Eq. (4-7) is used to compute the value of r to use in Eq. (4-4).

$$\bar{x} = x - x_p \tag{4-5}$$

$$\bar{y} = y - y_p \tag{4-6}$$

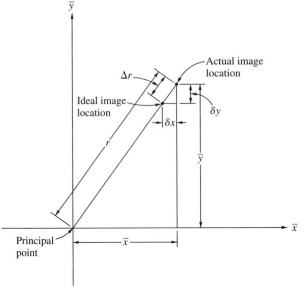

FIGURE 4-12
Relationship between radial lens distortion and corrections to x and y coordinates.

$$r = \sqrt{\bar{x}^2 + \bar{y}^2} \qquad (4\text{-}7)$$

After the radial lens distortion value of Δr is computed, its x and y components (corrections δx and δy) are computed and subtracted from $\bar{x}$ and $\bar{y}$, respectively. The δx and δy corrections are based on a similar-triangle relationship, as shown in Fig. 4-12. By similar triangles of that figure

$$\frac{\Delta r}{r} = \frac{\delta x}{\bar{x}} = \frac{\delta y}{\bar{y}}$$

from which

$$\delta x = \bar{x}\,\frac{\Delta r}{r} \qquad (4\text{-}8)$$

$$\delta y = \bar{y}\,\frac{\Delta r}{r} \qquad (4\text{-}9)$$

The corrected coordinates x_c and y_c are then computed by

$$x_c = \bar{x} - \delta x \qquad (4\text{-}10)$$

$$y_c = \bar{y} - \delta y \qquad (4\text{-}11)$$

Example 4-3. An older USGS camera calibration report specifies the calibrated focal length $f = 153.206$ mm and coordinates of the calibrated principal point as $x_p = 0.008$ mm and $y_p = -0.001$ mm. The report also lists mean radial lens distortion values given in columns (*a*) and (*b*) of the table below. Using these calibration values, compute the corrected coordinates for an image point having coordinates $x = 62.579$ mm, $y = -80.916$ mm relative to the fiducial axes.

	Mean radial distortion	
(*a*)	(*b*)	(*c*)
Field angle θ	Δr, mm	r, m
7.5°	0.004	0.0202
15°	0.007	0.0411
22.5°	0.007	0.0635
30°	0.001	0.0885
35°	−0.003	0.1073
40°	−0.004	0.1286

Solution. Compute r values (in meters) in column (*c*) by the following equation (see Fig. 4-13):

$$r = f \tan \theta$$

For example, for the field angle, $\theta = 7.5°$,

$$r = 0.153206 \tan 7.5° = 0.0202 \text{ m}$$

Using the least squares method presented in Sec. B-11 (see Example B-6), the following k values were computed.

$$k_1 = 0.2296 \qquad k_3 = 1018$$
$$k_2 = -35.89 \qquad k_4 = 12,100$$

Compute the distance from the principal point to the image point, using Eqs.(4-5), (4-6), and (4-7).

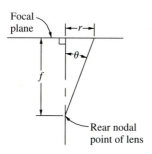
Focal plane

f

Rear nodal point of lens

FIGURE 4-13
Illustration of radial distance r as it relates to focal length f and field angle θ.

$$\bar{x} = x - x_p = 62.579 - 0.008 = 62.571 \text{ mm} = 0.062571 \text{ m}$$

$$\bar{y} = y - y_p = -80.916 - (-0.001) = -80.915 \text{ mm} = -0.080915 \text{ m}$$

$$r = \sqrt{0.062571^2 + (-0.080915)^2} = 0.1023 \text{ m}$$

Given this value for r and the k coefficients, compute Δr by Eq. (4-4):

$$\Delta r = (0.2296)(0.1023) + (-35.89)(0.1023)^3$$
$$+ (1018)(0.1023)^5 + (12,100)(0.1023)^7$$

$$= -0.0021 \text{ mm}$$

Compute δx and δy by Eqs. (4-8) and (4-9), respectively.

$$\delta x = 0.062571 \left(\frac{-0.0021}{0.1023} \right) = -0.0013 \text{ mm}$$

$$\delta y = -0.080915 \left(\frac{-0.0021}{0.1023} \right) = 0.0017 \text{ mm}$$

Compute corrected coordinates x_c and y_c by Eqs. (4-10) and (4-11), respectively.

$$x_c = 62.571 - (-0.0013) = 62.572 \text{ mm}$$

$$y_c = -80.915 - 0.0017 = -80.917 \text{ mm}$$

Lens design in modern aerial mapping cameras has evolved to such a level that symmetric radial lens distortion is of the same order of magnitude as decentering distortion, and camera calibration reports have been adapted to accommodate this change. For example, the mathematical model used in the current USGS calibration procedure, known as the Simultaneous Multi-camera Analytical Calibration (SMAC), computes both symmetric radial and decentering distortion parameters directly by least squares. Principal point coordinates and focal length are also determined in the solution. The USGS camera calibration report lists polynomial coefficients for symmetric radial lens distortion (k_0, k_1, k_2, k_3, k_4), and decentering distortion (p_1, p_2, p_3, p_4). It also gives calibrated principal point coordinates (x_p, y_p). To compute coordinates (x_c, y_c) corrected for these systematic errors, the following equations are used:

$$\delta x = \bar{x}(k_0 + k_1 r^2 + k_2 r^4 + k_3 r^6 + k_4 r^8) \tag{4-12}$$

$$\delta y = \bar{y}(k_0 + k_1 r^2 + k_2 r^4 + k_3 r^6 + k_4 r^8) \tag{4-13}$$

$$\Delta x = (1 + p_3 r^2 + p_4 r^4)[p_1(r^2 + 2\bar{x}^2) + 2p_2 \bar{x}\,\bar{y}] \tag{4-14}$$

$$\Delta y = (1 + p_3 r^2 + p_4 r^4)[2 p_1 \bar{x}\,\bar{y} + p_2(r^2 + 2\bar{y}^2)] \tag{4-15}$$

$$x_c = \bar{x} + \delta x + \Delta x \tag{4-16}$$

$$y_c = \bar{y} + \delta y + \Delta y \tag{4-17}$$

In Eqs. (4-12) through (4-17), $\bar{x}$ and $\bar{y}$ are coordinates of the image relative to the principal point as computed by Eqs. (4-5) and (4-6), respectively; r is the radial distance from the image to the principal point as computed by Eq. (4-7); k_0, k_1, k_2, k_3, and k_4 are coefficients of symmetric radial lens distortion from the calibration report; p_1, p_2, p_3, and p_4 are coefficients of decentering distortion from the calibration report; δx and δy are the symmetric radial lens distortion corrections to $\bar{x}$ and $\bar{y}$, respectively; and Δx and Δy are the decentering distortion corrections to $\bar{x}$ and $\bar{y}$, respectively.

Example 4-4. The parameters of a current USGS camera calibration report are given in the following table. Using these calibration values, compute the corrected coordinates for an image point having coordinates $x = -47.018$ mm, $y = 43.430$ mm relative to the fiducial axes.

Symmetric radial distortion parameters		Decentering distortion parameters		Calibrated principal point	
k_0	0.5493×10^{-4}	p_1	-0.7953×10^{-7} mm^{-1}	x_p	0.010 mm
k_1	-0.5984×10^{-8} mm^{-2}	p_2	0.1018×10^{-6} mm^{-1}	y_p	-0.001 mm
k_2	0.1053×10^{-12} mm^{-4}	p_3	0 mm^{-2}		
k_3	0 mm^{-6}	p_4	0 mm^{-4}		
k_4	0 mm^{-8}				

Solution. Compute $\bar{x}$, $\bar{y}$, and r by Eqs. (4-5), (4-6), and (4-7), respectively.

$$\bar{x} = -47.018 - 0.010 = -47.028 \text{ mm}$$

$$\bar{y} = 43.430 - (-0.001) = 43.431 \text{ mm}$$

$$r = \sqrt{(-47.028)^2 + 43.431^2} = 64.015 \text{ mm}$$

Compute symmetric radial lens distortion corrections δx and δy, using Eqs. (4-12) and (4-13), respectively.

$$\delta x = -47.028[0.5493 \times 10^{-4} - 0.5984 \times 10^{-8}(64.015)^2 + 0.1053 \times 10^{-12} (64.015)^4] = -0.0015 \text{ mm}$$

$$\delta x = 43.431[0.5493 \times 10^{-4} - 0.5984 \times 10^{-8}(64.015)^2 + 0.1053 \times 10^{-12} (64.015)^4] = 0.0014 \text{ mm}$$

Compute decentering distortion corrections Δx and Δy, using Eqs. (4-14) and (4-15), respectively.

$$\Delta x = 1\{(-0.7953 \times 10^{-7}) [64.015^2 + 2(-47.028)^2] + 2(0.1018 \times 10^{-6}) (-47.028) (43.431)\} = -0.0011 \text{ mm}$$

$$\Delta y = 1\{2(-0.7953 \times 10^{-7}) (-47.028) (43.431) + (0.1018 \times 10^{-6}) [64.015^2 + 2(43.431)^2]\} = 0.0011 \text{ mm}$$

Compute the corrected coordinates x_c and y_c, using Eqs. (4-16) and (4-17), respectively.

$$x_c = -47.028 - 0.0015 - 0.0011 = -47.031 \text{ mm}$$

$$y_c = 43.431 + 0.0014 + 0.0011 = 43.434 \text{ mm}$$

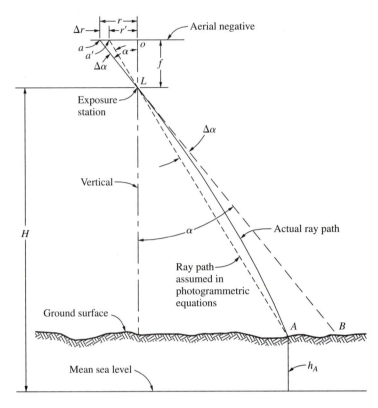

FIGURE 4-14
Atmospheric refraction in aerial photography.

4-14   CORRECTION FOR ATMOSPHERIC REFRACTION

It is well known that density (and hence the index of refraction) of the atmosphere decreases with increasing altitude. Because of this condition, light rays do not travel in straight lines through the atmosphere, but rather they are bent according to Snell's law (see Sec. 2-2), as shown in Fig. 4-14. The incoming light ray from point A of the figure makes an angle α with the vertical. If refraction were ignored, the light ray would appear to be coming from point B rather than from point A. Photogrammetric equations assume that light rays travel in straight paths, and to compensate for the known refracted paths, corrections are applied to the image coordinates.

In Fig. 4-14, if a straight path had been followed by the light ray from object point A, then its image would have been at a'. The angular distortion due to refraction is $\Delta\alpha$, and the linear distortion on the photograph is Δr. Refraction causes all imaged points to be displaced outward from their correct positions. The magnitude of refraction distortion increases with increasing flying height and with increasing α angle. Refraction distortion occurs radially from the photographic nadir point[2] (principal point of a vertical photo) and is zero at the nadir point. Atmospheric refraction in tilted photographs is treated in Sec. 10-14.

The usual approach to the atmospheric refraction correction is based on the assumption that change in the refractive index of air is directly proportional to change in height. Starting with the incident angle of the light ray at ground level, Snell's law can be solved continuously along the ray path for each infinitesimal change in angle due to refraction. When all the infinitesimal changes are summed, the total is proportional to the tangent of the incident angle. (Derivation of the relationship involves the solution of a differential equation, which is beyond the scope of this text.) The proportionality constant is based on the values of the refractive indices at ground level and at the camera position, which are related to elevation. The relationship that expresses the angular distortion $\Delta\alpha$ as a function of α is

$$\Delta\alpha = K \tan \alpha \tag{4-18}$$

In this equation, α is the angle between the vertical and the ray of light, as shown in Fig. 4-14, and K is a value which depends upon the flying height above mean sea level and the elevation of the object point. There are several different approaches to calculating a value for K, with most assuming a standard atmosphere. A convenient method, adapted from the *Manual of Photogrammetry,* is to compute K by

$$K = (7.4 \times 10^{-4})\,(H - h)\,[1 - 0.02(2H - h)] \tag{4-19}$$

In Eq. (4-19), H is the flying height of the camera above mean sea level in kilometers, and h is the elevation of the object point above mean sea level in kilometers. The units of K are degrees.

The procedure for computing atmospheric refraction corrections to image coordinates on a vertical photo begins by computing radial distance r from the principal point to the image, using Eq. (4-20). In this equation, the x and y image coordinates do not necessarily need to be related to the principal point since the error due to the assumption of vertical photography far overshadows any error which would be introduced.

$$r = \sqrt{x^2 + y^2} \tag{4-20}$$

Also from Fig. 4-14,

$$\tan \alpha = \frac{r}{f} \tag{4-21}$$

The values of K and $\tan \alpha$ from Eqs. (4-19) and (4-21), respectively, are then substituted into Eq. (4-18) to compute refraction angle $\Delta\alpha$.

$$\Delta\alpha = K \frac{r}{f} \tag{4-22}$$

---

[2] The photographic nadir point is defined in Sec. 10-2.

The radial distance r' from the principal point to the corrected image location can then be computed by

$$r' = f \tan(\alpha - \Delta\alpha) \tag{4-23}$$

The change in radial distance Δr is then computed by

$$\Delta r = r - r' \tag{4-24}$$

The x and y components of atmospheric refraction distortion corrections (δx and δy) can then be computed by Eqs. (4-8) and (4-9), using the values of x and y in place of $\bar{x}$ and $\bar{y}$, respectively. To compute corrected coordinates x' and y', the corrections δx and δy are subtracted from x and y, respectively.

Example 4-5. A vertical photograph taken from a flying height of 3500 m above mean sea level contains the image a of object point A at coordinates (with respect to the fiducial system) $x_a = 73.287$ mm and $y_a = -101.307$ mm. If the elevation of object point A is 120 m above mean sea level and the camera had a focal length of 153.099 mm, compute the x' and y' coordinates of the point, corrected for atmospheric refraction.

Solution. Compute r by Eq. (4-20).

$$r = \sqrt{73.287^2 + (-101.307)^2} = 125.036 \text{ mm}$$

Express $\tan \alpha$ by Eq. (4-21) and solve for α.

$$\tan \alpha = \frac{125.036}{153.099}$$

$$\alpha = 39.2386°$$

Compute K by Eq. (4-19).

$$K = (7.4 \times 10^{-4})(3.5 - 0.12)\{1 - 0.02[2(3.5) - 0.12]\} = 0.0022°$$

Compute $\Delta\alpha$ by Eq. (4-22).

$$\Delta\alpha = 0.0022° \left(\frac{125.036}{153.099}\right) = 0.0018°$$

Compute r' by Eq. (4-23).

$$r' = 153.099 \tan(39.2386° - 0.0018°) = 125.029 \text{ mm}$$

Compute Δr by Eq. (4-24).

$$\Delta r = 125.036 - 125.029 = 0.008 \text{ mm}$$

Compute δx and δy by Eqs. (4-8) and (4-9), respectively.

$$\delta x = 73.287\left(\frac{0.008}{125.036}\right) = 0.005 \text{ mm}$$

$$\delta y = -101.307\left(\frac{0.008}{125.036}\right) = -0.006 \text{ mm}$$

Subtract the corrections δx and δy from x and y, respectively to obtain corrected coordinates x' and y'.

$$x' = x - \delta x = 73.287 - 0.005 = 73.282 \text{ mm}$$

$$y' = y - \delta y = -101.307 - (-0.006) = -101.301 \text{ mm}$$

4-15 CORRECTION FOR EARTH CURVATURE

Traditionally, in analytical photogrammetry, corrections were commonly applied to measured photo coordinates to compensate for the effects of earth curvature. The rationale for this notion is that elevations of points are referenced to an approximately spherical datum (i.e., mean sea level) whereas photogrammetric equations assume that the zero-elevation surface is a plane. In addition, if horizontal coordinate systems such as state plane coordinates (see Sec. 5-6) are used in object space, the axis system is also curved, although it is curved primarily in only one direction (either east-west or north-south). It has long been recognized that the practice of making earth curvature corrections to measured photo coordinates is not theoretically correct. However, it has also long been accepted that using the correction generally leads to more accurate results than ignoring it, particularly in determining elevations.

The primary problem with the earth curvature correction is that because of the nature of map projection coordinates, correcting photo coordinates for earth curvature will degrade the accuracy of either X or Y object space coordinates, depending upon the map projection used. For example, when a UTM projection (see Sec. 5-6) is used, application of earth curvature corrections will yield more accurate elevations and Y values, but will degrade X values because the UTM projection does not curve in the X direction.

The proper approach, which avoids the need for any sort of earth curvature correction, employs a three-dimensional orthogonal object space coordinate system. One such coordinate system is the *local vertical coordinate system* described in Sec. 5-5. By using a local vertical coordinate system for a project, earth curvature ceases to be a distorting effect. Instead, the curvature of the earth will simply be a natural characteristic of the terrain, and it will be resolved in the same manner as any other topographic feature.

REFERENCES

Abdel-Aziz, Y. L.: "Asymmetrical Lens Distortion," *Photogrammetric Engineering and Remote Sensing,* vol. 41, no. 3, 1975, p. 337.

American Society of Photogrammetry: *Manual of Photogrammetry,* 4th ed., Bethesda, MD, 1980, chap. 9.

————: *Digital Photogrammetry: An Addendum to the Manual of Photogrammetry,* Bethesda, MD, 1996, chap 2.

Brown, D. C.: "Close-Range Camera Calibration," *Photogrammetric Engineering,* vol. 37, no. 8, 1971, p. 855.

————: "Unflatness of Plates as a Source of Systematic Error in Close-Range Photogrammetry," *Photogrammetria,* vol. 40, no. 9, 1986, p. 343.

Forrest, R. B.: "Refraction Compensation," *Photogrammetric Engineering,* vol. 40, no. 5, 1974, p. 577.

Fraser, C. S.: "Photogrammetric Measurement to One Part in a Million," *Photogrammetric Engineering and Remote Sensing,* vol. 58, no. 3, 1992, p. 305.

Fritz, L. W.: "A Complete Comparator Calibration Program," *Photogrammetria,* vol. 29, no. 4, 1973, p. 133.

Frost, R. M.: "Improved Well Positions for Geoscientific Applications: Exploiting NAPP Photographs with Digitizer and PC-Based Bundle Adjustment Program," *Photogrammetric Engineering and Remote Sensing,* vol. 61, no. 7, 1995, p. 927.

Gyer, M. S.: "Methods for Computing Photogrammetric Refraction Corrections for Vertical and Oblique Photographs," *Photogrammetric Engineering and Remote Sensing,* vol. 62, no. 3, 1996, p. 301.

Jaksic, Z.: "Deformations of Estar-Base Aerial Films," *Photogrammetric Engineering,* vol. 38, no. 3, 1972, p. 285.

Light, D. L.: "The New Camera Calibration System at the U.S. Geological Survey," *Photogrammetric Engineering and Remote Sensing,* vol. 58, no. 2, 1992, p. 185.

Oimoen, D. C.: "Evaluation of a Tablet Digitizer for Analytical Photogrammetry," *Photogrammetric Engineering and Remote Sensing,* vol. 53, no. 6, 1987, p. 601.

Scarpace, F. L., and P. R. Wolf: "Atmospheric Refraction," *Photogrammetric Engineering,* vol. 39, no. 5, 1973, p. 521.

Scholer, H.: "On Photogrammetric Distortion," *Photogrammetric Engineering and Remote Sensing,* vol. 41, no. 6, 1975, p. 761.

Tayman, W. P.: "User Guide for the USGS Aerial Camera Report of Calibration," *Photogrammetric Engineering and Remote Sensing,* vol. 50, no. 5, 1984, p. 577.

Wiley, A. G., and K. W. Wong: "Geometric Calibration of Zoom Lenses for Computer Vision Metrology," *Photogrammetric Engineering and Remote Sensing,* vol. 61, no. 1, 1995, p. 69.

Wolf, P. R.: "Trilaterated Photo Coordinates," *Photogrammetric Engineering,* vol. 35, no. 6, 1969, p. 543.

PROBLEMS

4-1. Assume that photo coordinates of points a and b of Fig. 4-1 are $x_a = 49.87$ mm, $y_a = 39.24$ mm, $x_b = 79.20$ mm, and $y_b = -62.81$ mm. Calculate photo distance ab and radial distances oa and ob.

4-2. Repeat Prob. 4-1 except that the photo coordinates are $x_a = -53.99$ mm, $y_a = -70.22$ mm, $x_b = -95.17$ mm, and $y_b = 23.64$ mm.

4-3. In Fig. 4-4, assume that $x_1 = -111.948$ mm, $y_1 = 0.010$ mm, $x_2 = -0.004$ mm, and $y_2 = -111.962$ mm. Calculate x_a and y_a if D_1 and D_2 are measured as 65.237 and 143.907 mm, respectively.

4-4. Repeat Prob. 4-3 except that $x_1 = -113.24$ mm, $y_1 = 0.00$ mm, $x_2 = 0.02$ mm, and $y_2 = -113.23$ mm. Also D_1 and D_2 are measured as 165.42 and 56.72 mm, respectively.

4-5. Repeat Prob. 4-3 except that D_1 and D_2 are measured as 127.0 and 97.6 mm, respectively.

4-6. Name and briefly describe the various systematic errors that may exist in photographic coordinates.

4-7. Calculate the acute angle of intersection of fiducial lines for a camera of the type shown in Fig. 4-1 if comparator measurements of the fiducial marks on a calibration "flash plate" were as follows:

Mark	X, mm	Y, mm
1	13.782	128.568
2	126.653	15.679
3	239.543	128.577
4	126.662	241.452

4-8. Repeat Prob. 4-7, except that the following flash plate measurements were taken:

Mark	X, mm	Y, mm
1	8.635	122.461
2	123.004	9.574
3	236.396	122.477
4	123.017	235.347

4-9. If the intersection of fiducial lines of the camera of Prob. 4-7 defines the principal point exactly, what are the x and y photo coordinates of the four fiducial marks in the photo system? Assume that the x and y photo coordinate axes are parallel to the comparator axes.

4-10. Repeat Prob. 4-9, except that it applies to the data of Prob. 4-8.

4-11. On a paper-print positive, the measured x distance between fiducials (1 and 3) was 226.4 mm and y between fiducials (2 and 4) was 225.0 mm. These x and y distances determined in camera calibration were 225.433 and 225.693 mm, respectively. Using the method of Example 4-2, calculate shrinkage-corrected coordinates of points a, b, and c whose coordinates were measured on the paper print as follows:

Point	x, mm	y, mm
a	20.3	−92.1
b	48.6	85.8
c	−111.1	−102.5

4-12. Repeat Prob. 4-11, except that the measured x distance on a paper-print positive between fiducials 1 and 3 was 222.75 mm and y between fiducials 2 and 4 was 222.49 mm; the calibrated distances between these same fiducials were 224.282 and 224.155 mm, respectively; and measured photo coordinates of points a, b, and c were as follows:

Point	x, mm	y, mm
a	14.63	−75.69
b	−78.08	41.81
c	−104.78	−58.42

4-13. A recent USGS camera calibration report yielded the parameters given in the following table. Using these calibration values, compute the corrected coordinates (to the nearest micrometer) for an image point having coordinates $x = 89.844$ mm and $y = -103.227$ mm relative to the fiducial axes.

Symmetric radial distortion parameters		Decentering distortion parameters		Calibrated principal point	
k_0	-0.1217×10^{-3}	p_1	0.1145×10^{-6} mm^{-1}	x_p	-0.008 mm
k_1	0.8252×10^{-8} mm^{-2}	p_2	0.5984×10^{-8} mm^{-1}	y_p	-0.004 mm
k_2	-0.1221×10^{-12} mm^{-4}	p_3	0 mm^{-2}		
k_3	0 mm^{-6}	p_4	0 mm^{-4}		
k_4	0 mm^{-8}				

4-14. Repeat Prob. 4-13, except that the coordinates, with respect to the fiducial system, of the image point are $x = 51.909$ mm and $y = 48.002$ mm.

4-15. The photo coordinates listed below have been corrected for film and lens distortions. The camera that took the photography had a focal length of 152.544 mm, the flying height above mean sea level (MSL) was 9410 m, and the average elevation of the terrain was 260 m above MSL. Calculate the photo coordinates (to the nearest micrometer) corrected for atmospheric refraction using the method of Example 4-5.

Point	x, mm	y, mm
a	28.738	49.211
b	57.820	-93.705
c	-117.232	-102.794

4-16. Repeat Prob. 4-15, except that the camera lens had a focal length of 88.792 mm, flying height above MSL was 3610 m, the average terrain elevation was 50 m above MSL, and the photo coordinates were as follows:

Point	x, mm	y, mm
a	59.238	74.281
b	-63.970	-113.444
c	103.296	-98.730

CHAPTER
5

OBJECT SPACE
COORDINATE
SYSTEMS

5-1 INTRODUCTION

Coordinate systems are a fundamental concept associated with spatial data. In the previous chapter, two-dimensional *image space* coordinates were discussed. In this chapter, three-dimensional *object space* coordinate systems are presented and described. Object space coordinate systems have always been important for specifying the relative positions of points in surveying, photogrammetry, and mapping. However, they have recently taken on added significance with the emergence and increasing importance of geographic information systems (GISs), which are heavily dependent upon coordinated data for their function.

Object space in photogrammetry refers to the three-dimensional region that encompasses the physical features imaged in the photographs. Most often object space relates to a portion of the earth's terrain and the natural and cultural features thereon; but it also can relate to other items such as celestial bodies, medical subjects, industrial machines, archeological objects, and many others. When mapping the earth's terrain and natural and cultural features, it is important that all mapped objects be accurately located with respect to an accepted geographic frame of reference. This is particularly important when spatial data from multiple sources are being integrated. If any of the spatial data sets are not accurately defined in an accepted frame of reference, then gaps, overlaps, and mismatches will occur. Several accepted reference systems exist: *geodetic, geocentric, local vertical,* and *map projection.* General concepts of these reference systems are discussed in this chapter, while mathematical equations which express the relationships between them are given in App. F.

5-2   CONCEPTS OF GEODESY

To a person standing in the middle of a prairie, the earth appears to be flat, and within a limited area, it is approximately so. Within this limited range, positions of points can be determined relative to a local plane surface without significant loss of accuracy. As the size of the area increases, however, the effects of the earth's curvature become significant and must be taken into account in order to maintain accuracy. Since photogrammetric projects often cover large areas, it is necessary to have quantitative knowledge of the earth's shape in order to determine accurate relative positions of points and to produce reliable maps.

The field of *geodesy* involves the study of the size, shape, gravity, rotation, and crustal movement of the earth. It is a highly refined science which provides the basis for earth-related reference coordinate systems. To understand its basis, three "reference surfaces" must be defined: the *physical earth,* the *geoid,* and the *ellipsoid*. These are illustrated in Fig. 5-1. The physical earth is exactly what the term implies. It contains mountains, valleys, plains, and ocean floors, and is the object which is to be mapped. Although generally viewed as a static object, the physical earth is subject to subtle crustal movements, which are accounted for in modern geodesy.

The *geoid* is an equipotential gravity surface, which is generally considered to be *mean sea level*. It can be imagined as the surface that the earth's seas would form if wind, wave, and tidal action ceased and the oceans were connected through the continents by narrow, frictionless canals. The geoid is a gently undulating surface which is everywhere perpendicular to the direction of gravity. These gentle undulations are due to gravity variations caused by the nonhomogeneous mass distribution of the earth. The shape of the geoid, in fact, results from the *net attraction,* comprised of gravity and the effect of the earth's rotation.

A reference *ellipsoid* is a mathematically defined surface which approximates the geoid either globally or in a large local area such as a continent. This surface is

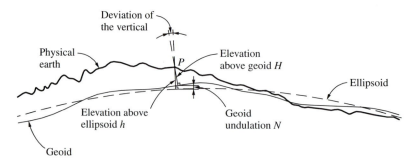

FIGURE 5-1
The three fundamental geodetic reference surfaces: physical earth, geoid, and ellipsoid.

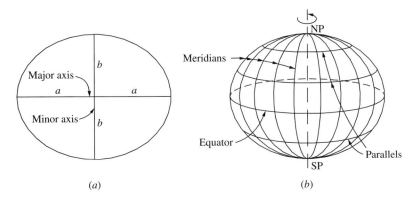

FIGURE 5-2
Definition of a reference ellipsoid. (*a*) Two-dimensional ellipse showing major and minor axes.
(*b*) Three-dimensional ellipsoid formed by rotation of ellipse about the minor axis.

formed by rotating a two-dimensional ellipse (shown in Fig. 5-2*a*) about its minor axis.
Rotation of an ellipse in this manner generates a three-dimensional ellipsoid, as shown
in Fig. 5-2*b*. This figure shows the (curved) lines which pass through the north and south
poles (NP and SP, respectively), known as *meridians,* and (curved) lines which are
parallel to the equator, called *parallels*. A meridian is formed by the intersection of the
ellipsoid with a plane containing the pole. A parallel, however, is formed by the intersection of the ellipsoid with a plane that is perpendicular to the pole (i.e., parallel to the
equator).

To define the size and shape of the reference ellipsoid, at least two constants are
required. They are derived from actual measurements made upon the earth. Generally,
the semimajor axis of the ellipsoid a and the flattening f are specified. From these two
defining constants, other ellipsoid parameters can be derived. The following equations
give relationships between the ellipsoid constants of a, the semimajor axis; b, the semiminor axis; f, the flattening; e, the first eccentricity; and e', the second eccentricity.

$$f = 1 - \frac{b}{a} \tag{5-1}$$

$$b = a(1 - f) \tag{5-2}$$

$$e = \frac{\sqrt{a^2 - b^2}}{a} \tag{5-3}$$

$$e^2 = f(2 - f) = \frac{a^2 - b^2}{a^2} \tag{5-4}$$

$$e' = \frac{\sqrt{a^2 - b^2}}{b} \tag{5-5}$$

$$e'^2 = \frac{e^2}{(1 - f)^2} = \frac{a^2 - b^2}{b^2} \tag{5-6}$$

TABLE 5-1
Parameters for select reference ellipsoids

Reference ellipsoid	Semimajor axis a	Flattening f
Clarke 1866	6,378,206.4 m	1/294.9786982
GRS80	6,378,137 m	1/298.25722210088
WGS84	6,378,137 m	1/298.257223563

The flattening f is a parameter for an ellipsoid (or ellipse) which quantifies how much it departs from a true sphere (or circle). The value of f for an ellipse can range from 0, which corresponds to a circle, to 1, which corresponds to a completely flattened ellipse, i.e., a straight line. The accepted value of f for the earth is roughly 0.0033, which implies that the earth is very nearly spherical. The first and second eccentricities e and e' are also parameters which quantify how much an ellipse departs from a true circle, with values near 0 denoting near circularity. Table 5-1 gives semimajor axis a and flattening f values for three commonly used reference ellipsoids.

As noted previously, the values for a and f are selected on the basis of geodetic measurements made in different locations of the world. In the past, reference ellipsoids were derived in order to fit the geoid in certain local regions, such as North America. The Clarke 1866 ellipsoid is an example of this type of local surface, which was a best-fit to the geoid in North America. More recently, given accurate global measurement technology such as GPS, Doppler satellite measurements, and very long baseline interferometry (VLBI), reference ellipsoids have been derived which give a best-fit to the geoid in a worldwide sense. The GRS80 and WGS84 ellipsoids are examples of worldwide reference surfaces.

5-3   GEODETIC COORDINATE SYSTEM

Geodetic coordinates for specifying point locations relative to the earth's surface are latitude ϕ, longitude λ, and height h. These coordinates all depend upon a reference ellipsoid for their basis. Latitude and longitude are horizontal components, while the vertical component is height. These three coordinates are illustrated in Fig. 5-3. This figure shows a point P with a line passing through it, perpendicular to the ellipsoid and extending to the polar axis. This line is called the *normal*. The longitude λ of a point is given by the angle in the plane of the equator from the *prime meridian* (usually the meridian through Greenwich, England) to the *local meridian* (meridian passing through the normal line). Values of longitude range from $-180°$ to $+180°$ with those west of the prime meridian being negative and those to the east being positive. The latitude (ϕ) of a point is the angle from the equatorial plane to the normal line. Values of latitude range from $-90°$ to $+90°$ with those north of the equator being positive and those to the south being negative. The latitude is more clearly illustrated in Fig. 5-4, which shows a plane section through the ellipsoid containing the local meridian and the normal.

As also illustrated in Fig. 5-4, height h is the distance from the surface of the ellipsoid to the point P, in the same direction as the normal. This value specifies the elevation of a point above the ellipsoid, also known as the *ellipsoid height*. The elevation

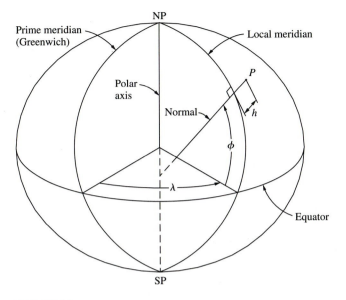

FIGURE 5-3
Geodetic coordinates of latitude ϕ, longitude λ, and height h.

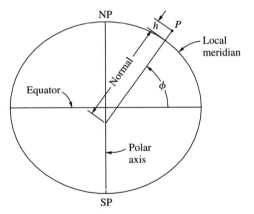

FIGURE 5-4
Illustration of latitude ϕ, normal, and height h in the plane of the local meridian.

of a point above the geoid H, also known as *orthometric height,* is commonly considered to be the *mean sea level* elevation. Figure 5-1 illustrates the relationship between these two height definitions. The difference between the two heights is referred to as the *geoid undulation* or *geoid height* and is indicated in Fig. 5-1 by the value N. The relationship between ellipsoid height h, orthometric height H, and geoid undulation N is specified in Eq. (5-7).

$$h \approx H + N \tag{5-7}$$

Figure 5-1 shows two lines through point P perpendicular to the ellipsoid and geoid, respectively. These lines intersect at an angle known as the *deviation of the vertical.* The deviation of the vertical can be determined by precise surveying techniques, and its value never exceeds 2 arc minutes anywhere on the earth. Therefore, although Eq. (5-7) is technically an approximation, it can be considered to be exact for practical purposes.

5-4 GEOCENTRIC COORDINATES

While geodetic coordinates $\phi\lambda h$ provide an earth-based definition for a point's three-dimensional position, they are related to a curved surface (reference ellipsoid). These coordinates are therefore nonorthogonal and as such are unsuitable for analytical photogrammetry, which assumes a rectangular or cartesian coordinate system. The geocentric coordinate system, on the other hand, is a three-dimensional *XYZ* cartesian system which provides an earth-centered definition of position, independent of any reference surface. This system has its *XY* plane in the plane of the equator with the *Z* axis extending through the north pole. The *X* axis is oriented such that its positive end passes through the prime meridian. Figure 5-5 illustrates the geocentric coordinate system and its relationship to geodetic coordinates.

The geocentric coordinate system is a convenient system for many worldwide geodetic applications such as satellite geodesy. Its use in photogrammetry and other mapping applications is not as advantageous since values of the coordinates are very large and have no obvious relationship to the cardinal directions in a local area. Also, the direction of the camera axis would be quantified relative to the earth's pole instead of

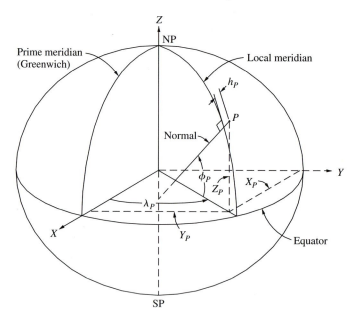

FIGURE 5-5
Relationship between geocentric and geodetic coordinates.

the local vertical. For these reasons, a special coordinate system, the local vertical system, with its origin in the local project area is generally used.

5-5   LOCAL VERTICAL COORDINATES

A local vertical coordinate system is a three-dimensional cartesian XYZ reference system which has its origin placed at a specific point within the project area. At this local origin, the Z axis extends straight up from the ellipsoid in the same direction as the normal at the origin. The positive X and Y axes are tangent to the ellipsoid and point to the east and north, respectively. Figure 5-6 shows the local vertical system and its relationship to geocentric and geodetic coordinates. In this figure, the position of the local origin is specified in terms of geodetic coordinates ϕ_o, λ_o, and h_o, with the last equal to zero. As shown in Fig. 5-6, the local origin has geocentric coordinates X_o, Y_o, and Z_o, and point P in the project area has local vertical coordinates X_{l_P}, Y_{l_P}, and Z_{l_P}.

Local vertical coordinates have the characteristic that X, Y, and Z values will correspond roughly to eastings, northings, and heights above the ellipsoid, respectively. However, the farther a point lies from the local origin, the less accurate these correspondences will be. This will have no detrimental effect on coordinates computed through analytical photogrammetry, since the local vertical coordinates can be accurately converted to useful geodetic coordinates or map projection eastings and northings. Appendix F contains detailed descriptions of these conversion procedures, complete with examples.

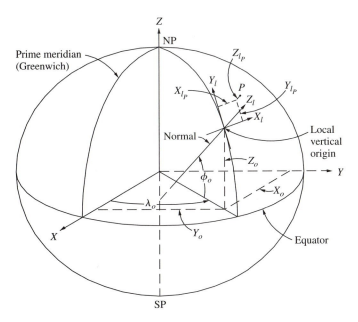

FIGURE 5-6
Local vertical coordinate system relative to geocentric and geodetic systems.

5-6   MAP PROJECTIONS

One of the fundamental products produced through photogrammetry is a map. A map, in general, consists of points, lines, arcs, symbols, or images which are placed on a flat, two-dimensional surface such as a sheet of paper or computer display. It is, in a manner of speaking, a scaled representation of what a human would see while walking from point to point in the mapped area. As such, it is preferable that the map present the viewpoint from directly overhead throughout the area. In mapping the earth's surface, it is impossible to achieve this overhead viewpoint on a two-dimensional medium without distortion because of the earth's curved shape. Map projections were created to accomplish this viewpoint with a carefully defined and understood amount of distortion.

While there are several types of map projections to choose from, those most often used in photogrammetric mapping applications are *conformal,* meaning that true shape is maintained. Two particular conformal map projections which will be discussed here are the *Lambert conformal conic* and the *transverse Mercator*. These map projections both employ the concept of the *developable surface*. A developable surface is a surface that may be three-dimensional in its natural form, but can be "unrolled" and laid flat. The developable surface is created which nearly coincides with the ellipsoid in the region being mapped. Points are then projected from the ellipsoid onto the developable surface or vice versa.

As its name implies, the Lambert conformal conic projection uses a cone as its developable surface. The axis of the cone is made to coincide with the minor axis of the ellipsoid and will pass through the ellipsoid along two parallels of latitude, called the *standard parallels*. Figure 5-7 shows a cone superimposed on the reference ellipsoid of the earth. Note that in the region between the standard parallels, the conic surface is below the ellipsoid; therefore, lines that are projected from the ellipsoid to the cone will be made shorter, and those outside the standard parallels will be made longer. This change in dimension can be quantified by a *scale factor* which is less than 1 between the standard parallels, greater than 1 outside, and exactly equal to 1 at the standard parallels. Since this scale factor varies in the north-south direction but remains the same in the east-west direction, the Lambert conformal conic projection is appropriate for areas of

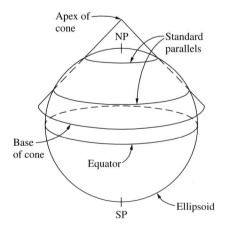

FIGURE 5-7
Cone used in the Lambert conformal conic projection.

limited extent north-south, but wide extent east-west. It is well-suited for an area such as the state of Tennessee.

The *XY* (easting, northing) coordinates of points within a Lambert conformal conic projection are based on the cone after it has been unrolled and laid flat. Figure 5-8 shows how the flattened cone is placed relative to the *XY* axes. In this figure, parallels of latitude form concentric circular arcs centered on the apex of the cone, and meridians appear as lines which are radial from the apex.

Different Lambert conformal conic projections can be set up for specific local areas or "zones." A zone is typically defined in terms of the extent of its latitude and longitude. When a Lambert conformal conic map projection is developed for a specific area, a *central meridian* is selected whose longitude is equal to that of the approximate center of the zone. The *origin* for the map projection is also selected. It lies on the central meridian at a location below the coverage of the zone. Figure 5-8 illustrates both the central meridian and the origin. Six parameters uniquely define a Lambert conformal conic map projection for a specific zone. Two of the parameters are the latitudes ϕ_1 and ϕ_2 of the two standard parallels depicted in Figs. 5-7 and 5-8. Two others are the latitude and longitude ϕ_o and λ_o, respectively, of the grid origin. The final two parameters are the *false easting* E_o and *false northing* N_o of the origin. These latter parameters, as shown in

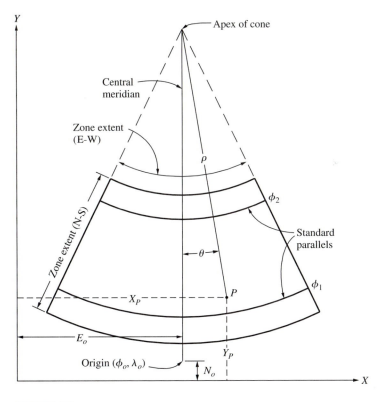

FIGURE 5-8
The Lambert cone unrolled and laid flat.

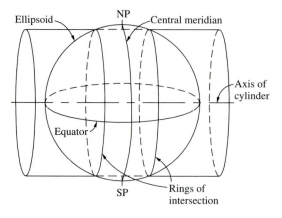

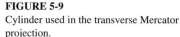

FIGURE 5-9
Cylinder used in the transverse Mercator
projection.

Fig. 5-8, are included so that all coordinates within the zone will be positive. A point P
having a particular latitude and longitude (ϕ_P, λ_P) will have corresponding map projection coordinates (X_P, Y_P).

Conversion of latitude and longitude $(\phi\lambda)$ of a point into map projection
coordinates (XY) is known as the *forward* or *direct* conversion. Computation of the
forward conversion for the Lambert conformal conic projection involves complex
mathematics which are given in Sec. F-5. Conversion from XY to $\phi\lambda$ is referred to as the
inverse conversion, procedures for which are also given in Sec. F-5.

Another commonly used map projection is the transverse Mercator, which employs a cylinder as its developable surface. The axis of the cylinder is defined so as to lie
in the plane of the equator, transverse to the minor axis of the ellipsoid. Figure 5-9 shows
a cylinder superimposed on the reference ellipsoid of the earth. The cylinder intersects
the reference ellipsoid along two rings which are nominally oriented in the north-south
direction. Note that in the region between the rings of intersection, the cylinder is below
the ellipsoid and therefore lines that are projected from the ellipsoid to the cylinder will
be made shorter, and those outside the rings will be made longer. This is similar to the
condition with regard to standard parallels of the Lambert conformal conic projection.
This change in dimension can also be quantified by a scale factor which is less than 1 between the rings of intersection, greater than 1 outside, and exactly equal to 1 at the
intersecting rings. Since this scale factor varies in the east-west direction but remains
approximately the same in the north-south direction, the transverse Mercator projection
is appropriate for areas of limited extent east-west, but with long extent north-south. It
is well suited for areas such as the states of Vermont and New Hampshire.

The XY (easting, northing) coordinates are based on the cylinder after it has been
unrolled and laid flat. Figure 5-10 shows how the flattened cylinder is placed relative to
the X and Y axes. In this figure, parallels of latitude and meridians of longitude appear
as lines which take the shapes of complex curves, symmetric about the central meridian.

As with the Lambert conformal conic projection, different transverse Mercator
projections can be set up in local areas or zones, also defined in terms of their ranges of
latitude and longitude. To develop a transverse Mercator map projection for a specific
area, a central meridian is selected in the approximate center of the zone. An origin is

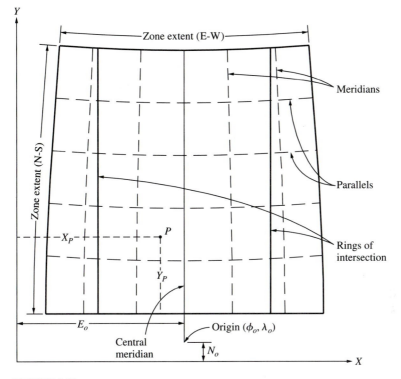

FIGURE 5-10
The transverse Mercator cylinder unrolled and laid flat.

also defined which lies on the central meridian at a location below the coverage of the zone (see Fig. 5-10). Five parameters uniquely define a transverse Mercator map projection for a specific zone. The first two parameters are the latitude and longitude, ϕ_o and λ_o, of the grid origin. A third parameter, k_o, is the scale factor along the central meridian. Finally, as shown in Fig. 5-10, a false easting E_o and false northing N_o of the origin are included to keep the coordinates positive. A point P having a particular latitude and longitude (ϕ_P, λ_P) will have corresponding map projection coordinates (X_P, Y_P).

The forward conversion of latitude and longitude $(\phi\lambda)$ of a point to XY coordinates for the transverse Mercator projection involves mathematical developments which are even more complex than those of the Lambert conformal conic projection (see Sec. F-6). The inverse conversion (from XY to $\phi\lambda$) is equally complex, and these procedures are also given in Sec. F-6.

Both the Lambert conformal conic and transverse Mercator projections are used in *state plane coordinate* (SPC) systems in the United States. These SPC systems were established to provide convenient local coordinate systems for surveying and mapping. In the SPC system, each state is divided into one or more zones chosen so that the maximum scale distortion is no more than 1 part in 10,000. To achieve this distortion limit, the north-south dimension of each Lambert conformal conic zone and the east-west dimension of each transverse Mercator zone are limited to approximately 254 kilometers (km). Each zone has its own unique set of defining parameters, with some states, such as Rhode Island, having a single zone, and other states, such as Alaska, having as many as 10 zones. (Note that one of the Alaska zones uses the oblique Mercator projection.)

Another common map projection system is the *universal transverse Mercator* (UTM) system. This system was established to provide worldwide coverage by defining 60 zones, each having a 6° longitude range. UTM zone 1 extends from 180° west longitude to 174° west longitude, with a central meridian of 177° west. Zone numbers increase to the east, at an equal spacing of 6° longitude. For example, zone 17 extends from 84° west to 78° west and has a central meridian of 81° longitude. The value of the scale factor along the central meridian k_o is equal to 0.9996 for every zone, resulting in a maximum scale distortion of 1 part in 2500. Each zone has its origin (ϕ_o, λ_o) at the intersection of the equator with the central meridian. The false easting for each zone is 500,000 m; for latitudes north of the equator, the false northing is 0 m, and for latitudes south of the equator, the false northing is 10,000,000 m.

The Lambert conformal conic and transverse Mercator map projections are by no means the only map projections available. Other projections which may be routinely encountered are the *polyconic, polar stereographic,* and *space oblique Mercator.* Although discussion of these projections is beyond the scope of this text, the reader is encouraged to consult references at the end of this chapter for details of these and other map projections.

5-7 HORIZONTAL AND VERTICAL DATUMS

A *datum* is a system of reference for specifying the spatial positions of points. These spatial positions are generally expressed in terms of two separate components, *horizontal* and *vertical.* Thus datums have also traditionally been of two kinds: horizontal and vertical. In a physical sense, a datum consists of a network of uniformly spaced control monuments whose positions have been determined through precise control surveys. These monuments serve as reference points for originating subordinate surveys of all types. Thus they provide the basis for specifying the relative positions of points both for surveying operations and for mapping purposes.

Common horizontal datums used in the United States include the *North American Datum of 1927* (NAD27), the *North American Datum of 1983* (NAD83), the *World Geodetic System of 1984* (WGS84), various statewide *high-accuracy reference networks* (HARNs), and the *International Terrestrial Reference Framework* (ITRF). They provide a means of relating horizontal coordinates derived through surveying and mapping processes to established coordinate reference systems.

Although the theoretical development of horizontal datums is rather complicated, conceptually they can be considered to be based on three primary components as minimum constraints: a reference ellipsoid, an origin, and an angular alignment. The North American Datum of 1927, for example, uses the Clarke 1866 ellipsoid as its reference surface which was a best-fit to the geoid in North America. The origin for NAD27 is a point known as "Meades Ranch," located in the state of Kansas. At this origin the geoid and ellipsoid, as well as their respective normals, were made to coincide; i.e., deviation of the vertical was assumed to be zero there. Based upon the best positional information available at the time, values of latitude and longitude were assigned and fixed for Meades Ranch, which by definition are exact. The angular alignment was established by fixing an azimuth to a nearby point named "Waldo" and by aligning the semiminor axis of the reference ellipsoid with the rotational axis of the earth. A large network of points, spanning the North American continent, were connected to this pair of points through high-accuracy control surveys. All points in the network were monumented to perpetuate their positions. Based on the latitude and longitude coordinates of Meades Ranch and the azimuth to Waldo, and the control survey data, coordinates of the other points were computed by a least squares adjustment. The resulting coordinates were then made available to the public so that surveyors and mappers throughout North America could establish coordinates relative to NAD27 by originating their surveys from monumented datum points in the vicinity.

As noted earlier, in a physical sense a horizontal datum such as NAD27 consists of the monument points along with their published coordinates. Since these coordinates were computed from a large number of measurements, each of which contains a certain amount of error, NAD27 likewise contains distortions due to these errors. In addition, subsequent surveys expanded NAD27 by connecting to previously established points in a piecemeal fashion. With advances in instrumentation, particularly accurate electronic distance-measuring devices, the distortions inherent in NAD27 began to cause difficulties in constraining the newer, more accurate measurements to the distorted system. To address this problem, the U.S. National Geodetic Survey created a new datum known as NAD83. This improved datum included virtually all points from NAD27 and its historic measurements, as well as nearly one-quarter million new points created from electronically measured distances and satellite observations. This entire system of combined measurements was adjusted by least squares. The values of computed coordinates were first published in 1986, 3 years later than planned, but nevertheless the system was called NAD83.

The reference ellipsoid of the NAD83 datum is the Geodetic Reference System of 1980 (GRS80) which was a worldwide best-fit to the geoid. The point of origin for NAD83 was made to coincide with the mass center of the earth, which was indirectly determined on the basis of satellite orbits. The semiminor axis of the reference ellipsoid was aligned with the rotational axis of the earth, and an internationally accepted basis for longitude was also used. Since this new datum included many new measurements and had a different set of constraints, points that were common to both datums now had two sets of horizontal coordinates. In some cases they differed by as much as 150 m. For this reason it is essential that the reference datum for any survey or map product be clearly indicated so as to avoid confusion.

The World Geodetic System of 1984 was established by the U.S. Department of Defense during the same time period that NAD83 was being developed. This datum is employed for all maps and charts produced for use by the U.S. armed forces. WGS84 is also the datum to which the broadcast ephemeris of the *Global Positioning System* (GPS) is referenced (see Sec. 16-7). Although it is defined in a manner similar to that of NAD83 (i.e., using a network of monument points and interconnecting measurements), it is different in one key aspect: It lacks a large network of easily accessible points having published coordinates. As such, this datum is not readily available for use by the general public.

At the same time that NAD83 was being completed, GPS was beginning to be widely used for geodetic surveys. Due to the exceptionally high accuracy of GPS, discrepancies were being revealed in the newly created NAD83. As use of GPS expanded, these discrepancies became a significant nuisance for geodesists, and newer, more accurate datums were sought. As a response, the National Geodetic Survey, in cooperation with individual states, began to establish high-accuracy reference networks (HARNs). These networks consisted of newly established points which were surveyed using high-accuracy GPS observations. The results were statewide networks of points whose relative positions are known to an accuracy of better than 1 part per million. The HARNs were individually connected to a high-accuracy worldwide network; therefore even though these networks have a high degree of internal consistency, there are discontinuities along the borders between states.

The previously mentioned horizontal datums are static systems; i.e., the coordinates of the monument points are based on a specific moment in time. It has been well established that the surface of the earth is dynamic, subject to crustal movements of several centimeters per year and perhaps more. In addition, the rotational axis of the earth is continually on the move at a slow but detectable rate. With current high-accuracy surveying techniques such as GPS, the dynamic nature of the earth's surface is readily detectable. In response, the International Earth Rotation Service established the *International Terrestrial Reference Frame,* or ITRF. The published values for this sophisticated datum consist of geocentric coordinates at a specific point in time (*epoch*) along with velocity vectors which can be used to determine the precise locations of the point at a later time. This ultraprecise datum is accurate to 1 part in 50,000,000, which corresponds to a precision of approximately 15 cm between any two points on the earth regardless of their locations! The ITRF is essentially a three-dimensional reference datum of high accuracy which is commonly used as a basis for precise GPS orbit determination. Periodically, WGS84 has been refined so as to closely coincide with the ITRF.

A vertical datum is a reference system for giving elevations of points relative to the geoid (i.e., orthometric heights). Two primary vertical datums are currently in use in the United States: the *National Geodetic Vertical Datum of 1929* (NGVD29) and the *North American Vertical Datum of 1988* (NAVD88).

The NGVD29 was established on the basis of tide gauge measurements at 26 stations on the North American coast, together with interconnecting leveling measurements through a vast network of benchmarks across the continent. At the time this datum was established, it was assumed that the geoid coincided with mean sea level as determined at the tide gaging stations. Elevations were therefore constrained to zero at these stations, and all benchmarks were adjusted to conform to this defined reference. Thus the NGVD29 can be considered to be a mean sea level datum.

The NGVD29 evolved in much the same way as the NAD27 in that many additional vertical surveys were connected to the network in local areas. This fact, in addition to distortions in the datum due to measurement errors and constraint to the tide gauging stations, led to a vertical datum which was not accurate enough. Also, the emerging use of GPS dictated the use of a vertical datum that more nearly corresponded to the geoid. Thus the NAVD88 was established. It is based on a worldwide gravity model which is, by definition, the geoid. The NAVD88 is more compatible with world-wide horizontal datums.

It is often necessary to convert (transform) points that have been referenced in one datum to another. Examples would be to transform from NAD27 to NAD83, or from NGVD29 to NAVD88. These transformations have become especially commonplace with the increasing use of geographic information systems. These systems often utilize information from different dates and different sources, and frequently the information is based on different reference coordinate systems. But the information must all be coordinated in a common reference system before being integrated for analysis and use in a GIS. A number of different mathematical procedures have been used for making these conversions. Unless the transformation procedure appropriately accounts for the distortions in the datums, however, errors on the order of several meters can result in the converted positions. To aid in making accurate horizontal datum conversions, the NGS has developed a program called NADCON. It can convert horizontal datum coordinates between NAD27 and NAD83 to an accuracy of approximately 15 cm with occasional errors as high as 50 cm. A related program called VERTCON, also available from NGS, performs vertical datum conversions between NGVD29 and NAVD88 to an accuracy of approximately 2 cm. Another useful program is GEOID99 which can be used to compute geoid undulation values N within the area encompassed by the NAVD88. These programs are available from the NGS on the World Wide Web.[1]

REFERENCES

American Society of Photogrammetry: *Manual of Photogrammetry,* 4th ed., Bethesda, MD, 1980, chaps. 8 and 9.

Bomford, G.: *Geodesy,* 4th ed., Clarendon Press, Oxford, 1980.

Boucher, C., and Z. Altamimi: "International Terrestrial Reference Frame," *GPS World,* vol. 7, no. 9, 1996, p. 71.

Cheves, M.: "NGS Releases New Geoid Model," *Professional Surveyor,* vol. 17, no. 1, 1997, p. 49.

Colvocoresses, A. P.: "The Gridded Map," *Photogrammetric Engineering and Remote Sensing,* vol. 63, no. 4, 1997, p. 377.

Doyle, F. J.: "Map Conversion and the UTM Grid," *Photogrammetric Engineering and Remote Sensing,* vol. 63, no. 4, 1997, p. 367.

Featherstone, W., and R. B. Langley: "Coordinates and Datums and Maps! Oh My!" *GPS World,* vol. 8, no. 1, 1997, p. 34.

National Imagery and Mapping Agency: "Department of Defense World Geodetic System 1984: Its Definition and Relationships with Local Geodetic Systems," NIMA Technical Report 8350.2, 3d ed., Bethesda, MD, 1997.

Snyder, J. P.: "Map Projections—A Working Manual," U.S. Geological Survey Professional Paper 1395, U.S. Geological Survey, Washington, 1987.

Stern, J. E.: "State Plane Coordinate System of 1983," NOAA Manual NOS NGS 5, National Oceanic and Atmospheric Administration, Rockville, MD, 1989.

Terry, N. G., Jr.: "Field Validation of the UTM Gridded Map," *Photogrammetric Engineering and Remote Sensing,* vol. 63, no. 4, 1997, p. 381.

---

[1] The Web site address of the NGS is http://www.ngs.noaa.gov.

Thompson, M. M.: *Maps For America,* 3d ed., U.S. Geological Survey, Washington, 1987.

Welch, R., and A. Homsey: "Datum Shifts for UTM Coordinates," *Photogrammetric Engineering and Remote Sensing,* vol. 63, no. 4, 1997, p. 371.

Wolf, P. R., and R. C. Brinker: *Elementary Surveying,* 9th ed., HarperCollins, New York, 1993.

PROBLEMS

5-1. List and briefly describe the three basic reference surfaces in geodesy.

— **5-2.** Using the values of a and f from Table 5-1, compute the values of the semiminor axis b, and the first eccentricity e for the Clarke 1866 ellipsoid. Express your answers to 10 significant figures.

5-3. Repeat Prob. 5-2, except the b and e values for WGS84 should be computed.

— **5-4.** The ellipsoid height h for a point was determined from GPS observation to be $+47.54$ m. If the geoid undulation N is equal to $+12.75$ m, what is the orthometric height H for the point?

5-5. Repeat Prob. 5-4, except that the ellipsoid height is -7.83 m and the geoid undulation is -26.74 m.

5-6. Briefly explain the difficulty associated with using the geocentric coordinate system in photogrammetric applications.

5-7. Illustrate and briefly describe how Z values in a local vertical coordinate system vary with distance from the local vertical origin for points having a constant ellipsoid height.

5-8. Briefly describe the variation in scale factor in the north-south direction for a Lambert conformal conic projection.

5-9. List the types of map projections that are used in the U.S. state plane coordinate system.

5-10. Based on the longitude of the area in which you live, determine the number of your local UTM zone.

5-11. Obtain the programs NADCON, VERTCON, and GEOID99 from the NGS World Wide Web site *http://www.ngs.noaa.gov*. Using the appropriate program, answer the following:

    (*a*) Using the values of latitude and longitude in NAD83 for a point in your local area, determine the corresponding values in NAD27, using NADCON.

    (*b*) Using the NGVD29 elevation of a point in your local area, determine the corresponding elevation in NAVD88, using VERTCON.

    (*c*) Using the same latitude and longitude as in part (*a*), determine the value of the geoid undulation using GEOID99.

CHAPTER 6

VERTICAL PHOTOGRAPHS

6-1 GEOMETRY OF VERTICAL PHOTOGRAPHS

As described in Chap. 1, photographs taken from an aircraft with the optical axis of the camera vertical or as nearly vertical as possible are called *vertical photographs*. If the optical axis is exactly vertical, the resulting photograph is termed *truly vertical*. In this chapter, equations are developed assuming truly vertical photographs. In spite of precautions taken to keep the camera axis vertical, small tilts are invariably present. For photos intended to be vertical, however, tilts are usually less than 1° and rarely exceed 3°. Photographs containing these small unintentional tilts are called *near-vertical* or *tilted photographs,* and for many practical purposes these photos may be analyzed using the relatively simple "truly vertical" equations of this chapter without serious error.

In this chapter, besides assuming truly vertical photographs, other assumptions are that the photo coordinate axis system has its origin at the photographic principal point and that all photo coordinates have been corrected for shrinkage, lens distortion, and atmospheric refraction distortion.

Figure 6-1 illustrates the geometry of a vertical photograph taken from an exposure station L. The *negative,* which is a reversal in both tone and geometry of the object space, is situated a distance equal to the focal length (distance $o'L$ in Fig. 6-1) above the rear nodal point of the camera lens. The *positive* may be obtained by direct

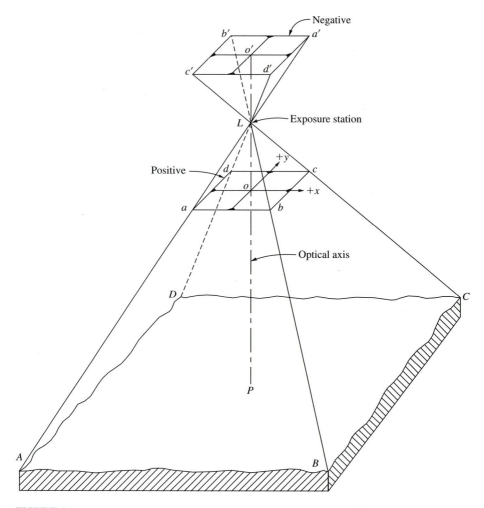

FIGURE 6-1
The geometry of a vertical photograph.

emulsion-to-emulsion "contact printing" with the negative. This process produces a reversal of tone and geometry from the negative, and therefore the tone and geometry of the positive are exactly the same as those of the object space. Geometrically the plane of a contact-printed positive is situated a distance equal to the focal length (distance oL in Fig. 6-1) below the front nodal point of the camera lens. The reversal in geometry from object space to negative is readily seen in Fig. 6-1 by comparing the positions of object points A, B, C, and D with their corresponding negative positions a', b', c', and d'. The correspondence of the geometry of the object space and the positive is also readily apparent. The photographic coordinate axes x and y, as described in Chap. 4, are shown on the positive of Fig. 6-1.

6-2 SCALE

Map scale is ordinarily interpreted as the ratio of a map distance to the corresponding distance on the ground. In a similar manner, the *scale of a photograph* is the ratio of a distance on the photo to the corresponding distance on the ground. Due to the nature of map projections, map scale is not influenced by terrain variations. An aerial photograph, however, is a perspective projection, and as will be demonstrated in this chapter, its scale varies with variations in terrain elevation.

Scales may be represented as *unit equivalents, unit fractions, dimensionless representative fractions,* or *dimensionless ratios.* If, for example, 1 inch (in) on a map or photo represents 1000 ft (12,000 in) on the ground, the scale expressed in the aforementioned four ways is

1. Unit equivalents: 1 in = 1000 ft
2. Unit fraction: 1 in/1000 ft
3. Dimensionless representative fraction: 1/12,000
4. Dimensionless ratio: 1:12,000

By convention, the first term in a scale expression is always chosen as 1. It is helpful to remember that a large number in a scale expression denotes a small scale, and vice versa; for example, 1:1000 is a larger scale than 1:5000.

6-3 SCALE OF A VERTICAL PHOTOGRAPH OVER FLAT TERRAIN

Figure 6-2 shows the side view of a vertical photograph taken over flat terrain. Since measurements are normally taken from photo positives rather than negatives, the negative has been excluded from this and other figures that follow in this text. The scale of a vertical photograph over flat terrain is simply the ratio of photo distance *ab* to corresponding ground distance *AB*. That scale may be expressed in terms of camera focal length *f* and flying height above ground *H'* by equating similar triangles *Lab* and *LAB* as follows:

$$S = \frac{ab}{AB} = \frac{f}{H'} \tag{6-1}$$

From Eq. (6-1) it is seen that the scale of a vertical photo is directly proportional to camera focal length (image distance) and inversely proportional to flying height above ground (object distance).

Example 6-1. A vertical aerial photograph is taken over flat terrain with a 152.4 mm-focal-length camera from an altitude of 1830 m above ground. What is the photo scale?

Solution. By Eq. (6-1),

$$S = \frac{f}{H'} = \frac{152.4 \text{ mm}}{1830 \text{ m}} = \frac{0.1524 \text{ m}}{1830 \text{ m}} = 1/12,\bar{0}00 = 1:12,\bar{0}00 \qquad (1 \text{ in}/10\bar{0}0 \text{ ft})$$

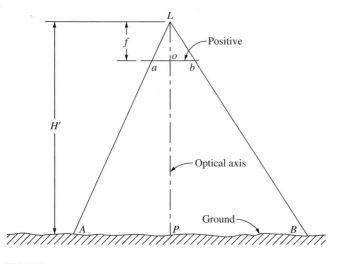

FIGURE 6-2
Two-dimensional view of a vertical photograph taken over flat terrain.

Note the use of the overbar in the solution of Example 6-1 to designate significant figures, as discussed in Sec. A-3.

6-4   SCALE OF A VERTICAL PHOTOGRAPH OVER VARIABLE TERRAIN

If the photographed terrain varies in elevation, then the object distance—or the denominator of Eq. (6-1)—will also be variable and the photo scale will likewise vary. For any given vertical photo scale increases with increasing terrain elevation and decreases with decreasing terrain elevation.

Suppose a vertical aerial photograph is taken over variable terrain from exposure station L of Fig. 6-3. Ground points A and B are imaged on the positive at a and b, respectively. Photographic scale at h, the elevation of points A and B, is equal to the ratio of photo distance ab to ground distance AB. By similar triangles Lab and LAB, an expression for photo scale S_{AB} is

$$S_{AB} = \frac{ab}{AB} = \frac{La}{LA} \tag{a}$$

Also, by similar triangles $LO_A A$ and Loa,

$$\frac{La}{LA} = \frac{f}{H - h} \tag{b}$$

Substituting Eq. (*b*) into Eq. (*a*) gives

$$S_{AB} = \frac{ab}{AB} = \frac{La}{LA} = \frac{f}{H - h} \tag{c}$$

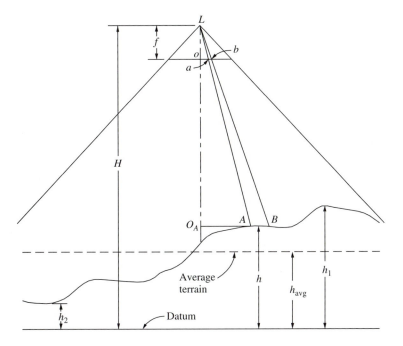

FIGURE 6-3
Scale of a vertical photograph over variable terrain.

Considering line *AB* to be infinitesimal, we see that Eq. (*c*) reduces to an expression of photo scale at a point. In general, by dropping subscripts, the scale at any point whose elevation above datum is *h* may be expressed as

$$S = \frac{f}{H - h} \tag{6-2}$$

In Eq. (6-2), the denominator $H - h$ is the object distance. In this equation as in Eq. (6-1), scale of a vertical photograph is seen to be simply the ratio of image distance to object distance. The shorter the object distance (the closer the terrain to the camera), the greater the photo scale, and vice versa. *For vertical photographs taken over variable terrain, there are an infinite number of different scales. This is one of the principal differences between a photograph and a map.*

6-5 AVERAGE PHOTO SCALE

It is often convenient and desirable to use an *average scale* to define the overall mean scale of a vertical photograph taken over variable terrain. Average scale is the scale at the average elevation of the terrain covered by a particular photograph and is expressed as

$$S_{\text{avg}} = \frac{f}{H - h_{\text{avg}}} \tag{6-3}$$

When an average scale is used, it must be understood that it is exact only at those points that lie at average elevation, and it is an approximate scale for all other areas of the photograph.

Example 6-2. Suppose that highest terrain h_1, average terrain h_{avg}, and lowest terrain h_2 of Fig. 6-3 are 610, 460, and 310 m above mean sea level, respectively. Calculate the maximum scale, minimum scale, and average scale if the flying height above mean sea level is 3000 m and the camera focal length is 152.4 mm.

Solution. By Eq. (6-2) (maximum scale occurs at maximum elevation),

$$S_{max} = \frac{f}{H - h_1} = \frac{152.4 \text{ mm}}{(3000 - 610) \text{ m}} = \frac{0.1524 \text{ m}}{2390 \text{ m}}$$

$$= 1/15,700 = 1:15,700 \qquad (1 \text{ in}/1310 \text{ ft})$$

15,682.414

and (minimum scale occurs at minimum elevation)

$$S_{min} = \frac{f}{H - h_2} = \frac{152.4 \text{ mm}}{(3000 - 310) \text{ m}} = \frac{0.1524 \text{ m}}{2690 \text{ m}}$$

$$= 1/17,700 = 1:17,700 \qquad (1 \text{ in}/1470 \text{ ft})$$

By Eq. (6-3)

$$S_{avg} = \frac{f}{H - h_{avg}} = \frac{152.4 \text{ mm}}{(3000 - 460) \text{ m}} = \frac{0.1524 \text{ m}}{2540 \text{ m}}$$

$$= 1/16,700 = 1:16,700 \qquad (1 \text{ in}/1390 \text{ ft})$$

In each of Eqs. (6-1), (6-2), and (6-3), it is noted that flying height appears in the denominator. Thus, for a camera of a given focal length, if flying height increases, object distance $H - h$ increases and scale decreases. Figures 6-4a through d illustrate this principle vividly. Each of these vertical photos was exposed using the very same 23-cm format and 152-mm-focal-length camera. The photo of Fig. 6-4a had a flying height of 460 m above ground, resulting in an average photo scale of 1:3000. The photos of Fig. 6-4b, c, and d had flying heights above average ground of 910 m, 1830 m, and 3660 m, respectively, producing average photo scales of 1:6000, 1:12,000, and 1:24,000, respectively.

FIGURE 6-4
Four vertical photos taken over Tampa, Florida, illustrating scale variations due to changing flying heights. Each photo was taken with a camera having a 23-cm square format and 152-mm-focal-length lens. In (a), flying height was 460 m above average ground, and average photo scale was 1:3000. In (b), (c), and (d), flying heights above average ground were 910 m, 1830 m, and 3660 m, respectively, and average photo scales were 1:6000, 1:12,000, and 1:24,000, respectively. (*Courtesy Aerial Cartographics of America, Inc.*)

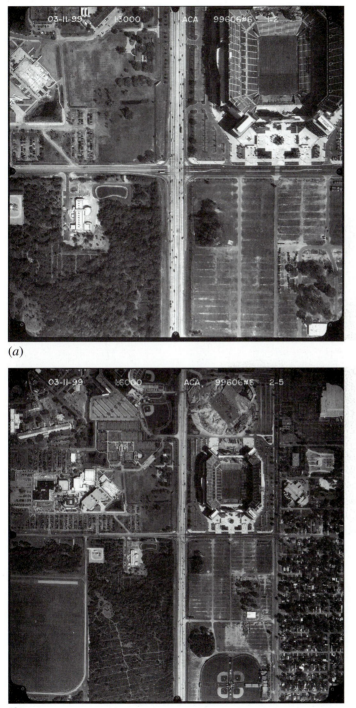

(a)

(b)

(*continued*)

(*c*)

(*d*)

FIGURE 6-4 (*concluded*)

6-6 OTHER METHODS OF DETERMINING SCALE OF VERTICAL PHOTOGRAPHS

In the previous sections, equations have been developed for calculating the scale of vertical aerial photographs in terms of camera focal length, flying height, and terrain elevation. There are, however, other methods of scale determination which do not require knowledge of these values.

A ground distance may be measured in the field between two points whose images appear on the photograph. After the corresponding photo distance is measured, the scale relationship is simply the ratio of the photo distance to the ground distance. The resulting scale is exact only at the elevation of the ground line, and if the line is along sloping ground, the resulting scale applies at approximately the average elevation of the two endpoints of the line.

Example 6-3. The horizontal distance AB between the centers of two street intersections was measured on the ground as 402 m. Corresponding line ab appears on a vertical photograph and measures 95.8 mm. What is the photo scale at the average ground elevation of this line?

Solution

$$S = \frac{ab}{AB} = \frac{95.8 \text{ mm}}{402 \text{ m}} = \frac{0.0958 \text{ m}}{402 \text{ m}} = \frac{1}{4200} = 1:42\overline{00} \qquad (1 \text{ in}/35\overline{0} \text{ ft})$$

The scale of a vertical aerial photograph may also be determined if a map covering the same area as the photo is available. In this method it is necessary to measure, on the photograph and on the map, the distances between two well-defined points that can be identified on both photo and map. Photographic scale can then be calculated from the following equation:

$$S = \frac{\text{photo distance}}{\text{map distance}} \times \text{map scale} \qquad (6\text{-}4)$$

Example 6-4. On a vertical photograph the length of an airport runway measures $16\overline{0}$ mm. On a map which is plotted at a scale of $1:24,000$, the runway scales 103 mm. What is the scale of the photograph at runway elevation?

Solution. From Eq. (6-4),

$$S = \frac{160 \text{ mm}}{103 \text{ mm}} \times \frac{1}{24,000} = 1/15,400 = 1:15,400 \qquad (1 \text{ in}/1290 \text{ ft})$$

The scale of a vertical aerial photograph can also be determined without the aid of a measured ground distance or a map if lines whose lengths are known by common knowledge appear on the photo. "Section lines" of a known 1-mile (1-mi) length, or a football field or baseball diamond, could be measured on the photograph, for example, and photographic scale could be calculated as the ratio of the photo distance to the known ground distance.

Example 6-5. What is the scale of a vertical aerial photograph on which a section line measures 151 mm?

Solution. The length of a section line is assumed to be 5280 ft. (Actually it can vary considerably from that value.) Photo scale is simply the ratio of the measured photo distance to the ground distance, or

$$S = \frac{151 \text{ mm}}{5280 \text{ ft}} \times \frac{1 \text{ ft}}{304.8 \text{ mm}} = 1:10,700 \qquad (1 \text{ in}/888 \text{ ft})$$

In each of the methods of scale determination discussed in this section, it must be remembered that the calculated scale applies only at the elevation of the ground line used to determine that scale.

6-7   GROUND COORDINATES FROM A VERTICAL PHOTOGRAPH

The ground coordinates of points whose images appear in a vertical photograph can be determined with respect to an arbitrary ground coordinate system. The arbitrary X and Y ground axes are in the same vertical planes as the photographic x and y axes, respectively, and the origin of the system is at the datum principal point (point in the datum plane vertically beneath the exposure station).

Figure 6-5 shows a vertical photograph taken at a flying height H above datum. Images a and b of the ground points A and B appear on the photograph, and their measured photographic coordinates are x_a, y_a, x_b, and y_b. The arbitrary ground coordinate axis system is X and Y, and the coordinates of points A and B in that system are X_A, Y_A, X_B, and Y_B. From similar triangles $La'o$ and $LA'A_o$, the following equation may be written:

$$\frac{oa'}{A_oA'} = \frac{f}{H - h_A} = \frac{x_a}{X_A}$$

from which

$$X_A = x_a \left(\frac{H - h_A}{f} \right) \tag{6-5}$$

Also, from similar triangles $La''o$ and $LA''A_o$,

$$\frac{oa''}{A_oA''} = \frac{f}{H - h_A} = \frac{y_a}{Y_A}$$

from which

$$Y_A = y_a \left(\frac{H - h_A}{f} \right) \tag{6-6}$$

Similarly, the ground coordinates of point B are

$$X_B = x_b \left(\frac{H - h_B}{f} \right) \tag{6-7}$$

$$Y_B = y_b \left(\frac{H - h_B}{f} \right) \tag{6-8}$$

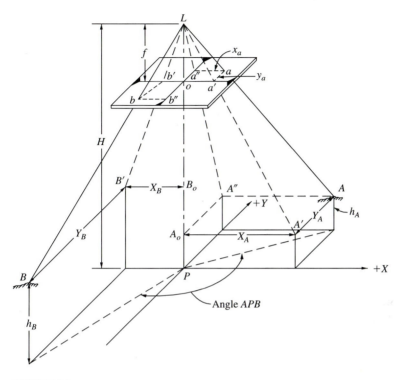

FIGURE 6-5
Ground coordinates from a vertical photograph.

Upon examination of Eqs. (6-5) through (6-8), it is seen that X and Y ground coordinates of any point are obtained by simply multiplying x and y photo coordinates by the inverse of photo scale at that point. From the ground coordinates of the two points A and B, the horizontal length of line AB can be calculated, using the pythagorean theorem, as

$$AB = \sqrt{(X_B - X_A)^2 + (Y_B - Y_A)^2} \qquad (6\text{-}9)$$

Also, horizontal angle APB may be calculated as

$$APB = 90° + \tan^{-1}\left(\frac{X_B}{Y_B}\right) + \tan^{-1}\left(\frac{Y_A}{X_A}\right) \qquad (6\text{-}10)$$

To solve Eqs. (6-5) through (6-8) it is necessary to know the camera focal length, flying height above datum, elevations of the points above datum, and photo coordinates of the points. The photo coordinates are readily measured, camera focal length is commonly known from camera calibration, and flying height above datum is calculated by methods described in Sec. 6-9. Elevations of points may be obtained directly by field measurements, or they may be taken from available topographic maps.

Example 6-6. A vertical aerial photograph was taken with a 152.4-mm-focal-length camera from a flying height of 1385 m above datum. Images a and b of two ground points A and B appear on the photograph, and their measured photo coordinates (corrected for shrinkage and distortions) are $x_a = -52.35$ mm, $y_a = -48.27$ mm, $x_b = 40.64$ mm, and $y_b = 43.88$ mm. Determine the horizontal length of line AB if the elevations of points A and B are 204 and 148 m above datum, respectively.

Solution. From Eqs. (6-5) through (6-8),

$$X_A = \frac{-52.35}{152.4}(1385 - 204) = -405.7 \text{ m}$$

$$Y_A = \frac{-48.27}{152.4}(1385 - 204) = -374.1 \text{ m}$$

$$X_B = \frac{40.64}{152.4}(1385 - 148) = 329.9 \text{ m}$$

$$Y_B = \frac{43.88}{152.4}(1385 - 148) = 356.2 \text{ m}$$

From Eq. (6-9),

$$AB = \sqrt{(329.9 + 405.7)^2 + (356.2 + 374.1)^2} = 1036 \text{ m}$$

Ground coordinates calculated by Eqs. (6-5) through (6-8) are in an arbitrary rectangular coordinate system, as previously described. If arbitrary coordinates are calculated for two or more "control" points (points whose coordinates are also known in an absolute ground coordinate system such as the state plane coordinate system), then the arbitrary coordinates of all other points for that photograph can be transformed to the ground system. The method of transformation used here is discussed in Secs. C-2 through C-5 (of App. C), and an example is given. Using Eqs. (6-5) through (6-8), an entire planimetric survey of the area covered by a vertical photograph can be made.

6-8   RELIEF DISPLACEMENT ON A VERTICAL PHOTOGRAPH

Relief displacement is the shift or displacement in the photographic position of an image caused by the relief of the object, i.e., its elevation above or below a selected datum. With respect to a datum, relief displacement is outward for points whose elevations are above datum and inward for points whose elevations are below datum.

The concept of relief displacement is illustrated in Fig. 6-6, which represents a vertical photograph taken from flying height H above datum. Camera focal length is f, and o is the principal point. The image of terrain point A, which has an elevation h_A above datum, is located at a on the photograph. An imaginary point A' is located vertically beneath A in the datum plane, and its corresponding imaginary image position is at a'. On the figure, both $A'A$ and PL are vertical lines, and therefore $A'AaLoP$ is a vertical plane. Plane $A'a'LoP$ is also a vertical plane which is coincident with $A'AaLoP$.

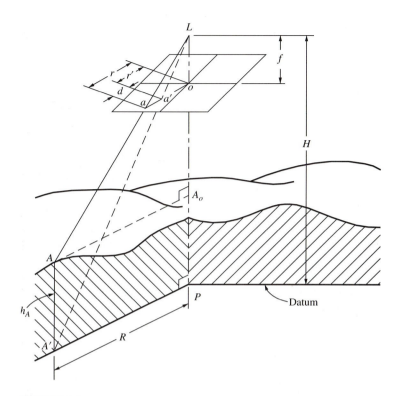

FIGURE 6-6
Relief displacement on a vertical photograph.

Since these planes intersect the photo plane along lines oa and oa', respectively, line aa' (relief displacement of point A due to its elevation h_A) is radial from the principal point.

An equation for evaluating relief displacement may be obtained by relating similar triangles. First consider planes Lao and LAA_o in Fig. 6-6:

$$\frac{r}{R} = \frac{f}{H - h_A} \qquad \text{Rearranging gives} \qquad r(H - h_A) = fR \qquad (d)$$

Also, from similar triangles $La'o$ and $LA'P$,

$$\frac{r'}{R} = \frac{f}{H} \qquad \text{or} \qquad r'H = fR \qquad (e)$$

Equating expressions (d) and (e) yields

$$r(H - h_A) = r'H$$

Rearranging the above equation, dropping subscripts, and substituting the symbol d for $r - r'$ gives

$$d = \frac{rh}{H} \qquad (6\text{-}11)$$

where d = relief displacement

h = height above datum of object point whose image is displaced

r = radial distance on photograph from principal point to displaced image
(The units of d and r must be the same.)

H = flying height above same datum selected for measurement of h

Equation (6-11) is the basic relief displacement equation for vertical photos. Examination of this equation shows that relief displacement increases with increasing radial distance to the image, and it also increases with increased elevation of the object point above datum. On the other hand, relief displacement decreases with increased flying height above datum. It has also been shown that relief displacement occurs radially from the principal point.

Figure 6-7 is a vertical aerial photograph which vividly illustrates relief displacement. Note in particular the striking effect of relief displacement on the tall buildings in the upper portion of the photo. Notice also that the relief displacement occurs radially from the center of the photograph (principal point). This radial pattern is also readily apparent for the relief displacement of all the other vertical buildings in the photo. The building in the center is one of the tallest imaged on the photo (as evidenced by the length of its shadow); however, its relief displacement is essentially zero due to its proximity to the principal point.

Relief displacement often causes straight roads, fence lines, etc., on rolling ground to appear crooked on a vertical photograph. This is especially true when such roads, fences, etc., occur near the edges of the photo. The severity of the crookedness will depend on the amount of terrain variation. Relief displacement causes some imagery to be obscured from view. Several examples of this are seen in Fig. 6-7; e.g., the street in the upper portion of the photo is obscured by relief displacement of several tall buildings adjacent to it.

Vertical heights of objects such as buildings, poles, etc., appearing on aerial photographs can be calculated from relief displacements. For this purpose, Eq. (6-11) is rearranged as follows:

$$h = \frac{dH}{r} \tag{6-12}$$

To use Eq. (6-12) for height determination, it is necessary that the images of both the top and bottom of the vertical object be visible on the photograph, so that d can be measured. Datum is arbitrarily selected at the base of the vertical object. Equation (6-12) is of particular import to the photo interpreter, who is often interested in relative heights of objects rather than absolute elevations.

Example 6-7. A vertical photograph taken from an elevation of 535 m above mean sea level (MSL) contains the image of a tall vertical radio tower. The elevation at the base of the tower is 259 m above MSL. The relief displacement d of the tower was measured as 54.1 mm, and the radial distance to the top of the tower from the photo center was 121.7 mm. What is the height of the tower?

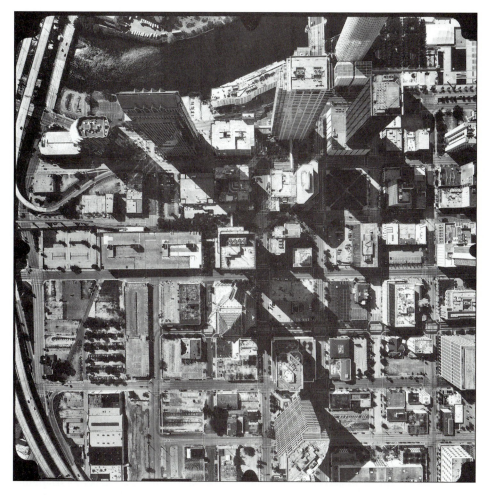

FIGURE 6-7
Vertical photograph of Tampa, Florida, illustrating relief displacements. (*Courtesy US Imaging, Inc.*)

Solution. Select datum at the base of the tower. Then flying height above datum is

$$H = 535 - 259 = 276 \text{ m}$$

By Eq. (6-12),

$$h = \frac{54.1(276)}{121.7} = 123 \text{ m}$$

Equation (6-11) may be used to calculate image displacements with respect to datum, and then corrected datum image positions may be located by laying off image

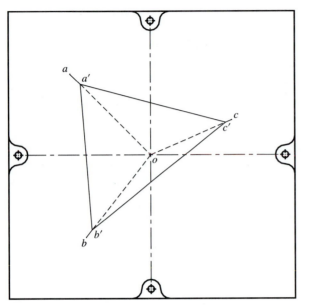

FIGURE 6-8
Relief displacement corrections laid off radially toward the principal point.

displacements along radial lines toward the principal point, as shown in Fig. 6-8. These "datum-corrected" images have true relative planimetric positions, just as they would be on a map plotted at photographic datum scale. Datum scale may be calculated by Eq. (6-1) using H above datum in the denominator. From these datum-corrected image positions, angles, lengths, and areas may be measured just as from a map.

Example 6-8. Figure 6-8 represents a vertical photo taken at a flying height of 1980 m above datum with a camera having a 152.4-mm focal length. On the photo, points a, b, and c are the imaged positions of the lot corners of a triangular parcel of land. Their radial distances from the principal point are 91.4 mm, 83.5 mm, and 70.1 mm, respectively. Corresponding ground elevations of points A, B, and C are 255 m, 183 m, and 137 m above MSL, respectively. Calculate relief displacements aa', bb', and cc' necessary to locate the datum positions a', b', and c' of the points, and calculate the datum scale.

Solution. By Eq. (6-11),

$$aa' = \frac{r_a h_A}{H} = \frac{91.4(255)}{1980} = 11.8 \text{ mm}$$

$$bb' = \frac{r_b h_B}{H} = \frac{83.5(183)}{1980} = 7.72 \text{ mm}$$

$$cc' = \frac{r_c h_C}{H} = \frac{70.1(137)}{1980} = 4.85 \text{ mm}$$

By Eq. (6-1),

$$\text{Datum scale} = \frac{f}{H} = \frac{152.4 \text{ mm}}{1980 \text{ m}} = \frac{0.1524 \text{ m}}{1980 \text{ m}} = 1:13,\overline{0}00$$

Note: Datum positions a', b', and c' in Fig. 6-8 are obtained by laying off distances aa', bb', and cc' along lines radial toward the principal point.

6-9 FLYING HEIGHT OF A VERTICAL PHOTOGRAPH

From previous discussion it is apparent that flying height above datum is an important quantity which is often needed for solving basic photogrammetric equations. Note, for example, that this parameter appears in scale, ground coordinate, and relief displacement equations. For rough computations, flying height may be taken from altimeter readings. Flying heights also may be obtained by using either Eq. (6-1) or Eq. (6-2) if a ground line of known length appears on the photograph. This procedure yields accurate flying heights for truly vertical photographs if the endpoints of the ground line lie at equal elevations. In general, the greater the difference in elevation of the endpoints, the greater the error in the computed flying height; therefore the ground line should lie on fairly level terrain. Accurate results can be obtained by this method, however, even though the endpoints of the ground line are at different elevations, if the images of the endpoints are approximately equidistant from the principal point of the photograph and on a line through the principal point.

Example 6-9. A vertical photograph contains the images of two fence line corners which identify adjacent section corners. The section line lies on fairly level terrain. Find the approximate flying height above the terrain if the camera focal length is 88.9 mm and the section line scales 94.0 mm on the photograph.

Solution. Assume the section line to be 5280 ft long. By Eq. (6-1),

$$\frac{94.0 \text{ mm}}{5280 \text{ ft}}\left(\frac{1 \text{ ft}}{304.8 \text{ mm}}\right) = \frac{88.9 \text{ mm}}{H'}\left(\frac{1 \text{ m}}{1000 \text{ mm}}\right)$$

from which

$$H' = 1520 \text{ m above terrain}$$

Accurate flying heights can be determined even though the endpoints of the ground line lie at different elevations, regardless of the locations of the endpoints in the photo. This procedure requires knowledge of the elevations of the endpoints of the line as well as of the length of the line. Suppose ground line AB has its endpoints imaged at a and b on a vertical photograph. Length AB of the ground line may be expressed in terms of ground coordinates, by the pythagorean theorem, as follows:

$$(AB)^2 = (X_B - X_A)^2 + (Y_B - Y_A)^2$$

Substituting Eqs. (6-5) through (6-8) into the previous equation gives

$$(AB)^2 = \left[\frac{x_b}{f}(H - h_B) - \frac{x_a}{f}(H - h_A)\right]^2 + \left[\frac{y_b}{f}(H - h_B) - \frac{y_a}{f}(H - h_A)\right]^2 \quad (6\text{-}13)$$

The only unknown in Eq. (6-13) is the flying height H. When all known values are inserted into the equation, it reduces to the quadratic form of $aH^2 + bH + c = 0$. The direct solution for H in the quadratic is

$$H = \frac{-b \pm \sqrt{b^2 - 4ac}}{2a} \quad (6\text{-}14)$$

Example 6-10. A vertical photograph was taken with a camera having a focal length of 152.3 mm. Ground points A and B have elevations 437.4 m and 445.3 m above sea level, respectively, and the horizontal length of line AB is 584.9 m. The images of A and B appear at a and b, and their measured photo coordinates are $x_a = 18.21$ mm, $y_a = -61.32$ mm, $x_b = 109.65$ mm, and $y_b = -21.21$ mm. Calculate the flying height of the photograph above sea level.

Solution. By Eq. (6-13),

$$(584.9)^2 = \left[\frac{109.65}{152.3}(H - 445.3) - \frac{18.21}{152.3}(H - 437.4)\right]^2$$
$$+ \left[\frac{-21.21}{152.3}(H - 445.3) + \frac{61.32}{152.3}(H - 437.4)\right]^2$$

Reducing gives

$$(584.9)^2 = (0.6004H - 268.3)^2 + (0.2634H - 114.1)^2$$

Squaring terms and arranging in quadratic form yields

$$0.4298H^2 - 382.3H - 257{,}100 = 0$$

Solving for H by Eq. (6-14) gives

$$H = \frac{382.3 \pm \sqrt{(-382.3)^2 - 4(0.4298)(-257{,}100)}}{2(0.4298)}$$

$$= \frac{382.3 \pm 766.9}{2(0.4298)} = 1337 \text{ m}$$

Note: The positive root was selected, since the negative root yields a ridiculous answer.

6-10   ERROR EVALUATION

Answers obtained in solving the various equations presented in this chapter will inevitably contain errors. It is important to have an awareness of the presence of these errors and to be able to assess their approximate magnitudes. Errors in computed answers are caused partly by random errors in measured quantities that are used in computation, and partly by the failure of certain assumptions to be met. Some of the more significant sources of errors in calculated values using the equations of this chapter are

1. Errors in photographic measurements, e.g., line lengths or photo coordinates
2. Errors in ground control
3. Shrinkage and expansion of film and paper
4. Tilted photographs where vertical photographs were assumed

Sources 1 and 2 can be minimized if precise, properly calibrated equipment and suitable caution are used in making the measurements. Source 3 can be practically eliminated by making corrections as described in Sec. 4-11. Magnitudes of error introduced by source 4 depend upon the severity of the tilt. Generally if the photos were intended to be vertical and if paper prints are being used, these errors are compatible with the other sources. If the photo is severely tilted, or if the highest accuracy is desired, analytical methods of Chap. 11 should be used. For the methods described in this chapter, errors caused by lens distortions and atmospheric refraction are relatively small and can generally be ignored.

A simple and straightforward approach to calculating the combined effect of several random errors is to use statistical error propagation, as discussed in Sec. A-4. This approach involves calculating rates of change with respect to each variable containing error and requires elementary differential calculus. As an example of this approach, assume that a vertical photograph was taken with a camera having a focal length of 152.4 mm. Assume also that a ground distance AB on flat terrain has a length of 1524 m and that its corresponding photo distance ab measures 127.0 mm. Flying height above ground may be calculated, using Eq. (6-1), as follows:

$$H' = f\left(\frac{AB}{ab}\right) = 152.4\left(\frac{1524}{127.0}\right) = 1829 \text{ m}$$

Now it is required to calculate the expected error dH' caused by errors in measured quantities AB and ab. This is done by taking derivatives with respect to each of these quantities containing error. Suppose that the error σ_{AB} in the ground distance is ± 0.50 m and that the error σ_{ab} in the measured photo distance is ± 0.20 mm. The rate of change of error in H' caused by the error in the ground length can be evaluated by taking the partial derivative $\partial H'/\partial AB$ as

$$\frac{\partial H'}{\partial AB} = \frac{f}{ab} = \frac{152.4 \text{ mm}}{127.0 \text{ mm}} = 1.200$$

In a similar manner, the rate of change of error in H' caused by the error in the measured image length can be evaluated by taking the partial derivative $\partial H'/\partial ab$ as

$$\frac{\partial H'}{\partial ab} = \frac{-f(AB)}{(ab)^2} = \frac{-152.4 \text{ mm} (1524 \text{ m})}{(127.0 \text{ mm})^2} = -14.40 \text{ m/mm}$$

A useful interpretation of these derivative terms is that an error of 1 m in ground distance AB will cause an error of approximately 1.2 m in the flying height, whereas an error of 1 mm in image distance *ab* will cause an error of approximately 14 m in the flying height. Substitution of these derivative terms into the error propagation Eq. (A-2) along with the error terms σ_{AB} and σ_{ab} gives

$$\sigma_{H'} = \sqrt{(1.200)^2(0.50 \text{ m})^2 + (-14.40 \text{ m/mm})^2(0.20 \text{ mm})^2}$$
$$= \sqrt{0.36 \text{ m}^2 + 8.29 \text{ m}^2} = \pm2.9 \text{ m}$$

Note that the error in H' caused by the error in the measurement of photo distance *ab* is the more severe of the two contributing sources. Therefore, to increase the accuracy of the computed value of H', it would be more beneficial to refine the measured photo distance to a more accurate value. Errors in computed answers using any of the equations presented in this chapter can be analyzed in the manner described above, and the method is valid as long as the contributing errors are small.

REFERENCES

American Society of Photogrammetry: *Manual of Photogrammetry,* 3d ed., Bethesda, MD, 1966, chap. 2.
————: *Manual of Photogrammetry,* 4th ed., Bethesda, MD, 1980, chap. 2.
Wolf, P. R., and C. G. Ghilani: *Adjustment Computations—Statistics and Least Squares in Surveying and GIS,* Wiley, New York, 1997.

PROBLEMS

Express answers for scale as dimensionless ratios and answers for distance in meters, unless otherwise specified.

6-1. The photo distance between two image points *a* and *b* on a vertical photograph is *ab*, and the corresponding ground distance is *AB*. What is the photographic scale at the elevation of the ground line?
 (*a*) *ab* = 3.78 in; *AB* = 779 ft
 (*b*) *ab* = 1.24 in; *AB* = 626 ft
 (*c*) *ab* = 5.83 in; *AB* = 5847 ft
 (*d*) *ab* = 4.65 in; *AB* = 9287 ft

6-2. Repeat Prob. 6-1, using the values of *ab* and *AB* indicated.
 (*a*) *ab* = 189.5 mm; *AB* = 474.6 m
 (*b*) *ab* = 148.3 mm; *AB* = 1151 m
 (*c*) *ab* = 195.7 mm; *AB* = 1377 m
 (*d*) *ab* = 120.0 mm; *AB* = 1.00 mi

6-3. On a vertical photograph a section line measures 73.7 mm. What is the photographic scale at the elevation of the section line?

6-4. On a vertical photograph a college football field measures 30.7 mm from goal line to goal line (100.0 yd). What is the scale of the photograph at the elevation of the football field?

6-5. A semitractor and trailer combination which is known to be 18 m long measures 10.2 mm long on a vertical aerial photo. What is the scale of the photo at the elevation of the semi?

6-6. Repeat Prob. 6-5, except that a railroad boxcar of 25.0-m known length measures 11.2 mm on the photo.

6-7. An interstate highway pavement of known 24.0-ft width measures 3.40 mm wide on a vertical aerial photo. What is the flying height above the pavement for this photo if the camera focal length was 152.4 mm?

6-8. In the photo of Prob. 6-7, a rectangular building near the highway has photo dimensions of 5.8 mm and 8.0 mm. What is the actual size of the structure?

6-9. Repeat Prob. 6-8, except that in the photo of Prob. 6-7 a bridge appears. If its photo length is 26.5 mm, what is the actual length of the bridge?

6-10. A vertical photograph was taken, with a camera having a 153.1-mm focal length, from a flying height 2280 m above sea level. What is the scale of the photo at an elevation of 430 m above sea level? What is the datum scale?

6-11. Aerial photographs are to be taken for highway planning and design. If a 152-mm-focal-length camera is to be used and if an average scale of 1 : 3000 is required, what should be the flying height above average terrain?

6-12. A vertical aerial photograph was taken from a flying height of $32\overline{0}0$ m above datum with a camera having a focal length of 209.45 mm. Highest, lowest, and average terrains appearing in the photograph are 2030 m, 940 m, and 1460 m, respectively. Calculate the maximum, minimum, and average photographic scales.

6-13. A vertical photograph was taken over the lunar surface from an altitude of 96.8 km with a 80.20-mm-focal-length camera. What is the actual diameter of a crater whose diameter on the photograph scales 10.63 mm? (Give answer in km.)

6-14. Vertical photography for military reconnaissance is required. If the lowest safe flying altitude over enemy defenses is 5000 m, what camera focal length is necessary to achieve a photo scale of 1 : 50,000?

6-15. A distance ab on a vertical photograph is 51.7 mm, and the corresponding ground distance AB is 1282 m. If the camera focal length is 88.95 mm, what is the flying height above the terrain upon which line AB is located?

6-16. Vertical photography at an average scale of 1 : $50\overline{0}0$ is to be acquired for the purpose of constructing a mosaic. What is the required flying height above average terrain if the camera focal length is 210.1 mm?

6-17. The distance on a map between two road intersections in flat terrain measures 49.3 mm. The distance between the same two points is 91.7 mm on a vertical photograph. If the scale of the map is 1 : 50,000, what is the scale of the photograph?

6-18. For Prob. 6-17, the intersections occur at an average elevation of 381 m above sea level. If the camera had a focal length of 209.6 mm, what is the flying height above sea level for this photo?

6-19. A section line scales 96.4 mm on a vertical aerial photograph. What is the scale of the photograph?

6-20. For Prob. 6-19, the average elevation of the section line is at 417 m above sea level, and the camera focal length is 152.4 mm. What would be the actual length of a ground line that lies at elevation 285 m above sea level and measures 57.9 mm on this photo?

6-21. A vertical aerial photo is exposed at 1950 m above mean sea level using a camera having an 88.9-mm focal length. A triangular parcel of land that lies at elevation 850 ft above sea level appears on the photo, and its sides measure 36.3 mm, 32.5 mm, and 24.1 mm, respectively. What is the approximate area of this parcel in acres?

6-22. On a vertical aerial photograph, a line which was measured on the ground to be 536 m long scales 29.4 mm. What is the scale of the photo at the average elevation of this line?

6-23. Points A and B are at elevations 317 m and 379 m above datum, respectively. The photographic coordinates of their images on a vertical photograph are $x_a = 68.27$ mm, $y_a = -32.37$ mm, $x_b = -87.44$ mm, and $y_b = 26.81$ mm. What is the horizontal length of line AB if the photo was taken from 4275 m above datum with a 152.35-mm-focal-length camera?

6-24. Images a, b, and c of ground points A, B, and C appear on a vertical photograph taken from a flying height of 2540 m above datum. A 153.16-mm-focal-length camera was used. Points A, B, and C have elevations of 395 m, 341 m, and 429 m above datum, respectively. Measured photo coordinates of the images are $x_a = -60.2$ mm, $y_a = 47.3$ mm, $x_b = 52.4$ mm, $y_b = 80.8$ mm, and $x_c = 94.1$ mm, and $y_c = -79.7$ mm. Calculate the horizontal lengths of lines AB, BC, and AC and the area within triangle ABC in hectares.

6-25. The image of a point whose elevation is 1475 ft above datum appears 53.87 mm from the principal point of a vertical aerial photograph taken from a flying height of $60\overline{0}0$ ft above datum. What would this distance from the principal point be if the point were at datum?

6-26. The images of the top and bottom of a utility pole are 129.8 mm and 125.2 mm, respectively, from the principal point of a vertical photograph. What is the height of the pole if the flying height above the base of the pole is 875 m?

6-27. An area has an average terrain elevation of 315 m above datum. The highest points in the area are 525 m above datum. If the camera focal plane opening is 23 cm square, what flying height above datum is required to limit relief displacement with respect to average terrain elevation to 5.0 mm? (*Hint:* Assume the image of a point at highest elevation occurs in the corner of the camera format.) If the camera focal length is 209.7 mm, what is the resulting average scale of the photography?

6-28. The datum scale of a vertical photograph taken from 915 m above datum is $1 : 6\overline{0}00$. The diameter of a cylindrical oil storage tank measures 6.87 mm at the base and 7.01 mm at the top. What is the height of the tank if its base lies at 247 m above datum?

6-29. Assume that the smallest discernible and measurable relief displacement that is possible on a vertical photo is 0.5 mm. Would it be possible to determine the height of a telephone utility box imaged in the corner of a 23-cm-square photo taken from 930 m above ground? (*Note:* Telephone utility boxes actually stand 1.2 m above the ground.)

6-30. If your answer to Prob. 6-29 is yes, what is the maximum flying height at which it would be possible to discern the relief displacement of the utility box? If your answer is no, at what flying height would the relief displacement of the box be discernible?

6-31. On a vertical photograph, images a and b of ground points A and B have photographic coordinates $x_a = -12.68$ mm, $y_a = 70.24$ mm, $x_b = 89.07$ mm, and $y_b = -92.41$ mm. The horizontal distance between A and B is 1207 m, and the elevations of A and B are 391 m and 418 m above datum, respectively. Calculate the flying height above datum if the camera had a 152.5-mm focal length.

6-32. Repeat Prob. 6-31, except that the horizontal distance AB is 1612 m and the camera focal length is 88.92 mm.

6-33. In Prob. 6-13, assume that the values given for focal length, photo distance, and flying height contain random errors of ±0.10 mm, ±0.05 mm, and ±0.30 km, respectively. What is the expected error in the computed diameter of the crater?

6-34. In Prob. 6-15, assume that the values given for focal length, photo distance, and ground length contain random errors of ±0.005 mm, ±0.50 mm, and ±0.30 m, respectively. What is the expected error in the computed flying height?

6-35. In Prob. 6-26, assume that the random error in each measured photo distance is ±0.10 mm and that the error in the flying height is ±2.0 m. What is the expected error in the computed height of the utility pole?

CHAPTER 7

STEREOSCOPIC VIEWING

7-1 DEPTH PERCEPTION

In our daily activities we unconsciously measure depth or judge distances to a vast number of objects about us through our normal process of vision. Methods of judging depth may be classified as either *stereoscopic* or *monoscopic*. Persons with normal vision (those capable of viewing with both eyes simultaneously) are said to have *binocular* vision, and perception of depth through binocular vision is called stereoscopic viewing. *Monocular* vision is the term applied to viewing with only one eye, and methods of judging distances with one eye are termed *monoscopic*. A person having normal binocular vision can, of course, view monocularly by covering one eye.

Distances to objects, or depths, can be perceived monoscopically on the basis of (1) relative sizes of objects, (2) hidden objects, (3) shadows, and (4) differences in focusing of the eye required for viewing objects at varying distances. Examples of the first two of these are shown in Fig. 7-1. Depth to the far end of the football field may be perceived, for example, on the basis of the relative sizes of the goalposts. The goalposts are actually the same size, of course, but one appears smaller because it is farther away. Also, the building is quickly judged to be a considerable distance away because it is partially hidden behind the football stadium.

Monoscopic methods of depth perception enable only rough impressions to be gained of distances to objects. With stereoscopic viewing, however, a much greater degree of accuracy in depth perception can be attained. Stereoscopic depth perception

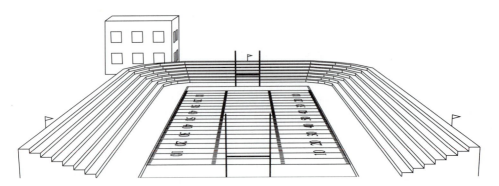

FIGURE 7-1
Depth perception by relative sizes and hidden objects.

is of fundamental importance in photogrammetry, for it enables the formation of a three-dimensional stereomodel by viewing a pair of overlapping photographs. The stereomodel can then be studied, measured, and mapped. An explanation of how this phenomenon is achieved is the subject of this chapter, and explanations of its use in measuring and mapping are given in the chapters that follow.

7-2   THE HUMAN EYE

The phenomenon of stereoscopic depth perception can be more clearly understood with the help of a brief description of the anatomy and physiology of the human eye. The human eye functions in much the same manner as a camera. As shown in Fig. 7-2, the eye is essentially a spherical organ having a circular opening called the *pupil*. The pupil is protected by a transparent coating called the *cornea*. Incident light rays pass through the cornea, enter the eye through the pupil, and strike the *lens*, which is directly behind the pupil. The cornea and lens refract the light rays according to Snell's law (see Sec. 2-2).

The lens of the eye is biconvex and is composed of a refractive transparent medium. It is suspended by many muscles, which enable the lens to be moved so that the *optical axis* of the eye can be aimed directly at an object to be viewed. As with a camera, the eye must satisfy the lens formula, Eq. (2-4), for each different object

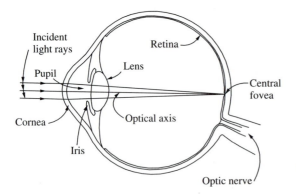

FIGURE 7-2
Cross section of the human eye.

distance. The eye's image distance is constant, however; therefore, to satisfy the lens formula for varying object distances, the focal length of the lens changes. When a distant object is viewed, the lens muscles relax, causing the spherical surfaces of the lens to become flatter. This increases the focal length to satisfy the lens formula and accommodate the long object distance. When close objects are viewed, a reverse procedure occurs. The eye's ability to focus for varying object distances is called *accommodation*.

As with a camera, the eye has a diaphragm called the *iris*. The iris (colored part of the eye) automatically contracts or expands to regulate the amount of light entering the eye. When the eye is subjected to intense light, the iris contracts, reducing the pupil aperture. When the intensity of light lessens, the iris dilates to admit more light.

The cornea partially refracts incident light rays before they encounter the lens. The lens refracts them further and brings them to focus on the *retina*, thereby forming an image of the viewed object. The retina is composed of very delicate tissue. The most important region of the retina is the *central fovea*, a small pit near the intersection of the optical axis with the retina. The central fovea is the area of sharpest vision. The retina performs a function similar to that performed by the emulsion of photographic film. When it is stimulated by light, the sense of vision is caused, which is transmitted to the brain via the *optic nerve*.

7-3   STEREOSCOPIC DEPTH PERCEPTION

With binocular vision, when the eyes fixate on a certain point, the optical axes of the two eyes converge on that point, intersecting at an angle called the *parallactic angle*. The nearer the object, the greater the parallactic angle, and vice versa. In Fig. 7-3, the optical

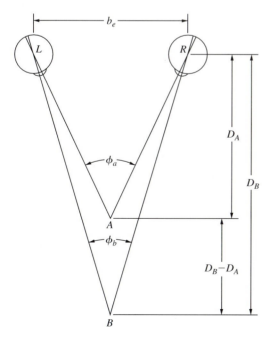

FIGURE 7-3
Stereoscopic depth perception as a function of parallactic angle.

axes of the two eyes L and R are separated by a distance b_e, called the *eye base*. For the average adult, this distance is between 63 and 69 mm, or approximately 2.6 in. When the eyes fixate on point A, the optical axes converge, forming parallactic angle ϕ_a. Similarly, when sighting an object at B, the optical axes converge, forming parallactic angle ϕ_b. The brain automatically and unconsciously associates distances D_A and D_B with corresponding parallactic angles ϕ_a and ϕ_b. The depth between objects A and B is $D_B - D_A$ and is perceived from the difference in these parallactic angles.

The ability of human beings to detect changes in parallactic angles, and thus judge differences in depth, is quite remarkable. Although it varies somewhat among individuals, the average person is capable of discerning parallactic angle changes of about 3 seconds of arc, but some are able to perceive changes as small as 1 second of arc. This means that photogrammetric procedures for determining heights of objects and terrain variations based on depth perception by comparisons of parallactic angles can be highly precise.

7-4   VIEWING PHOTOGRAPHS STEREOSCOPICALLY

Suppose that while a person is gazing at object A of Fig. 7-4, a transparent medium containing image marks a_1 and a_2 is placed in front of the eyes as shown. Assume further that the image marks are identical in shape to object A, and that they are placed on the

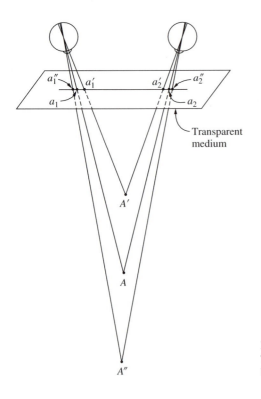

FIGURE 7-4
The apparent depth to the object A can be changed by changing the spacing of the images.

optical axes so that the eyes are unable to detect whether they are viewing the object or the two marks. Object A could therefore be removed without any noticeable changes in the images received on the retinas of the eyes. As shown in Fig. 7-4, if the image marks are moved closer together to, say, a_1', and a_2', the parallactic angle increases and the object is perceived to be nearer the eyes at A'. If the marks are moved farther apart to a_1'' and a_2'', the parallactic angle decreases and the brain receives an impression that the object is farther away, at A''.

The phenomenon of creating the three-dimensional or stereoscopic impression of objects by viewing identical images of the objects can be achieved photographically. Suppose that a pair of aerial photographs is taken from exposure stations L_1 and L_2 so that the building appears on both photos, as shown in Fig. 7-5. Flying height above ground is H', and the distance between the two exposures is B, the *air base*. Object points A and B at the top and bottom of the building are imaged at a_1 and b_1 on the left photo and at a_2 and b_2 on the right photo. Now, if the two photos are laid on a table and

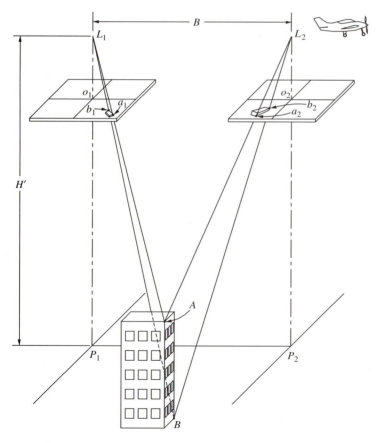

FIGURE 7-5
Photographs from two exposure stations with building in common overlap area.

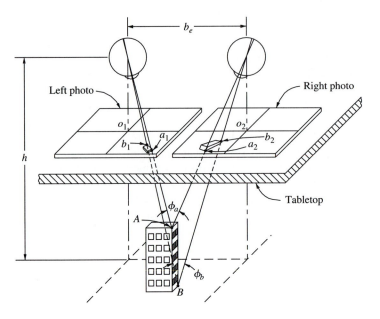

FIGURE 7-6
Viewing the building stereoscopically.

viewed so that the left eye sees only the left photo and the right eye sees only the right photo, as shown in Fig. 7-6, a three-dimensional impression of the building is obtained. The three-dimensional impression appears to lie below the tabletop at a distance h from the eyes. The brain judges the height of the building by associating depths to points A and B with the parallactic angles ϕ_a and ϕ_b, respectively. When the eyes gaze over the entire overlap area, the brain receives a continuous three-dimensional impression of the terrain. This is achieved by the continuous perception of changing parallactic angles of the infinite number of image points which make up the terrain. The three-dimensional model thus formed is called a *stereoscopic model* or simply a *stereomodel,* and the overlapping pair of photographs is called a *stereopair.*

7-5   STEREOSCOPES

It is quite difficult to view photographs stereoscopically without the aid of optical devices, although some individuals can do it. Besides being an unnatural operation, one of the major problems associated with stereoviewing without optical aids is that the eyes are focused on the photos, while at the same time the brain perceives parallactic angles which tend to form the stereomodel at some depth beyond the photos—a confusing situation, to say the least. These difficulties in stereoscopic viewing may be overcome through the use of instruments called *stereoscopes.*

There is a wide selection of stereoscopes serving a variety of special purposes. All operate in essentially the same manner. The *lens* or *pocket* stereoscope, shown in Fig. 7-7, is the least expensive and most commonly used stereoscope. It consists of two

FIGURE 7-7
Lens or pocket stereoscope.
(*Courtesy University of Florida.*)

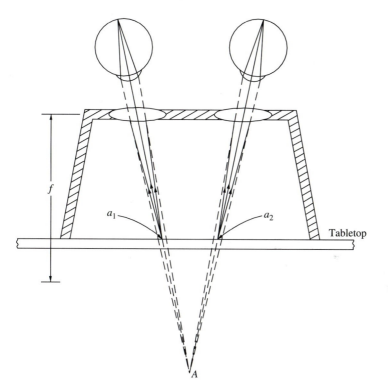

FIGURE 7-8
Schematic diagram of the pocket stereoscope.

simple convex lenses mounted on a frame. The spacing between the lenses can be varied to accommodate various eye bases. The legs fold or can be removed so that the instrument is easily stored or carried—a feature which renders the pocket stereoscope ideal for fieldwork. A schematic diagram of the pocket stereoscope is given in Fig. 7-8. The legs of the pocket stereoscope are slightly shorter than the focal length f of the lenses. When the stereoscope is placed over the photos, light rays emanating from points such as a_1 and a_2 on the photos are refracted slightly as they pass through each lens. (Recall from Chap. 2 that a bundle of light rays from a point exactly a distance f from a lens will be refracted and emerge through the lens parallel.) The eyes receive the refracted rays (shown dashed in Fig. 7-8), and on the basis of the eye focusing associated with these incoming rays, the brain receives the impression that the rays actually originate from a greater distance than that to the tabletop upon which the photos rest. This overcomes the difficulties noted above. The lenses also serve to magnify the images, thereby enabling details to be seen more clearly.

In using a pocket stereoscope, the photos are placed so that corresponding images are slightly less than the eye base apart, usually about 5 cm. For normal 23-cm-format photos taken with 60 percent end lap, the common overlap area of a pair of photos is a rectangular area about 14 cm wide, as shown crosshatched in Fig. 7-9a. If the photos are separated by 5 cm for stereoviewing with a pocket stereoscope, as shown in Fig. 7-9b, there is a rectangular area, shown double crosshatched, in which the top photo obscures the bottom photo, thereby preventing stereoviewing. To overcome this problem, the top photo can be gently rolled up out of the way to enable viewing the corresponding imagery of the obscured area.

The *mirror* stereoscope shown in Fig. 7-10 permits the two photos to be completely separated when viewed stereoscopically. This eliminates the problem of one photo obscuring part of the overlap of the other, and it also enables the entire width of the stereomodel to be viewed simultaneously. The operating principle of the mirror stereoscope is illustrated in Fig. 7-11. The stereoscope has two large wing mirrors and two smaller eyepiece mirrors, all of which are mounted at 45° to the horizontal. Light rays emanating from image points on the photos such as a_1 and a_2 are reflected from the mirror surfaces, according to the principles of reflection discussed in Sec. 2-2, and are received at the eyes, forming parallactic angle ϕ_a. The brain automatically associates the depth to point A with that parallactic angle. The stereomodel is thereby created beneath the eyepiece mirrors, as illustrated in Fig. 7-11.

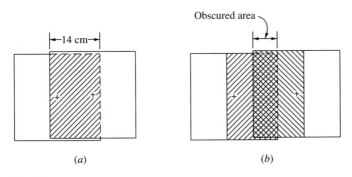

(a)   (b)

FIGURE 7-9
(a) The common overlap area of a pair of 23-cm-format photos taken with 60 percent end lap (corresponding images coincident). (b) Obscured area when photos are oriented for viewing with pocket stereoscope.

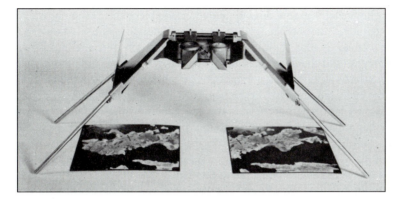

FIGURE 7-10
ST-4 mirror stereoscope. (*Courtesy LH Systems, LLC.*)

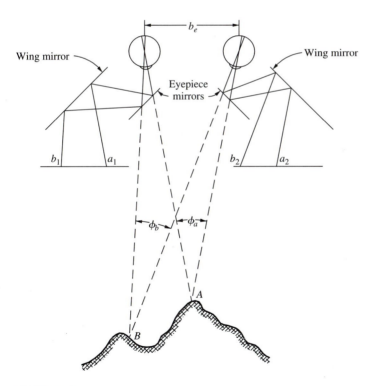

FIGURE 7-11
Operating principle of the mirror stereoscope.

Simple lenses are usually placed directly above the eyepiece mirrors as shown in Fig. 7-10. Their separation may be changed to accommodate various eye bases. The focal length of these lenses is also slightly greater than the length of the reflected ray path from photo to eyes, and therefore they serve basically the same function as the lenses of the pocket stereoscope. Mirror stereoscopes may be equipped with binoculars which fasten over the eyepiece mirrors. The binoculars, which may be focused individually to accommodate each eye, permit viewing images at high magnification—a factor which is especially important and useful in photo interpretation or in identifying small image points. High magnification, of course, limits the field of view so that the entire stereomodel cannot be viewed simultaneously. The stereoscope must therefore be moved about if all parts of the stereomodel are to be seen.

It is extremely important to avoid touching the first-surfaced mirrors of a mirror stereoscope. This is true because the hands contain oils and acids which can tarnish the coatings on the mirrors, rendering them useless. If the mirrors are accidently soiled by fingerprints, they should be cleaned immediately, using a soft cloth and lens-cleaning fluid.

A different type of stereoscope called the *zoom stereoscope* is shown in Fig. 7-12. A variety of these instruments are manufactured, affording a choice of special features such as continuous zoom magnification up to 120X, capability of rotating images optically (which permits convenient correction for crab or alignment), and differential enlargement so that two photos of different scales can be viewed stereoscopically. For direct stereoscopic viewing of film negatives, these stereoscopes may be obtained mounted on a light table and equipped with a special scanning mechanism. A reel of film and a take-up reel are mounted on either end of the table. By turning a crank, the frames are brought into position for viewing.

Other systems have been developed to facilitate stereoscopic viewing. These systems employ schemes such as colored filters, polarized light, or alternating imagery in

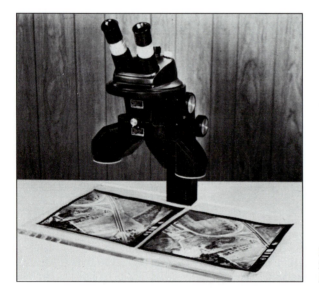

FIGURE 7-12
Zoom 95 stereoscope. (*Courtesy Bausch and Lomb Co.*)

order to achieve the stereoscopic effect. Sections 12-5 and 15-2 include a description of several of these approaches as they apply to stereoscopic plotting instruments.

7-6 THE USE OF STEREOSCOPES

Before you attempt to use a stereoscope, it is important to study the operator's manual if one is available. This is especially true for stereoscopes having more elaborate optical viewing systems. Also, the lenses and mirrors should be inspected and cleaned if necessary.

In stereoscopic viewing, it is important to orient the photos so that the left and right eyes see the left and right photos, respectively. If the photos are viewed in reverse, a *pseudoscopic view* results in which ups and downs are reversed; e.g., valleys appear as ridges and hills appear as depressions. This can be advantageous for certain work such as tracing drainage patterns, but normally the correct stereoscopic view is desired.

Accurate and comfortable stereoscopic viewing requires that the eye base, the line joining the centers of the stereoscope lenses, and the flight line all be parallel. Therefore, after the photos have been inspected and laid out so as to prevent a pseudoscopic view, the flight line is marked on both photos. For vertical photographs, the flight line is the line from the center of the left photo to the center of the right photo. In marking the flight line, the photo centers (principal points) are first located by joining opposite fiducial marks with straight lines. Principal points are shown at o_1 and o_2 on Fig. 7-13. *Corresponding principal points* (also called *conjugate principal points*), which are the locations of principal points on adjacent overlapping photos, are marked next. This may be done satisfactorily by carefully observing images immediately surrounding the principal points, and then marking the corresponding principal points by estimating their positions with respect to these surrounding images. The corresponding principal points are shown at o_1' and o_2' on Fig. 7-13.

The next step in orienting a pair of photos for stereoscopic viewing is to fasten the left photo down onto the table. Then the right photo is oriented so that the four points defining the flight line (o_1, o_2', o_1', and o_2) all lie along a straight line, as shown in Fig. 7-13. The right photo is retained in this orientation, and while being viewed through the stereoscope, it is moved sideways until the spacing between corresponding images produces a comfortable stereoscopic view. Normally the required spacing

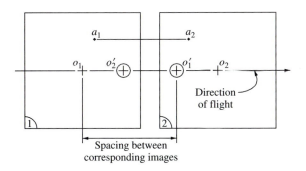

FIGURE 7-13
Pair of photos properly oriented for stereoscopic viewing.

between corresponding images is slightly more than 5 cm for a pocket stereoscope and about 25 cm for a mirror stereoscope.

It is not absolutely necessary to mark the flight lines and orient photos for stereoscopic viewing in the manner outlined above; in fact, for casual stereoviewing, the geometry shown in Fig. 7-13 is normally achieved by a trial method in which the photos are simply shifted in position until a clear stereoscopic view is obtained. If accuracy and eye comfort are considerations, however, orientation by the flight-line procedure is recommended.

As previously stated, comfortable stereoscopic viewing requires that the line joining the stereoscope lens centers be parallel with the flight line. Once the photos are properly oriented, the operator can easily align the stereoscope by simply rotating it slightly until the most comfortable viewing position is obtained. The operator should look directly into the centers of the lenses, thereby holding the eye base parallel with the flight line.

7-7   CAUSES OF Y PARALLAX

An essential condition which must exist for clear and comfortable stereoscopic viewing is that the line joining corresponding images be parallel with the direction of flight. This condition is fulfilled with the corresponding images a_1 and a_2 shown in Fig. 7-13. When corresponding images fail to lie along a line parallel to the flight line, *y parallax,* denoted by p_y, is said to exist. Any slight amount of y parallax causes eyestrain, and excessive amounts prevent stereoscopic viewing altogether.

If a pair of truly vertical overlapping photos taken from equal flying heights is oriented perfectly, then no y parallax should exist anywhere in the overlap area. Failure of any of these conditions to be satisfied will cause y parallax. In Fig. 7-14, for example, the photos are improperly oriented, and the principal points and corresponding principal points do not lie on a straight line. As a result, y parallax exists at both points a and b. This condition can be prevented by careful orientation.

In Fig. 7-15 the left photo was exposed from a lower flying height than the right photo, and consequently its scale is larger than the scale of the right photo. Even though the photos are truly vertical and properly oriented, y parallax exists at both points a and b due to variation in flying heights. To obtain a comfortable stereoscopic view, the y parallax can be eliminated by sliding the right photo upward transverse to the flight line when viewing point a and sliding it downward when viewing point b.

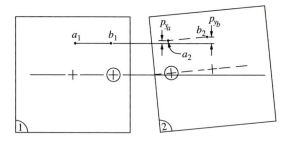

FIGURE 7-14
Here y parallax is caused by improper orientation of the photos.

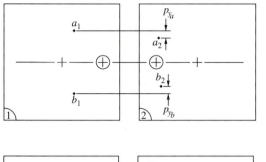

FIGURE 7-15
The *y* parallax is caused by variation in
flying height.

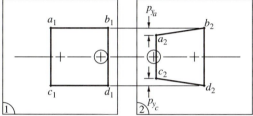

FIGURE 7-16
The *y* parallax is caused by tilt of the photos.

The effect of tilted photos is illustrated in Fig. 7-16. The left photo is truly vertical and shows positions of images *a* through *d* of a square parcel of property on flat terrain. The right photo was tilted such that the same parcel appears as a trapezoid. In this case, *y* parallax exists throughout the stereoscopic model as a result of the tilt, as indicated for points *a* and *c*. In practice, the direction of tilt is random, and therefore small *y* parallaxes from this source are likely to exist in variable amounts throughout most stereomodels. If it is the intent to obtain vertical photography from a constant flying height, however, these conditions are generally so well controlled that *y* parallaxes from these sources are seldom noticeable. Most serious *y* parallaxes usually occur from improper orientation of the photos, a condition which can be easily corrected.

7-8 VERTICAL EXAGGERATION IN STEREOVIEWING

Under normal conditions, the vertical scale of a stereomodel will appear to be greater than the horizontal scale; i.e., an object in the stereomodel will appear to be too tall. This apparent scale disparity is called *vertical exaggeration*. It is usually of greatest concern to photo interpreters, who must take this condition into account when estimating heights of objects, rates of slopes, etc.

Although other factors are involved, vertical exaggeration is caused primarily by the lack of equivalence of the *photographic base-height ratio, B/H',* and the corresponding *stereoviewing base-height ratio, b_e/h.* The term B/H' is the ratio of the *air base* (distance between the two exposure stations) to flying height above average ground, and b_e/h is the ratio of the *eye base* (distance between the two eyes) to the distance from the eyes at which the stereomodel is perceived. Figures 7-17*a* and *b*

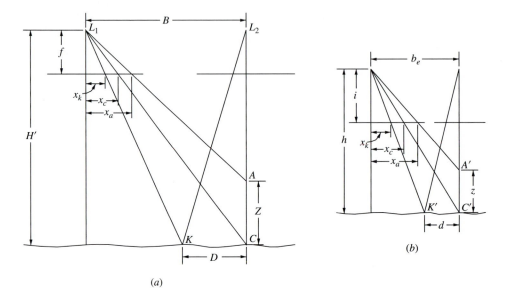

FIGURE 7-17
Simplistic diagrams for analyzing vertical exaggeration. (*a*) Geometry of overlapping aerial photography.
(*b*) Geometry of stereoscopic viewing of the photos of part (*a*).

depict, respectively, the taking of a pair of vertical overlapping photographs and the stereoscopic viewing of those photos. In Fig. 7-17*a*, the camera focal length is *f*, the air base is *B*, the flying height above ground is H', the height of ground object AC is Z, and the horizontal ground distance KC is D. In Fig. 7-17*a*, assume that Z is equal to D. In Fig. 7-17*b*, *i* is the image distance from the eyes to the photos, b_e is the eye base, *h* is the distance from the eyes to the perceived stereomodel, *z* is the stereomodel height of object $A'C'$, and *d* is the horizontal stereomodel distance $K'C'$. Note that while the ratio Z/D is equal to 1, the ratio z/d is greater than 1 due to vertical exaggeration.

An equation for calculating vertical exaggeration can be developed with reference to these figures. From similar triangles of Fig. 7-17*a*,

$$\frac{x_a}{B} = \frac{f}{H' - Z} \qquad \text{from which} \qquad x_a = \frac{Bf}{H' - Z} \qquad (a)$$

Also

$$\frac{x_c}{B} = \frac{f}{H'} \qquad \text{from which} \qquad x_c = \frac{Bf}{H'} \qquad (b)$$

Subtracting (*b*) from (*a*) and reducing gives

$$x_a - x_c = Bf \frac{Z}{(H')^2 - H'Z} \qquad (c)$$

Also from similar triangles of Fig. 7-17b,

$$\frac{x_a}{b_e} = \frac{i}{h - z} \qquad \text{from which} \qquad x_a = \frac{b_e i}{h - z} \qquad (d)$$

and

$$\frac{x_c}{b_e} = \frac{i}{h} \qquad \text{from which} \qquad x_c = \frac{b_e i}{h} \qquad (e)$$

Subtracting (e) from (d) and reducing yields

$$x_a - x_c = b_e i \frac{z}{h^2 - hz} \qquad (f)$$

Equating (c) and (f) gives

$$Bf \frac{Z}{(H')^2 - H'Z} = b_e i \frac{z}{h^2 - hz}$$

In the above equation, the values of Z and z are normally considerably smaller than the values of H' and h, respectively; thus

$$\frac{BfZ}{(H')^2} \approx \frac{b_e iz}{h^2} \qquad \text{from which} \qquad \frac{z}{Z} = \frac{fh}{H'i} \frac{Bh}{H'b_e} \qquad (g)$$

Also from similar triangles of Fig. 7-17a and b,

$$\frac{x_c - x_k}{D} = \frac{f}{H'} \qquad \text{from which} \qquad D = (x_c - x_k) \frac{H'}{f} \qquad (h)$$

and

$$\frac{x_c - x_k}{d} = \frac{i}{h} \qquad \text{from which} \qquad d = (x_c - x_k) \frac{h}{i} \qquad (i)$$

Dividing (i) by (h) and reducing yields

$$\frac{d}{D} = \frac{fh}{H'i} \qquad (j)$$

Substituting (j) into (g) and reducing gives

$$\frac{z}{Z} = \frac{d}{D} \frac{Bh}{H'b_e} \qquad (k)$$

In Eq. (k), if the term $Bh/(H'b_e)$ is equal to 1, there is no vertical exaggeration of the stereomodel. (Recall that Z is equal to D.) Thus an expression for the magnitude of vertical exaggeration V is given by

$$V \approx \frac{B}{H'} \frac{h}{b_e} \qquad (7\text{-}1)$$

From Eq. (7-1) it is seen that the magnitude of vertical exaggeration in stereoscopic viewing can be approximated by multiplying the B/H' ratio by the inverse of the b_e/h

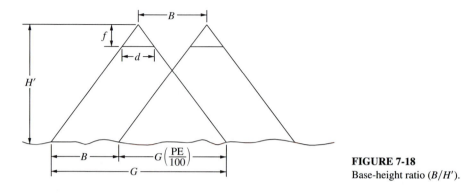

FIGURE 7-18
Base-height ratio (B/H').

ratio. An expression for the B/H' ratio can be developed with reference to Fig. 7-18. In this figure, G represents the total ground coverage of a vertical photo taken from an altitude of H' above ground. Air base B is the distance between exposures. From the figure,

$$B = G - G\frac{PE}{100} = G\left(1 - \frac{PE}{100}\right) \qquad (l)$$

In equation (l), PE is the percentage of end lap, which gives the amount that the second photo overlaps the first. Also by similar triangles of the figure,

$$\frac{H'}{G} = \frac{f}{d} \qquad \text{from which} \qquad H' = \frac{fG}{d} \qquad (m)$$

In Eq. (m), f is the camera focal length and d its format dimension. Dividing Eq. (l) by Eq. (m) and reducing gives

$$\frac{B}{H'} = \left(1 - \frac{PE}{100}\right)\frac{d}{f} \qquad (7\text{-}2)$$

The stereoviewing base-height ratio b_e/h is a somewhat difficult variable to measure, and it differs slightly among individuals. Repeated tests of many individuals, however, indicate that its value is approximately 0.15.

Example 7-1. Calculate the approximate vertical exaggeration for vertical aerial photos taken with a 152.4-mm-focal-length camera having a 23-cm-square format if the photos were taken with 60 percent end lap.

Solution. By Eq. (7-2),

$$\frac{B}{H'} = \left(1 - \frac{60}{100}\right)\frac{230}{152.4} = 0.60$$

By Eq. (7-1), assuming b_e/h to be 0.15,

$$V = 0.60\left(\frac{1}{0.15}\right) = 4.0 \qquad \text{(approx.)}$$

Note: If a 305-mm-focal-length camera had been used, the B/H' ratio would have been 0.30 and vertical exaggeration would have been reduced to 2.

REFERENCES

Ambrose, W. R.: "Stereoscopes with High Performance," *Photogrammetric Engineering,* vol. 31, no. 5, 1965, p. 822.

American Society of Photogrammetry: *Manual of Photogrammetry,* 4th ed., Bethesda, MD, 1980, chap. 10.

Collins, S. H.: "Stereoscopic Depth Perception," *Photogrammetric Engineering and Remote Sensing,* vol. 47, no. 1, 1981, p. 45.

Dalsgaard, J.: "Stereoscopic Vision—A Problem in Terrestrial Photogrammetry," *Photogrammetria,* vol. 34, no. 1, 1978, p. 3.

Gumbel, E. J.: "The Effect of the Pocket Stereoscope on Refractive Anomalies of the Eyes," *Photogrammetric Engineering,* vol. 30, no. 5, 1964, p. 795.

Howard, A. D.: "The Fichter Equation for Correcting Stereoscopic Slopes," *Photogrammetric Engineering,* vol. 34, no. 4, 1968, p. 386.

LaPrade, G. L.: "Stereoscopy—A More General Theory," *Photogrammetric Engineering,* vol. 38, no. 12, 1972, p. 1177.

————: "Stereoscopy—Will Dogma or Data Prevail?" *Photogrammetric Engineering,* vol. 39, no. 12, 1973, p. 1271.

Miller, C. I.: "Vertical Exaggeration in the Stereo Space Image and Its Use," *Photogrammetric Engineering,* vol. 26, no. 5, 1960, p. 815.

Myers, B. J., and F. P. Van der Duys: "A Stereoscopic Field Viewer," *Photogrammetric Engineering and Remote Sensing,* vol. 41, no. 12, 1975, p. 1477.

Nicholas, G., and J. T. McCrickerd: "Holography and Stereoscopy: The Holographic Stereogram," *Photographic Science and Engineering,* vol. 13, no. 6, 1969, p. 342.

Palmer, D. A.: "Stereoscopy and Photogrammetry," *Photogrammetric Record,* vol. 4, 1964, p. 391.

Raju, A. V., and E. Parthasarathi: "Stereoscopic Viewing of Landsat Imagery," *Photogrammetric Engineering and Remote Sensing,* vol. 43, no. 10, 1977, p. 1243.

Scheaffer, C. E.: "Stereoscope for Strips," *Photogrammetric Engineering,* vol. 34, no. 10, 1968, p. 1044.

Thayer, T. P.: "The Magnifying Single Prism Stereoscope: A New Field Instrument," *Journal of Forestry,* vol. 61, 1963, p. 381.

Yacoumelos, N.: "The Geometry of the Stereomodel," *Photogrammetric Engineering,* vol. 38, no. 8, 1972, p. 791.

PROBLEMS

7-1. What are some of the monocular methods of perceiving depth?

7-2. What is a parallactic angle?

7-3. Compare the advantages and disadvantages of the pocket and mirror stereoscopes.

7-4. Give a step-by-step procedure for orienting photos for stereoscopic viewing.

7-5. What is y parallax? What are the causes of y parallax in a stereomodel?

7-6. Prepare a table of B/H' ratios for camera focal lengths of 89, 152, 210, and 305 mm; camera format of 23-cm square; and end laps of 55, 60, and 65 percent.

7-7. Calculate the approximate vertical exaggeration in a stereomodel from photos taken with a 152-mm-focal-length camera having a 23-cm-square format if the photos are taken at 55 percent end lap.

7-8. Repeat Prob. 7-7, except that a 210-mm-focal-length camera was used, and end lap was 65 percent.

CHAPTER
8

STEREOSCOPIC PARALLAX

8-1 INTRODUCTION

Parallax is the apparent displacement in the position of an object, with respect to a frame of reference, caused by a shift in the position of observation. A simple experiment will serve to illustrate parallax. If a finger is held in front of the eyes, and while gazing at the finger the head is quickly shifted from side to side without moving the finger, the finger will appear to move from side to side with respect to objects beyond the finger, such as pictures on the wall. Rather than shifting the head, the same effect can be created by alternately blinking one's eyes. The closer the finger is held to the eyes, the greater will be its apparent shift. This apparent motion of the finger is parallax, and it is due to the shift in the position of observation.

If a person looked through the viewfinder of an aerial camera as the aircraft moved forward, images of objects would be seen to move across the field of view. This image motion is another example of parallax caused by shifting the location of the observation point. Again, the closer an object is to the camera, the more its image will appear to move.

An aerial camera exposing overlapping photographs at regular intervals of time obtains a record of positions of images at the instants of exposure. The change in position of an image from one photograph to the next caused by the aircraft's motion is termed *stereoscopic parallax, x parallax,* or simply *parallax.* Parallax exists for all images appearing on successive overlapping photographs. In Fig. 8-1, for example, images

164

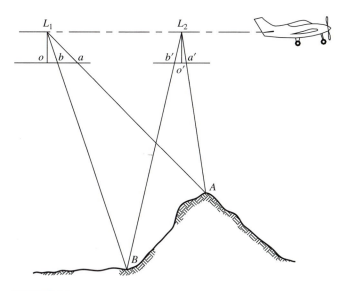

FIGURE 8-1
Stereoscopic parallax of vertical aerial photographs.

of object points A and B appear on a pair of overlapping vertical aerial photographs which were taken from exposure stations L_1 and L_2. Points A and B are imaged at a and b on the left-hand photograph. Forward motion of the aircraft between exposures, however, caused the images to move laterally across the camera focal plane parallel to the flight line, so that on the right-hand photo they appear at a' and b'. Because point A is higher (closer to the camera) than point B, the movement of image a across the focal plane was greater than the movement of image b; in other words, the parallax of point A is greater than the parallax of point B. This calls attention to two important aspects of stereoscopic parallax: (1) The parallax of any point is directly related to the elevation of the point, and (2) parallax is greater for high points than for low points. Variation of parallax with elevation provides the fundamental basis for determining elevations of points from photographic measurements. In fact, X, Y, and Z ground coordinates can be calculated for points based upon their parallaxes. Equations for doing this are presented in Sec. 8-6.

Figure 8-2 shows the two photographs of Fig. 8-1 in superposition. Parallaxes of object points A and B are p_a and p_b, respectively. Stereoscopic parallax for any point such as A whose images appear on two photos of a stereopair, expressed in terms of *flight-line* photographic coordinates, is

$$p_a = x_a - x_a' \tag{8-1}$$

In Eq. (8-1), p_a is the stereoscopic parallax of object point A, x_a is the measured photo coordinate of image a on the left photograph of the stereopair, and x_a' is the photo coordinate of image a' on the right photo. These photo coordinates *are not* measured with respect to the fiducial axis system which was described in Sec. 4-2. Rather, they are

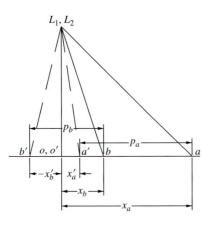

FIGURE 8-2
The two photographs of Fig. 8-1 are
shown in superposition.

measured with respect to the flight-line axis system described in Sec. 8-2. In Eq. (8-1) it is imperative that proper algebraic signs be given to measured photo coordinates to obtain correct values for stereoscopic parallax.

Figure 8-3 is a portion of a stereopair of vertical photographs taken over Washington, D.C., with a 152-mm-focal-length camera at a flying height of 1830 m above ground. On these photos, note how all images moved laterally with respect to the y axis from their positions on the left photo to their positions on the right photo. Note also how vividly the Washington Monument illustrates the increase in parallax with higher points; i.e., the top of the monument has moved farther across the focal plane than the bottom of the monument.

In Fig. 8-3 the Washington Monument affords an excellent example for demonstrating the use of Eq. (8-1) for finding parallaxes. The top of the monument has an x coordinate ($x_t = 4.32$ in $= 109.7$ mm) and an x' coordinate ($x'_t = 0.72$ in $= 18.3$ mm). By Eq. (8-1), the parallax $p_t = 4.32 - 0.72 = 3.60$ in $= 91.4$ mm. Also, the bottom of the monument has an x coordinate ($x_b = 3.96$ in $= 100.6$ mm) and an x' coordinate ($x'_b = 0.67$ in $= 17.0$ mm). Again by Eq. (8-1), $p_b = 3.96 - 0.67 = 3.29$ in $= 83.6$ mm.

8-2 PHOTOGRAPHIC FLIGHT-LINE AXES FOR PARALLAX MEASUREMENT

Since parallax occurs parallel to the direction of flight, the photographic x and x' axes for parallax measurement must be parallel with the flight line for each of the photographs of a stereopair. (Primed values denote the right-hand photo of a stereopair.) For a vertical photograph of a stereopair, the flight line is the line connecting the principal point and corresponding (conjugate) principal point. Principal points are located by intersecting the x and y fiducial lines. A monoscopic method of establishing corresponding principal points was described in Sec. 7-6. Stereoscopic methods are discussed in Sec. 8-4. The y and y' axes for parallax measurement pass through their respective principal points and are perpendicular to the flight line.

All photographs except those on the ends of a flight strip may have two sets of flight axes for parallax measurements—one to be used when the photo is the left photo of the

FIGURE 8-3
Overlapping vertical photographs taken over Washington, DC, illustrating stereoscopic parallax. (*Photos courtesy Ayres Associates, Inc.*)

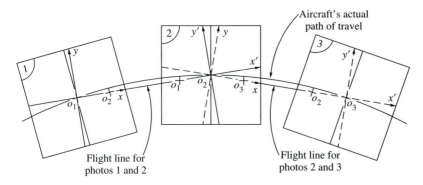

FIGURE 8-4
Flight-line axes for measurement of stereoscopic parallax.

stereopair and one when it is the right photo. An example is shown in Fig. 8-4, where photographs 1 through 3 were exposed as shown. Parallax measurements in the overlap area of photos 1 and 2 are made with respect to the solid xy axis system of photo 1 and the solid $x'y'$ system of photo 2. However, due to the aircraft's curved path of travel, the flight line of photos 2 and 3 is not in the same direction as the flight line of photos 1 and 2. The refore, parallax measurements in the overlap area of photos 2 and 3 must be made with respect to the dashed xy axis system on photo 2 and the dashed $x'y'$ system of photo 3. It is possible for the two axis systems to be coincident; however, this does not generally occur in practice. Henceforth in this chapter it is understood that photographic coordinates for parallax determination are measured with respect to the flight-line axis system.

8-3   MONOSCOPIC METHODS OF PARALLAX MEASUREMENT

Parallaxes of points on a stereopair may be measured either monoscopically or stereoscopically. There are certain advantages and disadvantages associated with each method. In either method the photographic flight line axes must first be carefully located by marking principal points and corresponding principal points.

The simplest method of parallax measurement is the monoscopic approach, in which Eq. (8-1) is solved after direct measurement of x and x' on the left and right photos, respectively. A disadvantage of this method is that two measurements are required for each point.

Another monoscopic approach to parallax measurement is to fasten the photographs down on a table or base material, as shown in Fig. 8-5. In this method the photographic flight lines o_1o_2 and $o'_1o'_2$ are marked as usual. A long straight line AA' is drawn on the base material, and the two photos are *carefully* mounted as shown so that the photographic flight lines are coincident with this line. Now that the photos are fastened down, the distance D between the two principal points is a constant which can be

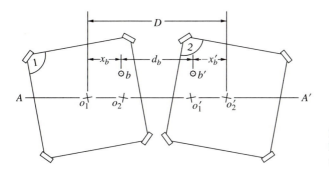

FIGURE 8-5
Parallax measurement using a simple scale.

measured. The parallax of point B is $p_b = x_b - x_b'$ (note that in Fig. 8-5 the x_b' coordinate is negative). However, by examining the figure, it is seen that parallax is also

$$p_b = D - d_b \tag{8-2}$$

With D known in Eq. (8-2), to obtain the parallax of a point it is necessary only to measure the distance d between its images on the left and right photos. The advantage is that for each additional point whose parallax is desired, only a single measurement is required. With either of these monoscopic methods of parallax measurement, a simple scale as described in Sec. 4-3 may be used, with the choice being based upon the desired accuracy.

8-4 PRINCIPLE OF THE FLOATING MARK

Parallaxes of points can be measured while viewing stereoscopically with the advantages of speed and accuracy. Stereoscopic measurement of parallax makes use of the principle of the *floating mark*. When a stereomodel is viewed through a stereoscope, two small identical marks etched on clear glass, called *half marks*, may be placed over the photographs—one on the left photo and one on the right photo, as illustrated in Fig. 8-6. The left mark is seen with the left eye and the right mark with the right eye. The half marks may be shifted in position until they fuse together into a single mark which appears to exist in the stereomodel and to lie at a particular elevation. If the half marks are moved closer together, the parallax of the half marks is increased and the fused mark will therefore appear to rise. Conversely, if the half marks are moved apart, parallax is decreased and the fused mark appears to fall. This apparent variation in the elevation of the mark as the spacing of half marks is varied is the basis for the term *floating mark.*

The spacing of the half marks, and hence the parallax of the half marks, may be varied so that the floating mark appears to rest exactly on the terrain. This produces the same effect as though an object of the shape of the half marks had existed on the terrain when the photos were originally taken. The floating mark may be moved about the stereomodel from point to point, and as the terrain varies in elevation, the spacing of the half marks may be varied to make the floating mark rest exactly on the terrain.

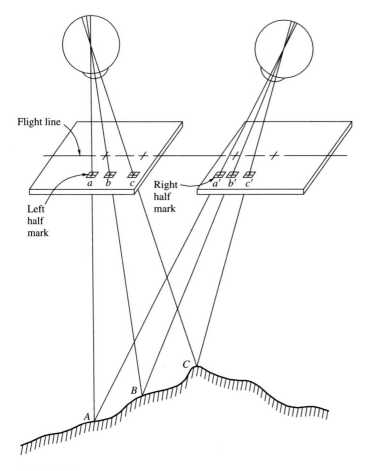

FIGURE 8-6
The principle of the floating mark.

Figure 8-6 demonstrates the principle of the floating mark and illustrates how the mark may be set exactly on particular points such as A, B, and C by placing the half marks at a and a', b and b', and c and c', respectively.

The principle of the floating mark can be used to transfer principal points to their corresponding locations, thereby marking the flight-line axes. In this procedure the principal points are first located as usual at the intersection of fiducial lines. Then by means of a point transfer device such as that shown in Fig. 8-7, the principal points are transferred to their corresponding locations. The point transfer device consists of two separate pieces containing identical half marks etched on glass. The left half mark of Fig. 8-7 is placed over one of the principal points, say, the left point o_1. It is held fixed in that position. Using a mirror stereoscope for viewing, the right half mark is placed on the right photo and moved about until a clear stereoscopic view of the floating mark is obtained and the fused mark appears to rest exactly on the ground. The right half mark,

FIGURE 8-7
Simpson point transfer device.
(*Courtesy Alan Gordon Enterprises, Inc.*)

which is hinged, is then raised to make way for lowering a hinged arm containing a pin. The pin comes down exactly on the position occupied by the right half mark. It is pressed into the photograph, thereby marking the corresponding principal point. This stereoscopic procedure is very accurate if carefully performed, and it has the advantage that discrete images near the principal points are not necessary, as they are with the monoscopic method. Imagine, e.g., the difficulty of monoscopically transferring a principal point that falls in the middle of a wheat field. This transfer could be readily done by the stereoscopic method, however.

A homemade version of the stereoscopic point transfer device described above consists of two small pieces of transparent plastic upon which identical crosses are printed. When a half mark is stereoscopically located over a corresponding principal point, its photographic position is marked by pinpricking through the center of the cross.

Once corresponding principal points have been marked, the *photo base b* can be determined. The photo base is the distance on a photo between the principal point and the corresponding principal point from the overlapping photo. Figure 8-8 is a vertical section through the exposure stations of a pair of overlapping vertical photos. By Eq. (8-1), the parallax of the left-photo ground principal point P_1 is $p_{o_1} = x_{o_1} - (-x'_{o_1}) = 0 - (-b') = b'$. (The x coordinate of o_1 on the left photo is zero.)

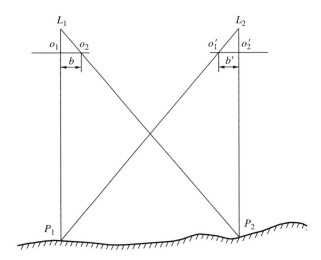

FIGURE 8-8
Parallax of the principal points.

Also, the parallax of the right-photo ground principal point P_2 is $p_{o_2} = x_{o_2} - (-x'_{o_2}) = b - 0 = b$. From the foregoing, it is seen that *the parallax of the left ground principal point is photo base b' measured on the right photo,* and *the parallax of the right ground principal point is photo base b measured on the left photo.* In areas of moderate relief, the values of b and b' will be approximately equal, and the photo base for the stereopair can be taken as the average of these two values.

8-5 STEREOSCOPIC METHODS OF PARALLAX MEASUREMENT

Through the principle of the floating mark, parallaxes of points may be measured stereoscopically. This method employs a stereoscope in conjunction with an instrument called a *parallax bar,* also frequently called a *stereometer.* A parallax bar consists of a metal rod to which are fastened two half marks. The right half mark may be moved with respect to the left mark by turning a micrometer screw. Readings from the micrometer are taken with the floating mark set exactly on points whose parallaxes are desired. From the micrometer readings, parallaxes or differences in parallax are obtained. A parallax bar is shown lying on the photos beneath a mirror stereoscope in Fig. 8-9.

When a parallax bar is used, the two photos of a stereopair are first *carefully* oriented for comfortable stereoscopic viewing, in such a way that the flight line of each photo lies precisely along a common straight line, as line AA' shown in Fig. 8-5. The photos are then fastened securely, and the parallax bar is placed on the photos. The left half mark, called the *fixed mark,* is unclamped and moved so that when the floating mark is fused on a terrain point of average elevation, the parallax bar reading is approximately in the middle of the run of the graduations. The fixed mark is then clamped, where it will remain for all subsequent parallax measurements on that particular stereopair. After the fixed mark is positioned in this manner, the right half mark, or *movable mark,* may be moved left or right with respect to the fixed mark (increasing or decreasing the parallax) as required to accommodate high points or low points without exceeding the run of the parallax bar graduations.

Figure 8-10 is a schematic diagram illustrating the operating principle of the parallax bar. After the photos have been oriented and the left half mark is fixed in position

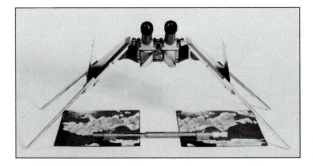

FIGURE 8-9
Wild ST-4 mirror stereoscope with binocular attachment and parallax bar. (*Courtesy LH Systems, LLC.*)

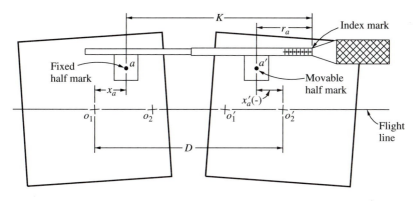

FIGURE 8-10
Schematic diagram of the parallax bar.

as just described, the *parallax bar constant* C for the setup is determined. For the setup, the spacing between principal points is a constant, denoted by D. Once the fixed mark is clamped, the distance from the fixed mark to the index mark of the parallax bar is also a constant, denoted by K. From Fig. 8-10, the parallax of point A is

$$p_a = x_a - x_a' = D - (K - r_a) = (D - K) + r_a$$

The term $(D - K)$ is C, the parallax bar constant for the setup. Also r_a is the micrometer reading. By substituting C into the above equation, the expression becomes

$$p_a = C + r_a \qquad (8\text{-}3)$$

To calculate the parallax bar constant, a micrometer reading is taken with the floating mark set on a selected point. The parallax of that point is also directly measured monoscopically and calculated using Eq. (8-1). Then with p and r for that point known, the value of C is calculated by using Eq. (8-3), as

$$C = p - r \qquad (8\text{-}4)$$

The parallax bar constant should be determined on the basis of micrometer readings and parallax measurements for two points. Then the mean of the two values may be adopted. Any two points may be selected for this purpose; however they should be clear, discrete images, and selected so that they lie on opposite sides of the flight line and approximately equidistant from the flight line. This minimizes error in parallaxes due to tilt and faulty orientation of the photos.

One of the advantages of measuring parallax stereoscopically is increased speed, for once the parallax bar constant is determined, the parallaxes of all other points are quickly obtained with a single micrometer reading for each point. Another advantage is increased accuracy. An experienced person using quality equipment and clear photos is generally able to obtain parallaxes to within approximately 0.03 mm of their correct values.

8-6   PARALLAX EQUATIONS

As noted earlier, X, Y, and Z ground coordinates can be calculated for points based upon the measurements of their parallaxes. Figure 8-11 illustrates an overlapping pair of vertical photographs which have been exposed at equal flying heights above datum. Images of an object point A appear on the left and right photos at a and a', respectively. The planimetric position of point A on the ground is given in terms of ground coordinates X_A and Y_A. Its elevation above datum is h_A. The XY ground axis system has its origin at the datum principal point P of the left-hand photograph; the X axis is in the same vertical plane as the photographic x and x' flight axes; and the Y axis passes through the datum principal point of the left photo and is perpendicular to the X axis. According to this definition, each stereopair of photographs has its own unique ground coordinate system.

By equating similar triangles of Fig. 8-11, formulas for calculating h_A, X_A, and Y_A may be derived. From similar triangles L_1oa_y and $L_1A_oA_y$,

$$\frac{Y_A}{H - h_A} = \frac{y_a}{f}$$

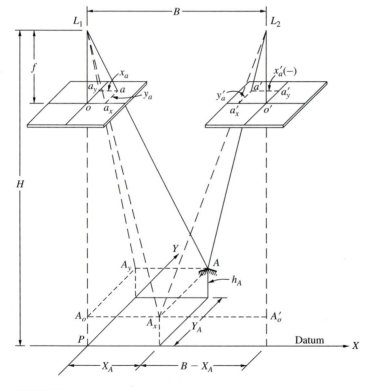

FIGURE 8-11
Geometry of an overlapping pair of vertical photographs.

from which

$$Y_A = \frac{y_a}{f}(H - h_A) \qquad (a)$$

and equating similar triangles $L_1 o a_x$ and $L_1 A_o A_x$, we have

$$\frac{X_A}{H - h_A} = \frac{x_a}{f}$$

from which

$$X_A = \frac{x_a}{f}(H - h_A) \qquad (b)$$

Also from similar triangles $L_2 o' a'_x$ and $L_2 A'_o A_x$,

$$\frac{B - X_A}{H - h_A} = \frac{-x'_a}{f}$$

from which

$$X_A = B + \frac{x'_a}{f}(H - h_A) \qquad (c)$$

Equating Eqs. (b) and (c) and reducing gives

$$h_A = H - \frac{Bf}{x_a - x'_a} \qquad (d)$$

Substituting p_a for $x_a - x'_a$ into Eq. (d) yields

$$h_A = H - \frac{Bf}{p_a} \qquad (8\text{-}5)$$

Now substituting Eq. (8-5) into each of Eqs. (b) and (a) and reducing:

$$X_A = B \frac{x_a}{p_a} \qquad (8\text{-}6)$$

$$Y_A = B \frac{y_a}{p_a} \qquad (8\text{-}7)$$

In Eqs. (8-5), (8-6), and (8-7), h_A is the elevation of point A above datum, H is the flying height above datum, B is the air base, f is the focal length of the camera, p_a is the parallax of point A, X_A and Y_A are the ground coordinates of point A in the previously defined unique arbitrary coordinate system, and x_a and y_a the photo coordinates of point a measured with respect to the flight-line axes on the left photo.

Equations (8-5), (8-6), and (8-7) are commonly called the *parallax equations*. They are among the most useful equations to the photogrammetrist. These equations enable a complete survey of the overlap area of a stereopair to be made, provided the focal length is known and sufficient ground control is available so the air base B and flying height H can be calculated.

Equations (8-6) and (8-7) yield X and Y ground coordinates in the unique arbitrary coordinate system of the stereopair, which is not related to any standard two-dimensional ground coordinate system. However, if arbitrary XY coordinates are determined using these equations for at least two points whose ground coordinates are also known in a standard two-dimensional coordinate system (e.g., state plane coordinates), then the arbitrary XY coordinates of all other points can be transformed into that ground system through a two-dimensional coordinate transformation, as described in App. C.

Example 8-1. A pair of overlapping vertical photographs was taken from a flying height of 1233 m above sea level with a 152.4-mm-focal-length camera. The air base was $39\overline{0}$ m. With the photos properly oriented, flight-line coordinates for points a and b were measured as $x_a = 53.4$ mm, $y_a = 50.8$ mm, $x'_a = -38.3$ mm, $y'_a = 50.9$ mm, $x_b = 88.9$ mm, $y_b = -46.7$ mm, $x'_b = -7.1$ mm, $y'_b = -46.7$ mm. Calculate the elevations of points A and B and the horizontal length of line AB.

Solution. By Eq. (8-1)

$$p_a = x_a - x'_a = 53.4 - (-38.3) = 91.7 \text{ mm}$$

$$p_b = x_b - x'_b = 88.9 - (-7.1) = 96.0 \text{ mm}$$

By Eq. (8-5),

$$h_A = H - \frac{Bf}{p_a} = 1233 - \frac{390(152.4)}{91.7} = 585 \text{ m above sea level}$$

$$h_B = H - \frac{Bf}{p_b} = 1233 - \frac{390(152.4)}{96.0} = 614 \text{ m above sea level}$$

By Eqs. (8-6) and (8-7),

$$X_A = B\frac{x_a}{p_a} = 390\left(\frac{53.4}{91.7}\right) = 227 \text{ m}$$

$$Y_A = B\frac{y_a}{p_a} = 390\left(\frac{50.8}{91.7}\right) = 216 \text{ m}$$

$$X_B = B\frac{x_b}{p_b} = 390\left(\frac{88.9}{96.0}\right) = 361 \text{ m}$$

$$Y_B = B\frac{y_b}{p_b} = 390\left(\frac{-46.7}{96.0}\right) = -190 \text{ m}$$

The horizontal length of line AB is

$$AB = \sqrt{(X_B - X_A)^2 + (Y_B - Y_A)^2} = \sqrt{(361 - 227)^2 + (-190 - 216)^2} = 427 \text{ m}$$

8-7   ELEVATIONS BY PARALLAX DIFFERENCES

Parallax differences between one point and another are caused by different elevations of the two points. While parallax Eq. (8-5) serves to define the relationship of stereoscopic parallax to flying height, elevation, air base, and camera focal length, parallax differences

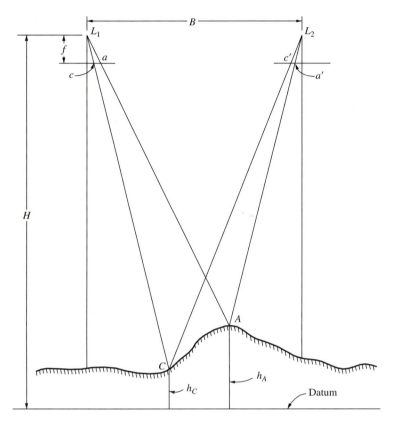

FIGURE 8-12
Elevations by parallax differences.

are more convenient for determining elevations. In Fig. 8-12, object point C is a control point whose elevation h_C above datum is known. The elevation of object point A is desired. By rearranging Eq. (8-5), parallaxes of both points can be expressed as

$$p_c = \frac{fB}{H - h_C} \tag{e}$$

$$p_a = \frac{fB}{H - h_A} \tag{f}$$

The difference in parallax $p_a - p_c$, obtained by subtracting Eq. (e) from Eq. (f) and rearranging, is

$$p_a - p_c = \frac{fB(h_A - h_C)}{(H - h_A)(H - h_C)} \tag{g}$$

Let $p_a - p_c$ equal Δp, the difference in parallax. By substituting $H - h_A$ from Eq. (f), and Δp into (g) and reducing, the following expression for elevation h_A is obtained:

$$h_A = h_C + \frac{\Delta p (H - h_C)}{p_a} \qquad (8\text{-}8)$$

Example 8-2. In Example 8-1, flight-line axis x and x' coordinates for the images of a vertical control point C were measured as $x_c = 14.3$ mm and $x'_c = -78.3$ mm. If the elevation of point C is 591 m above sea level, calculate the elevations of points A and B of that example, using parallax difference Eq. (8-8).

Solution. By Eq. (8-1),

$$p_c = x_c - x'_c = 14.3 - (-78.3) = 92.6 \text{ mm}$$

For point A,

$$\Delta p = p_a - p_c = 91.7 - 92.6 = -0.9 \text{ mm}$$

By Eq. (8-8),

$$h_A = h_C + \frac{\Delta p (H - h_C)}{p_a} = 591 + \frac{(-0.9)(1233 - 591)}{91.7}$$

$$= 585 \text{ m above sea level}$$

For point B,

$$\Delta p = p_b - p_c = 96.0 - 92.6 = 3.4 \text{ mm}$$

By Eq. (8-8),

$$h_B = h_C + \frac{\Delta p (H - h_C)}{p_b} = 591 + \frac{(3.4)(1233 - 591)}{96.0} = 614 \text{ m above sea level}$$

Note that these answers check the values computed in Example 8-1.

If a number of control points are located throughout the overlap area, use of Eq. (8-8) permits elevations of unknown points to be most accurately determined from the parallax difference of the nearest control point. This minimizes the effects of two primary errors—photographic tilt and imperfect alignment of the photos for parallax measurement.

8-8   APPROXIMATE EQUATION FOR HEIGHTS OF OBJECTS FROM PARALLAX DIFFERENCES

In many applications it is necessary to estimate heights of objects to a moderate level of accuracy. Utilizing parallax differences for height determination is particularly useful when application of relief displacement is not possible because either the feature is not vertical (e.g., a construction crane) or the base of the feature is obscured (e.g., trees in a forest). In a situation like this, a parallax difference can be determined between a point on the ground and the top of the feature. A fundamental assumption is, of course, that the point on the ground is at the same elevation as the base of the feature. In a large number of cases, this assumption is valid as long as only moderate accuracy is required.

A simplified equation for height determination can be obtained from Eq. (8-8) by choosing the vertical datum to be the elevation of the point on the ground that is used as the basis for the parallax difference. This makes h_C zero, and Eq. (8-8) simplifies to

$$h_A = \frac{\Delta p H}{p_a} \tag{8-9}$$

In Eq. (8-9), h_A is the height of point A above ground, $\Delta p = p_a - p_c$ is the difference in parallax between the top of the feature and the ground (p_c is the parallax of the ground), and H is the flying height above ground, since datum is at ground. If the heights of many features are needed in an area where the ground is approximately level, the photo base b can be utilized as the parallax of the ground point. In this case, Eq. (8-9) can be modified to

$$h_A = \frac{\Delta p H}{b + \Delta p} \approx \frac{\Delta p H}{b} \tag{8-10}$$

In Eq. (8-10), b is the photo base for the stereopair, $\Delta p = p_a - b$, and the other terms are as previously defined. Since parallax difference is generally small compared to b, Δp can be eliminated from the denominator of Eq. (8-10) with only slight loss of accuracy. For very low flying heights or in areas of significant relief, or both, the assumptions of Eq. (8-10) are not met; in these cases, Eq. (8-8) should be used. Equation (8-10) is especially convenient in photo interpretation where rough elevations, building and tree heights, etc., are often needed.

Example 8-3. The parallax difference between the top and bottom of a tree is measured as 1.3 mm on a stereopair of photos taken at 915 m above ground. Average photo base is 88.2 mm. How tall is the tree?

Solution. By Eq. (8-10),

$$h = \frac{1.3 \times 915}{88.2 + 1.3} = 13 \text{ m}$$

Or using the approximation gives

$$h = \frac{1.3 \times 915}{88.2} = 13 \text{ m}$$

Note that in this case, the approximation yields the same value as the more accurate equation.

8-9 MEASUREMENT OF PARALLAX DIFFERENCES

Parallax differences may be determined in any of the following ways:

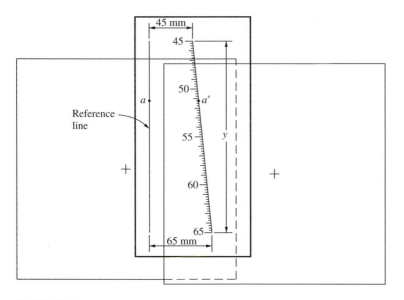

FIGURE 8-13
Parallax wedge.

1. By monoscopic measurement of parallaxes followed by subtraction
2. By taking differences in parallax bar readings
3. By *parallax wedge*

A parallax wedge, as illustrated in Fig. 8-13, consists of a piece of transparent film upon which are drawn two converging lines. The left line is a reference line while the line on the right contains graduations from which readings can be made. The spacing of the two lines depends on whether the parallax wedge will be used with a mirror stereoscope or a pocket stereoscope. For a pocket stereoscope the spacing should vary from about 65 mm at the bottom to about 45 mm at the top. This spacing accommodates the usual spacing between corresponding images when a stereopair is oriented for viewing with a pocket stereoscope, and it gives a possible range of about 20 mm in parallax differences that can be measured.

Suppose line spacings of a parallax wedge were exactly 65 mm at the bottom and 45 mm at the top, as shown in Fig. 8-13. If the total height y of the graduations was 200 mm, then for graduations spaced at 2.5-mm intervals along the line, each of the 80 graduations proceeding upward on the scale is 0.25 mm closer to the reference line than the next-lower graduation. Graduations numbered from 45 to 65 on the parallax wedge represent horizontal spacings from the reference line, with the smallest graduation interval representing a parallax difference of 0.25 mm even though the vertical spacing between consecutive lines is 2.5 mm.

When a parallax wedge is used, the photos are first carefully oriented as usual and secured. The parallax wedge is placed in the overlap area and viewed stereoscopically; the two lines of the parallax wedge will fuse and appear as a single *floating line* in areas where the spacing of the lines is slightly less than the spacing of corresponding photo images. The floating line will appear to split where the parallax of the lines is equal to that of the photo images. The position of the parallax wedge can be adjusted so that the floating line splits forming a wedge exactly at a point whose parallax is desired, and at that point a reading is taken from the scale. The parallax wedge reading at point *a* of Fig. 8-13, for example, is 51.25 mm. Parallax differences are obtained by simply taking differences in parallax wedge readings for different points.

An expedient means of producing a parallax wedge is to create a drawing using an available computer-aided drafting (CAD) program and to print the result on a transparency using a laser printer.

8-10   COMPUTING FLYING HEIGHT AND H, B
AIR BASE

To use parallax equations, it is generally necessary to compute the flying height and air base. Flying height may be calculated using the methods described in Sec. 6-9. For best results, the average of flying heights for the two photos of a stereopair should be used.

If the air base is known and if one vertical control point is available in the overlap area, flying height for the stereopair may be calculated by using Eq. (8-5).

> **Example 8-4.** An overlapping pair of vertical photographs taken with a 152.4-mm-focal-length camera has an air base of 548 m. The elevation of control point *A* is 283 m above sea level, and the parallax of point *A* is 92.4 mm. What is the flying height above sea level for this stereopair?
>
> *Solution.* By rearranging Eq. (8-5),
>
> $$H = h + \frac{Bf}{p} = 283 + \frac{548(152.4)}{92.4} = 1187 \text{ m above sea level}$$

If the flying height above datum is known and if one vertical control point is available in the overlap area, the air base for the stereopair may be calculated by using Eq. (8-5).

> **Example 8-5.** An overlapping pair of vertical photos was exposed with a 152.4-mm-focal-length camera from a flying height of 1622 m above datum. Control point *C* has an elevation of 263 m above datum, and the parallax of its images on the stereopair is 86.3 mm. Calculate the air base.
>
> *Solution.* By rearranging Eq. (8-5),

$$B = (H - h)\frac{p}{f} = (1622 - 263)\frac{86.3}{152.4} = 77\overline{0} \text{ m}$$

If a line of known horizontal length appears in the overlap area, then the air base can be readily calculated. The horizontal length of a line may be expressed in terms of rectangular coordinates, according to the pythagorean theorem, as

$$AB = \sqrt{(X_B - X_A)^2 + (Y_B - Y_A)^2}$$

Substituting Eqs. (8-6) and (8-7) into the above for the rectangular coordinates, gives

$$AB = \sqrt{\left(\frac{Bx_b}{p_b} - \frac{Bx_a}{p_a}\right)^2 + \left(\frac{By_b}{p_b} - \frac{By_a}{p_a}\right)^2}$$

Solving the above equation for B yields

$$B = \frac{AB}{\sqrt{(x_b/p_b - x_a/p_a)^2 + (y_b/p_b - y_a/p_a)^2}} \tag{8-11}$$

Example 8-6. Images of the endpoints of ground line AB, whose horizontal length is 650.47 m, appear on a pair of overlapping vertical photographs. Photo coordinates measured with respect to the flight axis on the left photo were $x_a = 33.3$ mm, $y_a = 13.5$ mm, $x_b = 41.8$ mm, and $y_b = -95.8$ mm. Photo coordinates measured on the right photo were $x_a' = -52.3$ mm and $x_b' = -44.9$ mm. Calculate the air base for this stereopair.

Solution. By Eq. (8-1),

$$p_a = x_a - x_a' = 33.3 - (-52.3) = 85.6 \text{ mm}$$

$$p_b = x_b - x_b' = 41.8 - (-44.9) = 86.7 \text{ mm}$$

By Eq. (8-9),

$$B = \frac{650.47}{\sqrt{(41.8/86.7 - 33.3/85.6)^2 + (-95.8/86.7 - 13.5/85.6)^2}} = 514 \text{ m}$$

8-11   ERROR EVALUATION

Answers obtained using the various equations presented in the chapter will inevitably contain errors. It is important to be aware of the presence of these errors and to be able to assess their magnitudes. Some of the sources of error in computed answers using parallax equations are as follows:

1. Locating and marking the flight lines on photos
2. Orienting stereopairs for parallax measurement
3. Parallax and photo coordinate measurements

4. Shrinkage or expansion of photographs
5. Unequal flying heights for the two photos of stereopairs
6. Tilted photographs
7. Errors in ground control
8. Other errors of lesser consequence such as camera lens distortion and atmospheric refraction distortion

A general approach for determining the combined effect of several random errors in computed answers is presented in Sec. A-4. This approach is demonstrated in the following example.

> **Example 8-8.** In the computation of the elevation of point A in Example 8-1, suppose that the random errors were ±2 m in H, ±2 m in B, and ±0.1 mm in p_a. Compute the resulting error in h_A due to the presence of these errors.
>
> **Solution.** The basic equation used was Eq. (8-5), and the derivatives in that equation taken with respect to each of the three error sources are
>
> $$\frac{\partial h_A}{\partial H} = 1 \qquad \frac{\partial h_A}{\partial B} = -\frac{f}{p_a} \qquad \frac{\partial h_A}{\partial p_a} = \frac{Bf}{p_a^2}$$
>
> Substituting the above expressions into Eq. (A-2) gives
>
> $$\sigma_{h_A} = \pm \sqrt{(1)^2\sigma_H^2 + \left(\frac{-f}{p_a}\right)^2 \sigma_B^2 + \left(\frac{Bf}{p_a^2}\right)^2 \sigma_{p_a}^2}$$
>
> Substituting numerical values into the above, we get
>
> $$\sigma_{h_A} = \pm \sqrt{(1)^2(2)^2 + \left(\frac{-152.4}{91.7}\right)^2(2)^2 + \left[\frac{390\,(152.4)}{(91.7)^2}\right]^2 (0.1)^2}$$
>
> $$= \pm \sqrt{4 + 11 + 0.5} = \pm 3.9 \text{ m}$$

Errors in computed answers using any of the equations of this chapter can be analyzed in the fashion described above. It is, of course, necessary to estimate the magnitude of the random errors in the measured variables used in the equations. It is more difficult to analyze errors caused by tilt in the photographs. The subject of tilted photographs is discussed in Chap. 10, and rigorous analytical methods are presented in Chap. 11. For the present, however, suffice it to say that for normal photography intended to be vertical, errors in parallax equation answers due to tilt are compatible with errors from the other sources that have been considered.

REFERENCES

American Society of Photogrammetry: *Manual of Photogrammetry,* 3d ed., Bethesda, MD, 1966, chap. 2.

————: *Manual of Photogrammetry,* 4th ed., Bethesda, MD, 1980, chap. 2.

Avery, T. E.: "Two Cameras for Parallax Height Measurements," *Photogrammetric Engineering,* vol. 32, no. 6, 1966, p. 576.

Bender, L. U.: "Derivation of Parallax Equation," *Photogrammetric Engineering,* vol. 33, no. 10, 1967, p. 1175.

Nash, A. J.: "Use a Mirror Stereoscope Correctly," *Photogrammetric Engineering,* vol. 38, no. 12, 1972, p. 1192.

Porter, G. R.: "Errors in Parallax Measurements and Their Assessment in Student Exercises," *Photogrammetric Record,* vol. 8, no. 46, 1975, p. 528.

Schut, G. H.: "The Determination of Tree Heights from Parallax Measurements," *Canadian Surveyor,* vol. 19, 1965, p. 415.

PROBLEMS

8-1. Calculate the stereoscopic parallaxes of points *A* through *D*, given the following measured flight-line axis coordinates. Which point is the highest in elevation? Which is lowest?

Point	x (left photo)	x′ (right photo)
A	59.9 mm	−27.2 mm
B	68.0 mm	−21.6 mm
C	99.6 mm	9.9 mm
D	100.4 mm	8.5 mm

8-2. Calculate the elevations of points *A* through *D* of Prob. 8-1 if the camera focal length is 152.62 mm, flying height above datum is 2469 m, and the air base is 1356 m.

8-3. A pair of overlapping vertical photographs is mounted for parallax measurement, as illustrated in Fig. 8-5. Distance *D* is measured as 263.4 mm. Calculate the stereoscopic parallaxes of the points whose measured *d* values are as follows. Which point is highest in elevation? Which is lowest?

Point	d
A	172.5 mm
B	179.6 mm
C	161.3 mm
D	167.6 mm

8-4. Repeat Prob. 8-3, except *D* was measured as 266.6 mm, and measured *D* values are as follows.

Point	d
A	170.2 mm
B	164.3 mm
C	176.0 mm
D	166.5 mm

8-5. Assume that point A of Prob. 8-3 has an elevation of 362.7 m above datum and that the photos were taken with an 88.74-mm-focal-length camera. If the air base is 1125 m, what are the elevations of points B, C, and D?

8-6. Assume that point A of Prob. 8-4 has an elevation of 375.0 m above datum and that the photos were taken with a 153.07-mm-focal-length camera. If the air base is 1832 m, what are the elevations of points B, C, and D?

8-7. From the information given for Probs. 8-1 and 8-2, calculate the horizontal ground length of line AC. Measured y coordinates on the left photo are $y_a = -59.2$ mm and $y_c = 101.9$ mm.

8-8. Repeat Prob. 8-7 except that the computations are for line BD. Measured y coordinates on the left photo are $y_b = 42.4$ mm and $y_d = -76.7$ mm.

8-9. From the data of Probs. 8-3 and 8-5, calculate the horizontal area of triangle ABC. Measured x and y flight-line axis coordinates of a, b, and c on the left photo were $x_a = -9.5$ mm, $y_a = 111.0$ mm, $x_b = 14.9$ mm, $y_b = -112.0$ mm, $x_c = 102.5$ mm, and $y_c = 47.5$ mm.

8-10. Distances b on the left photo and b' on the right photo of a pair of overlapping vertical photos are 90.3 mm and 89.8 mm, respectively. If the air base is 563 m and the camera focal length is 88.78 mm, which ground principal point is higher and by how much?

8-11. Repeat Prob. 8-10, except that b and b' are 92.8 mm and 91.3 mm, respectively, the air base is 635 m, and the camera focal length is 152.60 mm.

8-12. A pair of overlapping vertical photos is taken from a flying height of 1082 m above ground with a 152.46-mm-focal-length camera. The x coordinates on the left photo of the base and top of a certain tree are 81.5 mm and 84.3 mm, respectively. On the right photo these x' coordinates are -12.4 mm and -14.2 mm, respectively. Determine the height of the tree.

8-13. A pair of overlapping vertical photos is taken from a flying height of 1835 m above the base of a radio tower. The x coordinates on the left photo of the top and base of the tower were 96.5 mm and 90.5 mm, respectively. On the right photo these x' coordinates were -1.1 mm and -1.0 mm, respectively. What is the approximate height of the tower?

8-14. The air base of a pair of overlapping vertical photos was determined to be 757 m. The focal length of the camera was 152.35 mm. The image coordinates of point A, whose elevation is 282 m above datum, were determined on the left photo as $x_a = 3.3$ mm and on the right photo as $x_a' = -85.0$ mm. What is the flying height above datum for the stereopair?

8-15. Repeat Prob. 8-14, except that the air base was 1055 m, the camera focal length was 209.60 mm, and point A, whose elevation was 283 m above datum, had image coordinates of $x_a = 42.9$ mm on the left photo and $x_a' = -47.3$ mm on the right photo.

8-16. The images of two control points A and B appear in the overlap area of a pair of vertical photographs. The following photo coordinates and ground coordinates apply to points A and B. Calculate the air base of the stereopair, using Eq. (8-11).

Point	Left photo coordinates		Right photo coordinates		Ground coordinates	
	x, mm	y, mm	x', mm	y', mm	X, m	Y, m
A	26.4	-97.2	-65.1	-97.3	78,164.6	27,996.4
B	-19.4	34.2	-92.8	34.1	78,250.2	27,351.6

8-17. Repeat Prob. 8-16, except that the photo coordinates and ground coordinates for points A and B were as follows:

	Left photo coordinates		Right photo coordinates		Ground coordinates	
Point	x, mm	y, mm	x', mm	y', mm	X, m	Y, m
A	65.8	82.7	−33.9	82.7	102,055.8	35,781.1
B	41.8	−76.3	−50.2	−76.3	100,989.8	34,196.6

8-18. A pair of overlapping vertical photos was exposed with a camera having a 209.80-mm focal length. Calculate B and H from the following information on ground points D and E. (Hint: Set up Eq. (8-5) for point D and for point E, then solve simultaneously.)

Point	Elevation, m	Left photo coordinates	Right photo coordinates
D	587	$x_d = -17.4$ mm	$x'_d = -111.0$ mm
E	729	$x_e = 99.2$ mm	$x'_e = 1.63$ mm

8-19. Repeat Prob. 8-18, except that the camera focal length is 152.53 mm and the following information applies to points D and E.

Point	Elevation, m	Left photo coordinates	Right photo coordinates
D	547.1	$x_d = 70.9$ mm	$x'_d = -22.9$ mm
E	478.5	$x_e = 26.2$ mm	$x'_e = -64.0$ mm

8-20. A parallax wedge for use with a pocket stereoscope similar to that shown in Fig. 8-13 has height of graduations y equal to 120 mm. The lateral spacing between reference line and the graduated line is 45 mm at the top and 60 mm at the bottom. What is the vertical spacing of reference marks on the graduated line if the difference in parallax between adjacent graduations is 0.25 mm?

8-21. In Prob. 8-14, suppose that random errors were ±1 m in h and B, and ±0.1 mm in each of x_a and x'_a. What is the expected resultant error in the calculated value of H due to these random errors? (Assume the focal length to be error-free.)

8-22. In Prob. 8-15, suppose that random errors were ±1 m in h_A, ±2 m in B, and ±0.05 mm in both x_a and x'_a. What is the expected error in the calculated value of H due to these errors? (Assume the focal length to be error-free.)

8-23. In Prob. 8-12, assume that random errors existed in the amounts of ±2 m in H and ±0.2 mm for each of the measured photo coordinates. What is the expected error in the calculated height of the tree due to these random errors?

CHAPTER
9

ELEMENTARY METHODS OF PLANIMETRIC MAPPING FOR GIS

9-1 INTRODUCTION

This chapter describes elementary methods that can be used for compiling planimetric maps from vertical photographs and satellite images. These include (1) tracing with the use of reflection instruments, (2) *georeferencing* of digital imagery, (3) tracing of hardcopy imagery with a tablet digitizer, (4) performing heads-up digitizing, (5) preparing *photomaps,* and (6) constructing *mosaics.*[1] Each of these techniques is relatively uncomplicated to perform and generally requires simpler and less expensive equipment compared to the rigorous photogrammetric mapping techniques that are presented in later chapters. These methods can have definite utility, depending upon the extent and required accuracy of planimetric mapping to be accomplished. For map revision over limited areas, and in many applications of geographic information systems (GISs), particularly those involving natural resource mapping, the high accuracy afforded by

---

[1]In this context, compiling planimetric maps denotes the drawing of scaled diagrams which show planimetric features by means of lines and symbols or the production of images which portray planimetric features in picture form. Planimetric maps portray only horizontal position and give no information concerning elevations. The maps are prepared to some designated scale; thus all features are presumed to be shown in their true relative positions. Pure digital products such as computer-aided drafting (CAD) drawings and georeferenced images do not have a specific scale per se, but rather include explicit information regarding ground coordinates of features or pixels.

187

rigorous photogrammetric methods is often not necessary. By using appropriate elementary methods, substantial cost savings can be realized, while obtaining planimetric information at acceptable accuracy.

The choice among the above-cited methods depends upon the purpose of the map or image product, extent of the area to be mapped, required accuracy, available imagery, and budget constraints. As an example, it may be necessary to include a recently constructed road or shopping center which is not shown on an otherwise satisfactory existing planimetric map. It would be expensive, and unnecessary, to prepare a new map of the area if these features could be satisfactorily superimposed onto the existing map. This type of planimetric map revision can readily be done using procedures described in this chapter.

The accuracies that can be achieved in planimetric mapping by using these methods are generally of a lower order than those attainable with stereoplotting instruments (see Chap. 12) or orthophoto processing (see Chap. 15). However, for some work, especially if ample care is exercised, suitable results can be achieved.

9-2 PLANIMETRIC MAPPING WITH REFLECTION INSTRUMENTS

A number of instruments have been devised that enable an operator to view an image superimposed with a map manuscript. The basic operation of these instruments is fundamentally the same. Photo control[2] points which have been plotted to scale on the map manuscript are used as a basis for orienting the image. By adjusting image scale, rotation, translation, and in some cases tilt or nonuniform image stretch, the images of the photo control points can be made to approximately coincide with the plotted control points on the manuscript. The degree of coincidence depends on a number of factors including the amount of terrain relief. Once the device is oriented, additional image features can be directly traced onto the manuscript by virtue of the operator's simultaneous view of the oriented image and the map.

The zoom transfer scope (ZTS) of Fig. 9-1 is a versatile reflection instrument which is used primarily for planimetric map revision. With the ZTS, an aerial photograph or other hard-copy image is placed on a viewing stage and secured with clamps. Some ZTSs can accommodate imagery on either transparent film or paper, and some can function in either monoscopic or stereoscopic mode. The map manuscript is placed on a horizontal working surface which can be viewed simultaneously with the image. The instrument can accommodate large differences in scale from photo to map by means of zoom optics which provide continuous variations in magnification from 1 to 7X. In addition, the anamorphic optical system of the ZTS enables different magnification ratios to be applied in the x and y directions. The ZTS also has the capability of removing the effects of nonorthogonal *affine* distortion (see Sec. C-6) in the image by means of

---

[2]Photo control, as described in detail in Chap. 16, consists of any discrete objects whose images appear on the photo and whose ground coordinates are known. There are several methods for determining the ground coordinates, including scaling from an existing map, direct field survey, or photogrammetric techniques such as aerotriangulation (described in Chap. 17).

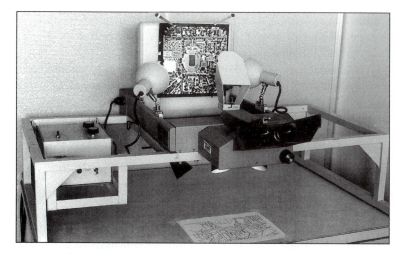

FIGURE 9-1
Zoom transfer scope. (*Courtesy University of Florida.*)

rotating prisms. These features facilitate adjusting photo images to coincide with map points or control points. Once the image and map are adjusted, the manuscript is taped down and the operator then traces the desired features.

Of course, photographs are perspective projections, and as described in Chap. 6, they contain scale variations due to tilt and relief. In compiling maps using reflection instruments, therefore, traced features will contain any scale variations that exist in the photo images. However, if relief is moderate, and relatively small areas are traced that lie within a perimeter of control points, the resulting map will be sufficiently accurate to satisfy many purposes.

9-3 GEOREFERENCING OF DIGITAL IMAGERY

A digital image as described in Sec. 2-13 can be obtained through a number of sources including satellite imaging sensors, digital cameras, and scanned aerial photographs. In its rudimentary form, a digital image bears no relationship to a ground coordinate reference system. Rather, its coordinate basis consists of integer column and row numbers which specify a pixel's location within a rectangular image array. *Georeferencing,* sometimes called *ground registration,* is a technique whereby a digital image is processed so that the columns and rows of the resulting product are aligned with north and east in a ground coordinate system. Some authors, particularly those in the remote-sensing field, refer to georeferencing as *rectification;* however, in the context of photogrammetry, the term *rectification* is reserved for the process of removing the effects of tilt from an aerial photograph. (See Secs. 10-8 through 10-13.)

The process of georeferencing an image involves two fundamental steps: (1) computing the parameters of a two-dimensional coordinate transformation (see App. C) which relates the digital image to the ground system and (2) filling an array, which is aligned with the ground coordinate system, with the appropriate digital numbers that quantify the brightness of the ground at the corresponding locations.

In the first step, a number of ground control points (GCPs) are selected which can be identified in the image, and for which accurate ground coordinates are available. The column and row image coordinates of each GCP are obtained and subsequently related to the ground coordinates. The simplest method for obtaining image coordinates of a GCP is to display the image on a computer screen. Then by using a pointing device such as a computer mouse, a cursor icon is guided to the location of the image point and a button is pressed to record the column and row. Many image manipulation programs are capable of this rudimentary operation.

The coordinate transformation in this first step of the georeferencing process *converts from ground coordinates (x and y) to image coordinates (X and Y)*. At first, the direction of this conversion (from ground to image) may seem backward. However, for reasons which will become clear in this section, the transformation must be performed in this manner. Furthermore, the use of lowercase x and y for ground coordinates and uppercase X and Y for image coordinates seems to be a reversal from the usual convention. Actually, this is appropriate when one looks at the uppercase/lowercase convention from a different point of view. In the coordinate transformation equations presented in App. C, lowercase variables are used for coordinates in the "from" system (the initial system), and uppercase variables are used for coordinates in the "to" system (the final system). For georeferencing, since the ground system is the "from" system, lowercase x and y are used for its coordinates; and since the image system is the "to" system, uppercase X and Y are used for its coordinates. Image coordinates of the common points will serve as control for the transformation, and the resulting parameters will give the relationship from ground to image coordinates. Any two-dimensional coordinate transformation could be used, but the conformal and affine are most often employed because of their convenience and suitability. For illustration purposes here, the two-dimensional conformal coordinate transformation will be used. Equations (C-11) express this transformation relationship, and for convenience they are repeated here as Eqs. (9-1). [Note that the variable names have been changed to relate to the coordinate system designations and specific conversion process (noted in italics) above.]

$$X = ax - by + T_X$$
$$Y = ay + bx + T_Y \tag{9-1}$$

In Eqs. (9-1), x and y are coordinates of points in the ground system; X and Y are coordinates of points in the image that have been obtained by converting from their column and row values; and a, b, T_X, and T_Y are parameters which are determined during this first step. Once computed, these parameters are used to transform coordinates of additional points from the ground system to the image system. The reason for this arrangement of the transformation will become apparent in the description of the second step, which follows.

During the second step of georeferencing, an image that is aligned with the ground coordinate system is produced. To understand this step, it is useful to visualize the ground as being divided into a rectangular grid of individual elements called *groundels* (ground elements), analogous to pixels (picture elements) of a digital image. The difference is that while pixels have no specific relationship to the ground, groundels are arranged at a nominal spacing in a grid which is parallel to the ground coordinate system. For each groundel, the x and y coordinates of its center point are transformed, based on the parameters computed in the first step, into corresponding image coordinates. The brightness value from the image at this corresponding location is then inserted into the groundel array. This involves the process of *resampling,* which is covered in App. E. After this process has been applied to each groundel, the georeferenced image is contained in the array of groundels.

To illustrate the above-described process of georeferencing, refer to Figs. 9-2*a* and *b*. Figure 9-2*a* represents an 8-pixel by 8-pixel digital image prior to georeferencing. Note that this image is represented as a square, nonrotated image, which is how it might appear if displayed on a computer screen. The solid-line grid of Fig. 9-2*b* represents an array of groundels which are nominally aligned with a ground coordinate system xy. Nominal ground coordinates are associated with the centers of the groundels, shown as small dots in Fig. 9-2*b*. The dashed lines represent the digital image as it would appear if properly aligned with the ground system. Four control points (1, 2, 3, and 4) having coordinates in the ground xy system appear in the image. Coordinates of these control points in both the image system of Fig. 9-2*a* and the ground system are

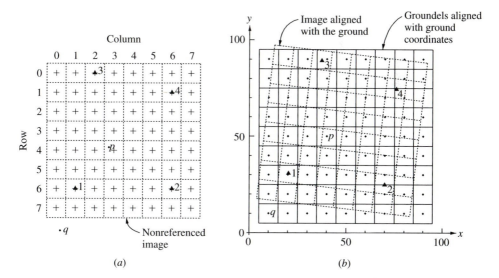

FIGURE 9-2
(*a*) Schematic representation of nonreferenced digital image. (*b*) Schematic representation of georeferenced image (solid lines) and orientation of image aligned with the ground (dashed lines).

TABLE 9-1
Ground and image coordinates for control points of Fig. 9-2

Point	Ground coordinates		Image coordinates	
	x, m	y, m	Column	Row
1	20.4	30.6	1	6
2	70.1	24.9	6	6
3	37.1	89.3	2	0
4	75.8	74.4	6	1

listed in Table 9-1. Note that image coordinates, which correspond to pixel centers (indicated by crosses in Fig. 9-2*a*), are specified to the nearest whole row and column, which is typical of most georeferencing applications. More elaborate techniques are available which can yield accurate fractional row and column coordinates; however, these techniques are not discussed here.

To prepare for the coordinate transformation which will yield the mathematical relationship between the image and ground coordinates, it is useful to convert the column and row coordinates of image pixels to the more conventional *XY* system. Note that in Fig. 9-2*a* the column coordinates increase from left to right, similar to conventional *X* coordinates; however the row coordinates increase from top to bottom, which is in the opposite direction from conventional *Y* coordinates. Because of this configuration, using the column coordinate as the abscissa and the row coordinate as the ordinate results in a "left-handed" coordinate system, as opposed to the more commonly used "right-handed" system, which is typical of ground coordinate systems. Transformation of a left-handed system to a right-handed system can cause mathematical inconsistencies in certain coordinate transformations, such as the conformal transformation. This inconsistency can be remedied by a number of approaches.

One approach is to use the row coordinate as the abscissa and the column coordinate as the ordinate. This results in a right-handed coordinate system which is rotated 90° clockwise. Another approach is to simply use a transformation that is able to account for the difference between left- and right-handed coordinate systems. The affine transformation has this capability by virtue of its separate *x* and *y* scale factors (see Sec. C-6), one of which will be positive and the other, negative. A third method is to convert the image coordinates from a left-handed system with the origin in the upper left corner to a right-handed system with its origin in the lower left corner. This can be accomplished by selecting the column coordinate as the abscissa and calculating ordinate values by subtracting row coordinates from the maximum row number.

Assume that the two-dimensional conformal transformation will be used to relate the ground coordinate system of Fig. 9-2*b* to the image coordinate system of Fig. 9-2*a*. Using image coordinates from Table 9-1, a right-handed *XY* coordinate system will be created using the third method just described. The maximum row number of the image is 7; therefore the original row coordinates will be subtracted from 7 to obtain *Y* coordinates. The resulting *XY* image coordinates are listed in Table 9-2.

TABLE 9-2
Column and row image coordinates of Table 9-1 converted to *XY* coordinates

Point	Original image coordinates		Converted image coordinates	
	Column	Row	X = Column	$Y = 7 -$ Row
1	1	6	1	1
2	6	6	6	1
3	2	0	2	7
4	6	1	6	6

To transform from the *xy* ground coordinate system to the *XY* converted image co-ordinate system, the following equations [of the form of Eqs. (9-1)] are written:

$$20.4a - 30.6b + T_X = 1 + V_{X_1}$$
$$30.6a + 20.4b + T_Y = 1 + V_{Y_1}$$
$$70.1a - 24.9b + T_X = 6 + V_{X_2}$$
$$24.9a + 70.1b + T_Y = 1 + V_{Y_2}$$
$$37.1a - 89.3b + T_X = 2 + V_{X_3}$$
$$89.3a + 37.1b + T_Y = 7 + V_{Y_3}$$
$$75.8a - 74.4b + T_X = 6 + V_{X_4}$$
$$74.4a + 75.8b + T_Y = 6 + V_{Y_4}$$

(9-2)

Using the least squares techniques presented in Secs. B-9 and C-5, the equations are solved for the most probable values of a, b, T_X, and T_Y, which are

$$a = 0.09921 \qquad T_X = -0.67$$
$$b = 0.01148 \qquad T_Y = -2.27$$

Once these transformation parameters have been computed, the georeferencing is completed by filling the groundel array with brightness values from the digital image. Referring to Fig. 9-2*b*, which contains a 9 × 9 groundel grid, this involves 81 separate applications of the above transformation, each followed by access of the brightness value from the image through resampling. For example, point *p* in Fig. 9-2*b* has ground coordinates of $x = 40$ and $y = 50$, and its image coordinates X and Y, by substitution into Eqs. (9-1), are

$$X = 40a - 50b + T_X = 40(0.09921) - 50(0.01148) + (-0.67) = 2.7$$
$$Y = 50a + 40b + T_Y = 50(0.09921) + 40(0.01148) + (-2.27) = 3.1$$

Expressing the location in column and row coordinates gives

$$\text{Column} = X = 2.7$$
$$\text{Row} = 7 - Y = 3.9$$

This column and row location is indicated by the position of point p in Fig. 9-2a. This position falls within the area of the pixel at column $= 3$ and row $= 4$, so the brightness associated with that particular image pixel could be copied into the corresponding groundel at $x = 40$ and $y = 50$. This is the *nearest-neighbor* method of resampling, which is described in App. E. Alternatively, *bilinear* or *bicubic* interpolation could be used for the resampling.

It is possible for some of the groundels to fall outside the limits of the digital image, as is the case with point q. From Fig. 9-2b, point q corresponds to the groundel at coordinates $x = 10$ and $y = 10$. Transforming these coordinates into the XY image system gives

$$X = 10a - 10b + T_X = 10(0.09921) - 10(0.01148) + (-0.67) = 0.2$$

$$Y = 10a + 10b + T_Y = 10(0.09921) + 10(0.01148) + (-2.27) = -1.2$$

(a)

FIGURE 9-3
(a) Nonreferenced satellite image. (b) Georeferenced image. (*Courtesy University of Florida.*)

Expressing this location in column and row coordinates gives

$$\text{Column} = X = 0.2$$
$$\text{Row} = 7 - Y = 8.2$$

Notice that since the row coordinate of 8.2 is beyond the maximum row number of 7, there is no brightness value from the image corresponding to this particular groundel. In this case, it is appropriate to use a default value of 0 for the groundel brightness value.

Figure 9-3 shows the result of georeferencing as applied to a satellite image. Figure 9-3*a* shows the nonreferenced image. Notice that even though the true directions of streets in the area correspond to cardinal directions (north, south, east, and west), the general directions of streets appear to be rotated approximately 10° in the image. Figure 9-3*b* shows the resulting georeferenced image where the directions of streets now properly correspond to cardinal directions.

When georeferencing digital imagery, it is beneficial to choose groundel dimensions that are consistent with the pixel dimensions of the image. For example, when georeferencing a satellite image with 10-m pixels, the groundel size should also be 10 m. This

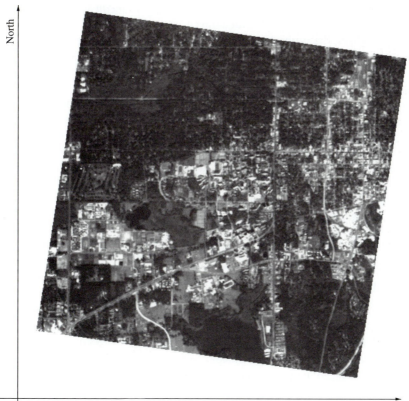

(*b*)

FIGURE 9-3 (*continued*)

is not an absolute necessity, but if the dimensions differ by more than a factor of about 1.5, more elaborate resampling methods may be necessary.

When choosing ground control points for georeferencing, it is important to select a well-distributed pattern of points, preferably with a point in each of the four corners as well as additional points spaced uniformly throughout the interior. It is particularly important to avoid having all control points in a small cluster in one area of the image. In addition, if the affine or projective transformation is used, it is critical that not all the control points lie along a straight line; otherwise, the transformation becomes ill-conditioned and yields poor results. Depending on the transformation used and the required geometric accuracy, the number of control points per image should be roughly 8 to 15. With a smaller number of control points, the accuracy will generally be lower and it will be more difficult to detect errors that may be present in the coordinates of one or more points. For the highest accuracy, a large number of control points is always desirable; however, beyond about 15 points, the marginal benefit of additional points seldom outweighs the extra effort involved.

The geometric accuracy of a georeferenced image can be assessed through evaluation of the residuals from the transformation, which should always be computed and checked. Large residuals may indicate errors in measured image or ground coordinates. Many computer software packages calculate the *root mean square* (RMS) of the residuals, which gives a nominal indication of accuracy. This value can be misleading, however, since the probability of an actual point's position being within the two-dimensional RMS error region is only about 40 percent. To increase the level of confidence to about 95 percent, the RMS value should be multiplied by 2.5.

It is possible to compute a transformation with small residuals but still have low accuracy in terms of absolute position. This would occur if, for example, the ground coordinates were scaled from a map that contains a significant systematic error. Perhaps the map was based on the NAD27 datum, but the resulting georeferenced image is assumed to be in the NAD83 datum (see Sec. 5-7). To disclose systematic errors of this type, it is advisable to perform a field test on a randomly selected sample of image points (checkpoints) using an independent source of higher accuracy, such as a global positioning system (GPS) survey. The magnitudes of discrepancies between coordinates as obtained from the georeferenced image and those obtained by the higher-accuracy method can be used to assess accuracy.

9-4 PLANIMETRIC MAPPING USING A TABLET DIGITIZER

For applications involving the use of hard-copy imagery, it is possible to digitize features from the image by using a tablet digitizer. A tablet digitizer, as described in Sec. 4-6 and illustrated in Fig. 4-6, is a device which is capable of measuring xy coordinates of points from flat media such as photographs and maps. To be used for this purpose, the tablet digitizer must be interfaced with a computer that has a graphical drawing program such as that of computer-aided drafting or geographic information system software.

The procedure involved in planimetric mapping with a tablet digitizer is similar to that of georeferencing. First, a series of control points whose *XY* coordinates are known in a ground coordinate system must be measured with the tablet digitizer, to obtain xy coordinates in the digitizer axis system. This enables a two-dimensional coordinate transformation to be computed which relates the xy digitizer coordinates to the ground XY system. Unlike the transformation associated with georeferencing, the arbitrary xy coordinates used on the right-hand side of transformation equations such as Eqs. (9-1) are those of the digitizer, and the XY coordinates on the left are in the ground system. (Once again, this is compatible with the convention of using lowercase for the "from" system and uppercase for the "to" system.) This enables digitizer coordinates of additional features (not control points) to be transformed to the ground system, and as such makes them suitable for mapping applications.

The resulting digital map consists of various point and line features with their positions represented by ground coordinates. Accuracy of the coordinates of these features depends upon a multitude of factors including terrain relief, accuracy of ground control, accuracy of the digitizer, and number of control points used. The type of coordinate transformation used is also important when digitizing aerial photographs. The projective transformation, described in Secs. 10-11 and C-8, is best for this application because it tends to compensate for tilt in the photos. An assessment of the accuracy can be made by evaluating residuals from the transformation and checkpoint discrepancies, as discussed in Sec. 9-3.

9-5 HEADS-UP DIGITIZING

Heads-up digitizing is a term used to describe the process where features, which are visible on a georeferenced digital image displayed on a computer screen, are digitized directly with a pointing device such as a mouse. The human operator views the digital image simultaneously with the digitized features on the computer screen. Since the operator's gaze remains on the computer screen throughout, the operation is referred to as *heads-up.*

Heads-up digitizing requires a computer that is capable of displaying a georeferenced image in the proper coordinate reference frame and enables simultaneous on-screen digitizing. Many GIS software packages are capable of this operation as well as some CAD packages. A key advantage of heads-up digitizing is that the points and lines associated with the digitized features are displayed as an overlay on the digital image, and if mistakes are made while pointing on the features, they can be readily detected and subsequently remeasured.

9-6 PHOTOMAPS

Photomaps are simply aerial photos that are used directly as planimetric map substitutes. The photos are usually brought to some desired average scale by enlargement or reduction by projection printing (see Sec. 2-12). Title information, place names, and other data may be superimposed on the photos in the same way that it is done on maps. Photomaps may be prepared from single aerial photos, or they may be made by piecing together two or more individual overlapping photos to form a single continuous composite picture. These composites are commonly referred to as *mosaics* and are described in Sec. 9-7.

Photomaps are similar to standard maps in many respects, but they have a number of definite advantages over maps. Photomaps show relative planimetric locations of a virtually unlimited number of objects, whereas features on maps—which are shown with lines and symbols—are limited by what was produced by the mapmaker. Photomaps of large areas can be prepared in much less time and at considerably lower cost than maps. Photomaps are easily understood and interpreted by people without photogrammetry backgrounds because objects are shown by their images. For this reason they are very useful in describing proposed construction or existing conditions to members of the general public, who may be confused by the same representations on a map.

Photomaps have the one serious disadvantage—they are not true planimetric representations. Rather, they are constructed from perspective photographs, which are subject to image displacements and scale variations. The most serious image displacements and scale variations are caused by variations in the terrain elevation and tilting of the camera axis. These displacements are generally most severe at the edges of the photographs. Of course, some small distortions result from shrinkage or expansion of the photo papers, camera lens imperfections, and atmospheric refraction, but these are generally negligible.

The effects of tilt can be eliminated by rectifying the photograph. (Rectification is described in Secs. 10-8 through 10-13.) Rectification does not remove the effects of topographic relief, however. Therefore the scale of the photomap is never constant throughout unless the terrain is perfectly flat. Relief displacements can be minimized by using a high flying height, while at the same time using a longer-focal-length camera to compensate for the decrease in scale. In measuring distances or directions from a photomap, it must be remembered that, due to image displacements, the scaled values will not be true. They are often used for qualitative studies only, and in that case slight planimetric inaccuracies caused by image displacements are of little consequence.

Because of their many advantages, photomaps are quite widely used. Their value is perhaps most appreciated in the field of planning, both in land-use planning and in planning for engineering projects. A photomap that shows an area completely and comprehensively can be rapidly and economically prepared. All critical features in the area that could affect the project can then be interpreted and taken into account. Alternative plans can be conveniently investigated, including considerations of soil types, drainage patterns, land-use and associated right-of-way costs, etc. As a result of this type of detailed study, the best overall plan is finally adopted.

Photomaps are valuable in numerous other miscellaneous areas besides planning. They are used to study geologic features, to inventory natural resources, to record the growth of cities and large institutions, to monitor construction activities at intervals of time, to record property boundaries, etc. They are also used as planimetric map substitutes for many engineering projects. Highway departments, for example, that are engaged in preparing plans for extensive construction projects frequently use photomaps to replace planimetric surveys. This not only eliminates much of the ground surveying but also does away with the office work associated with planimetric mapping. Design drawings and construction specifications are superimposed directly over the photomap. Used in that manner, these products have resulted in tremendous savings in time and cost, and they have yielded completely satisfactory results.

9-7 MOSAICS

If a single photo does not contain extensive enough coverage to serve as a photomap of an area, an aerial mosaic may be prepared. Traditionally, mosaics have been constructed manually from hard-copy paper prints, but recently digital mosaics prepared from scanned photographs have become more common. Whether accomplished manually or digitally, mosaics are constructed from a block of overlapping photographs which are trimmed and joined, much as cloth patches are stitched together to form a quilt. A special type of digital mosaic, known as a *composite orthophoto,* provides the most geometrically accurate image product available; however, its production is far more complex than that of a simple mosaic. Digital orthophotos are described in Sec. 15-8.

Aerial mosaics generally fall into three classes: *controlled, semicontrolled,* and *uncontrolled.* A controlled mosaic is the most accurate of the three classes. In the manual process, this type of mosaic is prepared from photographs that have been rectified and ratioed; i.e., all prints are made into equivalent vertical photographs that have the same nominal scale. In assembling the mosaic, image positions of common features on adjacent photos are matched as closely as possible. To increase the overall accuracy of the assembly, a plot of control points is prepared at the same scale as the ratioed photos. Then in piecing the photos together to form the mosaic, the control point images are also matched to their corresponding plotted control points to constrain the positions of the photos. Controlled digital mosaics have similar characteristics, but they are prepared in an analytical process similar to georeferencing, as discussed in Sec. 9-3. Along the edges between adjacent photos, images of features are aligned to the extent possible, although they will seldom line up exactly. These residual misalignments exist primarily because of relief displacements.

An uncontrolled mosaic is prepared by simply matching the image details of adjacent photos. There is no ground control, and aerial photographs that have not been rectified or ratioed are used. Uncontrolled mosaics are more easily and quickly prepared than controlled mosaics. They are not as accurate as controlled mosaics, but for many qualitative uses they are completely satisfactory.

Semicontrolled mosaics are assembled by utilizing some combinations of the specifications for controlled and uncontrolled mosaics. A semicontrolled mosaic may be prepared, for example, by using ground control but employing photos that have not been rectified or ratioed. The other combination would be to use rectified and ratioed photos but no ground control. Semicontrolled mosaics are a compromise between economy and accuracy. The mosaic of Fig. 9-4 is a semicontrolled mosaic prepared from nonrectified photos but assembled to fit U.S. Geological Survey quadrangle maps.

9-8 UNCONTROLLED DIGITAL MOSAICS

As noted above, a fundamental operation performed in the production of mosaics is the matching of images from adjacent photographs. In the manual process using hard-copy photographs, this is done through trial and error by shifting and rotating the photos slightly until optimal alignment is achieved. As previously mentioned, perfect alignment is seldom achieved due to relief displacements and tilt displacements of the

FIGURE 9-4
Semicontrolled mosaic showing the entire San Francisco Bay area. This mosaic consists of more than 2000 individual aerial photographs. (*Copyright photo, courtesy Pacific Resources, Inc.*)

images. With digital images, the matching procedure involves identifying *tie points* in the overlap areas between adjacent photos. To match a pair of photos, coordinates of the common tie points are measured in both photos, and a two-dimensional coordinate transformation is performed to define the geometric relationship between the photographs. This process is illustrated in Fig. 9-5, which shows two adjacent photographs in superposition. In this figure, the overlap area between photos 1 and 2 contains four tie points, *a*, *b*, *c*, and *d*. In the separated photographs shown at the bottom of the figure, each of these tie points has row and column coordinates in the digital image coordinate system of each photo. For instance, point *a* has coordinates C_{1_a}, R_{1_a} in photo 1 and C_{2_a}, R_{2_a} in photo 2. By selecting the coordinates of the tie points from photo 2 as control, a two-dimensional coordinate transformation can be computed that determines the parameters for transforming photo 1 coordinates to photo 2 coordinates. In the example of Fig. 9-5, since the tie points lie nearly in a straight line, a conformal transformation should be used because more complex transformations, such as affine or projective, become mathematically unstable when the control configuration is linear or nearly so.

After the transformation, the digital coverage of photo 1 can be extended to include the additional area covered by photo 2. This is done by increasing the number

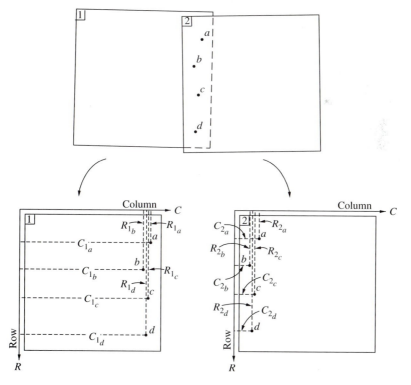

FIGURE 9-5
Use of tie points to join adjacent digital photographs.

of columns in the photo 1 array to create sufficient storage space to accommodate additional pixels from the photo 2 array. The brightness values associated with these additional columns are initially undefined. Then, in a process similar that of georeferencing, the coordinates of each pixel in the extended area of photo 1 are transformed to the coordinate system of photo 2, using the previously determined parameters. Brightness values are then resampled from photo 2 and stored in the extended portion of the photo 1 array. The result is a single image which incorporates the coverage of both photos.

When two photographs are attached by the foregoing method, a choice can be made as to which one is used to provide the brightness values in the overlap area. The simplest approach is to retain the values from the photo 1 array. This is generally not the best approach for two reasons. First, the extreme right edge of photo 1 typically contains the image of a side fiducial mark, which obscures the underlying terrain. Second, since relief displacements are more severe at greater distances from the principal point of a vertical photo, images of features along the seam between photos are more likely to be disjointed. To minimize these problems, the common area should be divided approximately into two halves, with one-half of the brightness values coming from each photograph. If a natural edge such as a street or fence line runs through the overlap area from top to bottom, it can be used to define the seam between the two photos. This will generally result in a less noticeable seam in the resulting image.

Large, uncontrolled digital mosaics can be constructed by repeating this attachment process for each photograph in a strip or block. As more digital photos are added, the area of the mosaic grows until the entire region is covered. Successive attachments of photos create an accumulation of errors and can cause the digital mosaic to become significantly warped. This effect can be reduced somewhat by starting in the middle of the block and working outward in all directions.

9-9   SEMICONTROLLED DIGITAL MOSAICS

As mentioned in Sec. 9-7, semicontrolled mosaics employ either rectified photos which can be joined (without control points) in the manner described in Sec. 9-8, or control points are used to constrain nonrectified photographs. The more likely situation associated with generation of a semicontrolled digital mosaic is to use nonrectified photographs in conjunction with a distributed configuration of control points. This situation is illustrated in Fig. 9-6. The figure shows 12 photographs (I through XII) arranged in a block formation with tie points (1 through 33) in the overlap areas between adjacent photos. Also, seven control points (A through G) are distributed about the block. Note that there is no single photograph with more than one control point; therefore, individual rectification of photographs is not possible. It is possible, however, to simultaneously transform points from all 12 photographs in a single least squares adjustment.

Suppose that the two-dimensional conformal transformation is used to perform the adjustment. Equations (9-1) express the conformal transformation relationship between coordinates x and y in an arbitrary system and coordinates X and Y in the control system. In Eqs. (9-1), a, b, T_X, and T_Y are the four parameters of the transformation.

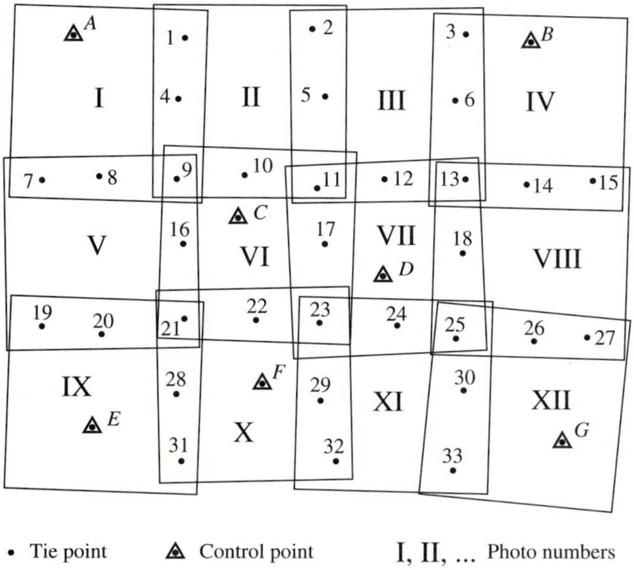

• Tie point   ▲ Control point   I, II, ... Photo numbers

FIGURE 9-6
Control and tie point configuration for a semicontrolled digital mosaic.

Since the block shown in Fig. 9-6 contains 12 individual photographs, 12 sets of four parameters (total of 48) are required to transform the individual photographs to the ground coordinate system. Calculation of the 48 unknown parameters requires a least squares adjustment involving two forms of observation equation pairs. The first type of observation equation pair is used for control points, and it has the same form as Eqs. (9-1) except that appropriate subscripts are included for the parameters and coordinates. As an example, note that in Fig. 9-6 control point C appears in photo VI. Digital photo coordinates x and y of the image of C in photo VI, which can be obtained from the row and column as described in Sec. 9-3, appear on the right-hand side of Eqs. (9-1); and X and Y ground coordinates of point C appear on the left. Also included in the right-hand side of Eqs. (9-1) are the parameters a, b, T_X, and T_Y which correspond to photo VI. Equations (9-3) express the observation equation pair which corresponds to the coordinates of control point C on photo VI, with appropriate subscripts.

$$X_C = a_{\mathrm{VI}} x_{C_{\mathrm{VI}}} - b_{\mathrm{VI}} y_{C_{\mathrm{VI}}} + T_{X_{\mathrm{VI}}}$$
$$Y_C = a_{\mathrm{VI}} y_{C_{\mathrm{VI}}} + b_{\mathrm{VI}} x_{C_{\mathrm{VI}}} + T_{Y_{\mathrm{VI}}}$$

(9-3)

Equations of this form are written for each control point and for each photo in which it appears. The second type of observation pair is for tie points. Since tie points have no ground coordinates associated with them, they cannot be used directly in Eqs. (9-1). Instead, the condition that the differences in ground coordinates X and Y for the common tie point based on the photo coordinates x and y in two adjacent photographs must equal zero, is enforced. An example of this condition is given by Eqs. (9-4), which express the relationship for tie point 1 between photos I and II.

$$a_1 x_{1_I} - b_1 y_{1_I} + T_{X_I} - (a_{II} x_{1_{II}} - b_{II} y_{1_{II}} + T_{X_{II}}) = 0$$
$$a_1 y_{1_I} + b_1 x_{1_I} + T_{Y_I} - (a_{II} y_{1_{II}} + b_{II} x_{1_{II}} + T_{Y_{II}}) = 0$$

(9-4)

A pair of equations of the form of Eqs. (9-4) is written for each occurrence of a tie point between a pair of photographs. Table 9-3 lists all the unique occurrences of tie points between adjacent photographs corresponding to Fig. 9-6. In this table, there are a total of 63 unique tie point connections between photographs, each of which contributes two equations of the form of Eqs. (9-4). Notice that tie points, such as point 1, which exist only on two photographs contribute only one pair of equations, whereas tie points, such as point 9, which exist on four photographs contribute six pairs of equations to the solution.

After all the equations have been formed, the unknown parameters (a, b, T_X, and T_Y) associated with each photo are computed by least squares. These parameters specify the geometric relationships between each photo and the ground system; however, the direction is the opposite of what is required. As described in Sec. 9-3 regarding georeferencing, coordinates of groundels must be transformed to the image coordinate system so that the brightness of the particular ground element can be determined. Since parameters a, b, T_X, and T_Y in Eqs. (9-2) transform from image coordinates to

TABLE 9-3
Connections between photos of Fig. 9-6 based on measured tie point coordinates

Point	Photo i	Photo j	Point	Photo i	Photo j	Point	Photo i	Photo j
1	I	II	13	III	VII	13	VII	VIII
4	I	II	13	III	VIII	18	VII	VIII
9	I	II	13	IV	VII	25	VII	VIII
7	I	V	13	IV	VIII	23	VII	X
8	I	V	14	IV	VIII	23	VII	XI
9	I	V	15	IV	VIII	24	VII	XI
9	I	VI	9	V	VI	25	VII	XI
2	II	III	16	V	VI	25	VII	XII
5	II	III	21	V	VI	25	VIII	XI
11	II	III	19	V	IX	25	VIII	XII
9	II	V	20	V	IX	26	VIII	XII
9	II	VI	21	V	IX	27	VIII	XII
10	II	VI	21	V	X	21	IX	X
11	II	VI	11	VI	VII	28	IX	X
11	II	VII	17	VI	VII	31	IX	X
3	III	IV	23	VI	VII	23	X	XI
6	III	IV	21	VI	IX	29	X	XI
13	III	IV	21	VI	X	32	X	XI
11	III	VI	22	VI	X	25	XI	XII
11	III	VII	23	VI	X	30	XI	XII
12	III	VII	23	VI	XI	33	XI	XII

ground, the equation must be rearranged. To do this, equations can be written in matrix form as

$$\begin{bmatrix} X \\ Y \end{bmatrix} = \begin{bmatrix} a & -b \\ b & a \end{bmatrix} \begin{bmatrix} x \\ y \end{bmatrix} + \begin{bmatrix} T_X \\ T_Y \end{bmatrix} \tag{9-5}$$

Solving Eq. (9-5) for the values of x and y results in Eq. (9-6).

$$\begin{bmatrix} x \\ y \end{bmatrix} = \begin{bmatrix} a & -b \\ b & a \end{bmatrix}^{-1} \begin{bmatrix} X - T_X \\ Y - T_Y \end{bmatrix} = \frac{1}{a^2 + b^2} \begin{bmatrix} a & b \\ -b & a \end{bmatrix} \begin{bmatrix} X - T_X \\ Y - T_Y \end{bmatrix} \tag{9-6}$$

By using the X and Y coordinates for each of the groundels in the area covered by the mosaic and the parameters of the corresponding photograph, Eq. (9-6) can be solved to obtain the position on the photo from which to extract the brightness values. The values are then resampled from the digital photo and placed in the groundel array. In areas of overlap between photographs, brightness values can be extracted from either photo. The choice can be made by the approach described in Sec. 9-8.

9-10 CONTROLLED DIGITAL MOSAICS

As mentioned in Sec. 9-7, controlled mosaics are made from rectified and ratioed photographs in conjunction with ground control. With digital mosaics, this process begins with photo rectification, as described in Sec. 10-13. In this process, at least four ground control points are identified in each photograph, and the two-dimensional projective transformation is used to georeference the image, using the process described in Sec. 9-3. Since each rectified digital photograph is georeferenced, the individual groundels will be directly related to ground coordinates. All that remains to create the controlled digital mosaic is to assemble the georeferenced images in a single groundel array that encompasses the full ground area of the mosaic. No further resampling or other significant processing is required, although choices must be made as to which brightness values to choose in overlapping areas.

An alternative approach can be used where there is insufficient ground control to rectify each digital photograph individually. It is a slight variation of the simultaneous coordinate transformation approach described in Sec. 9-9 which pertained to semi-controlled digital mosaics. The difference is that instead of using a two-dimensional conformal or affine transformation, the projective transformation is used. This involves solving for all the individual sets of eight projective transformation parameters for the entire block simultaneously. Since the transformation is projective, the photographs will be rectified and ratioed as well as adjusted to ground control in a single operation.

A simultaneous projective transformation is more complicated than either the conformal or the affine because the equations are nonlinear. The approach of simplifying the equations into a linear form as discussed in Sec. C-8 [see Eqs. (C-49)] cannot be used for tie points since the X and Y coordinates appear on both sides of the equation. Performing a solution of the nonlinear form of the projective equations requires a Taylor series expansion of Eqs. (C-48) as well as initial approximations for the unknown parameters. Equations (C-48) are repeated here as Eqs. (9-7) for convenience.

$$X = \frac{a_1x + b_1y + c_1}{a_3x + b_3y + 1}$$

$$Y = \frac{a_2x + b_2y + c_2}{a_3x + b_3y + 1}$$

(9-7)

For near-vertical photographs, a simultaneous affine transformation can be used to obtain initial approximations, because the parameters a_3 and b_3 of Eqs. (9-7) are approximately zero. This causes the denominators in these equations to reduce to unity, and thus the projective Eqs. (9-7) are equivalent to the affine equations [see Eqs. (C-22)]. When the initial approximations have been determined, the simultaneous projective transformation equations (which have been linearized by Taylor's series) can then be formed, and iterated until the solution converges. This results in a solution for the projective parameters which specifies the geometric relationship between each photograph and the ground coordinate system.

As was the case with the simultaneous conformal transformation method described in Sec. 9-9, the computed parameters describe the relationship in the reverse sense from that required. To reverse Eqs. (9-7) so that the x and y photo coordinates are isolated on the left requires substantial algebraic manipulation. The algebraic steps required to reverse these equations are left to the reader. Equations (9-8) express the necessary relationship in reverse.

$$x = \frac{(b_2 - b_3c_2)X + (b_3c_1 - b_1)Y + b_1c_2 - b_2c_1}{(a_2b_3 - a_3b_2)X + (a_3b_1 - a_1b_3)Y + a_1b_2 - a_2b_1}$$

$$y = \frac{(a_3c_2 - a_2)X + (a_1 - a_3c_1)Y + a_2c_1 - a_1c_2}{(a_2b_3 - a_3b_2)X + (a_3b_1 - a_1b_3)Y + a_1b_2 - a_2b_1}$$

(9-8)

Some final comments must be made regarding the simultaneous projective transformation solution. The least squares solution is often sensitive to large amounts of relief in the project area. Since the solution involves tie points for which there is no known height, corrections cannot be made for their relief displacements. If heights in object space of the selected tie points vary substantially from mean terrain, the projective equations can become ill-conditioned, which can cause the least squares solution to fail. In this case, the only recourse is to rectify each digital photograph individually and assemble them as mentioned at the beginning of this section. In addition to the difficulties caused by substantial relief, simultaneous projective transformations require more control points throughout the photo block as well as more tie points along the edges than either the simultaneous conformal or affine transformation.

REFERENCES

American Society of Photogrammetry: *Manual of Photogrammetry*, 4th ed., Bethesda, MD, 1980, chap. 15.
Derenyi, E. E.: "The Digital Transferscope," *Photogrammetric Engineering and Remote Sensing*, vol. 62, no. 6, 1996, p. 733.

Gao, J., and S. M. O'Leary: "The Role of Spatial Resolution in Quantifying SSC from Airborne Remotely Sensed Data," *Photogrammetric Engineering and Remote Sensing,* vol. 63, no. 3, 1997, p. 267.

Homer, C. G., R. D. Ramsey, T. C. Edwards, Jr., and A. Falconer: "Landscape Cover-Type Modeling Using a Multi-Scene Thematic Mapper Mosaic," *Photogrammetric Engineering and Remote Sensing,* vol. 63, no. 1, 1997, p. 59.

Jessiman, E. G., and M. R. Walsh: "An Approach to the Enhancement of Photomaps," *Canadian Surveyor,* vol. 30, no. 1, 1976, p. 11.

Kardoulas, N. G., A. C. Bird, and A. I. Lawan: "Geometric Correction of SPOT and Landsat Imagery: A Comparison of Map- and GPS-Derived Control Points," *Photogrammetric Engineering and Remote Sensing,* vol. 62, no. 10, 1996, p. 1173.

Marsik, Z.: "Use of Rectified Photographs and Differentially Rectified Photographs for Photo Maps," *Canadian Surveyor,* vol. 25, 1971, p. 567.

Novak, K.: "Rectification of Digital Imagery," *Photogrammetric Engineering and Remote Sensing,* vol. 58, no. 3, 1992, p. 339.

Steinwand, D. R., J. A. Hutchinson, and J. P. Snyder: "Map Projections for Global and Continental Data Sets and an Analysis of Pixel Distortion Caused by Reprojection," *Photogrammetric Engineering and Remote Sensing,* vol. 61, no. 12, 1995, p. 1487.

Vickers, E. W.: "Production Procedures for an Oversize Satellite Image Map," *Photogrammetric Engineering and Remote Sensing,* vol. 59, no. 2, 1993, p. 247.

PROBLEMS

9-1. Describe the various techniques available for elementary planimetric mapping using vertical photographs.

9-2. Describe the zoom transfer scope and its advantages in planimetric mapping by direct tracing.

9-3. Discuss ground control requirements necessary for georeferencing.

9-4. Briefly discuss the advantage of heads-up digitizing over mapping with a tablet digitizer.

9-5. Discuss the advantages and disadvantages of photomaps and aerial mosaics as compared to conventional line and symbol maps.

9-6. Outline some of the uses of photomaps.

9-7. Define the three main classes of mosaics.

9-8. A photomap is to be prepared by enlarging a 23-cm-square-format aerial photo to a size of an 80-cm square. If a Lear jet was used to obtain the photo and a 152-mm-focal-length camera was carried to an altitude of 13.2 km above ground for the exposure, what will be the resulting scale of the photomap, and how many hectares will it cover?

9-9. A mosaic to serve as a map substitute for locating utilities will be prepared from digital photographs scanned at a 25-micrometer (25-μm) pixel size. The photographs will be taken with a 305-mm-focal-length camera having a 23-cm-square format. If 75-cm-diameter manholes must span 10 pixels on the digital mosaic, what must be the flying height above average ground for the photography?

9-10. Repeat Prob. 9-9, except the camera lens focal length is 152 mm, the photos will be scanned at a 15-μm pixel size, and 10-cm-wide highway stripes must span 2 pixels on the digital mosaic.

CHAPTER 10

TILTED PHOTOGRAPHS

10-1 INTRODUCTION

In spite of level vials and other stabilizing equipment, in practice it is impossible to maintain the optical axis of the camera truly vertical. Unavoidable aircraft tilts cause photographs to be exposed with the camera axis tilted slightly from vertical, and the resulting pictures are called *tilted photographs*. If vertical photography is intended, the amount by which the optical axis deviates from vertical is usually less than 1° and it rarely exceeds 3°.

Six independent parameters called the *elements of exterior orientation* express the *spatial position* and *angular orientation* of a tilted photograph. The spatial position is normally given by X_L, Y_L, and Z_L, the three-dimensional coordinates of the exposure station in a ground coordinate system. Commonly Z_L is called H, the flying height above datum. Angular orientation is the amount and direction of tilt in the photo. Three angles are sufficient to define angular orientation, and in this book two different systems are described: the *tilt-swing-azimuth* (*t-s-α*) system and the *omega-phi-kappa* (*ω-φ-κ*) system. The omega-phi-kappa system possesses certain computational advantages over the tilt-swing-azimuth system and is therefore more widely used. The tilt-swing-azimuth system, however, is more easily understood and therefore is considered first.

10-2 ANGULAR ORIENTATION IN TILT, SWING, AND AZIMUTH

In Fig. 10-1, a tilted photograph is depicted showing the tilt-swing-azimuth angular orientation parameters. In the figure, L is the exposure station, and o is the principal point of the photo positive. Line Ln is a vertical line, n being the *photographic nadir point*, which occurs where the vertical line intersects the plane of the photograph. The extension of Ln intersects the ground surface at N_g, the *ground nadir point*, and intersects the datum surface at N_d, the *datum nadir point*. Line Lo is the camera optical axis; its extension intersects the ground at P_g, the ground principal point, and it intersects the

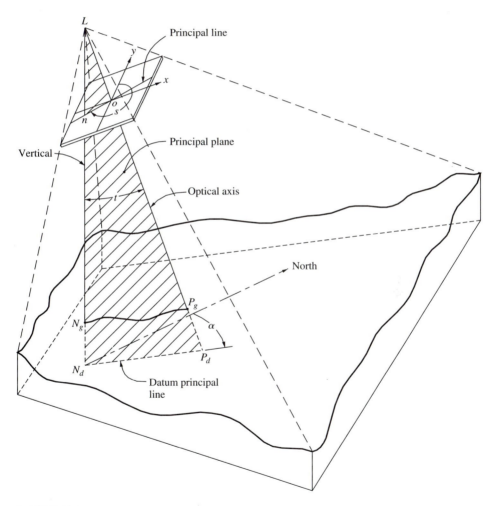

FIGURE 10-1
Geometry of a tilted photograph showing tilt-swing-azimuth angular orientation.

datum plane at P_d, the *datum principal point*. One of the orientation angles, *tilt,* is the angle t or nLo between the optical axis and the vertical line Ln. The tilt angle gives the magnitude of tilt of a photo.

Vertical plane Lno is called the *principal plane*. Its line of intersection with the plane of the photograph occurs along line *no,* which is called the *principal line*. The position of the principal line on the photo with respect to the reference fiducial axis system is given by s, the *swing* angle. *Swing* is defined as the clockwise angle measured in the plane of the photograph from the positive y axis to the downward or nadir end of the principal line, as shown in Fig. 10-1. The swing angle gives the direction of tilt on the photo. Its value can be anywhere between 0° and 360° or, alternatively, between −180° and 180°.

The third angular orientation parameter, α or *azimuth,* gives the orientation of the principal plane with respect to the ground reference axis system. Azimuth is the clockwise angle measured from the ground Y axis (usually north) to the *datum principal line* N_dP_d. It is measured in the datum plane or in a plane parallel to the datum plane, and its value can be anywhere between 0° and 360° or, alternatively, between −180° and 180°. The three angles of tilt, swing, and azimuth completely define the angular orientation of a tilted photograph in space. If the tilt angle is zero, the photo is vertical. Thus a vertical photograph is simply a special case of the general tilted photograph. For a vertical photo, swing and azimuth are undefined, but arbitrary definitions can be provided (see Sec. D-8).

10-3   AUXILIARY TILTED PHOTO COORDINATE SYSTEM

In the tilt-swing-azimuth system, certain computations require the use of an auxiliary $x'y'$ rectangular photographic coordinate system. This system, as shown in Fig. 10-2a, has its origin at the photographic nadir point n, and its y' axis coincides with the principal line (positive in the direction from n to o). Positive x' is 90° clockwise from positive y'. In solving tilted photo problems in the tilt-swing-azimuth system, photo coordinates are usually first measured in the fiducial coordinate system described in Sec. 4-2 and then converted to the auxiliary system numerically.

For any point in a tilted photo, the conversion from the xy fiducial system to the $x'y'$ tilted system requires (1) a rotation about the principal point through the angle θ and (2) a translation of origin from o to n. The rotation angle θ is defined as

$$\theta = s - 180° \tag{10-1}$$

The coordinates of image point a after rotation are x_a'' and y_a'', as shown in Fig. 10-2a. These coordinates, calculated in the same fashion as the E_C' and N_C' coordinates of Eqs. (C-2), are specified by

$$x_a'' = x_a \cos \theta - y_a \sin \theta$$

$$y_a'' = x_a \sin \theta + y_a \cos \theta$$

Auxiliary coordinate y_a' is obtained by adding the translation distance *on* to y_a''. From Fig. 10-2b, which is a side view of the principal plane, *on* is $f \tan t$. Since there is

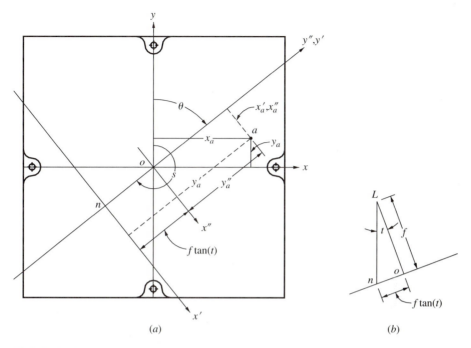

FIGURE 10-2
(a) Auxiliary $x'y'$ image coordinate system for a tilted photograph. (b) Principal plane of a tilted photo.

no x translation, $x_a' = x_a''$; and from the relationship between s and θ in Eq. (10-1), $\sin \theta = -\sin s$ and $\cos \theta = -\cos s$. Therefore, the coordinates x_a' and y_a' of a point a in the required auxiliary coordinate system are

$$x_a' = -x_a \cos s + y_a \sin s$$
$$y_a' = -x_a \sin s - y_a \cos s + f \tan t \tag{10-2}$$

In Eqs. (10-2), x_a and y_a are the fiducial coordinates of point a; f is the camera focal length; and t and s are the tilt and swing angles, respectively.

10-4   SCALE OF A TILTED PHOTOGRAPH

In Chap. 6 it was shown that scale variations in a vertical photograph are the result of variations in object distance (distance from camera to ground). The shorter the object distance, the larger the scale, and vice versa. For vertical photos, variations in object distances were caused only by topographic relief. In a tilted photograph, relief variations also cause changes in scale, but scale in various parts of the photo is further affected by the magnitude and angular orientation of the tilt. Figure 10-3a portrays the principal plane of a tilted photograph taken over a square grid on approximately flat ground. Figure 10-3b illustrates the appearance of the grid on the resulting tilted

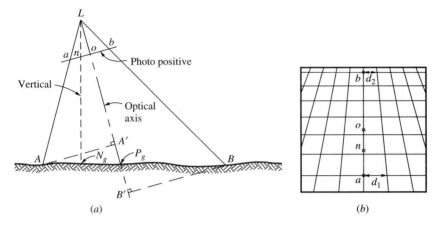

FIGURE 10-3
(*a*) Principal plane of a tilted photograph taken over approximately flat ground. (*b*) Image on the tilted photo of a square ground grid.

photograph. Due to tilt, object distance LA' in Fig. 10-3*a* is less than object distance LB', and hence a grid line near A would appear larger (at a greater scale) than a grid line near B. This is illustrated in Fig. 10-3*b*, where photo distance d_1 appears longer than photo distance d_2, yet both are the same length on the ground.

The scale at any point on a tilted photograph is readily calculated if tilt and swing for the photograph and the flying height of the photo and elevation of the point above datum are known. Figure 10-4 illustrates a tilted photo taken from a flying height H above datum; Lo is the camera focal length. The image of object point A appears at a on the tilted photo, and its coordinates in the auxiliary tilted photo coordinate system are x'_a and y'_a. The elevation of object point A above datum is h_A. Object plane $AA'KK'$ is a horizontal plane constructed a distance h_A above datum. Image plane $aa'kk'$ is also constructed horizontally. The scale relationship between the two parallel planes is the scale of the tilted photograph at point a because the image plane contains image point a and the object plane contains object point A. The scale relationship is the ratio of photo distance aa' to ground distance AA', and it may be derived from similar triangles $La'a$ and $LA'A$, and Lka' and LKA', as follows:

$$S_a = \frac{aa'}{AA'} = \frac{La'}{LA'} = \frac{Lk}{LK} \tag{a}$$

but

$$Lk = Ln - kn = \frac{f}{\cos t} - y'_a \sin t$$

also

$$LK = H - h_A$$

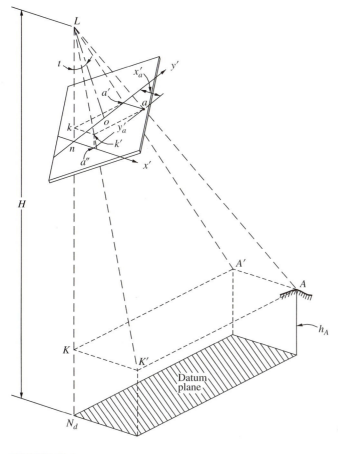

FIGURE 10-4
Scale of a tilted photograph.

Substituting Lk and LK into Eq. (a) and dropping subscripts, yields

$$S = \frac{f/\cos t - y' \sin t}{H - h} \qquad (10\text{-}3)$$

In Eq. (10-3), S is the scale on a tilted photograph for any point whose elevation is h above datum. Flying height above datum for the photo is H; f is the camera focal length; and y' is the coordinate of the point in the auxiliary system calculated by Eqs. (10-2). If the units of f and y' are millimeters and if H and h are meters, then the scale ratio is obtained in millimeters per meter. To obtain a dimensionless ratio, the right side of Eq. (10-3) must be multiplied by 1 m/1000 mm in that case. Examination of Eq. (10-3) shows that scale increases with increasing terrain elevation. If the photo is taken over level ground, then h is constant but scale still varies throughout the photograph with variations in y'.

Example 10-1. A tilted photo is taken with a 152.4-mm-focal-length camera from a flying height of 2266 m above datum. Tilt and swing are 2.53° and 218.2°, respectively. Point A has an elevation of 437 m above datum, and its image coordinates with respect to the fiducial axis system are $x_a = -72.4$ mm and $y_a = 87.1$ mm. What is the scale at point a?

Solution. By the second equation of Eqs. (10-2),

$$y_a' = 72.4 \sin 218.2° - 87.1 \cos 218.2° + 152.4 \tan 2.53° = 30.4 \text{ mm}$$

$$-44.77 - (-68.44) + 6.73$$

By Eq. (10-3),

$$S_a = \frac{152.4/\cos 2.53° - 30.4 \sin 2.53°}{2266 - 437} \left(\frac{1 \text{ m}}{1000 \text{ mm}} \right) = 1:12,1\overline{0}0 \quad (1 \text{ in}/101\overline{0} \text{ ft})$$

10-5 RELIEF DISPLACEMENT ON A TILTED PHOTOGRAPH

Image displacements on tilted photographs caused by topographic relief occur much the same as they do on vertical photos, except that *relief displacements on tilted photographs occur along radial lines from the nadir point.* (Relief displacements on a truly vertical photograph are also radial from the nadir point, but in that special case the nadir point coincides with the principal point.)

On a tilted photograph, image displacements due to relief vary in magnitude depending upon the flying height, height of object, amount of tilt, and location of the image in the photograph. Relief displacement is zero for images at the nadir point and increases with increased radial distances from the nadir.

As defined in Sec. 10-1, tilted photos are those that were intended to be vertical, but contain small unavoidable amounts of tilt. In practice, tilted photos are therefore so nearly vertical that their nadir points are generally very close to their principal points. Even for a photograph containing 3° of tilt taken with a 152-mm-focal-length camera, distance *on* is only about 8 mm. Relief displacements on tilted photos may therefore be calculated with satisfactory accuracy using Eq. (6-11), which applies to a vertical photograph. When this equation is used, radial distances *r* are measured from the principal point, even though theoretically they should be measured from the nadir point.

10-6 ANGULAR ORIENTATION IN OMEGA-PHI-KAPPA

As previously stated, besides the tilt-swing-azimuth system, angular orientation of a tilted photograph can be expressed in terms of three rotation angles: *omega, phi,* and *kappa*. These three angles uniquely define the angular relationships between the three axes of the tilted photo (image) coordinate system and the three axes of the ground (object) coordinate system.

Figure 10-5 illustrates a tilted photo in space. The ground coordinate system is *XYZ*. The tilted photo image coordinate system is *xyz* (shown dashed), and its origin is

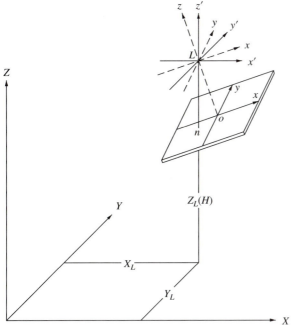

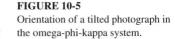

FIGURE 10-5
Orientation of a tilted photograph in
the omega-phi-kappa system.

at exposure station L. Consider an image coordinate system $x'y'z'$ with origin also at L and with its respective axes mutually parallel to the axes of the ground coordinate system, as shown in Fig. 10-5. As a result of three sequential rotations through the angles of omega, phi, and kappa, the $x'y'z'$ axis system can be made to coincide with the photographic xyz system. Each of the rotation angles omega, phi, and kappa is considered positive if counterclockwise when viewed from the positive end of the rotation axis.

The sequence of the three rotations is illustrated in Fig. 10-6. The first rotation, as illustrated in Fig. 10-6*a*, is about the x' axis through an angle omega. This first rotation creates a new axis system $x_1 y_1 z_1$. The second rotation phi is about the y_1 axis, as illustrated in Fig. 10-6*b*. As a result of the phi rotation, a new axis system $x_2 y_2 z_2$ is created. As illustrated in Fig. 10-6*c*, the third and final rotation is about the z_2 axis through the angle kappa. This third rotation creates the xyz coordinate system which is the photographic image system. Equations that express these three rotations are developed in Sec. C-7.

For any tilted photo there exists a unique set of angles omega, phi, and kappa that explicitly define the angular orientation of the photograph with respect to the reference ground coordinate system, provided appropriate ranges are maintained (see Sec. D-8). These three angles are related to the previously described tilt, swing, and azimuth angles; and if either set of three orientation angles is known for any photo, the other three can be determined as described in Sec. D-8. In the omega-phi-kappa system, as with the tilt-swing-azimuth system, the space position of any photo is given by the exposure station coordinates X_L, Y_L, and Z_L (or H).

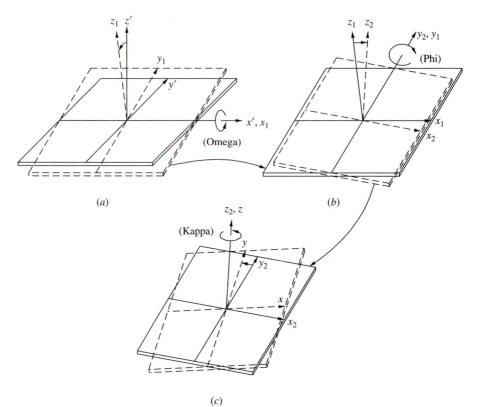

FIGURE 10-6
(a) Rotation about the x' axis through angle omega. (b) Rotation about the y_1 axis through angle phi.
(c) Rotation about the z_2 axis through angle kappa.

10-7 DETERMINING THE ELEMENTS OF EXTERIOR ORIENTATION

Many different methods, both graphical and numerical, have been devised to determine the six elements of exterior orientation of a single tilted photograph. In general, all methods require photographic images of at least three control points whose X, Y, and Z ground coordinates are known. Also, the calibrated focal length of the camera must be known. Today, only one method is still commonly in use: *space resection by collinearity*.

The method of space resection by collinearity is a purely numerical method which simultaneously yields all six elements of exterior orientation. Normally the angular values of omega, phi, and kappa are obtained in the solution, although the method is versatile and tilt, swing, and azimuth could also be obtained. Space resection by collinearity permits the use of redundant amounts of ground control; hence least squares computational techniques can be used to determine *most probable values* for the six elements. This method is a rather lengthy procedure requiring an iterative solution of nonlinear equations, but when it is programmed for a computer, a solution is easily obtained.

Space resection by collinearity involves formulating the *collinearity equations* for a number of control points whose X, Y, and Z ground coordinates are known and whose images appear in the tilted photo. The equations are then solved for the six unknown elements of exterior orientation which appear in them. The collinearity equations express the condition that for any given photograph the exposure station, any object point, and its corresponding image all lie on a straight line. Figure D-1 illustrates the collinearity condition for exposure station L, image point a, and object point A. The details involved in the numerical solution of space resection by collinearity are presented in Chap. 11.

10-8 RECTIFICATION OF TILTED PHOTOGRAPHS

Rectification is the process of making equivalent vertical photographs from tilted photo negatives. The resulting equivalent vertical photos are called *rectified photographs*. Rectified photos theoretically are truly vertical photos, and as such they are free from displacements of images due to tilt. They do, however, still contain image displacements and scale variations due to topographic relief. These relief displacements and scale variations can also be removed in a process called *differential rectification* or *orthorectification,* and the resulting products are then called *orthophotos.* Orthophotos are often preferred over rectified photos because of their superior geometric quality. Nevertheless, rectified photos are still quite popular because they do make very good map substitutes where terrain variations are moderate. This chapter discusses only the methods of rectification for the removal of the effects of tilt. Orthophoto production is described in Chap. 15.

Rectification is generally performed by any of three methods: analytical, optical-mechanical, and digital. Analytical rectification has the disadvantage that it can be applied only to individual discrete points, i.e., points that can be specifically identified so that their locations within the tilted photo can be measured. The resulting rectified photos produced by analytical methods are not really photos at all since they are not composed of photo images. Rather, they are plots of individual points in their rectified locations. The optical-mechanical and digital methods produce an actual picture in which the images of the tilted photo have been transformed to their rectified locations. Thus, the products of these two methods can be used in the production of photomaps and mosaics. In any of these rectification procedures, the rectified photos can be simultaneously *ratioed;* i.e., their average scales can be brought to some desired value different from that of the original photo. This is particularly advantageous if rectified photos are being made for the purpose of constructing a controlled mosaic, since all photos in the strip of block can be brought to a common scale. Thus, the resulting mosaic will have a more uniform scale throughout.

10-9 GEOMETRY OF RECTIFICATION

The fundamental geometry of rectification is illustrated in Fig. 10-7. This figure shows a side view of the principal plane of a tilted photo. When the exposure was made, the negative plane made an angle t with the datum plane. Rays from A and B were imaged at a' and b', respectively, on the negative, and their corresponding locations on the tilted photo are at a and b. The plane of an *equivalent vertical photo* is shown parallel to the

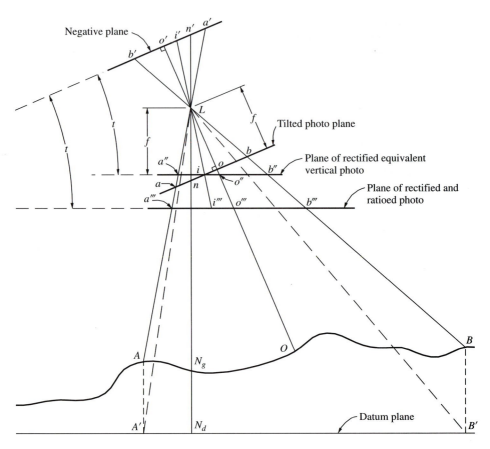

FIGURE 10-7
Principal plane of a tilted photograph showing the basic geometry of rectification.

datum plane and passing through i, the *isocenter* of the tilted photo[1]. The plane of a ratioed rectified photo is also shown. It is likewise parallel to the datum plane but exists at a level other than that of the equivalent vertical photo plane.

Methods of projecting points such as a and b either to a'' and b'' or to a''' and b''' are subjects of the following sections of this chapter. Figure 10-7 also illustrates, by virtue of lines LA' and LB', that although tilt displacements are removed in rectification, displacements due to relief are still present.

---

[1]An equivalent vertical photograph is an imaginary, truly vertical photo taken from the same exposure station as the tilted photo, with the same camera. The isocenter lies on the principal line of the tilted photo where the bisector of the tilt angle, constructed from the exposure station, intersects the tilted photo plane. A plane constructed through the isocenter, parallel to the datum principal plane, is the plane of an equivalent vertical photograph.

10-10 CORRECTION FOR RELIEF OF GROUND CONTROL POINTS USED IN RECTIFICATION

Since rectified and ratioed aerial photos retain the effects of relief, ground control points used in rectification must be adjusted slightly to accommodate their relief displacements. Conceptually, the process involves plotting ground control points at the locations that they will occupy in the rectified and ratioed photo. To this end, the positional displacements due to relief of the control points must be computed and applied to their horizontal positions in a radial direction from the exposure station, so that they will line up with points in the rectified photo. This procedure requires that the coordinates X_L, Y_L, and Z_L (or H) of the exposure station (which can be computed by space resection) and the X, Y, and Z (or h) coordinates for each ground control point be known. In addition, when the rectified photos are to be ratioed as well, it is convenient to select a plane in object space, at a specified elevation, to which the scale of the ratioed photo will be related. Generally, the elevation of this plane will be chosen as the elevation of average terrain h_{avg}.

The procedure is illustrated in Fig. 10-8, which shows the horizontal position of the exposure station L, denoted as a cross, and the unadjusted horizontal positions of four ground control points A, B, C, and D, denoted as circles. Also illustrated are radial distances r' from L to each of the points as well as the relief displacements d from the location in the plane of average terrain to the photo locations A', B', C', and D' of the control points, denoted as triangles. Since the elevations of these control points may be higher or lower than the average terrain, the relief displacements may be either outward (control point higher than average terrain) or inward (control point lower than average terrain). Note that in Fig. 10-8 the elevations of points B and D are less than average terrain elevation, and the elevations of points A and C are greater.

Determination of the coordinates of a displaced point (triangle) involves several steps. Initially, the value of r' is computed by Eq. (10-4) using the point's horizontal coordinates X and Y and the horizontal coordinates of the exposure station X_L and Y_L.

$$r' = \sqrt{(X - X_L)^2 + (Y - Y_L)^2} \tag{10-4}$$

Next, the relief displacement d is calculated by Eq. (10-5), which is a variation of Eq. (6-11):

$$d = \frac{r'(h - h_{avg})}{H - h} \tag{10-5}$$

In Eq. (10-5), d is the relief displacement, r' is the radial distance [computed by Eq. (10-4)], h is the height of the control point above datum, h_{avg} is the average terrain height in the tilted photo (also the height above datum of the ratioed photo plane), and H is the flying height above datum for the tilted photo. The units of all terms in the equation are those of the object space coordinate system, i.e., either meters or feet. Once the displacement d has been computed, the radial distance r to the displaced (image) location of the point can be computed by the following equation (be careful

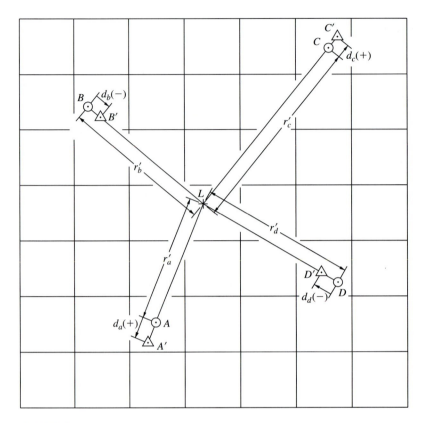

FIGURE 10-8
Plot of control points for rectification showing corrections made for relief displacements.

to use the proper algebraic sign of d):

$$r = r' + d \qquad (10\text{-}6)$$

Next, the azimuth α from the exposure location to the control point can be computed by

$$\alpha = \tan^{-1}\left(\frac{X - X_L}{Y - Y_L}\right) \qquad (10\text{-}7)$$

In Eq. (10-7), it is necessary to use the full-circle inverse tangent function so that the entire range of azimuth can be determined. Finally, the X' and Y' coordinates of the displaced (image) point are computed by

$$X' = X_L + r \sin \alpha \qquad (10\text{-}8)$$

$$Y' = Y_L + r \cos \alpha \qquad (10\text{-}9)$$

The resulting coordinates X' and Y' are appropriate for use in any of the methods of rectification.

10-11 ANALYTICAL RECTIFICATION

There are several methods available for performing analytical (numerical) rectification. Each of the analytical methods performs rectification point by point, and each requires that sufficient ground control appear in the tilted photo. Basic input required for the numerical methods, in addition to ground coordinates of control points, is x and y photo coordinates of all control points plus those of the points to be rectified. These are normally measured on a comparator. Due to the lengthy calculations required, numerical rectification is generally performed through use of a computer program.

Of the available methods of analytical rectification, the one that uses the two-dimensional projective transformation is the most convenient and is the only method discussed here. The transformation equations (C-48) are developed in Sec. C-8 and are repeated here as Eqs. (10-10), for convenience:

$$X = \frac{a_1 x + b_1 y + c_1}{a_3 x + b_3 y + 1}$$

$$Y = \frac{a_2 x + b_2 y + c_2}{a_3 x + b_3 y + 1}$$

(10-10)

In Eqs. (10-10), X and Y are ground coordinates, x and y are photo coordinates (in the fiducial axis system), and the a's, b's, and c's are the eight parameters of the transformation. The use of these equations to perform analytical rectification is a two-step process. First, a pair of Eqs. (10-10) is written for each ground control point. Four control points will produce eight equations, so that a unique solution can be made for the eight unknown parameters. It is strongly recommended that more than four control points be used, so that an improved solution can be arrived at by using least squares. An added benefit is that the redundant measurements may provide the ability of detecting mistakes in the coordinates, something which is not afforded by the unique solution using four control points.

Once the eight parameters have been determined, the second step of the solution can be performed—that of solving Eqs. (10-10) for each point whose X and Y rectified coordinates are desired. After rectified coordinates have been computed in the ground coordinate system, they can be plotted at the scale desired for the rectified and ratioed photo.

This analytical method is only rigorous if the ground coordinates X and Y of Eqs. (10-10) have been modified for relief displacements, as discussed in Sec. 10-10. If this is not done, a quasi-rectification results, although if the terrain is relatively flat and level, the errors will be minimal.

10-12 OPTICAL-MECHANICAL RECTIFICATION

In practice, the optical-mechanical method is widely used, although digital methods are rapidly surpassing this approach. The optical-mechanical method relies on instruments called *rectifiers*. They produce rectified and ratioed photos through the photographic process of projection printing (see Sec. 2-12); thus they must be operated in a darkroom.

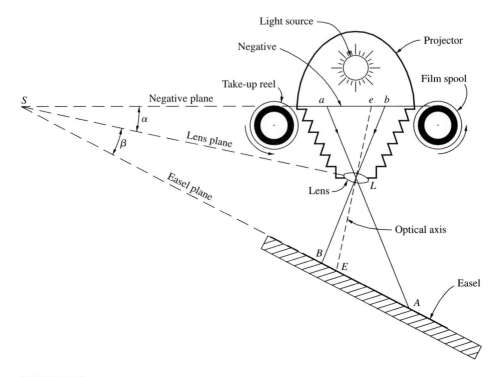

FIGURE 10-9
Schematic diagram of an optical-mechanical rectifier showing a side view of the principal plane.

As illustrated in Fig. 10-9, the basic components of a rectifier consist of a lens, a light source with reflector, a stage for mounting the tilted photo negative, and an easel which holds the photographic emulsion upon which the rectified photo is exposed. The instrument is constructed with controls so that the *easel plane, lens plane* (plane perpendicular to the optical axis of the rectifier lens), and *negative plane* can be tilted with respect to one another. Thus provision is made for varying angles α and β of Fig. 10-9. With some rectifiers the lens and easel planes can be tilted to vary these angles; however, most of these instruments do not have the capability of tilting the lens plane. Rather, the α and β angles are varied by tilting the negative and easel planes. This type of instrument is referred to as a *nontilting lens* rectifier.

So that rectified photos can be ratioed to varying scales, the rectifier must also have a magnification capability, and this is achieved by varying the projection distance (distance LE of Fig. 10-9 from the lens to the easel plane). To do this, however, and still maintain proper focus in accordance with Eq. (2-4), it is necessary to simultaneously vary the image distance (distance Le of Fig. 10-9 from the lens to the negative plane). The actual magnification that results along the axis of the rectifier lens is the ratio $LE/(Le)$, but it varies elsewhere in the photo due to the variable scale of the tilted photograph.

From Fig. 10-9 it can be seen that in rectification, projection distances vary depending upon the locations of points on the photo. To achieve sharp focus for all images in spite of this, the *Scheimpflug condition* must be satisfied. The Scheimpflug condition states that, in projecting images through a lens, if the negative and easel planes are not parallel, then the negative plane, lens plane, and easel plane must all intersect along a common line to satisfy the lens formula and achieve sharp focus for all images. Note that this condition is satisfied in Fig. 10-9 where these planes intersect at *S*.

In addition to being able to vary α, β, *LE*, and *Le*, rectifiers must allow for a rotation and shift of the negative, or a provision to tilt the easel plane in any direction and shift the negative. Rotation of the negative or tilt of the easel plane is necessary to place the principal plane of the photograph perpendicular to the easel plane of the rectifier, and the shift moves the principal point of the negative a specified distance away from the optical axis of the rectifier lens along the principal line. The amount of shift necessary and the values of α, β, *LE*, and *Le* are functions of the tilt angle, the magnification ratio, and the focal lengths of the rectifier lens and lens of the camera used to obtain the negative.

In practice, different techniques are employed to arrive at the proper rectifier settings. One method is to calculate numerically the elements of exterior orientation for each photo and from these values compute the required settings. Another possible method is to orient the photos in a stereoplotter, read the elements of exterior orientation, and from those values calculate the rectifier settings.

Rectifier settings that are actually needed depend upon the particular rectifier. If an *automatic rectifier* is used, for example, image and object distances are automatically held in the proper ratio to satisfy Eq. (2-4); and at the same time, a special mechanical device automatically satisfies the Scheimpflug condition regardless of the tilt angle or magnification ratio. With these instruments, fewer computations and settings are needed. An automatic rectifier is shown in Fig. 10-10.

Rather than compute rectifier settings, for near-vertical photography a trial-and-error method of orienting a rectifier which yields satisfactory results may be used. This is the procedure most often applied in practice. In this method four or more control points are plotted according to their ground control coordinates at the scale desired for the rectified photo. The plot is then placed on the easel, and, by trial and error, the rectifier controls are adjusted until the projected control point images coincide with their respective plotted points. This automatically creates proper orientation of the rectifier. Next the plot is removed from the easel, a sheet of photographic paper is placed there, and the rectified photo is exposed. Again, for a completely rigorous rectification to be achieved by using this method, the plotted control points must be corrected for their relief displacements in accordance with the method presented in Sec. 10-10; otherwise a quasi-rectification will result.

10-13 DIGITAL RECTIFICATION

Rectified photos can be produced by digital techniques that incorporate a photogrammetric scanner and computer processing. This procedure is a special case of the more general concept of georeferencing as presented in Chap. 9. The feature that distinguishes digital

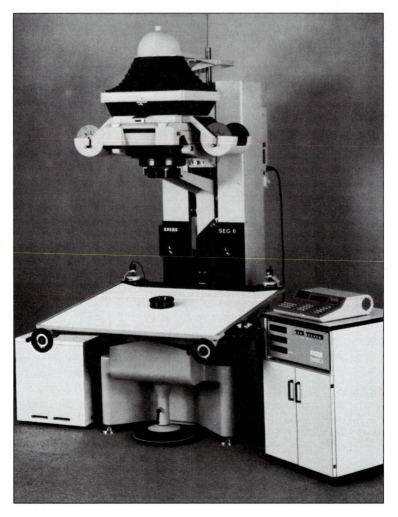

FIGURE 10-10
The SEG 6 automatic rectifier. (*Courtesy Carl Zeiss, Inc.*)

rectification from other georeferenced images is that rectification requires that a projective transformation be used to relate the image to the ground coordinate system, whereas georeferencing often uses simpler transformations such as the two-dimensional conformal or the two-dimensional affine transformation.

Figure 10-11*a* shows a portion of a scanned aerial photograph with a large amount of tilt (i.e., an oblique photo). Note that images in the foreground (near the bottom) are at a larger scale than those near the top. This photograph contains a generally rectangular street pattern, where the streets appear nonparallel due to the large amount of tilt. Figure 10-11*b* shows a digitally rectified version of the oblique photograph shown in Fig. 10-11*a*. Note that in this rectified image, the general street pattern

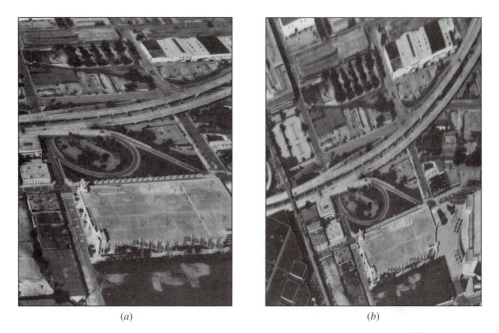

(a)   (b)

FIGURE 10-11
(a) Digital image of an oblique, nonrectified, photograph. (b) Image after digital rectification.

is approximately rectangular, although displacements remain due to relief in the terrain as well as buildings.

Three primary pieces of equipment needed for digital rectification are a photogrammetric scanner, computer, and plotting device capable of producing digital image output. Although a high-quality photogrammetric scanner is an expensive device, it is also highly versatile and can be used for many other digital photogrammetric operations. Low-accuracy desktop scanners can also be used for digital rectification; however, the geometric accuracy of the product will be substantially lower. The computer capability required is rather substantial, although not excessive. Off-the-shelf personal computers with large amounts of memory and disk space are more than adequate for digital rectification. Current plotting devices, generally large-format ink jet printers, produce image output of good quality, although the geometric accuracy and image resolution may be slightly less than those of photographic rectifiers. Image output from large-format ink jet plotters is currently a slow process, with large images taking perhaps 15 to 30 min. This drawback is sure to diminish as plotter technology improves.

10-14 ATMOSPHERIC REFRACTION IN TILTED AERIAL PHOTOGRAPHS

As mentioned in Sec. 4-14, atmospheric refraction in aerial photographs occurs radially from the nadir point. Now that parameters of tilted photographs have been described, a technique that corrects image coordinates in a tilted photo for the effects of atmospheric refraction is presented. In most practical situations, the assumption of vertical photography is sufficient when calculating atmospheric refraction. However for highest accuracy, when dealing with high-altitude photography (e.g., flying height greater than about 5000 m) and photographs with excessive tilts (e.g., tilts greater than about 5°), tilt should be considered when atmospheric refraction corrections are made.

Figure 10-12 illustrates a tilted photo containing image point a at coordinates x_a and y_a. The photographic nadir point n exists at coordinates x_n and y_n (in the fiducial axes coordinate system) which can be computed by the following equations. (Refer also to Fig. 10-2.)

$$x_n = on \sin s = f \tan t \sin s$$
$$y_n = on \cos s = f \tan t \cos s \qquad (10\text{-}11)$$

The angle between the vertical line Ln and the incoming light ray through point a is designated as α. Angle α can be computed by the application of the law of cosines to triangle Lna. (Refer again to Figs. 10-2 and 10-12.)

$$(na)^2 = (Ln)^2 + (La)^2 - 2(Ln)(La)\cos \alpha$$

Rearranging gives

$$\alpha = \cos^{-1}\left[\frac{(Ln)^2 + (La)^2 - (na)^2}{2(Ln)(La)}\right] \qquad (10\text{-}12)$$

where

$$na = \sqrt{(x_a - x_n)^2 + (y_a - y_n)^2} \qquad (10\text{-}13)$$

$$Ln = \frac{f}{\cos t} \qquad (10\text{-}14)$$

$$La = \sqrt{f^2 + r^2} = \sqrt{f^2 + x_a^2 + y_a^2} \qquad (10\text{-}15)$$

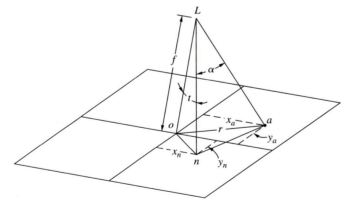

FIGURE 10-12
Diagram of a tilted photo with image point a whose coordinates will be corrected for atmospheric refraction.

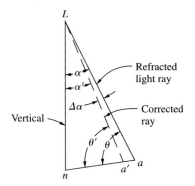

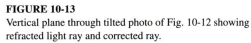

FIGURE 10-13
Vertical plane through tilted photo of Fig. 10-12 showing refracted light ray and corrected ray.

After angle α has been determined, the refraction angle $\Delta\alpha$ can be calculated by Eq. (4-18) by using a K value computed by Eq. (4-19).

Figure 10-13 shows triangle Lna along with the light ray La' which indicates the direction from the ground point that has been corrected for atmospheric refraction. From the figure, angle θ can be computed by the law of cosines as

$$\theta = \cos^{-1}\left[\frac{(La)^2 + (na)^2 - (Ln)^2}{2(La)(na)}\right] \tag{10-16}$$

Angle θ' can then be computed by

$$\theta' = \theta + \Delta\alpha \tag{10-17}$$

Application of the law of sines to triangle Lna' gives

$$\frac{\sin\theta'}{Ln} = \frac{\sin\alpha'}{na'}$$

Solving for na' gives

$$na' = Ln\frac{\sin\alpha'}{\sin\theta'} \tag{10-18}$$

where

$$\alpha' = \alpha - \Delta\alpha \tag{10-19}$$

and Ln and θ' are computed by Eqs. (10-14) and (10-17), respectively. The displacement aa' due to atmospheric refraction is then

$$aa' = na - na' \tag{10-20}$$

To compute the coordinate corrections δx and δy, displacement aa' must be applied in the direction of line an, as shown in Fig. 10-14. The direction of line an is specified by angle β as shown in the figure and can be computed by

$$\beta = \tan^{-1}\left(\frac{x_a - x_n}{y_a - y_n}\right) \tag{10-21}$$

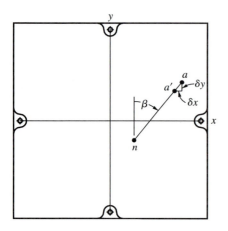

FIGURE 10-14
Photograph positions of image point
a and its new location *a'* after
correction for atmospheric refraction.

The corrections are then given by

$$\delta x = aa' \sin \beta$$
$$\delta y = aa' \cos \beta$$

(10-22)

Finally, the coordinates x'_a and y'_a for point *a* corrected for atmospheric refraction are computed by

$$x'_a = x_a - \delta x$$
$$y'_a = y_a - \delta y$$

(10-23)

Example 10-2. A tilted aerial photograph taken from a flying height of 4800 m above mean sea level contains the image *a* of object point *A* at fiducial coordinates $x_a = -64.102$ mm and $y_a = 83.220$ mm. The angles of tilt, swing, and azimuth for the photograph are 2.64°, 130.27°, and 5.15°, respectively. If the elevation of object point *A* is 140 m above mean sea level and the camera had a focal length of 152.794 mm, compute the *x'* and *y'* coordinates of the point, corrected for atmospheric refraction.

Solution. Compute the fiducial coordinates of the nadir point *n* by Eqs. (10-11).

$$x_n = 152.794 \tan 2.64° \sin 130.27° = 5.376 \text{ mm}$$

$$y_n = 152.794 \tan 2.64° \cos 130.27° = -4.554 \text{ mm}$$

Compute the values of *na*, *Ln*, and *La* by Eqs. (10-13), (10-14), and (10-15), respectively.

$$na = \sqrt{(-64.102 - 5.376)^2 + (83.220 + 4.554)^2} = 111.944 \text{ mm}$$

$$Ln = \frac{152.794}{\cos 2.64°} = 152.956 \text{ mm}$$

$$La = \sqrt{(152.794)^2 + (-64.102)^2 + (83.220)^2} = 185.420 \text{ mm}$$

Compute angle α by Eq. (10-12).

$$\alpha = \cos^{-1}\left[\frac{(152.956)^2 + (185.420)^2 - (111.944)^2}{2(152.956)(185.420)}\right] = 37.0933°$$

Compute K by Eqs. (4-19).

$$K = (7.4 \times 10^{-4})(4.8 - 0.14)\{1 - 0.02[2(4.8) - 0.14]\} = 0.0028°$$

Compute $\Delta\alpha$ by Eq. (4-18).

$$\Delta\alpha = 0.0028° \tan 37.0933° = 0.0021°$$

Compute θ by Eq. (10-16).

$$\theta = \cos^{-1}\left[\frac{(La)^2 + (na)^2 - (Ln)^2}{2(La)(na)}\right] = 55.4949°$$

Compute θ' and α' by Eqs. (10-17) and (10-19), respectively.

$$\theta' = 55.4949° + 0.0021° = 55.4970°$$

$$\alpha' = 37.0933° - 0.0021° = 37.0912°$$

Compute na' by Eq. (10-18).

$$na' = 152.956\,\frac{\sin 37.0912°}{\sin 55.4970°} = 111.935 \text{ mm}$$

Compute aa' and β by Eqs. (10-20) and (10-21), respectively.

$$aa' = 111.944 - 111.935 = 0.008 \text{ mm}$$

$$\beta = \tan^{-1}\left(\frac{-64.102 - 5.376}{83.220 + 4.554}\right) = -38.3634°$$

Compute δx and δy by Eqs. (10-22).

$$\delta x = 0.008 \sin(-38.3634°) = -0.005 \text{ mm}$$

$$\delta y = 0.008 \cos(-38.3634°) = 0.007 \text{ mm}$$

The corrected coordinates x'_a and y'_a are computed by Eqs. (10-23).

$$x'_a = -64.102 - (-0.005) = -64.097 \text{ mm}$$

$$y'_a = 83.220 - 0.007 = 83.213 \text{ mm}$$

Two final comments regarding refraction in a tilted photo are in order. First, the values of tilt and swing are not known until after an analytical solution is performed (see Chap. 11). However, photo coordinates are necessary to compute the analytical solution. Therefore, refinement for atmospheric refraction in a tilted photograph must be performed in an iterative fashion. Since analytical photogrammetric solutions are generally iterative due to the nonlinear equations involved, tilted photo refraction corrections can be conveniently inserted into the iterative loop.

Second, the foregoing discussion of atmospheric refraction in a tilted photograph assumes that the tilt angle is the angle between the optical axis and a vertical line. In analytical photogrammetry, however, a local vertical coordinate system is generally used for object space (see Sec. 5-5). As a result, tilt angles will be related to the direction of the local vertical Z axis, which is generally different from the true vertical direction. This effect, however, is negligible for practical situations. Unless a photograph is more than about 300 km from the local vertical origin, the effect from ignoring this difference in vertical direction will generally be less than 0.001 mm.

REFERENCES

American Society of Photogrammetry: *Manual of Photogrammetry,* 4th ed., Bethesda, MD, 1980, chaps. 2, 9, and 14.
Bomford, G.: *Geodesy,* 4th ed., Clarendon Press, Oxford, 1980.
Hallert, B.: "Quality of Exterior Orientation," *Photogrammetric Engineering,* vol. 32, no. 3, 1966, p. 464.
Jones, A. D.: "The Development of the Wild Rectifiers," *Photogrammetric Record,* vol. 2, 1966, p. 181.
Keller, M., and G. C. Tewinkel: *Space Resection in Photogrammetry,* ESSA Technical Report C&GS 32, U.S. Coast and Geodetic Survey, Washington, 1966.
Mugnier, C. J.: "Analytical Rectification Using Artificial Points," *Photogrammetric Engineering and Remote Sensing,* vol. 44, no. 5, 1978, p. 579.
Trachsel, A. F.: "Electro-Optical Rectifier," *Photogrammetric Engineering,* vol. 33, no. 5, 1967, p. 513.
Wilson, K. R., and J. Vlcek: "Analytical Rectification," *Photogrammetric Engineering,* vol. 36, no. 6, 1970, p. 570.

PROBLEMS

10-1. A particular tilted aerial photograph exposed with a 152.047-mm-focal-length camera has a tilt angle of 2.75° and a swing angle of 140.00°. On this photograph, what are the auxiliary x' and y' photo coordinates for points a and b, whose photo coordinates measured with respect to the fiducial axes are $x_a = 69.27$ mm, $y_a = -41.80$ mm, $x_b = -54.72$ mm, and $y_b = 106.38$ mm?

10-2. Repeat Prob. 10-1, except that the camera focal length is 88.907 mm, tilt angle is 1.92°, swing angle is 249.00°, $x_a = -62.41$ mm, $y_a = 76.80$ mm, $x_b = 98.55$ mm, and $y_b = -12.06$ mm.

10-3. Calculate photographic scale for image points a and b of Prob. 10-1 if flying height above datum was 2195 m and if elevations of points A and B were 264 m and 376 m above datum, respectively.

10-4. Calculate photographic scale for image points a and b of Prob. 10-2 if flying height above datum was 2682 m and if elevations of points A and B were 544 m and 572 m above datum, respectively.

10-5. Illustrate and briefly describe the six elements of exterior orientation.

10-6. Name and briefly describe the three different methods of performing rectification. Discuss some advantages and disadvantages of each.

10-7. A tilted aerial photograph is exposed at $X_L = 9475.2$ m, $Y_L = 8903.0$ m, and $Z_L = 1639.1$ m. A rectified and ratioed photograph is to be produced with an optical-mechanical rectifier using the trial-and-error approach. The following table lists X, Y, and Z coordinates of five control points which will control the rectification. If the plane of rectification (average terrain) is at $h_{avg} = 110.0$ m, compute the displaced image locations X' and Y' of the control points as illustrated in Fig. 10-8.

Point	X, m	Y, m	Z, m
1	8,819.8	9,580.4	114.2
2	10,324.6	9,279.9	120.8
3	9,055.4	8,603.9	101.5
4	10,185.0	8,075.3	116.7
5	9,483.7	7,993.3	102.5

10-8. Repeat Prob. 10-7, except that the tilted photograph is exposed at $X_L = 12{,}742.8$ m, $Y_L = 15{,}818.9$ m, and $Z_L = 1007.3$ m; average terrain is at $h_{avg} = 225.0$ m; and the X, Y, and Z coordinates of the control points are as listed in the following table.

Point	X, m	Y, m	Z, m
1	12,274.6	15,975.7	206.2
2	13,228.7	16,360.1	215.1
3	12,504.5	15,312.1	276.4
4	13,091.2	15,625.6	217.2
5	12,670.4	16,180.7	221.6

10-9. The following table lists measured photo coordinates for images of control points 1 through 5 of Prob. 10-7 as well as additional points 6 through 8. Using the analytical rectification technique discussed in Secs. 10-11 and C-8, determine the rectified coordinates of points 6, 7, and 8. Use the original X and Y coordinates from the table from Prob. 10-7 as control, not the X' and Y' coordinates adjusted for relief displacement.

Point	x, mm	y, mm
1	−60.39	72.53
2	90.56	37.94
3	−39.54	−25.41
4	70.81	−82.62
5	0.47	−86.76
6	−24.14	63.25
7	56.25	91.20
8	−12.99	−70.98

10-10. Repeat Prob. 10-9, except use X' and Y' coordinates, adjusted for relief displacement, from Prob. 10-7 as control for the rectification.

10-11. The following table lists measured photo coordinates for images of control points 1 through 5 of Prob. 10-8 as well as additional points 6, 7, and 8. Using the analytical rectification technique discussed in Secs. 10-11 and C-8, determine the rectified coordinates of points 6 through 8. Use the original X and Y coordinates from the table from Prob. 10-8 as control, not the X' and Y' coordinates adjusted for relief displacement.

Point	x, mm	y, mm
1	−91.96	24.32
2	87.26	99.08
3	−51.03	−113.13
4	65.84	−40.15
5	−17.17	65.20
6	50.93	−76.45
7	50.88	−25.41
8	−12.69	25.27

10-12. Repeat Prob. 10-11, except use X' and Y' coordinates, adjusted for relief displacement, from Prob. 10-8 as control for the rectification.

10-13. A tilted aerial photograph taken from a flying height of 3340 m above mean sea level contains the image a of object point A at fiducial coordinates $x_a = 18.227$ mm and $y_a = -102.499$ mm. The angles of tilt, swing, and azimuth for the photograph are 1.95°, −49.58°, and 139.44°, respectively. If the elevation of object point A is $h_A = 295$ m above mean sea level and the camera had a focal length $f = 152.013$ mm, compute the x' and y' coordinates of the point, corrected for atmospheric refraction in the tilted photo.

10-14. Repeat Prob. 10-13 except that the flying height is 6270 m above mean sea level, $x_a = 87.559$ mm, $y_a = 81.080$ mm, tilt = 4.97°, swing = −150.04°, azimuth = 47.22°, $h_A = 167$ m, and $f = 153.097$ mm.

CHAPTER
11

INTRODUCTION TO ANALYTICAL PHOTOGRAMMETRY

11-1 INTRODUCTION

Analytical photogrammetry is a term used to describe the rigorous mathematical calculation of coordinates of points in object space based upon camera parameters, measured photo coordinates and ground control. Unlike the elementary methods presented in earlier chapters, this process rigorously accounts for any tilts that exist in the photos. Analytical photogrammetry generally involves the solution of large, complex systems of redundant equations by the method of least squares. The concepts of analytical photogrammetry have existed for many years, but the process was not considered practical until relatively recently because of the heavy computational effort that it entails. The evolution of computer technology and photogrammetric software, however, has now made analytical photogrammetry a commonplace technique. Analytical photogrammetry forms the basis of many modern hardware and software systems, including: stereoplotters (analytical and softcopy), digital terrain model generation, orthophoto production, digital photo rectification, and aerotriangulation.

This chapter presents an introduction to some fundamental topics and elementary applications in analytical photogrammetry. The coverage here is limited to computations involving single photos and stereopairs. Later chapters in the book cover more advanced topics and applications in this subject. In particular, Chap. 17 describes analytical photogrammetry solutions for handling strips and blocks of photos.

11-2 IMAGE MEASUREMENTS

A fundamental type of measurement used in analytical photogrammetry is an x and y photo coordinate pair. These coordinates, generally measured in millimeters, *must be related to the principal point as the origin.* Since mathematical relationships in analytical photogrammetry are based on assumptions such as "light rays travel in straight lines" and "the focal plane of a frame camera is flat," various coordinate refinements may be required to correct measured photo coordinates for distortion effects that otherwise cause these assumptions to be violated. Techniques of photo coordinate refinement, as presented in Secs. 4-9 through 4-14 and Sec. 10-14, can be used to correct photo coordinates so that the assumptions will be valid.

A number of instruments and techniques are available for making photo coordinate measurements, as discussed in Secs. 4-5 through 4-8. To ensure results having the highest accuracy, these measurements must be made with great care. In many analytical photogrammetry methods, it is necessary to measure image coordinates of common object points that appear in more than one photograph. In these cases it is essential that the image of each object point be precisely identified between photos so that the measurements will be consistent.

11-3 CONTROL POINTS

In addition to measurement of image coordinates, a certain number of control points in object space are generally required for analytical photogrammetry. Object space coordinates of these control points, which may be either image-identifiable features or exposure stations of the photographs themselves, are generally determined via some type of field survey technique such as GPS. It is important that the object space coordinates be based on a three-dimensional cartesian system which has straight, mutually perpendicular axes. This often requires three-dimensional coordinate conversions which are described in Chap. 5 and App. F.

✳ 11-4 COLLINEARITY CONDITION

Perhaps the most fundamental and useful relationship in analytical photogrammetry is the *collinearity* condition. Collinearity, as described in App. D, is the condition that the exposure station, any object point, and its photo image all lie along a straight line in three-dimensional space. The collinearity condition is illustrated in Fig. 11-1, where L, a, and A lie along a straight line. Two equations express the collinearity condition for any point on a photo: one equation for the x photo coordinate and another for the y photo coordinate. The mathematical relationships are developed in App. D and expressed by Eqs. (D-5) and (D-6). They are repeated here for convenience.

$$x_a = x_o - f\left[\frac{m_{11}(X_A - X_L) + m_{12}(Y_A - Y_L) + m_{13}(Z_A - Z_L)}{m_{31}(X_A - X_L) + m_{32}(Y_A - Y_L) + m_{33}(Z_A - Z_L)}\right] \quad (11\text{-}1)$$

$$y_a = y_o - f\left[\frac{m_{21}(X_A - X_L) + m_{22}(Y_A - Y_L) + m_{23}(Z_A - Z_L)}{m_{31}(X_A - X_L) + m_{32}(Y_A - Y_L) + m_{33}(Z_A - Z_L)}\right] \quad (11\text{-}2)$$

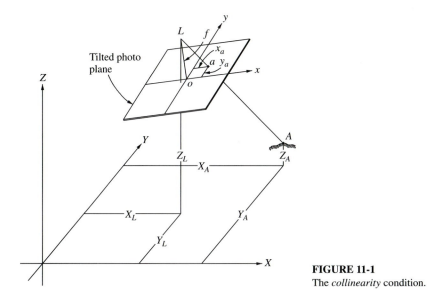

FIGURE 11-1
The *collinearity* condition.

In Eqs. (11-1) and (11-2), x_a and y_a are the photo coordinates of image point a; X_A, Y_A, and Z_A are object space coordinates of point A; X_L, Y_L, and Z_L are object space coordinates of the exposure station; f is the camera focal length; x_o and y_o are the coordinates of the principal point (usually known from camera calibration); and the m's (as described in Sec. C-7) are functions of three rotation angles, and most often omega, phi, and kappa are the angles employed.

The collinearity equations are nonlinear and can be linearized by using Taylor's theorem as described in Sec. D-4. The linearized forms are Eqs. (D-11) and (D-12), and they are also repeated here for convenience.

$$b_{11}\, d\omega + b_{12}\, d\phi + b_{13}\, d\kappa - b_{14}\, dX_L - b_{15}\, dY_L - b_{16}\, dZ_L$$
$$+ b_{14}\, dX_A + b_{15}\, dY_A + b_{16}\, dZ_A = J + v_{x_a} \tag{11-3}$$

$$b_{21}\, d\omega + b_{22}\, d\phi + b_{23}\, d\kappa - b_{24}\, dX_L - b_{25}\, dY_L - b_{26}\, dZ_L$$
$$+ b_{24}\, dX_A + b_{25}\, dY_A + b_{26}\, dZ_A = K + v_{y_a} \tag{11-4}$$

In Equations (11-3) and (11-4), v_{x_a} and v_{y_a} are residual errors in measured x_a and y_a image coordinates; $d\omega$, $d\phi$, and $d\kappa$ are corrections to initial approximations for the orientation angles of the photo; dX_L, dY_L, and dZ_L are corrections to initial approximations for the exposure station coordinates; and dX_A, dY_A, and dZ_A are corrections to initial values for the object space coordinates of point A. The b's and the J and K terms are described in Sec. D-4. Because higher order terms are ignored in linearization by Taylor's theorem, the linearized forms of the equations are approximations. They must therefore be solved iteratively, as described in App. D, until the magnitudes of corrections to initial approximations become negligible.

11-5   COPLANARITY CONDITION

Coplanarity, as illustrated in Fig. 11-2, is the condition that the two exposure stations of a stereopair, any object point, and its corresponding image points on the two photos all lie in a common plane. In the figure, for example, points L_1, L_2, a_1, a_2, and A all lie in the same plane. The coplanarity condition equation is

$$0 = B_X (D_1F_2 - D_2F_1) + B_Y (E_2F_1 - E_1F_2) + B_Z (E_1D_2 - E_2D_1) \qquad (11\text{-}5)$$

where
$$B_X = X_{L_2} - X_{L_1}$$
$$B_Y = Y_{L_2} - Y_{L_1}$$
$$B_Z = Z_{L_2} - Z_{L_1}$$
$$D = (m_{12})x + (m_{22})y - (m_{32})f$$
$$E = (m_{11})x + (m_{21})y - (m_{31})f$$
$$F = (m_{13})x + (m_{23})y - (m_{33})f$$

In Eq. (11-5), subscripts 1 and 2 affixed to terms D, E, and F indicate that the terms apply to either photo 1 or photo 2. The m's again are functions of the three rotation angles omega, phi, and kappa, as defined in Sec. C-7. One coplanarity equation may be written for each object point whose images appear on both photos of the stereopair. The coplanarity equations do not contain object space coordinates as unknowns; rather, they contain only the elements of exterior orientation of the two photos of the stereopair.

Like collinearity equations, the coplanarity equation is nonlinear and must be linearized by using Taylor's theorem. Linearization of the coplanarity equation is somewhat more difficult than that of the collinearity equations and is beyond the scope of this text. References cited at the end of this chapter explain these procedures, however. Coplanarity is not used nearly as extensively as collinearity in analytical photogrammetry.

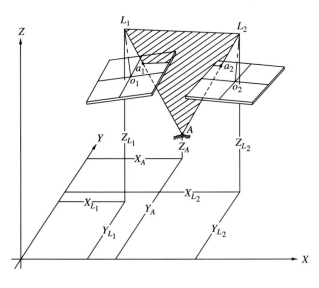

FIGURE 11-2
The *coplanarity* condition.

11-6 SPACE RESECTION BY COLLINEARITY

Space resection, as discussed in Sec. 10-7, is a method of determining the six elements of exterior orientation (ω, ϕ, κ, X_L, Y_L, and Z_L) of a photograph. This method requires a minimum of three control points, with known XYZ object space coordinates, to be imaged in the photograph. If the ground control coordinates are assumed to be known and fixed, then the linearized forms of the space resection collinearity equations for a point A are

$$b_{11}\, d\omega + b_{12}\, d\phi + b_{13}\, d\kappa - b_{14}\, dX_L - b_{15}\, dY_L - b_{16}\, dZ_L = J + v_{x_a} \quad (11\text{-}6)$$

$$b_{21}\, d\omega + b_{22}\, d\phi + b_{23}\, d\kappa - b_{24}\, dX_L - b_{25}\, dY_L - b_{26}\, dZ_L = K + v_{y_a} \quad (11\text{-}7)$$

In Eqs. (11-6) and (11-7), the terms are as defined in Secs. 11-4 and D-4. Two equations are formed for each control point, which gives six equations if the minimum of three control points is used. In this case a unique solution results for the six unknowns, and the residual terms on the right sides of Eqs. (11-6) and (11-7) will be zero. If four or more control points are used, more than six equations can be formed, allowing a least squares solution.

Since the collinearity equations are nonlinear, and have been linearized using Taylor's theorem, initial approximations are required for the unknown orientation parameters. For the typical case of near-vertical photography, zero values can be used as initial approximations for ω and ϕ. The value of Z_L (the height H above datum) can be computed using the method of Sec. 6-9. Since this method requires only two control points, several solutions are possible, using different pairs of control points. An improved approximation can be made by computing several values for H and taking the average. After H has been determined, ground coordinates from a vertical photograph, as described in Sec. 6-7, can be computed, using the measured x and y photo coordinates, focal length f, flying height H, and elevation of the object point Z [or h as it appears in Eqs. (6-5) and (6-6)]. A two-dimensional conformal coordinate transformation can then be performed, which relates the ground coordinates as computed from the vertical photo equations to the control values. The two-dimensional conformal coordinate transformation Eqs. (C-9), repeated here in a slightly different form, are used.

$$
\begin{aligned}
X &= ax' - by' + T_X \\
Y &= ay' + bx' + T_Y
\end{aligned}
\quad (11\text{-}8)
$$

In Eqs. (11-8), X and Y are ground control coordinates for the point; x' and y' are ground coordinates from a vertical photograph as computed by Eqs. (6-5) and (6-6); and a, b, T_X, and T_Y are the transformation parameters. A pair of equations of the type of Eqs. (11-8) can be written for each control point, and the four unknown parameters computed by least squares (see Secs. C-4 through C-5). The translation factors T_X and T_Y determined from this solution can then be used as initial approximations for X_L and Y_L, respectively. Rotation angle θ, which can be computed by Eq. (C-11), can be used as an approximation for κ.

By using these initial approximations in Eqs. (11-6) and (11-7), a least squares solution can be computed for the unknown corrections to the approximations. The solution is iterated until the corrections become negligible.

Example 11-1. A near-vertical aerial photograph taken with a 152.916-mm-focal-length camera contains images of four ground control points A through D. Refined photo coordinates and ground control coordinates (in a local vertical system) of the four points are listed in the following table. Calculate the exterior orientation parameters ω, ϕ, κ, X_L, Y_L, and Z_L for this photograph.

Point	Photo coordinates		Ground control coordinates		
	x, mm	y, mm	X, m	Y, m	Z, m
A	86.421	−83.977	1268.102	1455.027	22.606
B	−100.916	92.582	732.181	545.344	22.299
C	−98.322	−89.161	1454.553	731.666	22.649
D	78.812	98.123	545.245	1268.232	22.336

Solution

1. Determine initial approximations.
   a. Set $\omega = 0°$ and $\phi = 0°$.
   b. Use the method of Sec. 6-9, based on points A and B, to compute H. (The result is given below.)

$$H = 649.6 \text{ m}$$

   c. Use Eqs. (6-5) and (6-6) to compute ground coordinates from an assumed vertical photo for points A through D. (The results of these calculations are listed in the following table.)

Point	Ground coordinates from vertical photo	
	X, m	Y, m
A	354.356	−344.334
B	−413.993	379.804
C	−403.126	−365.565
D	323.295	402.511

   d. Compute two-dimensional conformal coordinate transformation parameters by a least squares solution using all four control points, and use the results for assigning initial approximations. (The results are:)

$$a = -0.22189$$
$$b = 0.97432$$
$$T_X = 1009.923 = X_L$$
$$T_Y = 1038.056 = Y_L$$
$$\kappa = \theta = \tan^{-1}\left(\frac{-0.22189}{0.97432}\right) = 102.83° \quad \text{(used full circle inverse tangent)}$$

2. Form the rotation matrix using Eqs. (C-33). Its elements are

$$M = \begin{bmatrix} -0.22205 & 0.97504 & 0.00000 \\ -0.97504 & -0.22205 & 0.00000 \\ 0.00000 & 0.00000 & 1.00000 \end{bmatrix}$$

3. Form the linearized observation equations. These are Eqs. (11-6) and (11-7). Note that a pair of these equations is formed for each control point, and thus a total of eight equations results. These can be represented in matrix form as

$$B\Delta = \varepsilon + V$$

where

$$B = \begin{bmatrix} b_{11_a} & b_{12_a} & b_{13_a} & -b_{14_a} & -b_{15_a} & -b_{16_a} \\ b_{21_a} & b_{22_a} & b_{23_a} & -b_{24_a} & -b_{25_a} & -b_{26_a} \\ b_{11_b} & b_{12_b} & b_{13_b} & -b_{14_b} & -b_{15_b} & -b_{16_b} \\ b_{21_b} & b_{22_b} & b_{23_b} & -b_{24_b} & -b_{25_b} & -b_{26_b} \\ b_{11_c} & b_{12_c} & b_{13_c} & -b_{14_c} & -b_{15_c} & -b_{16_c} \\ b_{21_c} & b_{22_c} & b_{23_c} & -b_{24_c} & -b_{25_c} & -b_{26_c} \\ b_{11_d} & b_{12_d} & b_{13_d} & -b_{14_d} & -b_{15_d} & -b_{16_d} \\ b_{21_d} & b_{22_d} & b_{23_d} & -b_{24_d} & -b_{25_d} & -b_{26_d} \end{bmatrix} \quad \Delta = \begin{bmatrix} d\omega \\ d\phi \\ d\kappa \\ dX_L \\ dY_L \\ dZ_L \end{bmatrix} \quad \varepsilon = \begin{bmatrix} J_a \\ K_a \\ J_b \\ K_b \\ J_c \\ K_c \\ J_d \\ K_d \end{bmatrix} \quad V = \begin{bmatrix} V_{x_a} \\ V_{y_a} \\ V_{x_b} \\ V_{y_b} \\ V_{x_c} \\ V_{y_c} \\ V_{x_d} \\ V_{y_d} \end{bmatrix}$$

Calculation of the coefficients corresponding to point A is shown as an example.

Compute r, s, and q terms for point A by Eqs. (D-7) and (D-8):

$$r = m_{11}(X_A - X_L) + m_{12}(Y_A - Y_L) + m_{13}(Z_A - Z_L)$$

$$= -0.22205(1268.102 - 1009.923) + 0.97504(1455.027 - 1038.056)$$
$$+ 0(22.606 - 649.614) = 349.233$$

$$s = m_{21}(X_A - X_L) + m_{22}(Y_A - Y_L) + m_{23}(Z_A - Z_L)$$

$$= -0.97504(1268.102 - 1009.923) - 0.22205(1455.027 - 1038.056)$$
$$+ 0(22.606 - 649.614) = -344.322$$

$$q = m_{31}(X_A - X_L) + m_{32}(Y_A - Y_L) + m_{33}(Z_A - Z_L)$$

$$= 0(1268.102 - 1009.923) + 0(1455.027 - 1038.056)$$
$$+ 1(22.606 - 649.614) = -627.008$$

Compute linearized collinearity equation coefficients for point A by Eqs. (D-11) and (D-12):

$$b_{11} = \frac{f}{q^2}[r(-m_{33}\,\Delta Y + m_{32}\,\Delta Z) - q(-m_{13}\,\Delta Y + m_{12}\,\Delta Z)]$$

$$= \frac{152.916}{(-627.008)^2}\{349.233[-1(1455.027 - 1038.056)$$

$$+\ 0(22.606 - 649.614)] + 627.008[-0(1455.027 - 1038.056)$$

$$+\ 0.97504(22.606 - 649.614)]\}$$

$$=\ -205.739$$

$$b_{12} = \frac{f}{q^2}[r(\cos\phi\,\Delta X + \sin\omega\,\sin\phi\,\Delta Y - \cos\omega\,\sin\phi\,\Delta Z)$$

$$-\ q(-\sin\phi\,\cos\kappa\,\Delta X + \sin\omega\,\cos\phi\,\cos\kappa\,\Delta Y$$

$$-\ \cos\omega\,\cos\phi\,\cos\kappa\,\Delta Z)]$$

$$= \frac{152.916}{(-627.008)^2}\{349.233[\cos(0)(1268.102 - 1009.923)$$

$$+\ \sin(0)\sin(0)(1455.027 - 1038.056)$$

$$-\ \cos(0)\sin(0)(22.606 - 649.614)]$$

$$+\ 627.008\ [-\sin(0)\cos(102.83°)(1268.102 - 1009.923)$$

$$+\ \sin(0)\cos(0)\cos(102.83°)(1455.027 - 1038.056)$$

$$-\ \cos(0)\cos(0)\cos(102.83°)(22.606 - 649.614)]\}$$

$$=\ 1.116$$

$$b_{13} = \frac{-f}{q}(m_{21}\,\Delta X + m_{22}\,\Delta Y + m_{23}\,\Delta Z)$$

$$= \frac{-152.916}{(-627.008)^2}[-0.97504(1268.102 - 1009.923)$$

$$-\ 0.22205(1455.027 - 1038.056) + 0(22.606 - 649.614)]$$

$$=\ -83.974$$

$$b_{14} = \frac{f}{q^2}(rm_{31} - qm_{11})$$

$$= \frac{152.916}{(-627.008)^2}[349.233(0) + 627.008(-0.22205)]$$

$$=\ -0.05415$$

$$b_{15} = \frac{f}{q^2}(rm_{32} - qm_{12})$$

$$= \frac{152.916}{(-627.008)^2}[349.233(0) + 627.008(0.97504)]$$

$$= 0.23779$$

$$b_{16} = \frac{f}{q^2}(rm_{33} - qm_{13}) = \frac{152.916}{(-627.008)^2}[349.233(1) + 627.008(0)]$$

$$= 0.13584$$

$$b_{21} = \frac{f}{q^2}[s(-m_{33}\,\Delta Y + m_{32}\,\Delta Z) - q\,(-m_{23}\,\Delta Y + m_{22}\,\Delta Z)]$$

$$= \frac{152.916}{(-627.008)^2}\{-344.322[-1(1455.027 - 1038.056)$$

$$+ 0(22.606 - 649.614)] + 627.008[-0(1455.027$$

$$- 1038.056) - 0.22205(22.606 - 649.614)]\}$$

$$= 89.799$$

$$b_{22} = \frac{f}{q^2}[s(\cos\phi\,\Delta X + \sin\omega\sin\phi\,\Delta Y - \cos\omega\sin\phi\,\Delta Z)$$

$$- q(\sin\phi\sin\kappa\,\Delta X - \sin\omega\cos\phi\sin\kappa\,\Delta Y + \cos\omega\cos\phi\sin\kappa\,\Delta Z)]$$

$$= \frac{152.916}{(-627.008)^2}\{-344.322[\cos(0)(1268.102 - 1009.923)$$

$$+ \sin(0)\sin(0)(1455.027 - 1038.056)$$

$$- \cos(0)\sin(0)(22.606 - 649.614)]$$

$$+ 627.008[\sin(0)\sin(102.83°)(1268.102 - 1009.923)$$

$$- \sin(0)\cos(0)\sin(102.83°)(1455.027 - 1038.056)$$

$$+ \cos(0)\cos(0)\sin(102.83°)(22.606 - 649.614)]\}$$

$$= -183.676$$

$$b_{23} = \frac{f}{q}(m_{11}\,\Delta X + m_{12}\,\Delta Y + m_{13}\,\Delta Z)$$

$$= \frac{152.916}{(-627.008)^2}[-0.22205(1268.102 - 1009.923)$$

$$+ 0.97504(1455.027 - 1038.056) + 0(22.606 - 649.614)]$$

$$= -85.172$$

$$b_{24} = \frac{f}{q^2}(sm_{31} - qm_{21})$$

$$= \frac{152.916}{(-627.008)^2}[-344.322(0) + 627.008(-0.97504)] = -0.23779$$

$$b_{25} = \frac{f}{q^2}(sm_{32} - qm_{22}) = \frac{152.916}{(-627.008)^2}[-344.322(0)$$

$$+ 627.008(-0.22205)] = -0.05415$$

$$b_{26} = \frac{f}{q^2}(sm_{33} - qm_{23}) = \frac{152.916}{(-627.008)^2}[-344.322(1)$$

$$+ 627.008(0)] = -0.13393$$

$$J = x_a - x_o + f\frac{r}{q} = 86.421 - 0 + 152.916\left(\frac{349.233}{-627.008}\right) = 1.249$$

$$K = y_a - y_o + f\frac{s}{q} = -83.977 - 0 + 152.916\left(\frac{-344.322}{-627.008}\right) = -0.003$$

Remaining coefficients, corresponding to points B, C, and D, are calculated in a similar manner.

Numerical values for the B matrix elements are

$$B = \begin{bmatrix} -205.739 & 1.116 & -83.974 & 0.05415 & -0.23779 & -0.13584 \\ 89.799 & -183.676 & -85.172 & 0.23779 & 0.05415 & 0.13393 \\ -229.270 & 11.238 & 92.682 & 0.05413 & -0.23768 & 0.16271 \\ 106.750 & -190.133 & 102.073 & 0.23768 & 0.05413 & -0.14774 \\ -196.473 & -102.705 & -89.144 & 0.05416 & -0.23781 & 0.15462 \\ -9.609 & -212.318 & 96.943 & 0.23781 & 0.05416 & 0.14218 \\ -178.404 & -93.117 & 97.990 & 0.05413 & -0.23769 & -0.12732 \\ -2.002 & -221.688 & -79.864 & 0.23769 & 0.05413 & -0.15622 \end{bmatrix}$$

Numerical values for the ε matrix elements are

$$\varepsilon = \begin{bmatrix} 1.249 \\ -0.003 \\ 1.157 \\ -0.100 \\ -1.379 \\ -0.017 \\ -1.052 \\ 0.133 \end{bmatrix}$$

4. Form and solve normal equations. Note that the solution conforms to Eq. (B-12), where Δ, B, and ε have been substituted for X, A, and L,

respectively. The solution is

$$\Delta = (B^\mathrm{T}B)^{-1}(B^\mathrm{T}\varepsilon)$$

and the detailed matrices in the solution are

$$
\Delta =
\begin{bmatrix}
184{,}878.9 & -321.6 & -1463.5 & 0.116 & 202.557 & -21.822 \\
-321.6 & 183{,}458.6 & -5662.4 & -201.986 & -0.116 & 5.587 \\
-1463.5 & -5662.4 & 66{,}639.7 & 9.024 & -2.312 & 0.000 \\
0.116 & -201.986 & 9.024 & 0.23781 & 0.00000 & -0.00365 \\
202.557 & -0.116 & -2.312 & 0.00000 & 0.23781 & -0.01438 \\
-21.822 & 5.587 & 0.000 & -0.00365 & -0.01438 & 0.16943
\end{bmatrix}^{-1}
$$

$$
\times
\begin{bmatrix}
-74.619 \\
247.729 \\
0.020 \\
0.00171 \\
0.00676 \\
-0.06946
\end{bmatrix}
$$

$$
=
\begin{bmatrix}
-0.00714 \\
0.02119 \\
-0.00059 \\
18.017 \\
6.049 \\
-1.127
\end{bmatrix}
$$

5. Add corrections. (*Note*: Angle corrections are in radians.)

$$\omega = 0° - 0.00714\left(\frac{180°}{\pi}\right) = -0.4093°$$

$$\phi = 0° + 0.02119\left(\frac{180°}{\pi}\right) = 1.2144°$$

$$\kappa = 102.83° - 0.00059\left(\frac{180°}{\pi}\right) = 102.7959°$$

$$X_L = 1009.923 + 18.017 = 1027.940$$

$$Y_L = 1038.056 + 6.049 = 1044.105$$

$$Z_L = 649.614 - 1.127 = 648.487$$

6. Second iteration: Repeat step 2, using updated approximations.

$$
M =
\begin{bmatrix}
-0.22143 & 0.97517 & -0.00227 \\
-0.97495 & -0.22132 & 0.02225 \\
0.02119 & 0.00714 & 0.99975
\end{bmatrix}
$$

7. Repeat step 3: Form matrices B and ε.

$$B = \begin{bmatrix} -208.944 & 1.559 & -83.926 & 0.05185 & -0.24241 & -0.13926 \\ 88.454 & -183.535 & -86.389 & 0.24423 & 0.05576 & 0.13033 \\ -225.520 & 10.925 & 92.541 & 0.05661 & -0.23338 & 0.15915 \\ 107.916 & -190.210 & 100.876 & 0.23138 & 0.05219 & -0.15085 \\ -199.865 & -103.710 & -89.124 & 0.05808 & -0.23982 & 0.15932 \\ -8.397 & -212.459 & 98.273 & 0.24395 & 0.05572 & 0.13848 \\ -175.370 & -92.117 & 98.085 & 0.05072 & -0.23586 & -0.12353 \\ -3.132 & -221.659 & -78.761 & 0.23164 & 0.05223 & -0.15988 \end{bmatrix}$$

$$\varepsilon = \begin{bmatrix} 0.032 \\ -0.051 \\ -0.040 \\ 0.041 \\ -0.049 \\ -0.037 \\ 0.051 \\ 0.038 \end{bmatrix}$$

8. Repeat step 4: Form and solve the normal equations.

$$\Delta = \begin{bmatrix} 184{,}767.9 & -190.1 & -55.9 & -0.306 & 202.510 & -22.387 \\ -190.1 & 183{,}499.5 & -5{,}665.2 & -202.007 & 0.097 & 7.169 \\ -55.9 & -5{,}665.2 & 66{,}670.9 & 8.657 & -1.204 & -0.175 \\ -0.306 & -202.007 & 8.657 & 0.23820 & -0.00029 & -0.00155 \\ 202.510 & 0.097 & -1.204 & -0.00029 & 0.23804 & -0.01370 \\ -22.387 & 7.169 & -0.175 & -0.00155 & -0.01370 & 0.16984 \end{bmatrix}^{-1} \begin{bmatrix} 3.151 \\ 0.681 \\ 4.930 \\ -0.00370 \\ 0.00034 \\ -0.04907 \end{bmatrix}$$

$$\Delta = \begin{bmatrix} -0.00003 \\ -0.00007 \\ 0.00008 \\ -0.083 \\ 0.009 \\ -0.289 \end{bmatrix}$$

9. Repeat step 5: Add corrections.

$$\omega = -0.4093° - 0.00003 \left(\frac{180°}{\pi} \right) = -0.4109°$$

$$\phi = 1.2144° - 0.00007 \left(\frac{180°}{\pi} \right) = 1.2101°$$

$$\kappa = 102.7959° + 0.00008 \left(\frac{180°}{\pi} \right) = 102.8003°$$

$$X_L = 1027.940 - 0.083 = 1027.857$$

$$Y_L = 1044.105 + 0.009 = 1044.114$$

$$Z_L = 648.487 - 0.289 = 648.197$$

Further iterations yield negligible corrections. Thus the final values for the six parameters of exterior orientation for the photo are those given in step 9.

As this example problem demonstrates, the computations involved in analytical photogrammetry, even in the most elementary problems, are too difficult and time-consuming for hand solution. However, the computations are readily programmed, and the solutions easily obtained, using a computer.

11-7 SPACE INTERSECTION BY COLLINEARITY

If space resection is used to determine the elements of exterior orientation for both photos of a stereopair, as described in the preceding section, then object point coordinates for points that lie in the stereo overlap area can be calculated. The procedure is known as *space intersection*, so called because corresponding rays to the same object point from the two photos must intersect at the point, as shown in Fig. 11-3. To calculate the coordinates of point A by space intersection, collinearity equations of the linearized form given by Eqs. (11-3) and (11-4) can be written for each new point, such as point A of Fig. 11-3. Note, however, that since the six elements of exterior orientation are known, the only remaining unknowns in these equations are dX_A, dY_A, and dZ_A. These are corrections to be applied to initial approximations for object space coordinates X_A, Y_A, and Z_A, respectively, for ground point A. The linearized forms of the space intersection equations for point A are

$$b_{14}\,dX_A + b_{15}\,dY_A + b_{16}\,dZ_A = J + v_{x_a} \tag{11-9}$$

$$b_{24}\,dX_A + b_{25}\,dY_A + b_{26}\,dZ_A = K + v_{y_a} \tag{11-10}$$

In Eqs. (11-9) and (11-10), the terms are as defined in Secs. 11-4 and D-4. Two equations of this form can be written for point a_1 of the left photo, and two more for point a_2 of the right photo; hence four equations result, and the three unknowns dX_A, dY_A, and dZ_A can be computed in a least squares solution. These corrections are added to the

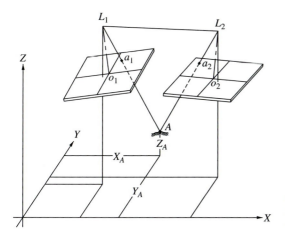

FIGURE 11-3
Space intersection with a stereopair of aerial photos.

initial approximations to obtain revised values for X_A, Y_A, and Z_A. The solution is then repeated until the magnitudes of the corrections become negligible.

Again because the equations have been linearized using Taylor's theorem, initial approximations are required for each point whose object space coordinates are to be computed. For these calculations, with normal aerial photography vertical photos can be assumed, and the initial approximations can be determined by using the parallax equations [Eqs. (8-5), (8-6), and (8-7)]. Note that because the X, Y, and Z coordinates for both exposure stations are known, for making these computations H can be taken as the average of Z_{L_1} and Z_{L_2}, and B is computed from

$$B = \sqrt{(X_{L_2} - X_{L_1})^2 + (Y_{L_2} - Y_{L_1})^2} \qquad (a)$$

The coordinates that result from Eqs. (8-6) and (8-7) are in the arbitrary system described in Sec. 8-6. (Let these coordinates be designated as x' and y'.) To convert them to the X and Y ground system, coordinate transformation Eqs. (11-8) can be used. For this transformation, the two exposure stations can serve as the control because their X and Y coordinates are known in the ground system, and their x' and y' coordinates in the parallax system are $x'_{L_1} = y'_{L_1} = y'_{L_2} = 0$, and $x'_{L_2} = B$.

11-8 ANALYTICAL STEREOMODEL

Aerial photographs for most applications are taken so that adjacent photos overlap by more than 50 percent. Two adjacent photographs that overlap in this manner form a *stereopair,* and object points that appear in the overlap area constitute a *stereomodel.* The mathematical calculation of three-dimensional ground coordinates of points in the stereomodel by analytical photogrammetric techniques forms an *analytical stereomodel.*

The process of forming an analytical stereomodel involves three primary steps: *interior orientation, relative orientation,* and *absolute orientation.* After these three steps are achieved, points in the analytical stereomodel will have object coordinates in the ground coordinate system. These points can then be used for many purposes, such as digital mapping, serving as control for orthophoto production, or DEM generation. The three orientation steps can be performed as distinct mathematical operations, or it is possible to combine them in a simultaneous solution. In this chapter, the three steps will be described as distinct, sequential operations.

11-9 ANALYTICAL INTERIOR ORIENTATION

Interior orientation for analytical photogrammetry is the step which mathematically recreates the geometry that existed in the camera when a particular photograph was exposed. This requires camera calibration information as well as quantification of the effects of atmospheric refraction. These procedures, commonly called *photo coordinate refinement,* are described in Secs. 4-11 through 4-14. The process begins with coordinates of fiducials and image points which have been measured by a comparator or related device. A two-dimensional (usually affine) coordinate transformation is used to relate the comparator coordinates to the fiducial coordinate system as well as to correct for film distortion. The lens distortion and principal-point information from camera calibration are then used to refine the coordinates so that they are correctly related to the principal point and free from lens distortion. Finally, atmospheric refraction corrections can be applied to the photo coordinates to complete the refinement and, therefore, finish the interior orientation.

11-10   ANALYTICAL RELATIVE ORIENTATION

Analytical relative orientation is the process of determining the relative angular attitude and positional displacement between the photographs that existed when the photos were taken. This involves defining certain elements of exterior orientation and calculating the remaining ones. The resulting exterior orientation parameters will not be the actual values that existed when the photographs were exposed; however, they will be correct in a "relative sense" between the photos. In analytical relative orientation, it is common practice to fix the exterior orientation elements ω, ϕ, κ, X_L, and Y_L of the left photo of the stereopair to zero values. Also for convenience, Z_L of the left photo (Z_{L_1}) is set equal to f, and X_L of the right photo (X_{L_2}) is set equal to the photo base b. (With these choices for Z_{L_1} and X_{L_2}, initial approximations for the unknowns are more easily calculated, as will be explained later.) This leaves five elements of the right photo that must be determined. Figure 11-4 illustrates a stereomodel formed by analytical relative orientation.

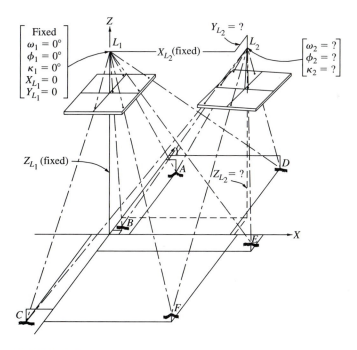

FIGURE 11-4
Analytical relative orientation of a stereopair.

Although the coplanarity condition equation can be used for analytical relative orientation, the collinearity condition is more commonly applied. In applying collinearity, each object point in the stereomodel contributes four equations: a pair of equations for the x and y photo coordinates of its image in the left photo, and a pair of equations for the x and y photo coordinates of its image in the right photo. In addition to the five unknown orientation elements, each object point adds three more unknowns which are their X, Y, and Z coordinates in the stereomodel. Thus each point used in relative orientation results in a net gain of one equation for the overall solution, and therefore at least 5 object points are required for a solution. If 6 or more points are available, an improved solution is possible through least squares. If 6 points were used for relative orientation, a system of 24 equations with 23 unknowns would result; and if 12 points were used, the system would consist of 48 equations and 41 unknowns.

$A =$

$d\omega_2$	$d\phi_2$	$d\kappa_2$	dY_{L_2}	dZ_{L_2}	dX_A	dY_A	dZ_A	dX_B	dY_B	dZ_B
0	0	0	0	0	$(b_{a_{14}})_1$	$(b_{a_{15}})_1$	$(b_{a_{16}})_1$	0	0	0
0	0	0	0	0	$(b_{a_{24}})_1$	$(b_{a_{25}})_1$	$(b_{a_{26}})_1$	0	0	0
0	0	0	0	0	0	0	0	$(b_{b_{14}})_1$	$(b_{b_{15}})_1$	$(b_{b_{16}})_1$
0	0	0	0	0	0	0	0	$(b_{b_{24}})_1$	$(b_{b_{25}})_1$	$(b_{b_{26}})_1$
0	0	0	0	0	0	0	0	0	0	0
0	0	0	0	0	0	0	0	0	0	0
0	0	0	0	0	0	0	0	0	0	0
0	0	0	0	0	0	0	0	0	0	0
0	0	0	0	0	0	0	0	0	0	0
0	0	0	0	0	0	0	0	0	0	0
0	0	0	0	0	0	0	0	0	0	0
0	0	0	0	0	0	0	0	0	0	0
0	0	0	0	0	0	0	0	0	0	0
$(b_{a_{11}})_2$	$(b_{a_{12}})_2$	$(b_{a_{13}})_2$	$(-b_{a_{15}})_2$	$(-b_{a_{16}})_2$	$(b_{a_{14}})_2$	$(b_{a_{15}})_2$	$(b_{a_{16}})_2$	0	0	0
$(b_{a_{21}})_2$	$(b_{a_{22}})_2$	$(b_{a_{23}})_2$	$(-b_{a_{25}})_2$	$(-b_{a_{26}})_2$	$(b_{a_{24}})_2$	$(b_{a_{25}})_2$	$(b_{a_{26}})_2$	0	0	0
$(b_{b_{11}})_2$	$(b_{b_{12}})_2$	$(b_{b_{13}})_2$	$(-b_{b_{15}})_2$	$(-b_{b_{16}})_2$	0	0	0	$(b_{b_{14}})_2$	$(b_{b_{15}})_2$	$(b_{b_{16}})_2$
$(b_{b_{21}})_2$	$(b_{b_{22}})_2$	$(b_{b_{23}})_2$	$(-b_{b_{25}})_2$	$(-b_{b_{26}})_2$	0	0	0	$(b_{b_{24}})_2$	$(b_{b_{25}})_2$	$(b_{b_{26}})_2$
$(b_{c_{11}})_2$	$(b_{c_{12}})_2$	$(b_{c_{13}})_2$	$(-b_{c_{15}})_2$	$(-b_{c_{16}})_2$	0	0	0	0	0	0
$(b_{c_{21}})_2$	$(b_{c_{22}})_2$	$(b_{c_{23}})_2$	$(-b_{c_{25}})_2$	$(-b_{c_{26}})_2$	0	0	0	0	0	0
$(b_{d_{11}})_2$	$(b_{d_{12}})_2$	$(b_{d_{13}})_2$	$(-b_{d_{15}})_2$	$(-b_{d_{16}})_2$	0	0	0	0	0	0
$(b_{d_{21}})_2$	$(b_{d_{22}})_2$	$(b_{d_{23}})_2$	$(-b_{d_{25}})_2$	$(-b_{d_{26}})_2$	0	0	0	0	0	0
$(b_{e_{11}})_2$	$(b_{e_{12}})_2$	$(b_{e_{13}})_2$	$(-b_{e_{15}})_2$	$(-b_{e_{16}})_2$	0	0	0	0	0	0
$(b_{e_{21}})_2$	$(b_{e_{22}})_2$	$(b_{e_{23}})_2$	$(-b_{e_{25}})_2$	$(-b_{e_{26}})_2$	0	0	0	0	0	0
$(b_{f_{11}})_2$	$(b_{f_{12}})_2$	$(b_{f_{13}})_2$	$(-b_{f_{15}})_2$	$(-b_{f_{16}})_2$	0	0	0	0	0	0
$(b_{f_{21}})_2$	$(b_{f_{22}})_2$	$(b_{f_{23}})_2$	$(-b_{f_{25}})_2$	$(-b_{f_{26}})_2$	0	0	0	0	0	0

Prior to solving the linearized collinearity equations, initial approximations for all unknown values must be determined. For photography that was intended to be vertical, values of zero are commonly used for initial estimates of ω_2, ϕ_2, κ_2, and Y_{L_2}. An initial value for Z_{L_2} may be selected equal to the value used for Z_{L_1}. If the constraints that were noted earlier are used for the parameters, that is, $\omega_1 = \phi_1 = \kappa_1 = X_{L_1} = Y_{L_1} = 0$, $Z_{L_1} = f$, and $X_{L_2} = b$, then the scale of the stereomodel is approximately equal to photo scale. Thus the x and y photo coordinates of the left photo are good approximations for X and Y object space coordinates, and zeros are good approximations for Z object space coordinates, respectively.

Suppose that the six points of Fig. 11-4 were used in analytical relative orientation. In matrix form, the system of 24 equations involving 23 unknowns could be expressed as follows:

$$_{24}A^{23}{}_{23}X^1 = {}_{24}L^1 + {}_{24}V^1 \tag{11-11}$$

where

dX_C	dY_C	dZ_C	dX_D	dY_D	dZ_D	dX_E	dY_E	dZ_E	dX_F	dY_F	dZ_F
0	0	0	0	0	0	0	0	0	0	0	0
0	0	0	0	0	0	0	0	0	0	0	0
0	0	0	0	0	0	0	0	0	0	0	0
0	0	0	0	0	0	0	0	0	0	0	0
$(b_{c_{14}})_1$	$(b_{c_{15}})_1$	$(b_{c_{16}})_1$	0	0	0	0	0	0	0	0	0
$(b_{c_{24}})_1$	$(b_{c_{25}})_1$	$(b_{c_{26}})_1$	0	0	0	0	0	0	0	0	0
0	0	0	$(b_{d_{14}})_1$	$(b_{d_{15}})_1$	$(b_{d_{16}})_1$	0	0	0	0	0	0
0	0	0	$(b_{d_{24}})_1$	$(b_{d_{25}})_1$	$(b_{d_{26}})_1$	0	0	0	0	0	0
0	0	0	0	0	0	$(b_{e_{14}})_1$	$(b_{e_{15}})_1$	$(b_{e_{16}})_1$	0	0	0
0	0	0	0	0	0	$(b_{e_{24}})_1$	$(b_{e_{25}})_1$	$(b_{e_{26}})_1$	0	0	0
0	0	0	0	0	0	0	0	0	$(b_{f_{14}})_1$	$(b_{f_{15}})_1$	$(b_{f_{16}})_1$
0	0	0	0	0	0	0	0	0	$(b_{f_{24}})_1$	$(b_{f_{25}})_1$	$(b_{f_{26}})_1$
0	0	0	0	0	0	0	0	0	0	0	0
0	0	0	0	0	0	0	0	0	0	0	0
0	0	0	0	0	0	0	0	0	0	0	0
0	0	0	0	0	0	0	0	0	0	0	0
$(b_{c_{14}})_2$	$(b_{c_{15}})_2$	$(b_{c_{16}})_2$	0	0	0	0	0	0	0	0	0
$(b_{c_{24}})_2$	$(b_{c_{25}})_2$	$(b_{c_{26}})_2$	0	0	0	0	0	0	0	0	0
0	0	0	$(b_{d_{14}})_2$	$(b_{d_{15}})_2$	$(b_{d_{16}})_2$	0	0	0	0	0	0
0	0	0	$(b_{d_{24}})_2$	$(b_{d_{25}})_2$	$(b_{d_{26}})_2$	0	0	0	0	0	0
0	0	0	0	0	0	$(b_{e_{14}})_2$	$(b_{e_{15}})_2$	$(b_{e_{16}})_2$	0	0	0
0	0	0	0	0	0	$(b_{e_{24}})_2$	$(b_{e_{25}})_2$	$(b_{e_{26}})_2$	0	0	0
0	0	0	0	0	0	0	0	0	$(b_{f_{14}})_2$	$(b_{f_{15}})_2$	$(b_{f_{16}})_2$
0	0	0	0	0	0	0	0	0	$(b_{f_{24}})_2$	$(b_{f_{25}})_2$	$(b_{f_{26}})_2$

$$
X = \begin{bmatrix} d\omega_2 \\ d\phi_2 \\ d\kappa_2 \\ dY_{L_2} \\ dZ_{L_2} \\ dX_A \\ dY_A \\ dZ_A \\ dX_B \\ dY_B \\ dZ_B \\ dX_C \\ dY_C \\ dZ_C \\ dX_D \\ dY_D \\ dZ_D \\ dX_E \\ dY_E \\ dZ_E \\ dX_F \\ dY_F \\ dZ_F \end{bmatrix}
\qquad
L = \begin{bmatrix} (J_a)_1 \\ (K_a)_1 \\ (J_b)_1 \\ (K_b)_1 \\ (J_c)_1 \\ (K_c)_1 \\ (J_d)_1 \\ (K_d)_1 \\ (J_e)_1 \\ (K_e)_1 \\ (J_f)_1 \\ (K_f)_1 \\ (J_a)_2 \\ (K_a)_2 \\ (J_b)_2 \\ (K_b)_2 \\ (J_c)_2 \\ (K_c)_2 \\ (J_d)_2 \\ (K_d)_2 \\ (J_e)_2 \\ (K_e)_2 \\ (J_f)_2 \\ (K_f)_2 \end{bmatrix}
\qquad
V = \begin{bmatrix} (v_{x_a})_1 \\ (v_{y_a})_1 \\ (v_{x_b})_1 \\ (v_{y_b})_1 \\ (v_{x_c})_1 \\ (v_{y_c})_1 \\ (v_{x_d})_1 \\ (v_{y_d})_1 \\ (v_{x_e})_1 \\ (v_{y_e})_1 \\ (v_{x_f})_1 \\ (v_{y_f})_1 \\ (v_{x_a})_2 \\ (v_{y_a})_2 \\ (v_{x_b})_2 \\ (v_{y_b})_2 \\ (v_{x_c})_2 \\ (v_{y_c})_2 \\ (v_{x_d})_2 \\ (v_{y_d})_2 \\ (v_{x_e})_2 \\ (v_{y_e})_2 \\ (v_{x_f})_2 \\ (v_{y_f})_2 \end{bmatrix}
$$

Note that Eq. (11-11) conforms to Eq. (B-10).

The terms of these matrices are from Eqs. (11-3) and (11-4), and the method of calculating each is explained in Sec. D-4. The subscripts a, b, c, d, e, and f correspond to the point names; subscript 1 refers to the left photo; and subscript 2 refers to the right photo. Note that the elements of the X matrix are listed at the tops of the columns of the A matrix for illustration purposes only.

Upon studying these matrices, particularly the A matrix, their systematic nature becomes apparent. The fact that many submatrices of zeros exist also indicates the relative ease with which partitioning could be applied in the solution. The following equations show the form of the A, X, L, and V matrices when partitioned by the standard approach.

$$A = \begin{bmatrix}
{}_2 0^5 & {}_2\ddot{B}_{1a}^3 & {}_2 0^3 & {}_2 0^3 & {}_2 0^3 & {}_2 0^3 & {}_2 0^3 \\
{}_2 0^5 & {}_2 0^3 & {}_2\ddot{B}_{1b}^3 & {}_2 0^3 & {}_2 0^3 & {}_2 0^3 & {}_2 0^3 \\
{}_2 0^5 & {}_2 0^3 & {}_2 0^3 & {}_2\ddot{B}_{1c}^3 & {}_2 0^3 & {}_2 0^3 & {}_2 0^3 \\
{}_2 0^5 & {}_2 0^3 & {}_2 0^3 & {}_2 0^3 & {}_2\ddot{B}_{1d}^3 & {}_2 0^3 & {}_2 0^3 \\
{}_2 0^5 & {}_2 0^3 & {}_2 0^3 & {}_2 0^3 & {}_2 0^3 & {}_2\ddot{B}_{1e}^3 & {}_2 0^3 \\
{}_2 0^5 & {}_2 0^3 & {}_2 0^3 & {}_2 0^3 & {}_2 0^3 & {}_2 0^3 & {}_2\ddot{B}_{1f}^3 \\
{}_2\dot{B}_{2a}^5 & {}_2\ddot{B}_{2a}^3 & {}_2 0^3 & {}_2 0^3 & {}_2 0^3 & {}_2 0^3 & {}_2 0^3 \\
{}_2\dot{B}_{2b}^5 & {}_2 0^3 & {}_2\ddot{B}_{2b}^3 & {}_2 0^3 & {}_2 0^3 & {}_2 0^3 & {}_2 0^3 \\
{}_2\dot{B}_{2c}^5 & {}_2 0^3 & {}_2 0^3 & {}_2\ddot{B}_{2c}^3 & {}_2 0^3 & {}_2 0^3 & {}_2 0^3 \\
{}_2\dot{B}_{2d}^5 & {}_2 0^3 & {}_2 0^3 & {}_2 0^3 & {}_2\ddot{B}_{2d}^3 & {}_2 0^3 & {}_2 0^3 \\
{}_2\dot{B}_{2e}^5 & {}_2 0^3 & {}_2 0^3 & {}_2 0^3 & {}_2 0^3 & {}_2\ddot{B}_{2e}^3 & {}_2 0^3 \\
{}_2\dot{B}_{2f}^5 & {}_2 0^3 & {}_2 0^3 & {}_2 0^3 & {}_2 0^3 & {}_2 0^3 & {}_2\ddot{B}_{2f}^3
\end{bmatrix}
\qquad
X = \begin{bmatrix}
\dot{\Delta}_2 \\ \ddot{\Delta}_A \\ \ddot{\Delta}_B \\ \ddot{\Delta}_C \\ \ddot{\Delta}_D \\ \ddot{\Delta}_E \\ \ddot{\Delta}_F
\end{bmatrix}
\qquad
L = \begin{bmatrix}
\varepsilon_{1a} \\ \varepsilon_{1b} \\ \varepsilon_{1c} \\ \varepsilon_{1d} \\ \varepsilon_{1e} \\ \varepsilon_{1f} \\ \varepsilon_{2a} \\ \varepsilon_{2b} \\ \varepsilon_{2c} \\ \varepsilon_{2d} \\ \varepsilon_{2e} \\ \varepsilon_{2f}
\end{bmatrix}
\qquad
V = \begin{bmatrix}
v_{1a} \\ v_{1b} \\ v_{1c} \\ v_{1d} \\ v_{1e} \\ v_{1f} \\ v_{2a} \\ v_{2b} \\ v_{2c} \\ v_{2d} \\ v_{2e} \\ v_{2f}
\end{bmatrix}$$

where

$$ {}_2 0^5 = \begin{bmatrix} 0 & 0 & 0 & 0 & 0 \\ 0 & 0 & 0 & 0 & 0 \end{bmatrix} \qquad {}_2 0^3 = \begin{bmatrix} 0 & 0 & 0 \\ 0 & 0 & 0 \end{bmatrix} $$

$$ {}_2\dot{B}_{2p}^5 = \begin{bmatrix} (b_{p11})_2 & (b_{p12})_2 & (b_{p13})_2 & (-b_{p15})_2 & (-b_{p16})_2 \\ (b_{p21})_2 & (b_{p22})_2 & (b_{p23})_2 & (-b_{p25})_2 & (-b_{p26})_2 \end{bmatrix} \qquad {}_2\ddot{B}_{ip}^3 = \begin{bmatrix} (b_{p14})_i & (b_{p15})_i & (b_{p16})_i \\ (b_{p24})_i & (b_{p25})_i & (b_{p26})_i \end{bmatrix} $$

$$ \dot{\Delta}_2 = \begin{bmatrix} d\omega_2 \\ d\phi_2 \\ d\kappa_2 \\ dY_{L_2} \\ dZ_{L_2} \end{bmatrix} \qquad \ddot{\Delta}_P = \begin{bmatrix} dX_P \\ dY_P \\ dZ_P \end{bmatrix} \qquad \varepsilon_{ip} = \begin{bmatrix} (J_p)_i \\ (K_p)_i \end{bmatrix} \qquad v_{ip} = \begin{bmatrix} (v_{x_p})_i \\ (v_{y_p})_i \end{bmatrix} $$

In the above submatrices, p is the point designation, and i is the photo designation. The prefixed subscript and postfixed superscript designate the number of rows and columns, respectively.

When the observation equations are partitioned in the above-described manner, the least squares solution $(A^T A)X = (A^T L)$ takes the following form.

$$\begin{bmatrix}
\dot{N}_2 & \overline{N}_a & \overline{N}_b & \overline{N}_c & \overline{N}_d & \overline{N}_e & \overline{N}_f \\
\overline{N}_a^T & \ddot{N}_a & {}_3 0^3 & {}_3 0^3 & {}_3 0^3 & {}_3 0^3 & {}_3 0^3 \\
\overline{N}_b^T & {}_3 0^3 & \ddot{N}_b & {}_3 0^3 & {}_3 0^3 & {}_3 0^3 & {}_3 0^3 \\
\overline{N}_c^T & {}_3 0^3 & {}_3 0^3 & \ddot{N}_c & {}_3 0^3 & {}_3 0^3 & {}_3 0^3 \\
\overline{N}_d^T & {}_3 0^3 & {}_3 0^3 & {}_3 0^3 & \ddot{N}_d & {}_3 0^3 & {}_3 0^3 \\
\overline{N}_e^T & {}_3 0^3 & {}_3 0^3 & {}_3 0^3 & {}_3 0^3 & \ddot{N}_e & {}_3 0^3 \\
\overline{N}_f^T & {}_3 0^3 & {}_3 0^3 & {}_3 0^3 & {}_3 0^3 & {}_3 0^3 & \ddot{N}_f
\end{bmatrix}
\begin{bmatrix}
\dot{\Delta}_2 \\ \ddot{\Delta}_A \\ \ddot{\Delta}_B \\ \ddot{\Delta}_C \\ \ddot{\Delta}_D \\ \ddot{\Delta}_E \\ \ddot{\Delta}_F
\end{bmatrix}
=
\begin{bmatrix}
\dot{K}_2 \\ \ddot{K}_A \\ \ddot{K}_B \\ \ddot{K}_C \\ \ddot{K}_D \\ \ddot{K}_E \\ \ddot{K}_F
\end{bmatrix}$$

where

$$\dot{N}_2 = \sum_{p=a}^{f} (\dot{B}_{2p}^{T} \dot{B}_{2p}) \qquad (5 \times 5 \text{ submatrix})$$

$$\overline{N}_p = \dot{B}_{2p}^{T} \ddot{B}_{2p} \qquad (5 \times 3 \text{ submatrix})$$

$$\ddot{N}_p = (\ddot{B}_{1p}^{T} \ddot{B}_{1p}) + (\ddot{B}_{2p}^{T} \ddot{B}_{2p}) \qquad (3 \times 3 \text{ submatrix})$$

$$\dot{K}_2 = \sum_{p=a}^{f} (\dot{B}_{2p}^{T} \varepsilon_{2p}) \qquad (5 \times 1 \text{ submatrix})$$

$$\ddot{K}_p = (\ddot{B}_{1p}^{T} \varepsilon_{1p}) + (\ddot{B}_{2p}^{T} \varepsilon_{2p}) \qquad (3 \times 1 \text{ submatrix})$$

$$_3 0^3 = \begin{bmatrix} 0 & 0 & 0 \\ 0 & 0 & 0 \\ 0 & 0 & 0 \end{bmatrix}$$

Special methods can be used to store and solve the partitioned normal equations; however, this subject is deferred until Chap. 17.

Example 11-2. A stereopair of near-vertical photographs is taken with a 152.113-mm-focal-length camera. Photo coordinates of the images of six points in the overlap area are listed in the following table. Perform analytical relative orientation of the stereopair, using the computer program provided.[1]

Point	Left photo coordinates		Right photo coordinates	
	x, mm	y, mm	x, mm	y, mm
a	−4.878	1.981	−97.920	−2.923
b	89.307	2.709	−1.507	−1.856
c	0.261	84.144	−90.917	78.970
d	90.334	83.843	−1.571	79.470
e	−4.668	−86.821	−100.060	−95.748
f	88.599	−85.274	−0.965	−94.319

Solution

1. With an ASCII text editor, create the following data file with a ".dat" extension:

```
152.113
a  -4.870    1.992   -97.920   -2.910
b  89.296    2.706    -1.485   -1.836
c   0.256   84.138   -90.906   78.980
d  90.328   83.854    -1.568   79.482
e  -4.673  -86.815  -100.064  -95.733
f  88.591  -85.269    -0.973  -94.312
```

---

[1] The program used in this example is titled "relor" and can be downloaded at the following Web site: http://www.surv.ufl.edu/wolfdewitt

The first line of data in the input file is the camera focal length. The information on each of the following lines, from left to right, consists of the point identification, its x and y photo coordinates on the left photo, and its x and y photo coordinates on the right photo.

2. Run the relor program to produce the following results:

```
Exterior orientation parameters:
Parameter  Left pho  Right pho  SD right
Omega(deg)   0.0000     2.4099   0.0171
Phi(deg)     0.0000     0.5516   0.0181
Kappa(deg)   0.0000    -0.2067   0.0084
XL           0.0000    91.9740
YL           0.0000    -1.7346   0.0545
ZL         152.1130   148.3015   0.0196


Object space coordinates:
point    X         Y         Z       SD-X    SD-Y    SD-Z
   a  -4.8352    1.9730    1.0888   0.0127  0.0107  0.0975
   b  89.0970    2.7047    0.3391   0.0464  0.0109  0.0813
   c   0.2542   83.5234    1.1159   0.0117  0.0522  0.1001
   d  89.2672   82.8667    1.7862   0.0469  0.0488  0.0809
   e  -4.6333  -86.0755    1.2917   0.0126  0.0555  0.1032
   f  89.3101  -85.9635   -1.2348   0.0491  0.0528  0.0866


Photo coordinate residuals:
point   xl-res    yl-res    xr-res    yr-res
   a  -0.0001   -0.0048    0.0001    0.0047
   b   0.0001    0.0048   -0.0001   -0.0047
   c   0.0001    0.0026   -0.0001   -0.0027
   d  -0.0001   -0.0026    0.0001    0.0027
   e  -0.0000    0.0023    0.0000   -0.0022
   f   0.0000   -0.0023   -0.0000    0.0022
 RMS   0.0001    0.0034    0.0001    0.0034

Standard error of unit weight: 0.0118
Degrees of freedom: 1
```

In the upper table of output, the units of omega, phi, and kappa are degrees, while X_L, Y_L, and Z_L are in millimeters. The rightmost column lists standard deviations in the computed unknowns, also in degrees and millimeters. The middle table lists stereomodel coordinates, and their computed standard deviations, in millimeters. The lower table lists x and y photo coordinates and their residuals for both photos. Notice that x photo coordinate residuals are nearly equal to zero for both the left and right photos. This is expected for a stereo pair displaced in the x direction because x parallax determines the Z coordinate in object space. The y photo coordinate residuals indicate the presence of y parallax, and their sizes are good indicators of the quality of the solution. Generally y coordinate residuals of under about 5 or 6 micrometers (μm) indicate a rather good solution.

11-11   ANALYTICAL ABSOLUTE ORIENTATION

For a small stereomodel such as that computed from one stereopair, analytical absolute orientation can be performed using a three-dimensional conformal coordinate transformation (see Sec. C-7). This requires a minimum of two horizontal and three vertical control points, but additional control points provide redundancy, which enables a least squares solution. In the process of performing absolute orientation, stereomodel coordinates of control points are related to their three-dimensional coordinates in a ground-based system. It is important for the ground system to be a true cartesian coordinate system, such as local vertical, since the three-dimensional conformal coordinate transformation is based on straight, orthogonal axes.

Once the transformation parameters have been computed, they can be applied to the remaining stereomodel points, including the X_L, Y_L, and Z_L coordinates of the left and right photographs. This gives the coordinates of all stereomodel points in the ground system.

Example 11-3. Ground coordinates in a local vertical system for three control points are listed in the table below. For the results of the analytical relative orientation of Example 11-2, perform analytical absolute orientation using a three-dimensional conformal coordinate transformation with the program provided.[2]

Point	X, m	Y, m	Z, m
C	9,278.062	10,482.868	59.741
E	9,269.903	9,922.635	69.799
F	9,580.264	9,927.325	66.109

Solution:

1.   With an ASCII text editor, create the following data file with a ".dat" extension:

```
C    0.2590    83.5132    1.1396   9278.062   10482.868   59.741
E   -4.6278   -86.0730    1.3111   9269.903    9922.635   69.799
F   89.3169   -85.9646   -1.2325   9580.264    9927.325   66.109
#
A   -4.8432    1.9676    1.0838
B   89.0758    2.7013    0.3937
D   89.2635   82.8498    1.8026
Lpho  0   0   152.113
Rpho 91.9825 -1.7057 148.3022
#
```

For the above input file, the first three lines relate to the control points. From left to right the data include the point identification; its *x*, *y*, and *z* stereomodel

---

[2]The program used in the example is titled "3dconf" and can be downloaded at the following Web site: http://www.surv.ufl.edu/wolfdewitt

coordinates; and its X, Y, and Z ground coordinates. The first # sign signifies that all control has been entered, and that the data following pertain to stereomodel points whose coordinates are to be transformed into the ground system. Each data line consists of the point identification, followed by its x, y, and z stereomodel coordinates. The second # sign completes the data.

2. Run the "3dconf" program to produce the following results:

```
Residuals:
Point   X res    Y res   Z res
   C    0.007    0.000 -0.000
   E   -0.006   -0.012   0.000
   F   -0.001    0.012 -0.000

Standard Error of Unit Weight: 0.01353

Final Results:
Param      Value     Stan.Err.
scale       3.30311   0.00009
omega      -0.9835d   0.0019d
  phi      -0.8849d   0.0035d
kappa       0.8159d   0.0015d
   Tx    9281.199     0.009
   Ty   10207.007     0.009
   Tz      60.760     0.009

Transformed Points:
Point    X          Y          Z      SDev.X SDev.Y SDev.Z
   A   9265.057  10213.334   63.983   0.009  0.009  0.010
   B   9575.214  10220.217   66.376   0.010  0.010  0.016
   D   9571.993  10484.979   66.436   0.013  0.013  0.023
Lpho   9273.439  10215.631  563.072   0.032  0.019  0.016
Rpho   9577.475  10214.188  555.203   0.032  0.019  0.021
```

The values listed in the top table of the output are residuals in the X, Y, and Z control point coordinates. The center table lists the seven parameters of the three-dimensional conformal coordinate transformation, and the lower table gives the transformed X, Y, and Z ground coordinates of noncontrol points, together with their computed standard deviations.

11-12   ANALYTICAL SELF-CALIBRATION

Analytical self-calibration is a computational process wherein camera calibration parameters are included in the photogrammetric solution, generally in a combined interior-relative-absolute orientation. The process uses collinearity equations that have been augmented with additional terms to account for adjustment of the calibrated focal length, principal-point offsets, and symmetric radial and decentering lens distortion. In addition, the equations might include corrections for atmospheric refraction as presented in Sec. 10-14. The common form of the augmented collinearity equations is given as

$$x_a = x_o - \bar{x}_a(k_1 r_a^2 + k_2 r_a^4 + k_3 r_a^6) - (1 + p_3^2 r_a^2)\left[p_1(3\bar{x}_a^2 + \bar{y}_a^2) + 2p_2\bar{x}_a\bar{y}_a\right] - f\frac{r}{q}$$

(11-12)

$$y_a = y_o - \bar{y}_a(k_1 r_a^2 + k_2 r_a^4 + k_3 r_a^6) - (1 + p_3^2 r_a^2)\left[2p_1\bar{x}_a\bar{y}_a + p_2(\bar{x}_a^2 + 3\bar{y}_a^2)\right] - f\frac{s}{q}$$

where x_a, y_a = measured photo coordinates related to fiducials

x_o, y_o = coordinates of the principal point

$\bar{x}_a = x_a - x_o$

$\bar{y}_a = y_a - y_o$

$r_a^2 = \bar{x}_a^2 + \bar{y}_a^2$

k_1, k_2, k_3 = symmetric radial lens distortion coefficients

p_1, p_2, p_3 = decentering distortion coefficients

f = calibrated focal length

r, s, q = collinearity equation terms as defined for Eqs. (D-7) and (D-8)

The interior orientation parameters x_o, y_o, f, k_1, k_2, k_3, p_1, p_2, and p_3 which appear in Eqs. (11-12) are included as unknowns in the solution, together with ω, ϕ, κ, X_L, Y_L, and Z_L for each photo and X_A, Y_A, and Z_A for each object point. These equations are of course nonlinear, and therefore Taylor's series is used to linearize them, and an iterative solution is made.

With the inclusion of the extra unknowns, it follows that additional independent equations will be needed to obtain a solution. Also, the numerical stability of analytical self-calibration is of serious concern. Merely including the additional parameters does not guarantee their solution. It is necessary to have special constraints and/or geometric configurations to ensure their solution. For example, with nominally vertical photography if the object points are at roughly the same elevation, then x_o, y_o, and f are strongly correlated with X_L, Y_L, and Z_L, respectively. Given this correlation, the solution may not produce satisfactory results. This problem can be overcome or at least alleviated if there is significant elevation variation in the object field, or by using highly convergent (nonvertical) photography, by making observations of the camera position (e.g., by airborne GPS), or by using redundant photographic coverage at different κ angles. In addition, to recover the lens distortion parameters, it is necessary to have many redundant object points whose images are well distributed across the format of the photograph.

REFERENCES

American Society of Photogrammetry: *Manual of Photogrammetry,* 4th ed., Falls Church, VA, 1980, chaps. 2 and 9.

Anderson, J. M., and C. Lee: "Analytical In-Flight Calibration," *Photogrammetric Engineering and Remote Sensing,* vol. 41, no. 11, 1975, p. 1337.

Brown, D. C.: "Close-Range Camera Calibration," *Photogrammetric Engineering,* vol. 37, no. 8, 1971, p. 855.

Eden, H. A.: "Point Transfer from One Photograph to Another," *Photogrammetric Record,* vol. 7, no. 41, 1973, p. 531.

Kenefick, J. F., M. S. Gyer, and B. F. Harp: "Analytical Self-Calibration," *Photogrammetric Engineering,* vol. 38, no. 11, 1972, p. 1117.

Leupin, M. M.: "Analytical Photogrammetry;" an Alternative to Terrestrial Point Determination," *Australian Surveyor,* vol. 28, no. 2, 1976, p. 73.

Maarek, A.: "Practical Numerical Photogrammetry," *Photogrammetric Engineering and Remote Sensing,* vol. 43, no. 10, 1977, p. 1295.

Smith, G. L.: "Analytical Photogrammetry Applied to Survey Point Coordination," *Australian Surveyor,* vol. 28, no. 5, 1977, p. 263.

Thompson, L. G.: "Determination of the Point Transfer Error," *Photogrammetric Engineering and Remote Sensing,* vol. 45, no. 4, 1979, p. 535.

Trinder, J. C.: "Some Remarks on Numerical Absolute Orientation," *Australian Surveyor,* vol. 23, no. 6, 1971, p. 368.

PROBLEMS

11-1. Describe two different conditions that are commonly enforced in analytical photogrammetry.

11-2. If 12 pass points are used in the analytical relative orientation of a stereopair, how many independent collinearity equations can be written?

11-3. A near-vertical aerial photograph taken with a 151.876-mm-focal-length camera contains images of five ground control points A through E. Refined photo coordinates and ground control coordinates in a local vertical system of the five points are listed in the following table. Calculate the exterior orientation parameters ω, ϕ, κ, X_L, Y_L, and Z_L for this photograph by space resection.

Point	Photo coordinates		Ground control coordinates		
	x, mm	*y*, mm	*X*, m	*Y*, m	*Z*, m
A	−53.845	65.230	6934.954	23,961.105	160.136
B	104.500	68.324	7860.202	23,941.563	152.653
C	4.701	−12.153	7261.078	23,491.497	142.208
D	−61.372	−79.559	6836.650	23,087.475	137.719
E	93.825	−62.060	7791.556	23,166.680	138.827

11-4. Repeat Prob. 11-3, except that the camera focal length is 152.401 mm, and the coordinates are as listed in the following table.

Point	Photo coordinates		Ground control coordinates		
	x, mm	*y*, mm	*X*, m	*Y*, m	*Z*, m
A	6.720	5.309	3261.671	4172.201	28.628
B	−75.094	58.105	3176.840	4220.876	26.471
C	99.467	77.810	3349.207	4248.519	25.899
D	−44.627	−69.133	3215.300	4096.584	30.008
E	51.001	−76.411	3309.646	4093.980	27.235

11-5. Orientation of a stereopair of aerial photographs taken with a 152.819-mm-focal-length camera resulted in the exterior orientation values listed in the following table. If refined photo coordinates for a particular point on photo 1 are $x_1 = 68.671$ mm and $y_1 = 17.768$ mm and those for the same point on photo 2 are $x_2 = -30.699$ mm and $y_2 = 14.656$ mm, compute the object space coordinates for the point by space intersection.

Parameter	Photo 1	Photo 2
ω	0.0000°	1.9153°
ϕ	0.0000°	−2.1785°
κ	0.0000°	−1.7248°
X_L	0.000 mm	99.070 mm
Y_L	0.000 mm	−3.265 mm
Z_L	152.819 mm	154.325 mm

11-6. Repeat Prob. 11-5 except that the camera focal length is 152.905 mm, the refined photo coordinates from the left photo are $x_1 = 33.099$ mm and $y_1 = -17.340$ mm, those from the right are $x_2 = -63.256$ mm and $y_2 = -25.911$ mm, and the orientation parameters are listed in the following table.

Parameter	Photo 1	Photo 2
ω	0.0000°	2.0875°
ϕ	0.0000°	−0.8180°
κ	0.0000°	−1.4795°
X_L	0.000 mm	93.285 mm
Y_L	0.000 mm	1.013 mm
Z_L	152.905 mm	152.167 mm

11-7. A stereopair of near-vertical photographs is taken with a 151.992-mm-focal-length camera. Photo coordinates of the images of six points in the overlap area are listed in the following table. Perform analytical relative orientation on the stereopair, using the computer program provided (see Ex. 11-2).

Point	Left photo coordinates		Right photo coordinates	
	x, mm	y, mm	x, mm	y, mm
A	−4.617	0.392	−84.078	−1.637
B	87.296	−0.309	7.322	−4.153
C	1.470	98.289	−75.107	94.664
D	71.917	73.563	−6.402	69.085
E	−2.274	−91.876	−84.680	−94.334
F	83.690	−100.003	1.120	−104.202

11-8. Repeat Prob. 11-7 except that the camera focal length is 153.219 mm, and photo coordinates of the images of 12 points in the overlap area are listed in the following table.

	Left photo coordinates		Right photo coordinates	
Point	**x, mm**	**y, mm**	**x, mm**	**y, mm**
A	0.979	2.194	−96.202	−5.590
B	4.295	6.768	−91.526	−1.073
C	91.252	5.175	−4.689	−5.971
D	87.353	0.153	−9.929	−10.908
E	−3.438	87.694	−97.285	79.217
F	−2.614	83.054	−95.869	74.701
G	89.914	84.645	−3.553	72.195
H	90.897	84.068	−0.899	71.556
I	2.181	−74.516	−100.467	−86.171
J	−0.790	−79.254	−103.638	−91.205
K	85.797	−75.263	−10.829	−89.604
L	86.453	−73.405	−12.521	−87.656

11-9. Ground coordinates in a local vertical system for three control points A, D, and F are listed in the table below. Using the results of the analytical relative orientation of Prob. 11-7, perform analytical absolute orientation with the program provided (see Ex. 11-3).

Point	**X, m**	**Y, m**	**Z, m**
A	4296.851	4233.813	108.354
D	4553.579	4473.130	123.791
F	4583.104	3900.089	112.791

11-10. Ground coordinates in a local vertical system for four control points E, H, J, and K are listed in the table below. Using the results of the analytical relative orientation of Prob. 11-8, perform analytical absolute orientation with the program provided (see Ex. 11-3).

Point	**X, m**	**Y, m**	**Z, m**
E	3723.452	6736.732	85.709
H	4010.237	6734.329	72.484
J	3731.260	6237.335	82.507
K	3999.126	6245.744	77.159

11-11. List the parameters of interior orientation that are included in the augmented collinearity equations used in analytical self-calibration.

11-12. Discuss the problems of numerical stability associated with analytical self-calibration.

11-13. Derive the partial derivatives of the interior orientation parameters of the augmented collinearity equations used in analytical self-calibration that are required for the Taylor's series linearization.

CHAPTER
12

STEREOSCOPIC
PLOTTING
INSTRUMENTS

12-1 INTRODUCTION

Stereoscopic plotting instruments (commonly called *stereoplotters* or simply "plotters") are instruments designed to provide rigorously accurate solutions for object point positions from their corresponding image positions on overlapping pairs of photos. A stereoplotter is essentially a three-dimensional digitizer, capable of producing accurate X, Y, and Z object space coordinates when properly oriented and calibrated. The fact that the photos may contain varying amounts of tilt is of no consequence in the resulting accuracy; in fact, many stereoplotters are capable of handling oblique or horizontal (terrestrial) photos. The primary uses of stereoplotters are compiling topographic maps and generating digital files of topographic information, and because these are the most widely practiced photogrammetric applications, the subject of stereoplotters is one of the most important in the study of photogrammetry.

The fundamental concept underlying the design of an early type of stereoplotter is illustrated in Fig. 12-1. In Fig. 12-1*a*, an overlapping pair of aerial photos is exposed. Transparencies or *diapositives,* as they are called, carefully prepared to exacting standards from the negatives, are placed in two stereoplotter projectors, as shown in Fig. 12-1*b*. This process is called *interior orientation*. With the diapositives in place, light rays are projected through them; and when rays from corresponding images on the left and right diapositives intersect below, they create a stereomodel (often simply called a *model*). In creating the intersections of corresponding light rays, the two projectors are oriented so

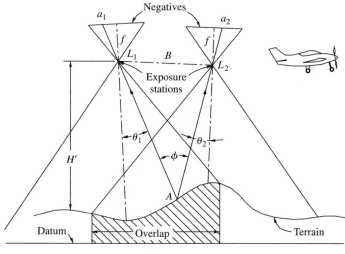

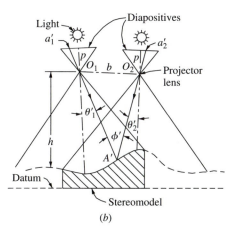

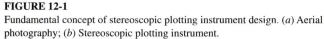

(b)

FIGURE 12-1
Fundamental concept of stereoscopic plotting instrument design. (a) Aerial
photography; (b) Stereoscopic plotting instrument.

that the diapositives bear the exact relative angular orientation to each other in the
projectors that the negatives had in the camera at the instants they were exposed. The
process is called *relative orientation* and creates, in miniature, a true three-dimensional
stereomodel of the overlap area. After relative orientation is completed, *absolute orientation* is performed. In this process the stereomodel is brought to the desired scale and
leveled with respect to a reference datum.

When orientation is completed, measurements of the model may be made and recorded, nowadays generally in digital, computer-compatible form. The position of any point is determined by bringing a reference mark (the floating mark) in contact with the model point. At the position of the reference mark, the three-dimensional coordinates (X, Y, and Z) are obtained through either an analog or a digital solution. Planimetric (X, Y) positions and elevations (Z) of points are thus obtained.

12-2   CLASSIFICATION OF STEREOSCOPIC PLOTTERS

A variety of stereoscopic plotting instruments are currently available, each with a different design. As an aid to understanding stereoplotters, it is helpful to classify them into groups having common characteristics. This classification is divided into four general categories: (1) *direct optical projection* instruments, (2) instruments with *mechanical or optical-mechanical projection,* (3) *analytical* stereoplotters, and (4) *softcopy* stereoplotters. The first-generation stereoplotters were of direct optical projection design, and although they are seldom used today, a description of their operation provides a good introduction to the subject of stereoplotters. These instruments create a true three-dimensional stereomodel by projecting transparency images through projector lenses, as illustrated in Fig. 12-1. The model is formed by intersections of light rays from corresponding images of the left and right diapositives. An operator is able to view the model directly and make measurements on it by intercepting projected rays on a viewing screen (platen).

Instruments of mechanical projection or optical-mechanical projection also create a true three-dimensional model from which measurements are taken. Their method of projection, however, is a simulation of direct projection of light rays by mechanical or optical-mechanical means. An operator views the diapositives stereoscopically directly through a binocular train.

Analytical stereoplotters form a stereomodel through a purely mathematical procedure which takes place in a computer. The mathematical basis behind these plotters was introduced in Chap. 11. As with mechanical plotters, an operator views the diapositives stereoscopically directly through a binocular train. The movements of the stereoscopic images are introduced by servomotors which are under computer control. Unlike direct projection or mechanical plotters, these versatile instruments are essentially unlimited in terms of the photographic geometry they can accommodate.

Softcopy instruments are the most recent innovation in the design of stereoplotters. Fundamentally, softcopy plotters operate in the same manner as analytical stereoplotters, except that instead of viewing film (hard-copy) diapositives through binocular optics, scanned (softcopy) photographs are displayed on a computer screen and viewed directly. Special viewing systems have been designed which enable the operator to view the left image with the left eye and the right image with the right eye.

Other methods of classifying stereoplotters are (1) classification by accuracy capability (i.e., first-, second-, or third-order plotters) and (2) classification according to whether an "approximate" solution or a "theoretically correct" solution is obtained. The first of these classifications is unsatisfactory because of difficulties in assessing true accuracy capabilities of various instruments. Plotting accuracy is not solely a function of the instrument but also depends upon other variables such as quality of photography and operator ability.

In the second of these classifications, instruments of the "approximate" category assume truly vertical photos and employ a parallax bar for measurement. These low-order instruments enable direct stereoscopic plotting, but the resulting map is a perspective projection, not an orthographic projection. For certain work, however, their accuracy is adequate. Plotters in the "theoretically correct" category are capable of creating a true stereomodel by means of interior, relative, and absolute orientation. This is the type of plotter discussed in this chapter.

It would be difficult, if not impossible, to describe each available instrument in detail in an entire book, let alone a single chapter. For the most part, descriptions are general without reference to specific plotters and without comparisons of available instruments. To emphasize and clarify basic principles, however, examples are made of certain instruments and in some cases pictures are given. Omission of other comparable stereoplotters is not intended to imply any inferiority of these instruments. Operator's manuals which outline the details of each of the different instruments are provided by the manufacturers. Comprehension of the principles presented in this chapter should provide the background necessary for understanding these manuals.

PART I
DIRECT OPTICAL PROJECTION STEREOPLOTTERS

12-3 COMPONENTS

The principal components of a typical direct optical projection stereoplotter are illustrated in the schematic diagram of Fig. 12-2. The numbered parts are the (1) *main frame,* which supports the projectors rigidly in place, thereby maintaining orientation of a stereomodel over long periods; (2) *reference table,* a large smooth surface which serves as the vertical datum to which model elevations are referenced and which also provides the surface upon which the manuscript map is compiled; (3) *tracing table,* to which the platen and tracing pencil are attached; (4) *platen,* the viewing screen which also contains the reference mark; (5) *guide rods,* which drive the illumination lamps, causing projected rays to be illuminated on the platen regardless of the area of the stereomodel being viewed; (6) *projectors;* (7) *illumination lamps;* (8) *diapositives;* (9) *leveling screws,* which may be used to tilt the projectors in absolute orientation; (10) *projector bar,* to which the projectors are attached; and (11) *tracing pencil,* which is located vertically beneath the reference mark on the platen.

Although direct optical projection plotters of different manufacturers vary somewhat in individual design and appearance, all are composed basically of the above parts. Reference to these parts will frequently be made in subsequent discussions in this chapter.

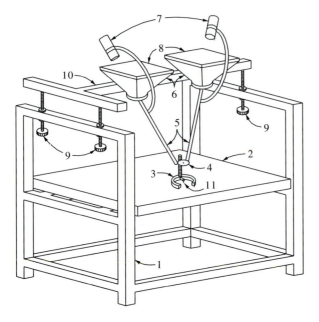

FIGURE 12-2
Principal components of a typical direct optical projection stereoplotter.

12-4   PROJECTION SYSTEMS

In the projection systems of direct optical projection stereoplotters, diapositives of a stereopair are placed in projectors and illuminated from above. Light rays are projected through the projector objective lenses and intercepted below on the reflecting surface of the platen. The projection systems of this type of stereoplotter require that the instruments be operated in a dark room.

Stereoplotter projectors are similar to ordinary slide projectors, differing primarily in their optical precision, physical size, and capability of adjustment in attitude relative to one another. Since projection takes place through an objective lens, the lens formula, Eq. (2-4), must be satisfied in order to obtain a sharply focused stereomodel. In terms of the stereoplotter symbols of Fig. 12-1, the lens formula is expressed as

$$\frac{1}{f'} = \frac{1}{p} + \frac{1}{h} \tag{12-1}$$

In Eq. (12-1), p is the *principal distance* of the projectors (distance from diapositive image plane to upper nodal point of the projector lens), h is the *projection distance* (distance from lower nodal point of the objective lens to the plane of optimum focus), and f' is the focal length of the projector objective lens. To obtain a clear stereomodel, intersections of projected corresponding rays must occur at a projection distance within the range of the *depth of field* of the projector lens (see Sec. 2-3).

To recreate the relative angular relationship of two photographs exactly as they were at the instants of their exposures (a process described in Sec. 12-8), it is necessary that the projectors have rotational and translational movement capabilities. These

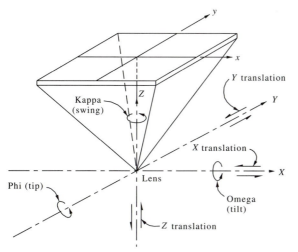

FIGURE 12-3
The six basic projector motions.

motions, six in number for each projector, are illustrated in Fig. 12-3. Three of the movements are angular rotations about each of three mutually perpendicular axes: *x* rotation, called *omega* or *tilt;* *y* rotation, called *phi* or *tip;* and *z* rotation, called *kappa* or *swing*. The origin of the axis system about which the rotations take place is at the upper nodal point of the projector lens, with the *x* axis being parallel to the projector bar. The other three movements are linear translations along each of the three axes. In general, projectors of direct optical projection stereoplotters have all three angular rotations; however, they do not necessarily have all three linear translations. As a minimum, though, they must have the *x* translation for changing the spacing between projectors.

Projectors of the type illustrated in Fig. 12-2 illuminate only a small area of the diapositive at a time. This type of illumination system consists of a small, narrow-angle light source which illuminates a circular area on the diapositive approximately 4 cm in diameter. When projected through the objective lens, an area slightly larger than the platen is illuminated in the model area. Driven by means of guide rods, the lamps swing above the diapositives, following the platen and illuminating it as it moves about the stereomodel. The Kelsh plotter shown in Fig. 12-4 utilizes this method of illumination. This instrument combines a nominal principal distance of 152 mm with an optimum projection distance of 760 mm, a combination which provides a nominal enlargement ratio of 5 from diapositive scale to model scale.

12-5 VIEWING SYSTEMS

The function of the viewing system of a stereoplotter is to enable the operator to view the stereomodel three-dimensionally. Stereoviewing is made possible by forcing the left eye to view only the overlap area of the left photo while the right eye simultaneously sees only the overlap area of the right photo. The different stereoviewing systems commonly used in direct optical projection plotters are (1) the *anaglyphic* system, (2) the *stereo-image alternator* (SIA), and (3) the *polarized-platen viewing* (PPV) system.

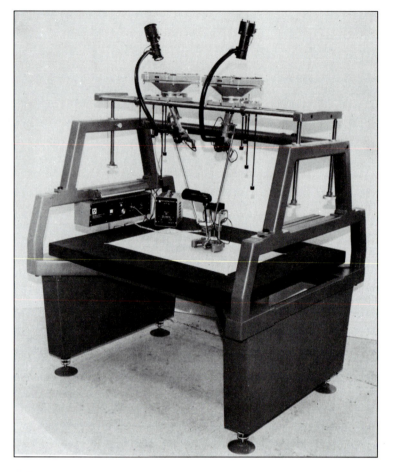

FIGURE 12-4
Kelsh stereoscopic plotting instrument. (*Courtesy of Kelsh Instrument Division, Danko Arlington, Inc.*)

An anaglyphic system uses filters of complementary colors, usually red and cyan (blue-green), to separate the left and right projections. Assume that a cyan filter is placed over the light source of the left projector while a red filter is placed over the right. Then, if the operator views the projected images while wearing a pair of spectacles having cyan glass over the left eye and red glass over the right eye, the stereomodel can be seen in three dimensions. The anaglyphic viewing system is simple and inexpensive; however, it precludes the use of color diapositives and causes considerable light loss, so that the model is not as bright as it could be if filters were not used.

The SIA system uses synchronized shutters to achieve stereoviewing. A shutter is placed in front of each projector lens. Also, a pair of eyepiece shutters, through which the operator must look, is situated in front of the platen. The shutters are synchronized so that the left projector and left eyepiece shutters are open simultaneously while the right projector and right eyepiece shutters are closed, and vice versa. An operator therefore sees only left projector images with the left eye and right projector images with the right eye. The shutters rotate at a rapid rate so that the operator is unaware of any discontinuity in the projection. An SIA system is shown attached to the Kelsh plotter pictured in Fig. 12-4.

The PPV system operates similarly to the anaglyphic system except that polarizing filters are used instead of colored filters. Filters of orthogonal polarity are placed in front of the left and right projectors, and the operator wears a pair of spectacles with corresponding filters on the left and right. In contrast to the anaglyphic system, the SIA and PPV systems both cause much less light loss, and both permit the use of color diapositives.

12-6   MEASURING AND TRACING SYSTEMS

A system for making precise measurements of the stereomodel is essential to every stereoplotter. Measurements may be recorded as direct tracings of planimetric features and contours of elevation, or they may be taken as X, Y, and Z model coordinates. One of the principal elements of the measuring system of a direct optical projection stereoplotter is a tracing table. The platen (see Fig. 12-2) contains a reference mark in its center, usually a tiny speck of light. The reference mark appears to float above the stereomodel if the platen is above the terrain; hence it is called the *floating mark*. The platen can be raised or lowered by turning a screw; the total vertical run of the threads is about 120 mm. Vertical movement of the platen is geared to a dial, and by varying gear combinations the dial can be made to display elevations directly in meters or feet for varying model scales.

A manuscript map, preferably of stable base material, is placed on top of the reference table, as illustrated in Fig. 12-4. The tracing table rests on the manuscript and is moved about manually in the X and Y directions. To plot the position of any point, the platen is adjusted in X, Y, and Z until the reference mark appears to rest exactly on the desired point in the model. A pencil point which is vertically beneath the reference mark is then lowered to record the planimetric position of the point on the map, and its elevation is read directly from the dial.

To trace a planimetric feature such as a creek, the pencil is lowered to the map and the tracing table is moved in the XY plane while the platen is moved up or down to keep the floating mark in contact with the stream. The pencil thereby records a continuous trace of the feature. Contours of elevation may also be traced by locking the dial at the elevation of the desired contour and then moving the tracing table about, keeping the floating mark in contact with the terrain. Tracing contours with a stereoplotter is a skill that takes years of practice to master.

12-7   INTERIOR ORIENTATION

As mentioned earlier, three steps are required to orient a stereoscopic plotter. The first, *interior orientation,* includes preparations necessary to recreate the geometry of the projected rays to duplicate exactly the geometry of the original photos; e.g., angles θ_1' and θ_2' of Fig. 12-1b must be exactly equal to angles θ_1 and θ_2 of Fig. 12-1a. This is necessary to obtain a true stereomodel. Procedures involved in interior orientation are (1) preparation of diapositives, (2) compensation for image distortions, (3) centering of diapositives in the projectors, and (4) setting off the proper principal distance in the projectors.

Diapositives are transparencies prepared on either optically flat glass or clear film base materials. They may be made either by *direct contact* printing or by *projection* printing. If contact-printed, their principal distances will be exactly equal to the focal length of the taking camera. For this reason, contact-printed diapositives can be used only in plotters whose accommodation range in principal distance includes the focal length of the taking camera. Contact printing creates true geometry as long as the principal distances of the projectors are set equal to the focal length of the taking camera. Projection-printed diapositives are seldom made, although they used to be required for old Multiplex and Balplex plotters which employed reduced-size diapositives.

In direct optical projection plotters, compensation for symmetric radial distortion of the lens of the taking camera may be accomplished in one of the following three ways: (1) elimination of the distortion with a "correction plate" in projection printing of the diapositives, followed by use of a distortion-free projector lens; (2) varying the projector principal distance by means of a cam, thereby reconstructing true geometry; and (3) use of a projector lens whose distortion characteristics negate the camera's distortion. With most modern aerial cameras, lens distortion is of such a low magnitude that correcting for it is often ignored.

Each diapositive must be centered in its projector so that the principal point is on the optical axis of the projector lens. Although this problem is solved slightly differently for each instrument, it is basically done by aligning fiducial marks of the diapositive with four calibrated collimation marks whose intersection locates the optical axis of the projector. Before diapositives are placed in the projectors, they are laid out so that their common area overlaps. They are then separated, rotated 180° about the Z axis, placed in the plate holders of the projectors, and centered. In projection, the imagery is rotated 180°, which once again makes the common areas of projected images overlap.

The final step in interior orientation is to set the diapositive principal distance on the projectors. This was unnecessary for some early plotters such as the Multiplex and Balplex, whose principal distances were permanently fixed and whose diapositives were prepared to conform with these principal distances. For other optical projection plotters, the principal distance may be varied by adjusting either graduated screws or a graduated ring to raise or lower the diapositive image plane. These projectors are designed to accommodate a certain nominal principal distance, and their range from that value is small; e.g., the Kelsh plotter pictured in Fig. 12-4 is designed for 152-mm photography, and its range in principal-distance accommodation is from only 150 to 156 mm.

12-8 RELATIVE ORIENTATION

Imagine the camera frozen in space at the instants of exposure of two photographs of a stereopair. The two negatives in the camera would then bear a definite position and attitude relationship relative to each other. In relative orientation, this relative position and attitude relationship is recreated for the two diapositives by means of movements imparted to the projectors.

The condition that is fulfilled in relative orientation is that each model point and the two projection centers form a plane in miniature just like the plane that existed for the corresponding ground point and the two exposure stations. This condition is illustrated by corresponding planes $A'O_1O_2$ and AL_1L_2 of Fig. 12-1. Also, parallactic angle ϕ' for any point of the stereomodel of Fig. 12-1 must equal the point's original parallactic angle ϕ. The implication of the foregoing condition of relative orientation is that projected rays of all corresponding points on the left and right diapositives must intersect at a point, and this is the basis of the systematic relative orientation procedures described below.

Since relative orientation is unknown at the start, the two projectors are first positioned relative to each other by estimation. Usually, if near-vertical photography is being used, the projectors are set so that the diapositives are nearly level and so that their X axes lie along a common line. Also, the projectors are adjusted so that their Y and Z settings (distances from the projector bar in the Y and Z directions) are equal. When the projector lamps are first turned on, corresponding light rays will not intersect, and their projected images may appear on the platen, as shown in Fig. 12-5a. Since the X component p_x of the mismatch of images is a function of the elevation of the point, it can be removed by raising or lowering the platen. The remaining Y component p_y, as shown in Fig. 12-5b, is called *y parallax,* and it must be removed or *cleared* for all points in the stereomodel in order to obtain a relatively oriented model.

Rather than attempt to clear *y* parallax at all points in the model, conventional procedure clears five standard points (plus a sixth for a check) located in the model, as shown in Fig. 12-6. If these five points are cleared of *y* parallax, then the entire model should be cleared. Points 1 and 2 are vertically beneath projector I (left projector) and projector II (right projector), respectively. Points 3 and 5 are on the Y axis through projector I, and points 4 and 6 are on the Y axis through projector II. Points 1, 2, 3, and 4 roughly form a square, as do points 1, 2, 5, and 6.

Before relative orientation procedures are described, it may be helpful to consider the movement of projected images in the model area caused by each of the six projector motions that were illustrated in Fig. 12-3. These image movements are illustrated in Figs. 12-7a through *f.* Figure 12-7a shows that an X translation imparts only X

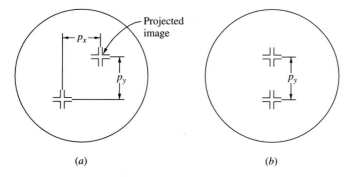

(a) (b)

FIGURE 12-5
(a) Observing both *y* parallax and *x* parallax on the platen. (b) After *x* parallax is removed, only *y* parallax remains.

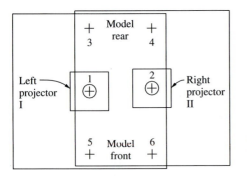

FIGURE 12-6
Stereomodel locations of six points
conventionally used in relative orientation.

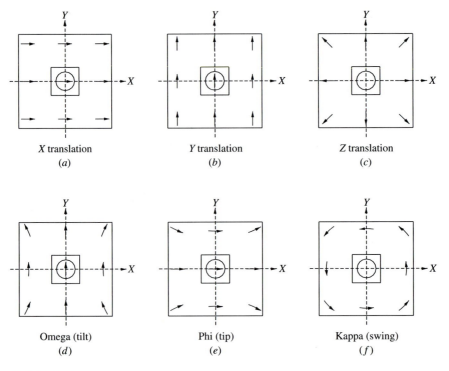

FIGURE 12-7
Movement of projected images in the model caused by the six projector motions.

movement to all projected images, and therefore no y parallax could be cleared using this motion. In Fig. 12-7b, Y translation causes equal Y movement of all projected images; hence any point in the stereomodel could be cleared of y parallax by using this motion. Fig. 12-7c shows that all points move radially outward from the projection center if a projector is translated upward in Z. Image movement would be radially inward for a downward Z translation. With the use of Z translation, Y parallax could be cleared for any points except those along the X axis. In Fig. 12-7d, omega rotation (tilt) causes Y

movement of all projected images. This motion could therefore be used to clear y parallax anywhere in the model. As shown in Fig. 12-7e, a phi rotation (tip) causes an X movement for all projected images, but it also imparts a Y component for points not on the X or Y axes. And finally, in Fig. 12-7f, a kappa rotation (swing) imparts Y components to points everywhere in the model except along the Y axis. A clear grasp of these image movements will help to explain why y parallax can be removed and orientation achieved by using the following procedures.

Two different systematic procedures for relative orientation are discussed in this text: the *two-projector* method, or what is commonly called the *swing-swing* method; and the *one-projector* method. The steps involved in the two-projector method are as follows (refer to Figs. 12-6 and 12-7):

a. Clear p_y at 1 with kappa (swing) of projector II.

b. Clear p_y at 2 with kappa (swing) of projector I.

c. Clear p_y at 3 with phi (tip) of projector II.

d. Clear p_y at 4 with phi (tip) of projector I.

e. Observe p_y at 5, remove this y parallax, and overcorrect by introducing one-half of the original p_y in the opposite direction with omega (tilt) of either projector.[1]

f. Repeat steps *a* through *e* until no p_y exists at 5.

g. Check 6 for p_y.

In analyzing the above steps, note that step *b* does not introduce y parallax at point 1, so both points 1 and 2 are clear following this step. Also step *c* introduces no y parallax at either point 1 or 2, so points 1, 2, and 3 are clear following this step. Furthermore, step *d* introduces no y parallax at point 1, 2, or 3, so points 1 through 4 are clear following this step. Step *e* introduces y parallax at all points, but if steps *a* through *d* are repeated, the y parallax at point 5 will be greatly diminished. Several repetitions of steps *a* through *e* may be required to finally clear point 5, however. If points 1 through 5 are cleared, point 6 should also be clear; and if it is not, this is generally an indication that the other five points are not truly clear.

The two-projector method described above requires rotations of both projectors. There are situations in which it is necessary to retain the orientation of one projector, say, the left, and to perform relative orientation with motions of the right projector only. An example of this is the orienting of a strip of three or more photos on a projector bar holding more than two projectors. After the first two photos are

---

[1] The actual amount of y parallax overcorrection that must be introduced into point 5 is a function of camera focal length f and distance y (measured on the original photo) from model point 1 to model point 5. It can be computed by the formula $0.5(1 + f^2/y^2) - 1$. For 152-mm-focal-length photography and a y distance of approximately 110 mm, the value is $\frac{1}{2}$; hence step *e* above applies specifically to those very commonly encountered conditions. Amounts of overcorrection for other situations can be calculated, however, and appropriate y parallax introduced accordingly. Note that overcorrection values calculated by the above formula are exact only for level terrain, but good approximations result for variable terrain. The inexactness of the step helps to explain the need to perform several iterations in relative orientation to finally clear all y parallax.

oriented, the third must be oriented to the second without upsetting the orientation of the second, etc. The one-projector method outlined in the following steps may be used in this situation. All motions are imparted to the right-hand projector. (Refer to Fig. 12-6 for point locations.)

a. Clear p_y at 2 with Y translation.

b. Clear p_y at 1 with kappa (swing).

c. Clear p_y at 4 with Z translation.

d. Clear p_y at 3 with phi (tip).

e. Observe p_y at 6, remove this y parallax, and overcorrect by introducing one-half of the original p_y in the opposite direction with omega (tilt).

f. Repeat steps a through e until no p_y exists at 6.

g. Check 5 for p_y.

12-9   ABSOLUTE ORIENTATION

After relative orientation is completed, a true three-dimensional model of the terrain exists. Although the horizontal and vertical scales of the model are equal, that scale is unknown and must be fixed at the desired value. Also the model is not yet level with respect to datum. Selecting model scale and fixing the model at that scale, and leveling the model are the purposes of absolute orientation.

12-9.1   Selecting Model Scale

Model scale is fixed within certain limits by the scale of the photography and by the characteristics of the particular stereoplotter. By comparing the geometry of Fig. 12-1a and b, model scale is seen to be the ratio of the sizes of triangles AL_1L_2 and $A'O_1O_2$. Equating these similar triangles, model scale may be expressed as

$$S_m = \frac{b}{B} = \frac{h}{H'} \tag{12-2}$$

In Eq. (12-2), S_m is model scale, b is the model air base, B is the photographic air base, h is plotter projection distance, and H' is the flying height above ground. From Chap. 6 it will be recalled that flying height above ground and camera focal length fix photo scale according to the relationship

$$S_p = \frac{f}{H'} \tag{12-3}$$

In Eq. (12-3), S_p is photo scale and f is the camera focal length. By substituting Eq. (12-3) into Eq. (12-2), the following convenient equation results for computing the optimum model scale (given photo scale):

$$S_m = \frac{h}{f} S_p \tag{12-4}$$

In Eq. (12-4), the term h/f is the ratio of projection distance to camera focal length, and it is also the enlargement ratio from photo scale to model scale. From the foregoing it is apparent, then, that for a given photo scale, *optimum model scale* is fixed for a particular stereoplotter by its *optimum projection distance*. A small aperture is used in the objective lenses of projectors of direct optical projection stereoplotters, so that a large depth of field results. A satisfactory model can therefore be obtained for a rather wide range in projection distance, and accordingly some flexibility is afforded in selecting the model scale. The range in projection distance also makes it possible to accommodate topographic relief in the stereomodel. The actual model scale that is adopted should be chosen near optimum model scale, but it should be rounded to one of the commonly used scales such as 1:1000 or 1 in/100 ft.

When model scale has been adopted, an initial model air base (spacing between projectors) is set off. This is most conveniently done prior to relative orientation, so that model scale after relative orientation is close to the required model scale. An initial model base can be obtained by multiplying the photo base (see Sec. 8-4) by the actual enlargement ratio S_m/S_p. The photo base can be measured directly from the photos, or it can be computed on the basis of percentage of end lap.

12-9.2 Scaling the Model

If a preliminary model base is calculated as described above and the projectors are set accordingly, then after relative orientation the stereomodel will be near the required scale. As shown in Fig. 12-8, model scale is changed by varying the model base. If the Y and Z settings of the two projectors are equal, then model base is composed only of an X component called b_x, and model scale is varied by simply changing the model base by Δb_x, as shown in Fig. 12-8a.

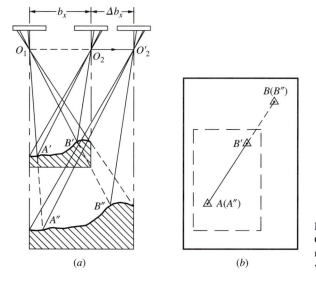

FIGURE 12-8
Changing model scale by adjusting model base. (*a*) Cross-sectional view; (*b*) plan view.

A minimum of two horizontal control points is required to scale a stereomodel. These points are plotted at adopted model scale on the manuscript map as points A and B of Fig. 12-8b. The manuscript is then positioned under the model, and with the floating mark set on one model point, such as A', the manuscript is moved until map point A is directly under the plotting pencil. The floating mark is then set on model point B'. While the manuscript is held firmly with a fingertip at point A, it is rotated until map line AB is collinear with model line $A'B'$. If model line $A'B'$ is shorter than map line AB, as is the case in Fig. 12-8b, model scale is too small and must be increased by increasing the model base until new model line $A''B''$ is equal in length to map line AB. The model base may be set to the required value by trial and error, or Δb_x may be calculated directly from

$$\Delta b_x = b_x \left(\frac{AB}{A'B'} - 1 \right) \tag{12-5}$$

In Eq. (12-5), AB and $A'B'$ are scaled from the manuscript in any convenient units. If the algebraic sign of Δb_x is negative, model scale is too large and b_x must be reduced by Δb_x. Once the model is scaled, it is recommended that a third horizontal control point be checked to guard against possible mistakes.

If the Y and Z settings of the projectors are not equal following relative orientation, as would be the case if the one-projector method were used, there will be b_x, b_y, and b_z components in the model air base, as shown in Fig. 12-9. In this situation if it were necessary, for example, to increase the model scale, projector II would have to be moved to II' along the baseline joining I and II; otherwise, y parallax would be introduced. This requires introducing Δb_x, Δb_y, and Δb_z components in the scaling process. First the Δb_x component is introduced (which of course introduces y parallax in the model). Then the Δb_y component is introduced by applying a sufficient amount of b_y motion to projector II to clear y parallax for a point near 2 (see Fig. 12-6). Finally the Δb_z component is introduced by applying a sufficient amount of b_z motion to projector II to clear y parallax for a point near 4 or 6 of Fig. 12-6.

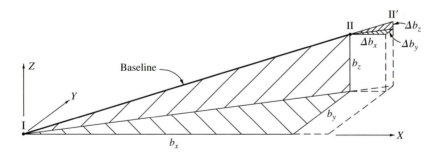

FIGURE 12-9
Model base containing b_x, b_y, and b_z components. (Note that an increase of b_x by Δb_x also requires introducing Δb_y and Δb_z components.)

12-9.3 Leveling the Model

The final step in absolute orientation is to level the model. This procedure requires a minimum of three vertical control points distributed in the model so that they form a large triangle. As a practical matter, four points, one near each corner of the model, should be used. A fifth point near the center of the model is also desirable. Before proceeding with leveling, it is imperative that the proper gears be inserted into the tracing table so that elevation changes recorded on the dial that occur with up and down motions of the platen are consistent with the required model scale.

A model with a vertical control point near each corner that has not yet been leveled is shown in Fig. 12-10. Note that there are two components of tilt in the model, an X component (also called Ω) and a Y component (also called Φ). The amount by which the model is out of level in each of these components is determined by reading model elevations of the vertical control points and comparing them with their known values. Assuming four vertical control points in a model, as points A through D of Fig. 12-10, the following systematic procedure would be applied in leveling the model:

a. Set the floating mark on model point A, and index the tracing table dial to read the control elevation of that point.
b. Read the model elevation of control point D.
c. From the difference between model elevation and control elevation, determine whether the model is X-tilted up or down toward the rear. (If model elevation is higher than control elevation, the model is tilted up in the rear.)
d. Introduce a corrective X tilt.
e. Repeat steps *a* through *d* until the model is level in the direction from A to D.

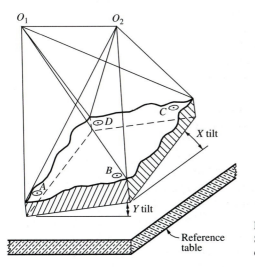

FIGURE 12-10
Stereomodel that is not level (note X and Y components of tilt).

f. Reindex the tracing table dial to read the control elevation of point *A* with the floating mark set on model point *A*.

g. Read the model elevation of control point *B*.

h. From the difference between model elevation and control elevation, determine whether the model is *Y*-tilted up or down to the right.

i. Introduce a corrective *Y* tilt.

j. Repeat steps *f* through *i* until the model is level in that direction.

k. Check point *D* to see if its model elevation still conforms to its control elevation. If points *A* and *D* are not on a line parallel to the stereoplotter *Y* axis, it will likely not conform. If it does not conform, the above steps should be repeated until the model elevations of points *A*, *D*, and *B* all agree with their control elevations.

l. Check point *C* to see if its model elevation conforms to its control elevation. If it does not, this could be an indication of model deformation caused by inaccuracies of relative orientation, or it could disclose a blunder in one or more of the vertical control points.

There are different methods available for introducing the corrective *X* and *Y* tilts. The choice is primarily dependent upon the particular design of the stereoplotter. Some instruments are designed so that their reference tables may be tilted in the *X* and *Y* directions to make them parallel with the model datum. With other stereoplotters, such as that of Fig. 12-4, corrective tilts can be introduced by turning the leveling screws to rotate the projector bar. This procedure is illustrated in Fig. 12-11. Either *X* or *Y* tilts can be introduced in this way, but if the instrument has four leveling screws, only one or the other should be done. If both are performed, all four leveling screws will not rest firmly on their supports afterward. A third method of introducing corrective tilts consists of tilting the projectors. As shown in Fig. 12-12*a*, to introduce a corrective *X* tilt, equal amounts of omega rotation (tilt) are applied to each projector. First a trial amount is introduced into the left projector, which of course introduces *y* parallax throughout the model. But if the *Y* and *Z* settings of the projectors are equal, this *y* parallax is entirely removed by introducing an equal amount of omega rotation to the right projector. If the operator clears *y* parallax for any point in the stereomodel when introducing omega to the right projector, the entire model will again be clear. In applying a corrective *Y* tilt, a trial amount of phi rotation is applied in equal amounts to both projectors. This has the effect of introducing *y* parallax in the model because it creates a b_z base component, as shown in Fig. 12-12*b*. To compensate for this condition, either the left or right projector must be raised or lowered, the exact amount being just enough to clear *y* parallax for a point in one corner of the model.

The leveling operation will disturb the previously established model scale, especially if large corrective tilts are required. Also, it is likely that absolute orientation will slightly upset relative orientation. Therefore it is not practical to labor at great lengths with either relative or absolute orientation the first time through. Rather, quick orientations should be performed at first, followed by careful refinements the second time through. When orientation is completed, the manuscript map should be firmly secured to the reference table in preparation for map compilation.

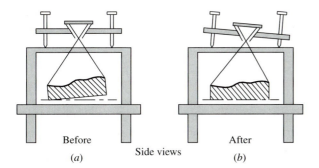

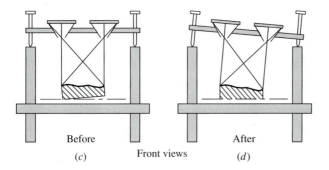

FIGURE 12-11
(*a*) and (*b*) Correcting *X* tilt of a model by *X* tilt of projector bar.
(*c*) and (*d*) Correcting *Y* tilt of a model by *Y* tilt of projector bar.

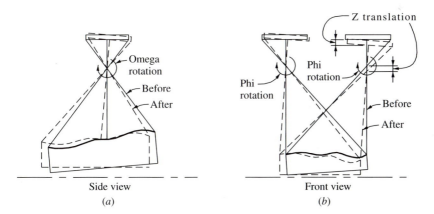

FIGURE 12-12
(*a*) Correcting *X* tilt of a model with equal omega rotations of both projectors. (*b*) Correcting
Y tilt of a model with equal phi rotations of both projectors followed by a *Z* translation.

12-10   MECHANICAL PROJECTION INSTRUMENTS

Mechanical projection stereoscopic plotting instruments simulate direct optical projection of light rays by means of two precisely manufactured, metal *space rods*. These instruments are preferred over direct optical projection plotters due to their versatility, higher accuracy, overall stability, and the fact that they need not be operated in a dark room. Although their use is diminishing today, many mechanical plotters are still used in production of photogrammetric maps. As with direct optical projection instruments, the many slightly different designs of mechanical projection stereoplotters cannot be described in detail in this text. Rather, general principles of mechanical projection plotters are explained. Manuals provided by manufacturers outline the details of operation for the specific instruments.

The basic principles of mechanical projection are illustrated in the simplified diagram of Fig. 12-13. Diapositives are placed in *carriers* and illuminated from above. The carriers are analogous to projectors of direct optical projection instruments. Two space rods are free to rotate about *gimbal joints* O' and O'', and they can also slide up and down through these joints. The space rods represent corresponding projected light rays, and the gimbal joints are mechanical projection centers, analogous to the objective lenses of projectors of direct optical projection stereoplotters. The model exposure stations are

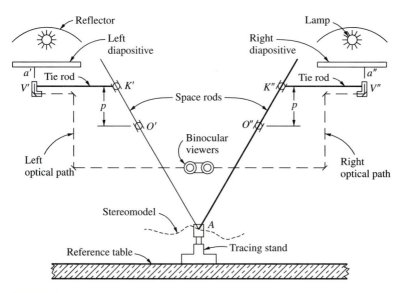

FIGURE 12-13
Basic principles of mechanical projection.

therefore represented by O' and O'', and the distance $O'O''$ is the model air base. Joints O' and O'' are fixed in position except that their spacing can be changed, either physically or theoretically, during absolute orientation to obtain correct model scale.

The viewing system consists of two individual optical trains of lenses, mirrors, and prisms. The two optical paths are illustrated by dashed lines in Fig. 12-13. An operator looking through binocular eyepieces along the optical paths sees the diapositives directly and perceives the stereomodel. Objective lenses V' and V'' are situated in the optical trains directly beneath the diapositives. The lenses are oriented so that viewing is orthogonal to the diapositives; consequently, the diapositive image planes (emulsion surfaces) can lie on the upper side of the diapositive glass with no refraction error being introduced because no rays pass obliquely through the glass. A reference half mark is superimposed on the optical axis of each of the lenses V' and V''. Movement is imparted to the lenses from the space rods by means of tie rods connected at another set of gimbal joints K' and K''. These joints are fixed in vertical position, and the vertical distance from lower gimbal joints O' and O'' to their corresponding upper gimbal joints K' and K'' is the principal distance p. During interior orientation, this distance is set equal to the principal distance of the diapositives.

The space rods intersect at the *tracing stand*. Manually moving the tracing stand imparts movement to the space rods, which in turn impels the viewing system and makes it possible to scan the diapositives. By manipulating the tracing stand in the X, Y, and Z directions, the optical axes of lenses V' and V'' can be placed on corresponding images such as a' and a''. This will occur when the reference half marks fuse into a single mark that appears to rest exactly on the model point. If orientation of the instrument has been carefully performed, with the floating mark fused on points a' and a'', the space rods have the same orientation that the incoming light rays from terrain point A had at the time of photography, and the intersection of space rods locates that model point. Each additional model point is located in this same manner.

By geometric comparison, the mechanical projection system illustrated in Fig. 12-13 is exactly the same as direct optical projection. Diapositives are placed in the carriers with their overlapping areas toward the outside. In scanning the diapositives, if the tracing stand is moved right, the viewing lenses move left, and vice versa. Also, if the tracing stand is pushed backward, the viewing lenses move forward, and vice versa. When the tracing stand screw is turned to raise the space rod intersection, the parallactic angle increases and the viewing lenses move apart, a manifestation of the increased x parallax which exists for higher model points.

The carriers of most mechanical projection stereoplotters are capable of rotations and translations, either directly or indirectly. These carrier motions are used to perform relative and absolute orientation. Many mechanical projection instruments are oriented by using exactly the same procedures that were described for direct optical projection stereoplotters. Others that do not possess all six rotations and translations use slight variations from these basic orientation procedures.

An example of a stereoplotter utilizing mechanical projection is the Galileo *Stereosimplex G-7*, pictured in Fig. 12-14. The operation of this plotter is similar in principle to that of the hypothetical instrument illustrated in Fig. 12-13, except that the optical system is fixed and the space rods impel the diapositive carriers for scanning the model

FIGURE 12-14
Galileo Stereosimplex G-7. (*Courtesy Galileo Corporation of America.*)

in the X and Y directions. The X and Y scanning motions are introduced manually by using a pantograph handle with one hand, while Z motions are introduced by turning a foot disk. Movements in X and Y are transmitted to the plotting table at the desired enlargement or reduction ratio via the pantograph. Each carrier of the instrument has omega, phi, and kappa rotations, and its common phi and common omega (which tilt both projectors simultaneously by equal amounts) facilitate leveling models. Its range in principal-distance accommodation is continuous from 85 to 310 mm, and its projection distance range is from 155 to 460 mm.

PART III
ANALYTICAL PLOTTERS

12-11 INTRODUCTION

Analytical plotters represent the next stage in stereoplotter evolution. Their development was made possible by advances in computers, digital encoders, and servosystems. By combining computerized control with precision optical and mechanical components, analytical plotters enable exact mathematical calculations to define the nature of the stereomodel. They are also easily interfaced with computer-aided drafting (CAD) systems, which facilitates map editing and updates. These instruments, with their digital output capability, are ideal for compiling data for use in geographic information systems.

The essential capabilities of an analytical plotter are (1) to precisely measure x and y photo coordinates on both photos of a stereopair and (2) to accurately move to defined x and y photo locations. These operations are carried out under direct computer control. Digital encoders provide the ability to measure x and y photo coordinates, with the output from the encoders being read by the computer. Servomotors, which respond to signals from the controlling computer, allow the photographs to be moved to the defined locations.

12-12 SYSTEM COMPONENTS AND METHOD OF OPERATION

An analytical plotter incorporates a pair of precision comparators which hold the left and right photos of a stereopair. Figure 12-15 shows a schematic diagram of a single comparator which employs a lead screw for measurement and movement. The plate carrier, which holds the photograph, moves independently in the X and Y directions as the servomotors turn the corresponding lead screws. Digital encoders attached to the shafts of the motors sense the position of a point on the photo relative to an index mark (half mark) which is fixed.

Figure 12-16 illustrates how the primary components of an analytical stereoplotter interact. The heart of the system is the *controller computer* which interfaces with the *operator controls, servomotors, encoders,* and the *CAD computer.* The controller computer accepts input from the operator controls and calculates left and right plate positions from these inputs. It then operates the servomotors to move the plates, stopping when the encoders indicate that the correct positions have been reached.

Analytical plotters form neither an *optical* nor a *mechanical* model, as do the stereoplotters described in preceding sections of this chapter. Rather, they compute a *mathematical* model based on the principles of analytical photogrammetry presented in Chap. 11. This mathematical model is established through numerical versions of interior, relative, and absolute orientation. Although the exact methods used to accomplish these orientations may vary among instruments, the fundamental approach is the same. Typically, the orientation software is a distinct module, separate from the data collection (mapping) software. This allows flexibility in the choice of available data collection software for a given analytical plotter.

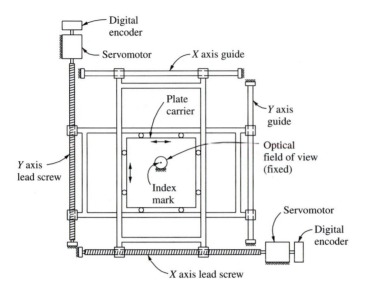

FIGURE 12-15
Schematic representation of a comparator from an analytical plotter.

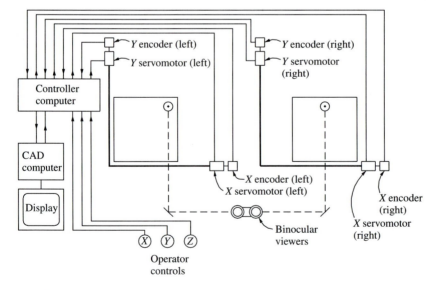

FIGURE 12-16
Schematic diagram of components and operation of an analytical plotter.

FIGURE 12-17
Zeiss P-3 analytical plotter. (*Courtesy Carl Zeiss, Inc.*)

Although the development of the softcopy stereoplotter has reduced the demand for analytical plotters somewhat, they are still widely used. Two popular analytical stereoplotters are the Zeiss P-3, and the Leica SD3000, shown in Figs. 12-17 and 12-18, respectively.

12-13 ADVANTAGES OF ANALYTICAL PLOTTERS

Because they have no optical or mechanical limitations in the formation of their mathematical models, analytical plotters have great versatility. They can handle any type of photography, including vertical, tilted, low oblique, convergent, high oblique, panoramic, and terrestrial photos. They can also be used with various types of satellite imagery. In addition, they can accommodate photography from any focal-length camera, and in fact they can simultaneously use two photos of different focal lengths to form a model.

In comparison with analog plotters, analytical plotters can provide results of superior accuracy for three fundamental reasons. First, because they do not form model points by intersecting projected light rays or mechanical space rods, optical and mechanical errors from these sources are not introduced. Second, analytical plotters can effectively correct for any combination of systematic errors caused by camera lens distortions, film shrinkage or expansion, atmospheric refraction, and, if necessary, earth curvature. Third, in almost every phase of their operation, they can take advantage of redundant observations and incorporate the method of least squares into the solution of the equations.

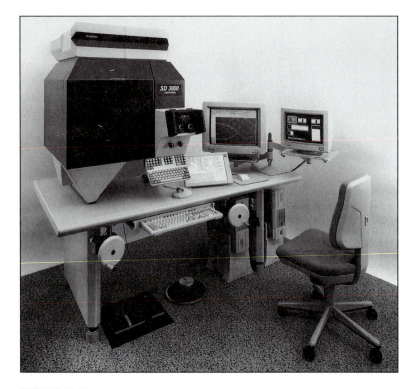

FIGURE 12-18
Leica SD3000 analytical plotter. (*Courtesy LH Systems, LLC.*)

12-14 ANALYTICAL PLOTTER ORIENTATION

As is necessary with all analog stereoplotters, interior, relative, and absolute orientation are also required for analytical plotters prior to going into most modes of operation. In all phases of orientation, and operation for that matter, a dialogue is maintained between operator and instrument. The system's computer directs the operator through each of the various steps of operation, calling for data when needed and giving the operator opportunities to make certain decisions.

Prior to using an analytical plotter, its measuring system should be calibrated using a precise grid plate. This plate consists of a flat piece of glass upon which an array of fine crosses has been etched. The coordinates of the crosses on a high-quality grid plate are known to within 1 or 2 μm. The grid plate is placed on the plate carrier, and the coordinates of each of the crosses are measured. A two-dimensional coordinate transformation is then computed which relates the raw plate coordinates of the analytical plotter to the accurate xy system of the grid plate. An analytical plotter should be recalibrated approximately every year as part of routine maintenance or anytime the instrument is moved, to ensure accurate results.

The orientation and operation of all analytical plotters are quite similar. The following subsections describe, in general, the three orientation steps for analytical plotters.

12-14.1   Interior Orientation

As mentioned in Sec. 12-7, interior orientation consists of (1) preparation of diapositives, (2) compensation for image distortions, (3) centering of diapositives, and (4) setting off the proper principal distance. Steps 2 and 4 simply amount to accessing the appropriate lens distortion coefficients and camera focal length from a data file. By using these parameters in the appropriate photogrammetric equations, these steps of interior orientation are readily accomplished. Diapositives used with analytical plotters are typically contact prints made on film transparencies, although some plotters can accommodate paper (opaque) prints. The individual diapositives of a stereopair are placed on the left and right plate carriers and held in place by glass covers. Centering of the diapositives is accomplished by measuring the X and Y plate coordinates (based on the encoders) of the fiducials of each photo. This phase of operation is aided by computer-activated servomotors, which automatically drive the measuring mark to the vicinity of the fiducials. A fine pointing is then made by the operator. As few as two fiducials can be measured, but more are recommended and up to eight should be measured if they are available, to increase redundancy. Individual two-dimensional coordinate transformations (see App. C) are then calculated for the left and right photos. This establishes the relationship between the XY plate coordinates and the xy calibrated fiducial coordinates, and at the same time it compensates for film shrinkage or expansion. A choice of coordinate transformations is available, but usually the affine or projective types are used (see Secs. C-6 and C-8). Residuals for this solution will be displayed so that the operator can either accept them or remeasure one or more fiducials. When the solution is accepted, the interior orientation parameters are stored in the computer.

12-14.2   Relative Orientation

Relative orientation is achieved by numerical techniques as described in Sec. 11-10. The operator measures the left and right photo coordinates of at least six pass points to provide observations for the least squares relative orientation solution. In areas of imagery that contain discrete features such as manhole covers and sidewalk intersections, the operator sets the individual left and right index marks on the images of the feature in each photo to measure their coordinates. In open areas such as grass-covered fields, the operator adjusts the floating mark while viewing in stereo until it appears to rest exactly on the ground. Once the raw plate coordinates are observed for all pass points, they are transformed to the calibrated fiducial system, and then lens distortion and atmospheric refraction corrections are applied, resulting in refined coordinates. Numerical relative orientation is then calculated, resulting in values for the exterior orientation parameters $(\omega, \phi, \kappa, X_L, Y_L, Z_L)$ for both photos. The computations are performed using least squares if more than five points are involved in the solution. Again the residuals will be displayed, and the operator has the option of discarding certain points, adding others, or accepting the solution. When relative orientation is accepted, the orientation parameters are stored in the computer. At this point the operation of the analytical plotter changes from a two-dimensional to a three-dimensional mode. The X, Y, and Z operator controls provide model coordinate input to the controller computer which then drives the plates to the appropriate image locations.

12-14.3   Absolute Orientation

Absolute orientation is commonly achieved by a three-dimensional conformal coordinate transformation. While viewing in stereo, the operator places the floating mark on the images of the ground control points which appear in the model and records the XYZ model coordinates for each. The computer then accesses a previously established data file which contains the ground coordinates for the control points, thus enabling a three-dimensional conformal coordinate transformation to be calculated. For absolute orientation a minimum of two horizontal and three well-distributed vertical control points are required, just as for absolutely orienting analog plotters. More than the minimum is recommended, however, so that a least squares solution can be made. Once again, residuals are displayed, and the operator has the option of discarding or remeasuring individual points. After the solution is accepted, the transformation parameters are stored, the analytical stereoplotter is fully oriented, and it is ready to be used as a three-dimensional mapping tool.

12-15   THREE-DIMENSIONAL OPERATION OF ANALYTICAL PLOTTERS

After the steps of orientation are complete, the parameters from each step are available for real-time control of the plotter. The operator supplies input of X, Y, and Z ground coordinates through hand (and/or foot) controls, and the instrument responds by calculating theoretical left and right plate coordinates for this ground position. The servomotors then move the left and right plates to these coordinates so that the floating mark appears at the corresponding location in the model. At first glance, this sequence of operation seems to be out of order. Users of analytical plotters often incorrectly assume that their hand control inputs directly move the index marks to positions in the photos and that ground XYZ coordinates are then computed. This confusion may arise due to the real-time nature of plotter operation.

   The top-down flow of this real-time operation is illustrated in Fig. 12-19. In this figure, computational steps are indicated with solid boxes, inputs represented with dashed boxes, and instrument movements indicated with ellipses. The process begins when the operator adjusts the XYZ controls which provide the primary inputs. The controlling computer then calculates the corresponding XYZ model coordinates by applying parameters of the three-dimensional conformal coordinate transformation, which was calculated in the absolute orientation step. Note that the three-dimensional transformation was computed going from ground system to model.

   After the XYZ model coordinates have been computed, the computer solves the collinearity equations [see Eqs. (11-1) and (11-2)] for x and y coordinates on the left

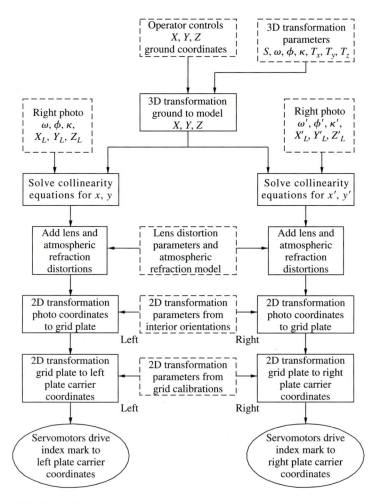

FIGURE 12-19
Flowchart showing the real-time, three-dimensional operation of an analytical
stereoplotter.

photo and x' and y' coordinates on the right. The calculations are made using the XYZ
model coordinates as well as the exterior orientation parameters ω, ϕ, κ, X_L, Y_L, and Z_L
of the left photo and ω', ϕ', κ', X'_L, Y'_L, and Z'_L of the right. The photo coordinates directly computed from the collinearity equations contain no distortions and may there-
fore be considered ideal coordinates. The actual images in the photographs, however, do
not occupy these ideal positions due to lens distortion and atmospheric refraction.
Therefore, when the servomotors move the plates, in order for the actual images in the
photos to correctly coincide with the index marks in the binoculars, distortions must be
added to the ideal coordinates computed by the collinearity equations.

The next step is to transform from the photo coordinate system to the calibrated grid plate coordinate system. This calculation is based on the two-dimensional coordinate transformation parameters which were computed during interior orientation. The final step is to transform from grid plate coordinates to the plate carrier coordinate system. This calculation relies on the two-dimensional coordinate transformation parameters which were computed during grid plate calibration. With the plate carrier coordinates known for both photographs, the servomotors are instructed to drive the left and right index marks to these locations. All these calculations can be performed in a fraction of a millisecond on a modern computer, giving a real-time response. The only lag in the system is due to the finite speed of the servomotors. However, unless fast, abrupt movements are made with the operator controls, the entire process happens smoothly and with instantaneous response.

The *XYZ* stereoplotter controls are directly linked to the CAD system inputs. The operator could, in fact, draw on the CAD system using these controls without looking into the plotter, although the drawn entities would have no relationship to the ground if this were done. By looking into the plotter, the operator can see the movement of the floating mark as it responds to *XYZ* input. The operator guides the floating mark in *X*, *Y*, and *Z* until it appears to rest directly on the feature of interest. At that point, the final position of the operator controls defines the *XYZ* ground position of the feature, and it is then recorded into the CAD system.

12-16   MODES OF USE OF ANALYTICAL PLOTTERS

The modes of use of analytical plotters vary only slightly with the different instruments. Generally all can perform planimetric and topographic mapping. In this case the operator simply traces out features and contours by keeping the floating mark in contact with each feature. The ground coordinates from the operator controls are directly transmitted to the CAD computer for subsequent off-line editing and plotting, or for the development of digital elevation models. "Layered information" needed for the databases of geographic information systems can also be conveniently compiled and stored.

Analytical plotters can also be operated in a profiling mode. In this operation *X*, *Y*, and *Z* coordinates can be measured along predetermined lines in the model. The locations and directions of desired profile lines can be input to the computer in terms of *XY* ground coordinates, and the machine will automatically drive the measuring mark to the corresponding locations while an operator monitors the floating mark to keep it in contact with the ground. The *XYZ* plotter output is continuously recorded by the CAD computer. Profile data can be used directly for various engineering applications such as investigating earthwork quantities along proposed highway centerlines, or they can be used to generate terrain elevation data for subsequent digital orthophoto production (see Sec. 15-8).

Another common application of analytical plotters is in aerotriangulation. Here various modes of operation are possible, from independent model triangulation to simultaneous block adjustment (see Chap. 17). When the aerotriangulation is completed, the diapositives can be used in other stereoplotters or photogrammetric scanners, or they can be reset in the analytical plotter based upon the now available ground control information. As a result of aerotriangulation, the parameters of relative and absolute orientation become known. Analytical plotters have a "reset" capability which enables reading in these known parameters, whereupon an instantaneous relative and absolute orientation can be achieved automatically, thus bypassing the measurements otherwise needed to determine those two steps. In fact, if for any reason a model that was once set should need to be reset to obtain additional data, the orientation parameters from the first setting can be recalled and the system will automatically perform relative and absolute orientation. Anytime photos are moved, however, interior orientations must be redone since the positions of the photographs on the plate carriers will have changed.

Analytical plotters can also be used simply as monocomparators or stereocomparators, where image coordinates are measured and recorded for use in aerotriangulation. Many analytical stereoplotters contain a software module which performs aerotriangulation. Alternatively, photo coordinate data can be recorded in a computer disk file in various formats, suitable for any of several commercially available aerotriangulation software programs.

PART IV
SOFTCOPY PLOTTERS

12-17 INTRODUCTION

The latest stage in the evolution of stereoplotters is that of softcopy plotters. Advances in the computational speed, memory capacity, disk storage, and display monitors of computer systems along with accurate, high-resolution photogrammetric scanners have led to these fully computerized systems. The fundamental operation of a softcopy plotter is the same as that of an analytical plotter except that instead of employing servomotors and encoders for point measurement, softcopy systems rely on digital imagery. Softcopy plotters can perform all the operations of an analytical plotter and, due to their design, can perform a wealth of digital image processing routines as well.

Softcopy stereoplotters have had a tremendous impact on the practice of photogrammetry. By replacing costly optical and mechanical components with digital display manipulation, softcopy plotters have become less expensive and more versatile than analytical plotters. Because of their importance in the current practice of photogrammetric mapping, softcopy stereoplotters are covered in detail in Chap. 15. For now, only a general overview is presented to serve as a comparison with other types of plotters.

12-18 GENERAL OVERVIEW OF SOFTCOPY PLOTTERS

The essential component of a softcopy plotter is a computer with a high-resolution graphics display. The computer must be capable of manipulating large digital images efficiently and must be able to display left and right photos of a stereopair simultaneously.

At the same time, a special configuration is required so that the operator's left eye views only the left photo and the right eye views only the right photo. Several approaches are available to perform this task, which are discussed in greater detail in Sec. 15-2.

Operator input to the softcopy system is similar to that of analytical stereoplotters. Controls must be available which provide continuous input of the primary X, Y, and Z coordinates, in addition to a standard keyboard for routine data entry. These controls can range from a cursor for XY input (similar to that of a tablet digitizer) and a thumbwheel for Z input, to more traditional handwheel controls for X and Y and a footwheel for Z input.

Manual use of a softcopy plotter is most similar to that of an analytical stereoplotter. Orientations can be performed by placing the measuring mark (a single pixel or small pattern of pixels on the display) on necessary points, followed by analytical calculations being performed to compute the orientation. Once oriented, the softcopy plotter can be used in the three-dimensional mode to measure (digitize) topographic features.

In addition to the manual uses available with softcopy plotters, these systems offer many automatic features not found on analytical plotters. One of the most useful automatic capabilities is the ability to perform routine point measurement by computer

FIGURE 12-20
Intergraph Image Station Z softcopy plotter. (*Courtesy Intergraph, Inc.*)

processing, requiring little or no operator input. This capability, discussed in greater detail in Secs. 15-6 and 15-7, can significantly speed up the process of orienting the plotter, as well as assist in collecting digital elevation model information, profiles, and cross-sections. Another convenience offered by softcopy plotters is *vector superimposition,* in which topographic map features (lines, points, etc.) are superimposed on the digital photos as they are being digitized. This capability, also found on a few analytical stereoplotters, allows operators to keep their gaze in one place rather than constantly moving back and forth from photograph display to CAD display.

Figure 12-20 shows the Image Station Z digital photogrammetric workstation. With this system, the operator wears special goggles while viewing the computer display for proper stereoviewing. Operator control of *XYZ* inputs is achieved through a handheld digitizing cursor with a thumbwheel. The digitizing tablet is built into the lower surface of the table.

REFERENCES

American Society of Photogrammetry: *Manual of Photogrammetry,* 4th ed., Falls Church, VA, 1980, chaps. 11 and 12.

American Society for Photogrammetry and Remote Sensing: *Digital Photogrammetry: An Addendum to the Manual of Photogrammetry,* Bethesda, MD, 1996, chaps. 6 and 10.

Case, J. B.: ASP-DTM Symposium, *Photogrammetric Engineering and Remote Sensing,* vol. 44, no. 12, 1978, p. 1477.

Collins, S. H.: "Terrain Parameters Directly from a Digital Terrain Model," *Canadian Surveyor,* vol. 29, no. 5, 1975, p. 507.

Danko, J. O., Jr.: "Color, the Kelsh, and the PPV," *Photogrammetric Engineering,* vol. 38, no. 1, 1972, p. 83.

Dorrer, E.: "Software Aspects in Desk-Top Computer Assisted Stereoplotting," *Photogrammetria,* vol. 33, no. 1, 1977, p. 1.

Dowman, I. J.: "A Working Method for the Calibration of Plotting Instruments Using Computers," *Photogrammetric Record,* vol. 7, no. 42, 1973, p. 662.

―――: "Model Deformation—An Interactive Demonstration," *Photogrammetric Engineering and Remote Sensing,* vol. 43, no. 3, 1977, p. 303.

Ghosh, S. K.: *Analytical Photogrammetry,* 2d ed., Pergamon Press, New York, 1988, chap. 2.

Graham, L. N., Jr., K. Ellison, Jr., and C. S. Riddell: "The Architecture of a Softcopy Photogrammetry System," *Photogrammetric Engineering and Remote Sensing,* vol. 63, no. 8, 1997, p. 1013.

Helava, U. V.: "The Analytical Plotter—Its Future," *Photogrammetric Engineering and Remote Sensing,* vol. 43, no. 11, 1977, p. 1361.

Mikhail, E. M.: "From the Kelsh to the Digital Photogrammetric Workstation, and Beyond," *Photogrammetric Engineering and Remote Sensing,* vol. 62, no. 6, 1996, p. 680.

Mundy, S. A.: "Evaluation of Analytical Plotters for the Commercial Mapping Firm," *Photogrammetric Engineering and Remote Sensing,* vol. 50, no. 4, 1984, p. 457.

Olaleye, J., and W. Faig: "Reducing the Registration Time for Photographs with Non-Intersecting Cross-Arm Fiducials on the Analytical Plotter," *Photogrammetric Engineering and Remote Sensing,* vol. 58, no. 6, 1992, p. 857

Petrie, G., and M. O. Adam: "The Design and Development of a Software Based Photogrammetric Digitising System," *Photogrammetric Record,* vol. 10, no. 55, 1980, p. 39.

Salsig, G.: "Calibrating Stereo Plotter Encoders," *Photogrammetric Engineering and Remote Sensing,* vol. 51, no. 10, 1985, p. 1635.

Whiteside, A. E., and C. W. Matherly: "Recent Analytical Stereoplotter Developments," *Photogrammetric Engineering,* vol. 38, no. 4, 1972, p. 373.

Zarzycki, J. M.: "An Integrated Digital Mapping System," *Canadian Surveyor,* vol. 32, no. 4, 1978, p. 443.

PROBLEMS

12-1. List and briefly describe the four main categories of stereoplotters.

12-2. Describe the basic differences between stereoplotters with direct optical projection and instruments with mechanical or optical-mechanical projection.

12-3. Three basic orientation steps are necessary prior to using a stereoplotter. Name them and give the objective of each.

12-4. What is the focal length of the lens in a Kelsh projector whose principal distance and optimum projection distance are 152 and 760 mm, respectively?

12-5. Discuss the different viewing systems used in direct optical projection stereoplotters.

12-6. Outline the steps of interior orientation.

12-7. What are the advantages and disadvantages of using film diapositives in stereoplotters as opposed to glass diapositives?

12-8. Draw a sketch of the stereoscopic neat model, and label the six points where y parallax is conventionally removed in relative orientation.

12-9. Outline the steps of the one-projector method of relative orientation.

12-10. Briefly describe the process of absolute orientation for a direct optical projection stereoplotter. Why must the steps be repeated?

12-11. Briefly describe the working components of an analytical stereoplotter.

12-12. Explain how the process of interior orientation for an analytical stereoplotter differs from that of an analog stereoplotter.

12-13. Repeat Prob. 12-12, except explain it for relative orientation.

12-14. Repeat Prob. 12-12, except explain it for absolute orientation.

12-15. Discuss the advantages of analytical plotters over analog plotters.

12-16. Compare softcopy plotters and analytical plotters, i.e., discuss their similarities and differences.

CHAPTER
13

TOPOGRAPHIC MAPPING AND SPATIAL DATA COLLECTION

13-1 INTRODUCTION

Mapping, and/or spatial data collection for GIS databases, can generally be categorized into either *planimetric* or *topographic* methods. Planimetric methods involve determining only the horizontal positions of features. The locations of these features are normally referenced in an *XY* coordinate system that is based upon some selected map projection (see Chap. 5). Topographic methods, on the other hand, include not only the location of planimetric details but also provide elevation information. Until recently, elevations have typically been represented by contours, which are lines connecting points of equal elevation. Now, however, elevations are often given in digital form; that is, Z coordinates are listed for a network of X, Y locations. This is particularly the case when data are being collected for GIS applications. This digital representation of elevations is called a *digital elevation model* (DEM), and this subject is discussed further in Sec. 13-6.

The purpose of the mapping, or spatial data collection, will normally dictate the level of detail required in representing features and the required level of accuracy as well. If the mapping is being done photogrammetrically by stereoscopic plotter, this in turn will dictate the flying height for the aerial photos, and hence also photo scale and map compilation scale. For example, maps used for the preliminary planning of a highway or transmission line would normally be compiled from high-altitude aerial photos which have a small scale. This would yield a small compilation scale, suitable for showing only general features and elevations with moderate accuracy. A map used for the design, construction, and operation of a utility, on the other hand, would generally be compiled from low-altitude photos, having a much larger scale. Thus compilation scale would be larger, and features and elevations would be shown in much greater detail and to a higher level of accuracy.

The concept of map scale and its relationship to level of detail that can be shown was well understood throughout the era when hard-copy maps were being compiled manually. That was the case because as map scale decreased, the number of features and contours that could be shown also had to decrease; otherwise, the map would have become congested to the point of being unreadable. Now, however, with the advent of digital mapping systems, the concept of mapping scale versus level of detail that can be shown is often not so obvious. With these systems, map products created from digital data can technically be plotted at any scale, and congestion can readily be removed by simply using the system's zoom feature. Nonetheless, the intended purpose of the mapping will still guide the level of detail presented, and this in turn will dictate photographic scale, compilation scale, and hence the positional accuracy of features shown.

13-2   DIRECT COMPILATION OF PLANIMETRIC FEATURES BY STEREOPLOTTER

In Chap. 12, fundamental operations of stereoscopic plotters were discussed. Regardless of the type of plotter employed, interior, relative, and absolute orientation (see Secs. 12-7 through 12-9) must be performed in order. Then compilation of the planimetric and/or topographic features within the "stereoscopic neat model" (see Sec. 18-7) can begin. Either of two basic methods may be employed in compiling topographic information with stereoplotters: (1) direct tracing from stereomodels to create hard-copy maps or (2) compiling digital files of topographic data from which computer-generated maps are prepared. The first method is described in this section, while the second is covered in Secs. 13-4 through 13-6.

Prior to the existence of digital mapping systems, photogrammetric map compilation was accomplished exclusively by direct tracing of topographic features and contours from stereomodels. This process, although still occasionally performed, now has almost completely given way to digital mapping. In the direct tracing process, to locate a "point" feature such as a utility pole, the floating mark is placed on the object in the stereomodel, and the corresponding location of the tracing pencil marks the point's map location. "Linear" features such as roads and streams, and "area" features such as lakes and buildings, are drawn by placing the floating mark on the feature at some starting point, and then tracing the feature by moving the floating mark continuously along or around the object. The corresponding movement of the tracing pencil on the map sheet simultaneously locates the feature. While tracing features in this manner, it is absolutely essential that the floating mark be always kept in contact with the feature being traced; otherwise, its planimetric position will be incorrect. If a feature is not visible on one photo (or both), the floating mark cannot be properly set. Often when buildings are traced, the base of the building cannot be seen, and generally the roof edges will be traced. Slender vertical features such as utility poles can be plotted at either the top or the base, for if the floating mark is truly in contact with the pole, both will correspond to the same point.

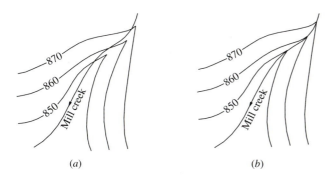

(a)   (b)

FIGURE 13-1
(a) Inconsistencies between planimetry and contour locations.
(b) Consistent renditions of planimetry and contours.

As a general rule in direct compilation, planimetric details are traced first, followed by contouring. This is so because natural and cultural features have a very significant effect on the location and appearance of contours (e.g., the V's of contours crossing streams must peak in the stream and contours cross roads at right angles). An example of carelessly compiled contours or planimetry is shown in Fig. 13-1a, while its corresponding correct rendition is shown in Fig. 13-1b. Also in direct compilation, it is advisable to plot all features of a kind at once (e.g., all roads), before proceeding to another feature, as this reduces the likelihood of omissions.

13-3 DIRECT COMPILATION OF CONTOURS BY STEREOPLOTTER

Direct compilation of contours is a difficult operation that takes considerable skill and experience on the part of the operator. To trace contours, the floating mark is set at the elevation of a desired contour and is moved within the model until the mark coincides with the apparent terrain surface in the model. This will occur at any point where the desired contour elevation exists. The operator then moves the floating mark about the stereomodel as necessary to keep it constantly in contact with the terrain, while the tracing pencil simultaneously follows the movement and records the contour. During the tracing of a contour, the operator must continuously look ahead to properly determine the direction in which to proceed.

In direct tracing of contours, it is recommended that all contours be completely compiled within a local area of the model, rather than attempt to trace a single contour across the entire model. To facilitate accurate contouring, the operator should become acquainted with the general shape of the terrain within the entire model before proceeding. This is usually most conveniently accomplished by viewing prints of the stereopair in three dimensions, using a stereoscope. Generally, contouring can be approached in much the same way as the assembly of a jigsaw puzzle. Easier areas and prominent features are compiled first, and the more difficult detail is filled in later. It can be helpful to initially locate and trace apparent drainage lines (ditches, swales, etc.) in order to guide the contouring operation. If necessary, these drainage lines can be subsequently erased during the map editing stage.

Some terrain areas such as flat expanses, regions of shadow, and areas of minimal image texture present particular difficulty when tracing contours. In these areas, it is sometimes necessary to determine "spot elevations" and interpolate contours from them. This is helpful because spot elevations can be read to significantly greater accuracy than direct contour tracing. In areas covered with trees or tall vegetation, it may be impossible to plot continuous contours. In these areas broken contours (contour lines plotted in open areas only) may be drawn; otherwise, spot elevations can be plotted and contours interpolated. Densely vegetated areas occasionally must be field-surveyed.

13-4   REPRESENTING TOPOGRAPHIC FEATURES IN DIGITAL MAPPING

Features that are typically depicted in topographic mapping are many and varied. In general, however, in digital mapping any feature can be represented as consisting of (1) a *point,* (2) a *line* (straight or curved), or (3) a *polygon* (closed shape). Horizontal and vertical control points, utility poles, and fire hydrants are examples of features that can be represented by points. Roads, railroads, streams, etc. can be shown as lines, while parcels of different land ownership or land-use types, large buildings, etc., can be represented by polygons.

As noted in the preceding section, map compilation scale dictates the level of detail of the features to be included. This in turn may affect the manner of representing a particular feature. For example, on a 1:24,000-scale map, an 8-m-wide road would be plotted only 0.3 mm wide, and therefore it would be appropriately represented as a single line. On the other hand, the same road shown on a 1:600-scale map would be 13 mm wide, and therefore both edges of the road could be clearly shown without congestion.

In digital mapping, code identifiers are used to keep track of the many different features that can be digitized from a stereomodel. Normally all features within one category are given a specific identifying code (number or letter). Then within that category, individual features may be further labeled with an additional identifier. Many different coding systems have been developed for categorizing map features based on type. These systems, while fundamentally similar, differ according to the primary purpose of the map product. Special considerations may also be required to fit the needs of individual projects. As an example, suppose an operator is digitizing all structures within a stereomodel, and in the mapping system being used, this category is identified by the number 3. Now within that category, buildings might be identified with the letter *A*, bridges with *B*, dams with *C*, retaining walls with *D*, etc. Thus a bridge would be identified with a feature code of *3B*. The following feature categories (with typical individual features listed within each category) are presented only as a general guide. The list is not all-inclusive and may need to be supplemented on a project-specific basis.

1. *Nonimage data:* coordinate grid, property boundaries, horizontal and vertical control points
2. *Streets, highways, and related features:* edge of pavement, centerlines, curbs, medians, barrier walls, shoulders, guardrails, parking lots, alleys, driveways
3. *Other transportation:* railroad tracks, abandoned railroads, airport runways and taxiways, unpaved roads, trails, sidewalks, port facilities, locks
4. *Structures:* buildings, bridges, water towers, dams, fence lines, stadiums, retaining walls, antennas, concrete slabs, swimming pools
5. *General land use:* cemeteries, parks and recreation areas, agricultural areas, quarries, stockpiles, landfills
6. *Natural features:* lakes, ponds, rivers, streams, beaches, islands, wetlands, wooded areas, individual trees, bushes and shrubs, meadows
7. *Terrain elevation:* contours, spot elevations
8. *Drainage features:* ditches and swales, retention basins, culverts, headwalls, catch basins, curb inlets, storm sewer manholes, flared end sections, berms
9. *Utilities:* utility poles, power lines, telephone lines, transmission lines, substations, transformers, fire hydrants, gate-valve covers, sanitary manholes
10. *Signs and lights:* traffic signals, streetlights, billboards, street signs
11. *Project-specific:* varies

13-5 DIGITIZING PLANIMETRIC FEATURES FROM STEREOMODELS

In recent years, given the development of computers and digitizing systems and the rapid growth and widespread use of geographic information systems (GISs), photogrammetric mapping procedures have changed dramatically. Whereas previous to that time, most maps were being traced directly from stereomodels and reproduced in hard copy by some type of printing process, today the vast majority of maps are being developed using computers and topographic data that have been compiled by digitizing from stereomodels. Many advantages accrue from the digital method of extracting topographic information. Perhaps the foremost advantage is that databases of geographic information systems require data in digital form; and in most cases, if proper formats and procedures are employed during the process of digitizing stereomodels, the resulting files can be entered directly without further modification into a GIS database. Other advantages are that digitized data afford greater flexibility of use. As examples, computerized mapping systems enable maps to be viewed on a screen, or they can be rapidly plotted in hard-copy form without the tedium of manual drafting. Furthermore, maps can be plotted using any desired scale and/or contour interval, and the topographic data can be instantaneously transmitted electronically to remote locations. Because the information is collected according to feature category or "layer" in the digitizing process, individual layers can be plotted and analyzed separately, or overlaid with other layers for analyses. These are just a few of the advantages—there are many others.

The process of digitizing planimetric features from stereomodels is fundamentally the same, whether the operator is using a digitized *mechanical projection stereoplotter* (see Part II of Chap. 12), an *analytical plotter* (see Part III of Chap. 12), or a *softcopy plotter* (see Part IV of Chap. 12 and Chap. 15). Of course, just as was the case of mapping by direct tracing from stereomodels, in digitizing data from stereomodels the instrument must be completely and accurately oriented prior to commencement of the digitizing process. To digitize an object within a stereomodel, the operator must bring the floating mark in contact with that object, enter the feature code, and then either push a button or depress a foot pedal. This causes the feature code and its X, Y, Z coordinates to be instantaneously stored in the computer. When digitizing linear or area features, a number of points along the line or polygon are digitized. This can be done very rapidly, since after the first point on any specific feature is identified and digitized, no further feature codes need be entered on that feature. On these types of features, most operators will pause momentarily when the button or foot pedal is depressed to ensure that the floating mark is on the desired object when the digitizing occurs. CAD systems associated with modern plotters provide for grouping of similar features into a specific layer in the drawing file. Generally, features are compiled in descending order of prominence. Point features that would appear smaller than about 1 mm at the intended map scale should be represented by appropriate symbols. Any notes or labels that the compiler feels are necessary to avoid later misidentification should also be made. Again, a set of paper prints and a stereoscope are essential to the compiler as an aid in identifying features.

13-6   DIGITAL ELEVATION MODELS AND INDIRECT CONTOURING

An alternative to the direct tracing of contours is to create a digital elevation model (DEM) and interpolate the contours from it. A digital elevation model is a discrete representation of a topographic surface. In some literature a DEM is referred to as a *digital terrain model* (DTM) or a *digital terrain elevation model* (DTEM). Two common forms of DEMs are *regular grid* and *triangulated irregular network* (TIN).

In a regular grid DEM, spot elevations are determined for a uniformly spaced array of ground cells or groundels. (See Sec. 9-3.) The elevations (Z values) are digitally stored in a computer array. Figure 13-2 shows a three-dimensional representation of a rectangular DEM. In this figure, lines are drawn to connect the center points of the groundel array, which, by virtue of their varying Z values, appear as undulating terrain. Since the z values are stored as an array of numbers, they are compatible for computer processing using various algorithms. These algorithms include contour generation, volume and slope calculation, orthophoto generation (see Sec. 13-7), and depiction of three-dimensional views. Computer algorithms applied to rectangular arrays of elevations tend to be very straightforward in their implementation. They readily lend themselves to investigations such as frequency analysis and finite element analysis.

A disadvantage associated with regular grid DEMs is that for large areas of level or uniformly sloping terrain, a great deal of computer memory is wasted storing highly redundant elevation information. For example, if a groundel spacing of 1 m is used for a flat area of terrain 100 meters square, then $100 \times 100 = 10,000$ storage locations will

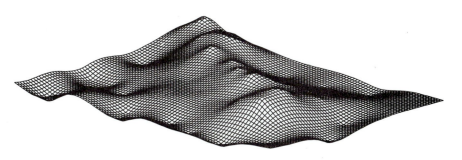

FIGURE 13-2
Three-dimensional representation of a rectangular DEM.

be used to represent the elevations. Since the terrain is flat, all 10,000 values will be the same. More complex data structures such as *quadtrees* can be used in lieu of arrays to reduce storage requirements of regular grids; however, their implementation is far more complicated.

A TIN affords a more efficient representation of terrain in terms of data storage. With a TIN, spot elevations can be acquired at *critical points,* which are high points, low points, locations of slope changes, ridges, valleys, and the like. Once these spot elevations have been collected, together with their X, Y coordinates, lines are constructed between the points forming a system of triangles covering the surface. The triangles can be assumed to be planar facets which are sloped in three-dimensional space. Each triangle side has a uniform slope, and the system of triangles is called a TIN model. The method generally used to construct the TIN is known as a *Delaunay triangulation.* In a Delaunay triangulation, the lines are drawn between points in closest proximity to each other, without any lines intersecting. The resulting set of triangles has the property that for each triangle, the circle that passes through the three vertices contains no vertices of any other triangles.

A further generalization of a TIN allows the inclusion of *breaklines*. Breaklines are lines which have constant slope and are used where there are discontinuities in the terrain surface, such as streams, retaining walls, and pavement edges. Within the TIN, a breakline will form the sides of two adjacent triangles and will have no other lines crossing it. By using breaklines in a TIN, the terrain surface can be more faithfully represented, and contours generated from the TIN will give a more accurate portrayal of the surface features.

Figure 13-3 illustrates the concept of using a TIN as a DEM in order to determine contours. Figure 13-3*a* shows a contour map, which for purposes of illustration is assumed to accurately represent a particular terrain area. Note that the figure includes a hill near the left edge, a depression at the upper left, a steady rise to the upper right, and a stream (*AB*) which runs from the upper left to the lower right. Figure 13-3*b* shows a TIN (dashed lines) formed from a sample of 20 spot elevations (shown as black dots at all triangle vertices) along with the set of interpolated contours (solid lines). The contours are formed by interpolating along the sides of the triangles and connecting the interpolated points with straight lines. (Smooth curves can also be used to approximate the straight-line contours.) Although the figure gives a nominal depiction of the hill, depression, and rise, the contours associated with the stream (*AB*) are not properly

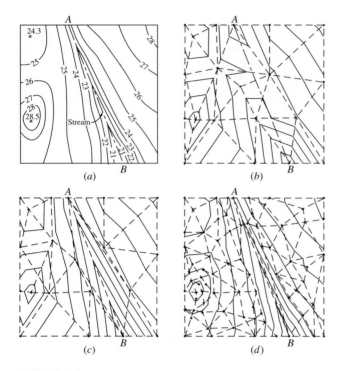

FIGURE 13-3

(*a*) Contour map showing accurate representation of terrain in an area. (*b*) Contours of the same area based on a TIN created from 20 data points, but without a breakline along stream *AB*. (Note the erroneous contour representations, especially in the area of the stream.) (*c*) Contours of the same area based on a TIN with 20 data points, but identifying stream line *AB* as a breakline. (Note the improvement in the contours in the area of the stream.) (*d*) Contours of the same area based on a TIN created from 72 data points, and with stream line *AB* identified as a breakline. [Note how well these contours agree with the accurate representation of (*a*).]

represented. In the TIN shown in Fig. 13-3*c*, the original 20 spot elevations are still shown, but a breakline has been added to connect the endpoints (*A* and *B*) of the stream. Note that in this figure, none of the TIN lines cross the stream as they did in Fig. 13-3*b*. As such, the contours associated with the stream give a much better representation. An even more accurate terrain representation is shown in Fig. 13-3*d*. In this figure, 72 spot elevations are used, in addition to the stream breakline. Note that the shapes and spacings of the contours in this figure are much closer to those in Fig. 13-3*a*, although some minor inconsistencies remain.

Because certain computational algorithms require the use of regular grids, it is sometimes necessary to convert a TIN to a regular grid DEM. This can be done by interpolating elevations between the TIN vertices at the centers of each grid cell. Various methods are available for performing this interpolation, including *nearest neighbor, inverse distance weighting, moving least squares,* and *Kriging.* As its name implies, the nearest-neighbor method simply assigns the elevation of a grid cell to that of the nearest spot elevation in the TIN. In the inverse distance weighting approach, elevations at

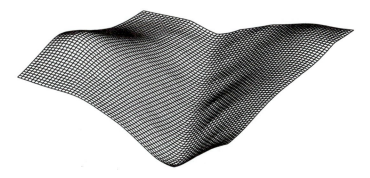

FIGURE 13-4
Three-dimensional view of a regular grid that has been interpolated from the
data points of Fig. 13-3*d* by Kriging.

the grid cell locations are computed by a weighted average of spot elevations, where
each weight is inversely proportional to the horizontal distance between the grid cell
center and the spot elevation point. By this approach, spot elevations that are nearer to
the grid cell will contribute more heavily to the weighted average. In the moving least
squares approach, a local polynomial surface $z = f(x, y)$ is fitted to a small set of points
nearest the grid cell being interpolated. Once the polynomial coefficients have been
computed by least squares, the z value (elevation) is computed at the given x, y of the
grid cell. The procedure is then repeated for every additional grid cell, with different
polynomial surfaces being computed for each. Kriging is a statistical method which
involves spatial correlation of points. The spatial correlation is specified a priori through
the use of a *variogram* which quantifies the influence a spot elevation will have on
points being interpolated. Proper definition of the variogram is important for generating
an accurate interpolation. Figure 13-4 shows a three-dimensional view of a regular grid
which has been interpolated by Kriging, using the spot elevations (and points along the
breakline) of Fig. 13-3*d*.

13-7 ORTHOPHOTO GENERATION

An orthophoto is a photograph showing images of objects in their true orthographic positions. Orthophotos are therefore geometrically equivalent to conventional line and
symbol planimetric maps which also show true orthographic positions of objects. The
major difference between an orthophoto and a map is that an orthophoto is composed of
images of features, whereas maps utilize lines and symbols plotted to scale to depict features. Because they are planimetrically correct, orthophotos can be used as maps for
making direct measurements of distances, angles, positions, and areas without making
corrections for image displacements. This, of course, cannot be done with perspective
photos. As discussed in Chaps. 20 and 21, orthophotos are widely used in connection
with geographic information systems, where they serve as planimetric frames of reference for performing analyses, and are also used for generating layers of information for
databases.

Orthophotos are produced from perspective photos (usually aerial photos) through a process called *differential rectification*, which eliminates image displacements due to photographic tilt and terrain relief. The process is essentially the same as standard rectification described in Secs. 10-12 and 10-13, except that it is performed independently at myriad individual, tiny surface patches or differential elements. In this way, rather than rectify the photograph to some average scale (which does not correct for relief displacements), each differential element is rectified to a common scale. Prior to the age of digital photogrammetry, complicated optical-mechanical devices were employed to produce orthophotos from film diapositives. Several of these are described in the second edition of this book. Modern orthophotos, however, are produced digitally, using scanned photographs and a DEM as the fundamental input information. Details of the method of producing digital orthophotos are given in Sec. 15-8.

Orthophotomaps prepared from orthophotos offer significant advantages over aerial photos and line and symbol maps because they possess the advantages of both. On one hand, orthophotos have the pictorial qualities of air photos because the images of an unlimited number of ground objects can be recognized and identified. Furthermore, because of the planimetric correctness with which images are shown, measurements may be taken directly from orthophotos just as from line maps. When digital orthophotos are stored in a computer in their numeric form, they can be incorporated with a regular grid DEM having the same ground resolution to produce three-dimensional pictorial views. This is achieved by assigning the brightness values of the orthophoto pixels to corresponding cells of the DEM and manipulating the view angle. The result is a three-dimensional view of the terrain with the image draped over the top, as shown in Fig. 13-5.

FIGURE 13-5
Three-dimensional pictorial view of terrain obtained by combining a DEM and orthophoto.

13-8   MAP EDITING

After map manuscripts have been compiled with a stereoplotter, they are passed on to an editor who provides finishing touches. Since modern stereoplotter manuscripts are compiled in a CAD environment, the resulting digital files can be used directly by the editor at a computer workstation. The editor checks the manuscript for completeness and the use of proper symbols and labels. Endpoints of lines which delineate polygon features are forced to close properly, and buildings are squared up. Lines which continue between adjacent manuscript sheets are edge-matched so that they will be continuous. Proper line types (dashed, solid, etc.) are assigned along with appropriate line thickness. Depending upon project requirements and the features in the mapped area, additional edits may be performed. In a CAD environment, many of the editing functions can be achieved through automated processing, particularly edge-matching, closing polygons, and building squaring. After the manuscripts have been edited, the map should be checked to verify its compliance with the appropriate accuracy standards.

When the digital manuscript has been edited to its final form, the original stereoplotter manuscripts form a continuous and seamless composite map of the project area. The corresponding digital CAD file can be saved on a tape or compact disk for subsequent delivery to the client. When individual map sheets are required, the editor will extract areas of the digital manuscript in a "cookie-cutter" fashion and place them in an appropriate border and title block. A north arrow, scale bar, legend, notations, etc., are then placed on the map which can be subsequently plotted at the appropriate scale.

13-9   ELEVATION DATA COLLECTION BY LASER MAPPING SYSTEMS

A recent innovation in topographic mapping devices is the laser scanner, or LIDAR (*li*ght *d*etection *a*nd *r*anging). It has resulted from advances in miniaturization of electronics and integration of different measuring devices. The system consists primarily of a laser scanner, inertial navigation system, GPS receiver, and controlling and data-recording computers. The system is flown in an airborne platform and is capable of obtaining measurements of X, Y, Z coordinates of a very large number of points on the earth's surface. Figure 13-6 shows a laser mapping system mounted in an aircraft. At the lower right in the figure is the laser device mounted directly over a hole in the floor of the aircraft. The control electronics and GPS receiver are located on the chassis at the left.

The system operates on a principle similar to that of the flying spot scanner, described in Sec. 3-14.3. As the aircraft flies along its trajectory, a laser pulse is transmitted to the terrain below and its return signal is detected a split second later. By accurately measuring the time delay between the transmission and return, and applying the speed of light, the distance from the scanner to the terrain point can be determined. At the same time, the accompanying inertial navigation device periodically records the attitude angles (roll, pitch, and yaw; or omega, phi, and kappa) of the aircraft. In addition, the interfaced GPS receiver is periodically recording the X, Y, Z positions of the antenna. After the signals from the three devices are processed and combined, the laser

FIGURE 13-6
Laser mapping system mounted in an
aircraft. (*Courtesy University of Florida.*)

pulses essentially define vector displacements (distances and directions) from specific
points in the air, to points on the ground. Since the laser device can generate pulses at
a very fast rate (thousands of pulses per second), the result is a dense pattern of
measured X, Y, Z points on the terrain. In addition, by measuring the intensities of the
return pulses, as well as their delays, a point-sampled image of the terrain can be
obtained.

The dense pattern of points collected by the laser scanner constitutes spot elevations which can then be stored and manipulated as a DEM. From this DEM, contour maps
and other topographic products can be produced. Currently, elevation accuracies of laser
scanners are on the order of 10 to 15 cm, and this accuracy is likely to improve with further research and development. Figure 13-7a shows a shaded relief map of a landfill site
produced from a laser mapping system. In this figure, notice that the laser scanner has
recorded spot elevations not only on the terrain, but on tree canopies as well. Computer
processing routines are available which can essentially remove these artifacts from the
DEM, leaving an accurate representation of the bare ground. Figure 13-7b shows an
image of the same area formed from the return intensities of the laser pulses. This image
is similar to a digital aerial photograph where the laser return intensities are analogous
to pixel brightness values. Note that while the image may not have the clarity and resolution of an aerial photograph, it has no shadows since the laser is an active light source.
As laser mapping systems become more common, they promise to become highly
effective topographic mapping tools.

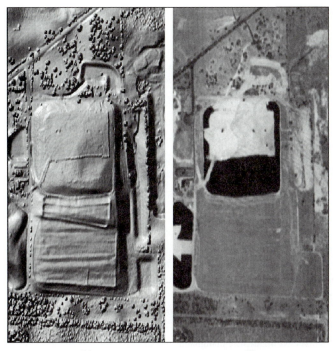

(a) (b)

FIGURE 13-7
(a) Shaded relief map of a landfill site produced from a laser mapping
system. (b) Image obtained from the intensities of returned laser pulses.
(*Courtesy University of Florida.*)

REFERENCES

American Society for Photogrammetry and Remote Sensing: *Digital Photogrammetry: An Addendum to the Manual of Photogrammetry,* Bethesda, MD, 1996, chaps. 6 and 9.

Benjamin, S., and L. Gaydos: "Spatial Resolution Requirements for Automated Cartographic Road Extraction," *Photogrammetric Engineering and Remote Sensing,* vol. 56, no. 1, 1990, p. 93.

Carter, J. R.: "Digital Representations of Topographic Surfaces," *Photogrammetric Engineering and Remote Sensing,* vol. 54, no. 11, 1988, p. 1577.

Cowen, D. J.: "GIS versus CAD versus DBMS: What Are the Differences?" *Photogrammetric Engineering and Remote Sensing,* vol. 54, no. 11, 1988, p. 1551.

Doytsher, Y., and B. Shmutter: "A New Approach to Monitoring Data Collection in Photogrammetric Models," *Photogrammetric Engineering and Remote Sensing,* vol. 54, no. 6, 1988, p. 715.

Flood, M., and B. Gutelius: "Commercial Implications of Topographic Terrain Mapping Using Scanning Airborne Laser Radar," *Photogrammetric Engineering and Remote Sensing,* vol. 63, no. 4, 1997, p. 327.

Fortune, S. J.: "A Sweepline Algorithm for Voronoi Diagrams," *Algorithmica,* vol. 2, 1987, p. 153.

Preparata, F. P., and M. I. Shamos, *Computational Geometry,* Springer-Verlag, New York, 1985.

Ramey, B. S.: "U.S. Geological Survey National Mapping Program: Digital Mapmaking Procedures for the 1990s," *Photogrammetric Engineering and Remote Sensing,* vol. 58, no. 8, 1992, p. 1113.

Shrestha, R., W. E. Carter, M. Lee, P. Finer, and M. Sartori: "Airborne Laser Swath Mapping: Accuracy Assessment for Surveying and Mapping Applications," *Surveying and Land Information Systems,* vol. 59, no. 2, 1999, p. 83.

U.S. Army Corps of Engineers: *Photogrammetric Mapping,* EM 1110-1-1000, Hyattsville, MD, 1993.

Wang, J., G. J. Robinson, and K. White: "A Fast Solution to Local Viewshed Computation Using Grid-Based Digital Elevation Models," *Photogrammetric Engineering and Remote Sensing,* vol. 62, no. 10, 1996, p. 1157.

PROBLEMS

13-1. Explain the basic differences between planimetric mapping and topographic mapping.

13-2. Discuss the relationships between map scale and the level of detail that can be shown on a map.

13-3. Describe the process of using a stereoscopic plotter to create a hard-copy map by directly tracing planimetric features.

13-4. Describe the process of using a stereoscopic plotter to create a hard-copy map by directly tracing contours.

13-5. Why are feature codes necessary in digital mapping?

13-6. Describe the general process of digitizing planimetric features from stereomodels.

13-7. Cite several advantages of digital mapping over direct tracing of hard-copy maps.

13-8. Describe the following terms: (*a*) DEM and (*b*) TIN.

13-9. What are the slope characteristics that exist for the sides of all triangles of a TIN model?

13-10. What is Delaunay triangulation?

13-11. Describe breaklines, and discuss their importance in digital mapping.

13-12. How do orthophotos differ from normal aerial photos?

13-13. Discuss the similarities and differences between orthophotomaps and line and symbol maps.

13-14. What is the purpose of the map editing process? Name some specific conditions that are examined in the process.

13-15. Describe the operating principles of the LIDAR system of mapping.

CHAPTER
14

FUNDAMENTAL PRINCIPLES OF DIGITAL IMAGE PROCESSING

14-1 INTRODUCTION

Digital image processing in general involves the use of computers for manipulating digital images in order to improve their quality and/or modify their appearance. In digital image processing, the digital number (see Sec. 2-13) of each pixel in an original image is input to a computer, with its inherent row and column location. The computer operates on the digital number according to some preselected mathematical function or functions, and then stores the results in another array which represents the new or modified image. When all pixels of the original image have been processed in this manner and stored in the new array, the result is a new digital image.

Many different types of digital image processes can be performed. One type falls under the general heading of *preprocessing operations*. These are generally aimed at correcting for distortions in the images which stem from the image acquisition process, and they include corrections for such conditions as scanner or camera imperfections and atmospheric refraction. These procedures are discussed in Chap. 4. Another type of digital image processing, called *image enhancement,* has as its goal the improvement of the visual quality of images. Image enhancement makes interpretation and analysis of images easier, faster, and more accurate; and thus it can significantly improve the quality of photogrammetric products developed from digital images, and reduce the cost of producing them. Digital orthophotos in particular benefit significantly from the improved image quality that results from image enhancements.

A third type of digital image processing, called *image classification,* attempts to replace manual human visual analysis with automated procedures for recognizing and identifying objects and features in a scene. Image classification processes have been widely used in a host of different interpretation and analysis applications, as well as in the production of a variety of thematic maps. An important new application is their use in automated soft copy mapping systems (see Chap. 15). A final type of digital image processing, *data merging,* combines image data for a certain geographic area with other geographically referenced information in the same area. The procedures may overlay multiple images of the same area taken at different dates—a technique which is very useful in identifying changes over time, such as monitoring a forest fire or following the spread of a disease in a certain tree species. The procedures can also combine image data with nonimage data such as DEMs, land cover, and soils. These types of digital image processing are extremely important in the operation of geographic information systems.

In this chapter, some concepts that are fundamental to digital image processing are presented, and basic procedures for a few of the more common types of digital image processes are introduced. Examples are given to illustrate their effects.

14-2   THE DIGITAL IMAGE MODEL

To comprehend the subject of digital image processing, an understanding of the concept of the digital image model is useful. In a mathematical sense, a digital image model can be considered as a mathematical expression which yields image density as a function of x, y (row, column) position. This is analogous to a regular grid DEM which yields elevation of a terrain surface as a function of X, Y position.

Digital images are imperfect renditions of the object scenes they portray. Their imperfections stem from a host of sources, including the imaging system, signal noise, atmospheric scatter, and shadows. The primary degradation of an image is a combined systematic blurring effect resulting from aberrations of the lens, resolution of the recording medium (e.g., film), and, to some extent, atmospheric scatter. The effect of these combined factors can be specified in terms of the *point-spread function.* A point-spread function can be thought of as the blurred image that would result from a perfect point source of light such as a star. This point spread function can be represented as a mathematical expression which models the systematic image imperfection, and which was applied to the ideal image through a process known as *convolution.* (The subject of convolution in image processing is discussed in Sec. 14-5, and examples are given.)

The cumulative effect of the systematic image degradations can be represented in a general sense by

$$I(x, y) = O(x, y) * P(x, y) + N(x, y) \tag{14-1}$$

In Eq. (14-1), $I(x, y)$ is the actual image model as represented by the digital image, $O(x, y)$ is the theoretical ideal image model of the object, $P(x, y)$ is the point-spread function, $N(x, y)$ is signal noise, and * (asterisk) indicates the convolution operation. One goal of image processing is to negate the effects of image noise and the point-spread function to recover the theoretical ideal image.

While there are techniques known as *optimal filters* that can negate these detrimental effects to some extent, their implementation is complex and requires an in-depth knowledge of signal processing theory. On the other hand, simpler methods are also available which can at least partially compensate for these effects. Some of these methods can reduce image noise, and others can reduce the blurring caused by the point-spread function through edge enhancement.

14-3 SPATIAL FREQUENCY OF A DIGITAL IMAGE

Section 2-13 gave a general description of digital images and described the processes by which they are produced. It introduced the concepts of pixels as discrete samples of the image at a specific geometric resolution, as well as the quantization of brightness values. These are important concepts in understanding digital image processing. Another notion that is key to the comprehension of digital image processing is that of *spatial frequency* (see Sec. 3-13). Objects (e.g., terrain features) have a virtually unlimited range of spatial frequencies. Imagine, for example, an ideal image of a cornfield, where the rows are nominally spaced 1 m apart. On a small-scale image (e.g., high-altitude air photo), the variations from light to dark to light, etc. (corresponding to corn, soil, corn, etc.) might be barely discernible. The spatial frequency at object scale associated with this variation would be 1 cycle per meter. A medium-scale image (e.g., low-altitude air photo) of the same cornfield might reveal variations from light to dark between individual leaves of the plants. If the leaves were nominally spaced 10 cm (0.1 m) apart, the corresponding spatial frequency would be 1 cycle per 0.1 m, or 10 cycles per meter. A large-scale image (e.g., close-up photo taken with a handheld camera) might reveal variations associated with individual kernels on an ear of corn. If the kernels were nominally spaced 5 mm (0.005 m) apart, the corresponding spatial frequency would be 1 cycle per 0.005 m, or 200 cycles per meter. If an individual kernel were viewed under a microscope, even higher spatial frequencies would be discernible. The point of this example is that a multitude of different frequencies exists in objects in nature.

Although there are many different frequencies in a natural scene, most of the higher frequencies will be lost when a digital image is acquired. This occurs whether the image is exposed directly by a digital sensor or is obtained by scanning a film-based photograph. In the process of obtaining digital images, brightness variations that occur within an individual pixel will be averaged together, thus attenuating high frequencies. This is a consequence of the physical size of the pixel being sampled. The highest frequency that can be represented in a digital image is the *Nyquist frequency,* which is one-half the sampling frequency. As an example, in a satellite image with pixels having a 10-m ground resolution, the sampling frequency is 1 sample per 10 m, or 0.1 sample per meter. The highest spatial frequency (Nyquist frequency) that can be represented in this image is therefore 0.05 cycle per meter, or 1 cycle per 20 m. Note that 1 cycle corresponds to two samples.

The relationship between the variations of digital numbers in a digital image and the spatial frequencies they represent, can be precisely quantified in a mathematical sense. Often, certain characteristics of a digital image are more logically described in terms of spatial frequency than by digital numbers in an image. In sections that follow, explanations will be given which should help to clarify this relationship. However, a full understanding of the concept will require substantial study in the area of signal processing theory, which is beyond the scope of this text.

14-4   CONTRAST ENHANCEMENT

Contrast enhancement is a commonly employed type of digital image processing operation. As discussed in Sec. 2-13, when each pixel of a scene is quantized, a digital number that corresponds to the brightness at its location is assigned. In an area where the overall scene brightness is low (e.g., a shadow region), the majority of the pixels will have low values. In such cases, the contrast between features will be diminished. To get an impression of the overall brightness values in an image, it is useful to produce a *histogram* of the digital numbers. The histogram is a plot of digital number on the abscissa versus number of occurrences on the ordinate. As an example, if a digital image has 1907 pixels having a digital number (DN) of 29, the histogram will show an ordinate of 1907 at the abscissa of 29. Figure 14-1*a* shows a small portion of a digital image which is in the shadow of a building, and Fig. 14-1*b* shows its corresponding histogram of digital numbers. The image contains a sidewalk and a bench which are barely noticeable because of the low contrast caused by the shadow. Notice from the histogram that the majority of digital numbers falls in the range of 20 to 65, even though the available range is 0 to 255. This indicates that when the image is displayed, the contrast among features in this area will be low, as is certainly evident in Fig. 14-1*a*.

A simple method of increasing the contrast is to apply a *linear stretch*. With this operation, the digital numbers within a certain range (e.g., from 20 to 65) are linearly

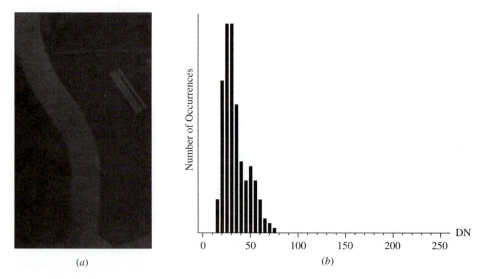

(*a*) (*b*)

FIGURE 14-1
(*a*) Digital image of an area in a shadow. (*b*) Histogram of digital numbers from the image.

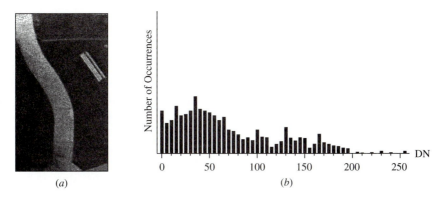

(a)   (b)

FIGURE 14-2
(a) Digital image from Fig. 14-1a after linear stretch contrast enhancement. (b) Histogram of digital numbers from the image after linear stretch.

expanded to the full available range of values (e.g., from 0 to 255). The effect of a linear stretch on the image of Fig. 14-1a is illustrated in Fig. 14-2a. Its associated histogram is shown in Fig. 14-2b. Notice in this histogram that the digital numbers have been expanded (stretched) so as to encompass the full available range (0 to 255). As can be seen in Fig. 14-2a, the effect of this linear stretch is enhanced contrast, making the sidewalk and bench more apparent.

A more complex method of increasing the contrast is to apply *histogram equalization*. The concept behind histogram equalization is to expand the digital numbers in a nonlinear fashion across the available range so that the values are more evenly distributed. The effect of histogram equalization on the image of Fig. 14-1a is illustrated in Fig. 14-3a. Its associated histogram is shown in Fig. 14-3b. Notice from this histogram that the numbers are more evenly distributed, particularly at the high end of the digital number range. Note also that the histogram-equalized image of Fig. 14-3a has even

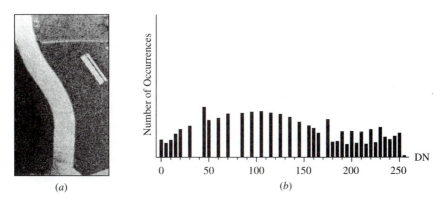

(a)   (b)

FIGURE 14-3
(a) Digital image from Fig. 14-1a after histogram equalization contrast enhancement.
(b) Histogram of digital numbers from the image after histogram equalization.

greater contrast than that of the linear stretched image, making the sidewalk and bench even more pronounced.

14-5 SPECTRAL TRANSFORMATIONS

As discussed in Sec. 14-3, digital images capture information at various spatial frequencies. Digital images in their native form exist in the *spatial domain.* The term *spatial domain* refers to the concept that the positions of pixels in the image relate directly to positions in the two-dimensional space of the image. Through certain mathematical operations, images can be converted from the spatial domain to the *frequency domain,* and vice versa. A digital image which has been converted to the frequency domain has pixels whose positions (row and column) relate to frequency instead of spatial locations. Images in the frequency domain no longer have visibly recognizable features as spatial domain images do. They are an abstraction of the original spatial image; however, they do contain the entire information content of the original. In the frequency domain, the spatial frequencies inherent in the native image are represented by measures of amplitude and phase for the various frequencies present in the image.

The most commonly used method for converting a digital image to the frequency domain (or vice versa) is the *discrete Fourier transform* (DFT). This process is roughly analogous to conversion between geodetic and geocentric coordinates (see Sec. F-3). Converting from geodetic to geocentric coordinates changes the representation of a point's position from latitude, longitude, and height; to X, Y, and Z. In an analogous fashion, the DFT converts the brightness values from a (spatial domain) digital image to a set of coefficients for sine and cosine functions at frequencies ranging from 0 to the Nyquist frequency. In an additional step, amplitude and phase information can be obtained from these sine and cosine coefficients. A more detailed explanation of the DFT requires the use of complex numbers, which is beyond the scope of this text.

The discrete Fourier transform has become a useful tool for analyzing digitally sampled periodic functions in one, two, or even more dimensions. (Note that, as described in Sec. 14-2, a digital image is a two-dimensional function.) A particularly efficient implementation of the discrete Fourier transform is the aptly named *fast Fourier transform* (FFT). Since its discovery and implementation in the 1960s, it has enabled discrete Fourier transforms to be computed at a greatly enhanced speed. This greatly enhanced computational speed has led to wider use of the Fourier transform as an image processing tool.

Figure 14-4*a* shows a digital satellite image of a portion of Gainesville, Florida, with its Fourier transform (amplitudes only) shown in Fig. 14-4*b*. In Fig. 14-4*b*, greater amplitudes correspond to brighter tones, and lesser amplitudes correspond to darker tones. This Fourier transform exhibits a star-shaped pattern which is typical of two-dimensional transforms of urban scenes. The lowest spatial frequencies are represented at the center of the star with frequency increasing with increasing distance from the center. It can be seen from Fig. 14-4*b* that lower frequencies have greater magnitudes than higher frequencies. The figure also shows discernible spikes radiating from the center. Those radiating in vertical and horizontal directions are due to discontinuities associated with the edges of the image, whereas the spikes oriented approximately 10° counterclockwise are due to the edges associated with the primary streets and buildings in the image.

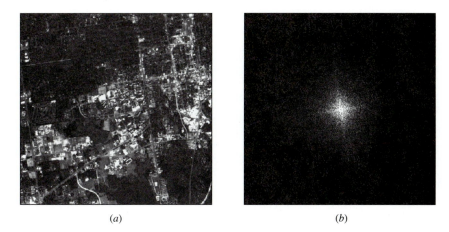

(a)   (b)

FIGURE 14-4
(a) Digital satellite image of Gainesville, Florida. (b) Fourier transform. (*Courtesy University of Florida.*)

Certain operations are easier to accomplish in the frequency domain than in the spatial domain. For example, assume a digital sensor had an electrical problem which caused an interference pattern in the image as illustrated in Fig. 14-5a. Conversion of this contaminated image from the spatial to the frequency domain reveals high amplitudes for the particular frequencies which contributed to the interference, as shown in Fig. 14-5b. The coefficients corresponding to the sine and cosine terms for those frequencies can be changed from the high values to zero values as illustrated in Fig. 14-5c, thus eliminating the interference in the frequency domain image. To view the image, an inverse Fourier transform would then be applied to the modified frequency image to convert back to the spatial domain, as illustrated in Fig. 14-5d. Since the systematic interference was eliminated in the frequency domain, it is no longer present in the spatial domain image. The interference removal was not perfect, however, as can be seen in the subtle interference patterns which remain in Fig. 14-5d.

As mentioned above, certain mathematical operations are easier to compute in the frequency domain than the spatial domain. Another such operation is *convolution*. This process involves two functions: the signal (e.g., digital image) and the response function. An example of a response function is the mathematical representation of the effect that a real imaging system imparts to an ideal image to produce the actual image (i.e., the point-spread function). The two functions are convolved, which will apply the "smearing" effect of the response function to the signal. Depending on the form of the response function, the convolution will have the effect of filtering out certain high frequencies from the image. In the absence of noise, the inverse operation *deconvolution* can be applied to an image that has been affected by a known response function, to reconstruct the high frequencies which were lost. Deconvolution may be done to remove the blurring effect of the point-spread function (see Sec. 14-2) in order to recover the ideal image from an imperfect real image. Deconvolution should be done with caution, however, since it can be highly sensitive to image noise.

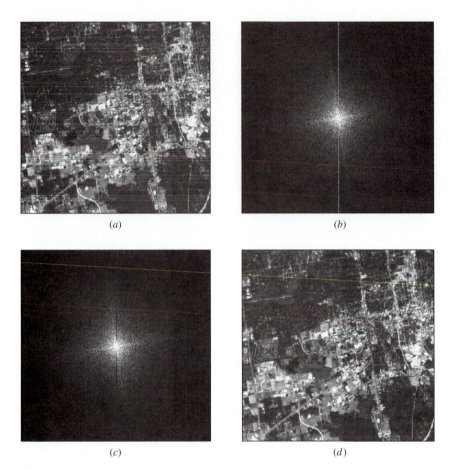

FIGURE 14-5
(*a*) Gainesville image with interference noise. (*b*) Fourier transform. (*c*) Fourier transform with high-frequency interference eliminated. (*d*) Cleaned image after inverse Fourier transform of (*c*). (*Courtesy University of Florida.*)

Another operation which is easy to compute in the frequency domain is *correlation*. Correlation also involves two functions, which can both be assumed to be signals (e.g., digital images). Correlation provides a measure of the similarity of two images by comparing them while superimposed, as well as shifting an integral number of pixels left, right, up, or down, all with one simultaneous operation. Correlation can be useful in matching patterns from one image that correspond to those of another image. Pattern matching provides the backbone of softcopy mapping systems (see Chap. 15). A drawback associated with correlation is that it is sensitive to noise in the image, and as such, other techniques are generally used for matching images.

Fourier transforms, while probably the best known, are not the only image transformations available. Other useful transformations are the *wavelet transform* and the

Hough transform. Wavelet transforms are useful for purposes such as reducing the data storage requirements for digital images (image compression) as well as many others. The use of wavelets has become very popular in digital signal analysis. The Hough transform is primarily used for joining discontinuous edges in an image. It can be highly useful for reconstructing linear and circular features from raster images, where the edges have breaks in them. Details of these transforms are beyond the scope of this text; however, references cited at the end of this chapter can be consulted for further information.

14-6 MOVING WINDOW OPERATIONS

As mentioned in Sec. 14-5, the Fourier transform is highly effective for performing convolution of a digital image with a response function. A simpler method known as the *moving window* can also be used for performing convolution. The moving window is particularly useful when the response function is highly localized. For example, a two-dimensional response function may consist of a 3×3 submatrix of nonzero terms contained within a large array which has zero terms everywhere else. When convolving using the Fourier transform, the image and response function must have the same dimensions, which can lead to inefficiency when the vast majority of terms in the response function are equal to zero. In cases such as this, the moving window can be used efficiently.

The moving window operation has two primary inputs: the original digital image and a localized response function known as a *kernel*. The kernel consists of a set of numbers in a small array, which usually has an odd number of rows and columns. The kernel is conceptually overlaid with a same-size portion of the input image (image window), and a specific mathematical operation is carried out. The value resulting from this operation is placed in the center of the corresponding location in the output image, the kernel is shifted by one column (or row), and another value is computed. This procedure is repeated until the entire output image has been generated. Figure 14-6 illustrates one step of the procedure. In this figure, the 3×3 kernel is centered on pixel 3, 4 (row = 3, column = 4) in the input image. After the convolution is performed, the result is placed in pixel 3, 4 of the output (convolved) image. The kernel is then shifted so that it is centered on pixel 3, 5, and the convolution is repeated. When convolving pixels at the edge of the original image, the kernel will extend outside the image boundary. Under the assumption that the image is a periodic function, the kernel values that extend beyond the image will "wrap-around" so as to overlay with image pixels at the opposite side of the image. The wrap-around effect can be visualized by imagining exact copies of the image to exist above, to the right, below, to the left, and diagonally away from each of the four corners of the original image. This is the same result that would have occurred if the Fourier transform had been used to perform the convolution.

The mathematical operation of convolution is accomplished through a series of multiplications and additions. For example, the convolution result depicted in Fig. 14-6 is computed by

$$C_{34} = k_1 \times I_{23} + k_2 \times I_{24} + k_3 \times I_{25} + k_4 \times I_{33} + k_5 \times I_{34}$$
$$+ k_6 \times I_{35} + k_7 \times I_{43} + k_8 \times I_{44} + k_9 \times I_{45} \tag{14-2}$$

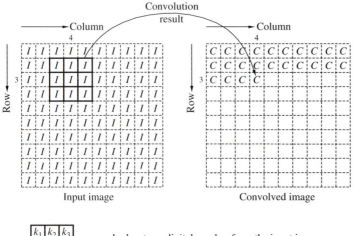

I - denotes a digital number from the input image
C - denotes a digital number from the convolved image

FIGURE 14-6
Moving-window image convolution.

In Eq. (14-2), the k values are the individual kernel elements, the I values are digital numbers from the input image with the appropriate row and column indicated by their subscripts, and C_{34} is the convolution result which is placed at row = 3, column = 4 in the output image.

Perhaps the simplest form of convolution kernel is one that computes the average of the nearby pixels. This type of convolution is known as a *low-pass filter*, so called because the averaging operation attenuates high frequencies, allowing low frequencies to be passed on to the convolved image. The kernel values used in a simple 3 × 3 low-pass filter are shown in Eq. (14-3).

$$K = \begin{bmatrix} \frac{1}{9} & \frac{1}{9} & \frac{1}{9} \\ \frac{1}{9} & \frac{1}{9} & \frac{1}{9} \\ \frac{1}{9} & \frac{1}{9} & \frac{1}{9} \end{bmatrix} \tag{14-3}$$

In this equation, K is the kernel matrix, which in this case has all elements equal to $\frac{1}{9}$. Convolution of this kernel matrix with a corresponding 3 × 3 image submatrix is equivalent to computing a weighted average of the digital numbers within the submatrix. Notice that the sum of the elements in this kernel matrix equals 1, which is appropriate when computing a weighted average. In fact, elements of a kernel can be considered to be the weights used in computing a weighted average. If the sum of the elements of the kernel matrix does not equal 1, the convolution result can be divided by the kernel's sum before placement into the output image.

Example 14-1. Given the following convolution kernel K and the matrix of digital numbers D from a 5×5 image, compute the convolved image C, using the moving window approach.

$$K = \begin{bmatrix} 0.1 & 0.1 & 0.1 \\ 0.1 & 0.2 & 0.1 \\ 0.1 & 0.1 & 0.1 \end{bmatrix}$$

$$D = \begin{bmatrix} 84 & 79 & 43 & 74 & 76 \\ 81 & 77 & 45 & 75 & 78 \\ 50 & 48 & 39 & 42 & 44 \\ 80 & 77 & 46 & 73 & 72 \\ 82 & 83 & 47 & 71 & 70 \end{bmatrix}$$

Solution

1. Create the periodic wrap-around effect by adding an extra row and column from the imaginary copies of the image surrounding the original image.

$$D' = \begin{bmatrix} 70 & 82 & 83 & 47 & 71 & 70 & 82 \\ 76 & 84 & 79 & 43 & 74 & 76 & 84 \\ 78 & 81 & 77 & 45 & 75 & 78 & 81 \\ 44 & 50 & 48 & 39 & 42 & 44 & 50 \\ 72 & 80 & 77 & 46 & 73 & 72 & 80 \\ 70 & 82 & 83 & 47 & 71 & 70 & 82 \\ 76 & 84 & 79 & 43 & 74 & 76 & 84 \end{bmatrix}$$

2. Convolve at each 3×3 position of the moving window. Calculation of the first two elements will be shown as examples.

$$c_{11} = 0.1 \times 70 + 0.1 \times 82 + 0.1 \times 83 + 0.1 \times 76 + 0.2 \times 84$$
$$+ \; 0.1 \times 79 + 0.1 \times 78 + 0.1 \times 81 + 0.1 \times 77$$

$$c_{11} = 79.4 = 79 \text{ (rounded to nearest integer)}$$

$$c_{12} = 0.1 \times 82 + 0.1 \times 83 + 0.1 \times 47 + 0.1 \times 84 + 0.2 \times 79$$
$$+ \; 0.1 \times 43 + 0.1 \times 81 + 0.1 \times 77 + 0.1 \times 45$$

$$= 70.0 = 70 \text{ (rounded to nearest integer)}$$

the remaining values are

$$C = \begin{bmatrix} 79 & 70 & 64 & 65 & 77 \\ 70 & 62 & 57 & 59 & 68 \\ 66 & 59 & 56 & 56 & 64 \\ 69 & 63 & 57 & 58 & 66 \\ 79 & 70 & 64 & 64 & 75 \end{bmatrix}$$

The result of a low-pass filter convolution is illustrated in Figs. 14-7a and b. Figure 14-7a shows the original image of cars in a parking lot. A low-pass filter was applied

(a)   (b)

FIGURE 14-7
(a) Parking lot image. (b) Parking lot image after applying a low-pass filter.

using a 5 × 5 convolution kernel which computed a simple average, resulting in the image of Fig. 14-7b. Notice that the image of Fig. 14-7b now appears somewhat blurred since the highest-frequency detail has been filtered out from the original image. By applying kernels of larger size (say, 7 × 7 or 9 × 9) a wider range of high-frequency information will be filtered from the image. The ultimate limit in kernel size would be the same as that of the original image. Such a convolution would result in all pixels of the output image having the same value—the overall average of digital numbers from the input image. In that case, detail at all frequencies will have been filtered out, leaving a uniformly gray image.

The intentional blurring of an image as the result of applying a low-pass filter may seem like a counterproductive operation. After all, high-frequency detail assists in identifying features and measuring the positions of their edges. The method could serve as a simple high-frequency noise filter, but there are better methods available for this purpose. However, another use for low-pass filtering is to employ the process as a precursor to a *high-pass* operation. By performing a pixel-by-pixel subtraction of the low-passed image from the original image, the result will give the high-frequency detail which was filtered out in the low-pass operation. Since subtraction can yield negative numbers which are inconvenient to deal with in a digital image, the absolute value can be taken after each subtraction. Figure 14-8 shows the high-pass filtered result from subtracting the digital numbers of Fig. 14-7b from those of Fig. 14-7a and taking the absolute value. As is apparent from the figure, this simple high-pass filter can be used to detect edges in the original image.

Moving window operations can be used to perform other useful operations, such as noise filtering. One such noise filter method is known as *median filtering*. In this approach, a convolution is not performed in the normal fashion. Rather, a 3 × 3 moving window is passed through the input image, and the 9 pixels in the immediate neighborhood are extracted at each step. The nine digital numbers are then sorted, and the median

FIGURE 14-8
Parking lot image after application of a simple high-pass filter.

(middle) value is placed in the corresponding location in the output image. By using the middle value rather than the average, the median filter will not be sensitive to any extremely high or low value which may be the result of image noise. The result of median filtering is shown in Figs. 14-9a and b. Figure 14-9a shows the same image of a parking lot shown in Fig. 14-7a, except that for purposes of this example, random "salt and pepper" noise has been added. After the median filter operation is applied, the image of Fig. 14-9b results. Note that the noise has indeed been eliminated, at the expense of a subtle loss of high-frequency information. Median filters are useful for removing many forms of random noise, but are not as effective at removing systematic noise.

Another class of operations which uses the moving window approach is edge detection. Earlier in this section, a simple form of edge detection based on high-pass filtering was presented. Other methods exist for edge detection, and two specific ones are presented here: the laplacian and Sobel operators.

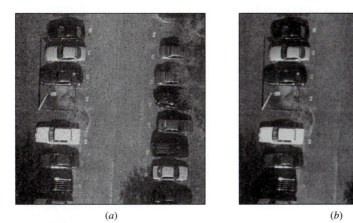

(a)   (b)

FIGURE 14-9
(a) Parking lot image with noise. (b) Parking lot image after application of median filter.

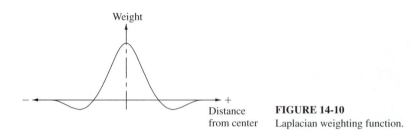

Weight

Distance
from center

FIGURE 14-10
Laplacian weighting function.

The laplacian operator is a convolution using a kernel which has values corresponding to the shape of the function shown in Fig. 14-10, except that it extends in two dimensions. In Fig. 14-10, note that at the center, the laplacian starts with a positive value and drops off to negative values at a certain distance from the center. The effect of this type of kernel is to amplify the differences between neighboring digital numbers which occur at an edge. If the kernel values (weights) are chosen so that their sum is equal to zero, the laplacian will result in a high-pass operation, with the lower frequency information filtered out. The result will be an image which contains edge information at the corresponding high frequency. By using a larger laplacian kernel which is flatter and wider, edge information at a particular frequency will be passed to the output image, resulting in a *bandpass* image. A bandpass image is an image which contains frequencies within a specific range only. Frequencies outside the range (band) are not present in the bandpass image. Note that since the laplacian kernel contains negative values, the convolution result can be negative, so the absolute value should be taken if the result is to be displayed.

The results of two laplacian convolutions on the parking lot image of Fig. 14-7a are shown in Figs. 14-11a and b. The convolution kernel used for Fig. 14-11a has the form shown in Eq. (14-4), and the kernel used for Fig. 14-11b has the form shown in Eq. (14-5). Note that the sum of the elements for each of these kernels is equal to zero.

$$K = \begin{bmatrix} -1 & -1 & -1 \\ -1 & 8 & -1 \\ -1 & -1 & -1 \end{bmatrix} \tag{14-4}$$

$$K = \begin{bmatrix} -1 & -2 & -3 & -2 & -1 \\ -2 & 2 & 4 & 2 & -2 \\ -3 & 4 & 8 & 4 & -3 \\ -2 & 2 & 4 & 2 & -2 \\ -1 & -2 & -3 & -2 & -1 \end{bmatrix} \tag{14-5}$$

Notice from Fig. 14-11a that the laplacian operation has enhanced the highest frequencies while filtering out all other frequencies. The resulting edges correspond to the highest frequencies and barely contain recognizable information. Since a wider laplacian kernel was used to produce Fig. 14-11b, the edges that were detected correspond to frequencies that are slightly lower than those of Fig. 14-11a. Notice that the outlines of the cars are readily discernible, as well as the lamppost and its shadow.

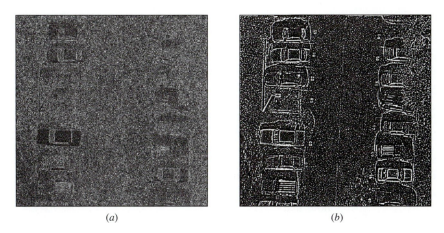

(a) (b)

FIGURE 14-11

(a) Parking lot image after convolution with 3×3 laplacian kernel. (b) Parking lot image after convolution with 5×5 laplacian kernel.

The laplacian edge detector is not sensitive to the orientation of the edge that it detects. A different edge detection operator, the Sobel, is capable of not only detecting edges, but determining their orientation as well. This is accomplished by using a pair of kernels (an x kernel and a y kernel) in the convolution. The forms of the x and y kernels are given in Eqs. (14-6) and (14-7), respectively.

$$K_x = \begin{bmatrix} -1 & 0 & +1 \\ -2 & 0 & +2 \\ -1 & 0 & +1 \end{bmatrix} \tag{14-6}$$

$$K_y = \begin{bmatrix} +1 & +2 & +1 \\ 0 & 0 & 0 \\ -1 & -2 & -1 \end{bmatrix} \tag{14-7}$$

Convolution by the two kernels at a specific window location gives the x component of the edge, S_x, and the y component of the edge, S_y. From these two convolution results, the magnitude and direction of the edge can be computed by Eqs. (14-8) and (14-9), respectively.

$$\text{Magnitude} = \sqrt{S_x^2 + S_y^2} \tag{14-8}$$

$$\text{Direction} = \tan^{-1}\left(\frac{S_y}{S_x}\right) \tag{14-9}$$

Application of the Sobel edge detection operation to the parking lot image results in the magnitude image shown in Fig. 14-12. Notice that the edges defining the cars and their shadows are well defined, as are the edges defining the lamppost and its shadow.

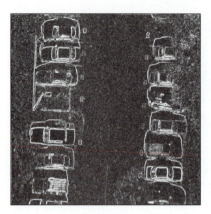

FIGURE 14-12
Parking lot image after application of
Sobel edge detection.

From the examples given on this and the previous page, it can be seen that the size and composition of the kernels applied in digital image processing have a major bearing upon the results achieved, and they also create many interesting possibilities.

14-7 MULTISCALE REPRESENTATION

In the preceding section, bandpass filters were discussed. Images that result from a bandpass filter contain edge information that corresponds only to the range of frequencies in the bandpass image. By using filters which pass lower ranges of frequencies, low-frequency (small-scale) edges can be detected. The effect of detecting edges at different scales (frequencies) can be seen in Figs. 14-13*a*, *b*, and *c*. Figure 14-13*a* is the result of performing a simple average convolution with a 3 × 3 kernel on the parking lot image, and subtracting the convolved image from the original. Note that this image contains a great deal of large-scale (high-frequency) edge information.

Figure 14-13*b* is the result of two convolutions, followed by a difference. The first convolution is the same as that used to produce Fig. 14-13*a*, i.e., a simple average convolution with a 3 × 3 kernel. Assume that the original parking lot image is image I and the result of the first convolution gives image I_3. The second convolution is performed on image I_3 using a simple average convolution with a 5 × 5 kernel, resulting in image $I_{3,5}$. Finally, image $I_{3,5}$ is subtracted from image I_3, resulting in Fig. 14-13*b*. Notice that this figure contains edge information at a smaller scale (lower frequency) than that of Fig. 14-13*a*. Notice also that the edges corresponding to the roof rack on the vehicle at the lower left are still discernible.

The last image, Fig. 14-13*c*, is the result of three convolutions followed by a difference. Starting with image $I_{3,5}$ which has been convolved twice, an additional convolution is performed using a 7 × 7 kernel (simple average), resulting in image $I_{3,5,7}$. Finally, $I_{3,5,7}$ is subtracted from image $I_{3,5}$, resulting in Fig. 14-13*c*. In this figure, edges of a still smaller scale (lower frequency) have been detected. Notice that in this figure, the edges associated with the roof rack have essentially disappeared.

This method of obtaining multiscale representations of edges is rather simplistic, but serves as an introduction to the concept. Other approaches such as those that employ

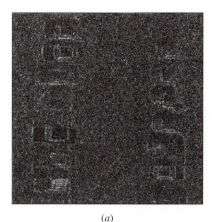

(a)

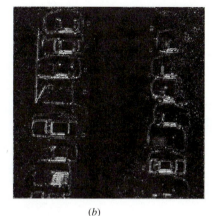

(b)

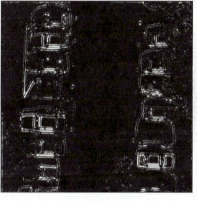

(c)

FIGURE 14-13
(a) Large-scale edge information from parking lot image. (b) Edge information at a smaller scale.
(c) Edge information at a still smaller scale.

the wavelet transform can produce multiscale representations with better resolution. The wavelet transform is often used to extract multiscale features (edges) for use in feature recognition and pattern matching.

Another concept relating to image scale is that of the *image pyramid*. An image pyramid is formed by successively convolving an image with a *gaussian* kernel, with each convolution producing a half-resolution copy of the previous image. A gaussian kernel is one which has weights that correspond to the shape of a normal (gaussian) distribution or bell-shaped curve. The series of images thus produced can be visualized as a stack of image layers forming a pyramid, as shown in Fig. 14-14. This figure shows an image pyramid formed from the parking lot image, with the original image at the bottom of the pyramid and successive half-resolution copies going up the pyramid.

FIGURE 14-14
Image pyramid formed from parking lot image.

An image is formed by convolving the original image with a gaussian kernel, such as that shown in Eq. (14-10).

$$K = \begin{bmatrix} 0.0025 & 0.0125 & 0.0200 & 0.0125 & 0.0025 \\ 0.0125 & 0.0625 & 0.1000 & 0.0625 & 0.0125 \\ 0.0200 & 0.1000 & 0.1600 & 0.1000 & 0.0200 \\ 0.0125 & 0.0625 & 0.1000 & 0.0625 & 0.0125 \\ 0.0025 & 0.0125 & 0.0200 & 0.0125 & 0.0025 \end{bmatrix} \qquad (14\text{-}10)$$

Notice that the weights in this kernel are largest in the center and fall off gradually away from the center. This pattern of weights mimics the behavior of a gaussian distribution in two dimensions.

The convolution is performed in the usual fashion, with the exception that instead of the moving window shifting one row or column at a time, it is shifted two rows or columns. This results in a convolved image having one-half as many rows and columns as the original. This reduction in number of pixels is offset by the fact that each pixel in the convolved image represents an area twice the width and height of the original pixels. Each successive convolution is performed on the previously convolved image, resulting in the series of half-resolution copies.

Image pyramids can be used for many purposes. One particularly important use is for multiresolution image matching. By matching images in upper layers of the pyramid, the location of the match can be predicted in lower layers within a couple of pixels, which avoids searching through the entire full-resolution image to find a matching feature. Another use is for quick display of an image while zooming in or out. By first constructing an image pyramid, an image zoom operation can be accomplished by accessing different layers of the pyramid. These are just two possible uses for image pyramids, and there are many others.

14-8   SUMMARY

The discussion of digital image processing methods presented in this chapter is only a brief treatment of the subject. These methods dealt primarily with edge detection, contrast enhancement, noise removal, and multiscale representations. A wealth of other digital image processing methods are available for dealing with these problems as well as many others. An in-depth understanding of digital image processing requires a great deal of study and experimentation. Those who are interested in learning more about this subject are directed to the references listed below. Further information can be gathered by consulting the bibliographies contained in these documents.

REFERENCES

American Society of Photogrammetry: *Manual of Remote Sensing,* vols. 1 and 2, 2d ed., Bethesda, MD, 1983.

American Society for Photogrammetry and Remote Sensing: *Digital Photogrammetry, An Addendum to the Manual of Photogrammetry,* Bethesda, MD, 1996.

————: Special Image Processing Issue, *Photogrammetric Engineering and Remote Sensing,* vol. 55, no. 9, 1989.

————: Special Image Processing Issue, *Photogrammetric Engineering and Remote Sensing,* vol. 56, no. 1, 1990.

————: Special Issue on Geostatistics and Scaling of Remote Sensing and Spatial Data, *Photogrammetric Engineering and Remote Sensing,* vol. 65, no. 1, 1999.

Atkinson, P., and P. Curran: "Choosing an Appropriate Spatial Resolution for Remote Sensing Investigations," *Photogrammetric Engineering and Remote Sensing,* vol. 63, no. 12, 1997, p. 1345.

Baxes, G.: *Digital Image Processing,* Prentice-Hall, Englewood Cliffs, NJ, 1984.

Burt, P. J.: "The Pyramid as a Structure for Efficient Computation," *Multiresolution Image Processing and Analysis,* ed. A. Rosenfeld, Springer-Verlag, Berlin, 1984.

Carnahan, W., and G. Zhou: "Fourier Transform Techniques for the Evaluation of the Thematic Mapper Line Spread Function," *Photogrammetric Engineering and Remote Sensing,* vol. 52, no. 5, 1986, p. 639.

Chavez, P.: "Radiometric Calibration of Landsat Thematic Mapper Multispectral Images," *Photogrammetric Engineering and Remote Sensing,* vol. 55, no. 9, 1989, p. 1285.

————: "Image Based Atmospheric Corrections—Revisited and Improved," *Photogrammetric Engineering and Remote Sensing,* vol. 62, no. 9, 1996, p. 1025.

Chui, C. K.: *An Introduction to Wavelets,* Academic Press, San Diego, CA, 1992.

Crippen, R.: "A Simple Spatial Filtering Routine for the Cosmetic Removal of Scan-Line Noise from Landsat TM P-Tape Imagery," *Photogrammetric Engineering and Remote Sensing,* vol. 55, no. 3, 1989, p. 327.

Ekstrom, M.: *Digital Image Processing Techniques,* Academic Press, New York, 1984.

Ekstrom, M., and A. McEwen: "Adaptive Box Filters for Removal of Random Noise from Digital Images," *Photogrammetric Engineering and Remote Sensing,* vol. 56, no. 4, 1990, p. 453.

Hord, R.: *Digital Image Processing of Remotely Sensed Data,* Academic Press, New York, 1982.

Horgan, G.: "Wavelets for SAR Image Smoothing," *Photogrammetric Engineering and Remote Sensing,* vol. 64, no. 12, 1998, p. 1171.

Hunt, B. R., T. W. Ryan, and F. A. Gifford: "Hough Transform Extraction of Cartographic Calibration Marks from Aerial Photography," *Photogrammetric Engineering and Remote Sensing,* vol. 59, no. 7, 1993, p. 1161.

Jensen, J.: *Introductory Digital Image Processing: A Remote Sensing Perspective,* Prentice-Hall, Englewood Cliffs, NJ, 1986.

Kardoulas, N., A. C. Bird, and A. I. Lawan: "Geometric Correction of SPOT and Landsat Imagery: A Comparison of Map and GPS Derived Control Points," *Photogrammetric Engineering and Remote Sensing,* vol. 62, no. 10, 1996, p. 1173.

Light, D.: "Film Cameras or Digital Sensors? The Challenge," *Photogrammetric Engineering and Remote Sensing,* vol. 62, no. 3, 1996, p. 285.

Lillesand, T., and R. Kiefer: *Remote Sensing and Image Interpretation,* 4th ed., Wiley, New York, 1999.

Lim, J. S.: *Two-Dimensional Signal and Image Processing,* Prentice-Hall, Englewood Cliffs, NJ, 1990.

Moik, J. G.: *Digital Processing of Remotely Sensed Images,* NASA SP-431, Government Printing Office, Washington, 1980.

Muller, J.: *Digital Image Processing in Remote Sensing,* Taylor & Francis, London, 1986.

Novack, K.: "Rectification of Digital Imagery," *Photogrammetric Engineering and Remote Sensing,* vol. 58, no. 3, 1992, p. 339.

Pratt, W. K.: *Digital Image Processing,* 2d ed., Wiley, New York, 1991.

Press, W. H., S. A. Teukolsky, W. T. Vetterling, and B. P. Flannery: *Numerical Recipes in C,* Cambridge University Press, New York, 1994.

Richards, J.: *Remote Sensing Digital Image Analysis: An Introduction,* Springer-Verlag, New York, 1986.

Thome, K., B. Markham, J. Barker, P. Slater, and S. Bigger: "Radiometric Calibration of Landsat," *Photogrammetric Engineering and Remote Sensing,* vol. 63, no. 7, 1997, p. 853.

Walsh, S., J. W. Cooper, I. E. Von Essen, and K. R. Gallager: "Image Enhancement of Landsat Thematic Mapper Data and GIS Integration for Evaluation of Resource Characteristics," *Photogrammetric Engineering and Remote Sensing,* vol. 56, no. 8, 1990, p. 1135.

PROBLEMS

14-1. Give a brief definition of a digital image model.

14-2. Define spatial frequency.

14-3. What is the Nyquist frequency? Why is it important?

14-4. Briefly describe the linear stretch method of contrast enhancement.

14-5. Give a brief description of the histogram equalization method of contrast enhancement.

14-6. Briefly describe the relationship between the spatial domain and frequency domain for a digital image.

14-7. Give a brief description of the discrete Fourier transform.

14-8. What is the fast Fourier transform and why is it important?

14-9. Given the following convolution kernel K and the matrix of digital numbers D from a 5×5 image, compute the convolved image C, using the moving window approach.

$$K = \begin{bmatrix} 0.05 & 0.05 & 0.05 \\ 0.05 & 0.60 & 0.05 \\ 0.05 & 0.05 & 0.05 \end{bmatrix}$$

$$D = \begin{bmatrix} 78 & 69 & 59 & 58 & 61 \\ 56 & 61 & 53 & 64 & 63 \\ 39 & 46 & 49 & 72 & 78 \\ 52 & 49 & 41 & 61 & 57 \\ 70 & 67 & 81 & 77 & 72 \end{bmatrix}$$

14-10. Repeat Prob. 14-9, except use the following convolution kernel K.

$$K = \begin{bmatrix} -1 & -1 & -1 \\ -1 & 9 & -1 \\ -1 & -1 & -1 \end{bmatrix}$$

14-11. Briefly explain the difference between high-pass and bandpass filters.

14-12. Outline a process by which edges at different scales can be extracted from a digital image.

14-13. Briefly explain how an image pyramid is constructed.

CHAPTER
15

PRINCIPLES OF SOFTCOPY PHOTOGRAMMETRY

15-1 INTRODUCTION

Softcopy systems are the most recent development in photogrammetric stereoplotters. They operate with digital images as opposed to film diapositives used with analytical and other stereoplotters. Sections 12-17 and 12-18 provided a brief overview of softcopy stereoplotters. In this chapter, many of the details associated with this type of stereoplotter are presented.

Development of softcopy stereoplotters occurred over nearly the same time period as that of analytical plotters. Many of the fundamental concepts were first developed in the late 1950s and early 1960s. Research progressed through the years until in the late 1980s the first commercially available softcopy plotter, the Kern DSP1, was introduced. Continuing advances in computer hardware, digital imaging systems and scanners, and computer algorithm development have-led to the current generation of softcopy stereoplotters. Development is sure to continue at a steady pace, with the promise of even more capable and user-friendly systems in the future.

15-2 SYSTEM HARDWARE

The fundamental hardware requirement for a softcopy stereoplotter is a high-powered computer. The computer must have not only substantial processing speed, but also large amounts of random access memory and mass storage (hard disk space). Some systems even employ multiple processors to increase computational speed. This high speed and the large memory capacity are necessary because digital image files used in softcopy plotters may be as large as several hundred megabytes or more. In addition, some form of archival storage such as optical disks or high-capacity tapes is required for off-line storage of digital images. The system also requires operator controls for X, Y, and Z position within the stereomodel. On a softcopy system, the X and Y controls are usually implemented as some type of computer "mouse," while the Z control is typically a small wheel which can be rotated by the operator's thumb. Some systems can even be fitted with older-style handwheels and a foot wheel, so as to be more familiar to operators who were trained on traditional stereoplotters.

For the computer to serve as a stereo workstation, additional hardware is required to enable the operator to see the left image with the left eye and the right image with the right eye. Various approaches are available to provide stereoviewing capability, such as polarizing filters, alternating shutters, and split screen with a stereoscope. In the polarizing filter approach, a computer monitor is fitted with an active polarizing screen, while the operator wears a simple pair of spectacles consisting of orthogonally polarized filters. The active polarizing screen has the capability of alternating the orientation of its polarity between horizontal and vertical, 120 times per second [120 hertz (Hz)]. The computer display has the capability of alternately displaying two images (left and right) at the same rate (120 Hz). The spectacles are constructed so that the filter over one eye is oriented vertically and the filter over the other eye is oriented horizontally. At a particular instant, the left image is displayed on the monitor while the polarity of the screen is set to that of the left filter of the spectacles, as depicted in Fig. 15-1a. Since the polarity of the left filter of the spectacles has the same orientation as that of the screen, the left image passes through to the operator's left eye, while the right eye sees nothing since its filter is orthogonal to that of the screen. A split second ($\frac{1}{120}$ s) later, the right image is displayed on the monitor and the screen switches polarity to match that of the right filter of the spectacles, as depicted in Fig. 15-1b. At this instant, the operator's right

Left image displayed

Right image displayed

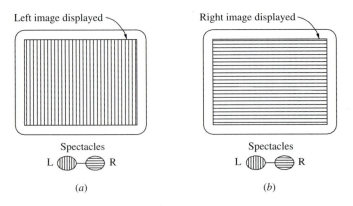

Spectacles

L ⬚ R   Spectacles

L ⬚ R

(a)   (b)

FIGURE 15-1
Stereoviewing principle of the active polarizing screen display.

eye sees the right image on the monitor, and the left eye sees nothing. This alternating display of left and right images with synchronized alternation of the screen polarity happens so rapidly that the operator is unaware of the effect, yet sees the left image with the left eye and the right image with the right eye. The DPW770 system by LH Systems, shown in Fig. 15-2, employs a viewing system of this type.

A second approach to stereoviewing uses a display monitor which shows alternating left and right images at 120 Hz, along with special viewing glasses that have liquid crystal display (LCD) masks which alternate at the same rate. The LCD glasses receive signals transmitted from an infrared device, mounted on top of the computer monitor, which controls the alternating left and right masking functions. At a particular instant, the left image is displayed on the monitor, and at the same time, the LCD mask over the right eye turns opaque while the LCD mask over the left eye is clear, as illustrated in Fig. 15-3a. Since the right eye is blocked and the left eye is unobstructed, the operator sees the left image with the left eye. A split second later, the right image is displayed on the monitor, and at the same time the LCD mask over the left eye turns opaque while the LCD mask over the right eye is clear, as illustrated in Fig. 15-3b. This causes the operator to see the right image with the right eye. When this is repeated at a rate of 120 Hz, the operator is unaware of the alternating images, yet the proper stereoview is created. The PHODIS system by Carl Zeiss, Inc., which is shown in Fig. 15-4, uses this method of stereo image display, as does the Intergraph Image Station Z of Fig. 12-20.

A third method of stereoviewing for softcopy plotters employs a split-screen display and mirror stereoscope. The mirror stereoscope is mounted in front of the

FIGURE 15-2
Leica DPW 770 Digital Photogrammetric Workstation. (*Courtesy LH Systems, LLC.*)

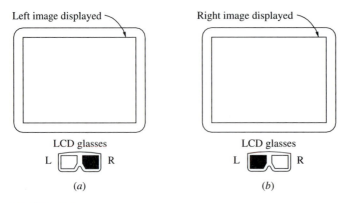

FIGURE 15-3
Stereoviewing principle of the alternating LCD shutter display.

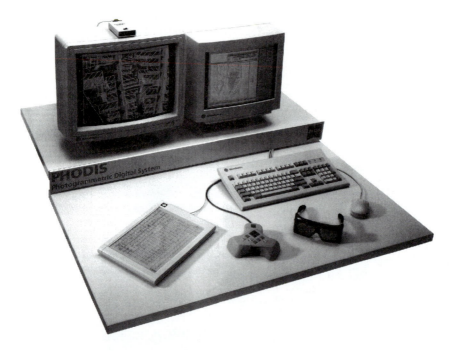

FIGURE 15-4
Zeiss PHODIS ST digital stereo workstation. (*Courtesy Carl Zeiss, Inc.*)

computer monitor, and the operator views the monitor through it. The left image is displayed on the left side of the monitor, and the right image is displayed on the right, as shown in Fig. 15-5. Since the mirror stereoscope diverts the optical paths to the operator's eyes, the left eye sees only the left image and the right eye sees only the right image.

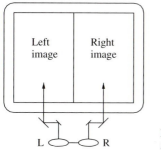

FIGURE 15-5
Stereoviewing principle of the split-screen display.

Each of the aforementioned methods of stereoviewing has its advantages. The split-screen method tends to be the least expensive implementation of the three. It can work with a standard computer display using a 60-Hz refresh rate as opposed to the other methods which require a 120-Hz refresh rate. An advantage of the polarizing screen method is that more than one person can view the stereo display at one time, with extra sets of inexpensive polarized spectacles. The alternating-shutter approach shares the advantage of more than one person viewing the stereo display at a time, although this requires multiple sets of LCD glasses, which are more expensive than simple polarized spectacles. In a production environment this advantage is minor, except perhaps for training novice operators. The polarizing screen and alternating-shutter methods also allow the operator more freedom of movement compared to the split-screen approach, where the operator's eyes must be directly in front of the stereoscope.

15-3 IMAGE MEASUREMENTS

As is the case with all stereoplotters, manual image measurements are accomplished through operator control of a floating mark. On a softcopy stereoplotter, a floating mark consists of left and right half marks which are superimposed on the left and right images, respectively. An individual half mark consists of a single pixel, or small pattern of pixels in the shape of a dot, cross, or more complex shape. The pixel(s) of the half mark is (are) set to brightness values which give a high contrast with the background image. When the operator moves the X, Y, or Z control, the positions of the individual half marks move with respect to the background image. Once the floating mark visually coincides with a feature of interest, the operator can press a button or foot pedal to record the feature's position.

Two approaches are used for half mark movement: a fixed mark with a moving image or a fixed image with a moving mark. The first approach, which mimics the action of an analytical stereoplotter, keeps the individual half marks at the same physical position on the screen while the image pans across the screen under operator control. This implementation places a high demand on the computer display in terms of data transfer rate, thus requiring a high-speed computer graphics adapter. The second approach works by keeping the left and right images in a fixed position on the display while the individual half marks move in response to the operator's input. While this approach places far less demand on the display adapter, it creates additional problems. One problem is that when the half marks approach the edge of the image area, the images must be reloaded and redisplayed so that the half marks can be placed at the center of the viewing area. This discontinuous image shift can be quite disruptive to the operator, since it requires a corresponding shift in the direction of the operator's gaze. Another problem, which is more subtle, is that the operator's eyes must follow the floating mark as it moves, which can cause eye strain. With the fixed half mark approach, the operator's gaze remains essentially fixed while the images move around it.

A key advantage afforded by softcopy plotters is their ability to make point measurements automatically. This is accomplished through some form of pattern-matching technique, wherein a small subarray of digital numbers from the left image is matched with a corresponding subarray from the right image. Finding the matching position is equivalent to manually setting the floating mark so that it appears to rest directly on a feature in the stereomodel. Various methods are available for performing the pattern-matching operation, and some of these are discussed in Sec. 15-6.

15-4   ORIENTATION PROCEDURES

Softcopy plotters are oriented in a similar fashion to that of analytical plotters; i.e., the same three steps of interior, relative, and absolute orientation must be performed. The key difference between orientations of softcopy and analytical plotters is that softcopy systems allow for greater automation in the process. Interior orientation, which primarily consists of pointing on the fiducial marks, can be done directly under operator control, or by use of pattern-matching methods. Systems that use pattern matching attempt to find the positions of the fiducial marks by matching a standard image of the fiducial, sometimes called a *template,* with a corresponding subarray from the image. Once all fiducials have been located, a two-dimensional transformation can be computed to relate image coordinates (row and column) to the fiducial axis system.

Relative orientation can also be greatly assisted by automatic pattern matching. Small subarrays from the left image in the standard pass point locations are matched with corresponding subarrays from the right image. Once a sufficient number of pass points have been matched (generally at least six points), a relative orientation can be computed. Accuracy of the relative orientation can be improved by matching additional pass points, thus providing greater redundancy in the least squares solution.

Absolute orientation is much less amenable to automation than interior or relative orientation. In absolute orientation, three-dimensional measurements must be made on the positions of ground control points which appear in the model. Since ground control points can have varying shapes and can appear anywhere within the model, they are more difficult to locate by automatic pattern-matching techniques. In these cases, manual pointing on the control points is usually done.

One situation where absolute orientation can be automated occurs when a block aerotriangulation has previously been performed on the digital images. As a result of the aerotriangulation, exterior orientation parameters will have been determined for each photo. Having known exterior orientation parameters essentially defines the absolute orientation, thus no additional measurements need to be taken.

15-5 EPIPOLAR GEOMETRY

As mentioned in Sec. 15-3, automatic methods can be used to set the floating mark on a feature. This involves pattern matching, which can be a computationally intensive task. To reduce the computational burden, it is helpful to constrain the search area so as to avoid unnecessary calculations. By exploiting the principle of *epipolar* geometry, search regions can be constrained along a single line.

As described in Sec. 11-5, coplanarity is the condition in which the left and right camera stations, an object point, and the left and right images of the object point lie in a common plane. If relative orientation is known for a given stereopair, the coplanarity condition can be used to define epipolar lines. This situation is illustrated in Fig. 15-6. The figure shows the intersection of the *epipolar plane* (any plane containing the two exposure stations and an object point, in this instance plane L_1AL_2) with the left and right photo planes. The resulting lines of intersection are the *epipolar lines*. They are important because, given the left photo location of image a_1, its corresponding point a_2 on the right photo is known to lie along the right epipolar line. Based solely upon the location of image point a_1, object point A could be located anywhere along line L_1A, at an arbitrary Z elevation. For example, if object point A were located at position A', its image would still be at point a_1 in the left photo. Based on an assumed object point position of A', the corresponding location of its image on the right photo, a_2' can be calculated by the collinearity equations (see App. D). A small subarray of digital numbers at the location of point a_2' can be compared with a corresponding subarray at the location of point a_1. Since these two locations do not both correspond to images of object point A, the patterns would not match, and another point on the right epipolar line would be tried. The search continues along the epipolar line until it zeros in on the corresponding image at a_2, where the patterns match. The coordinates of image points a_1 and a_2 can then be used to determine the three-dimensional object space coordinates of point A.

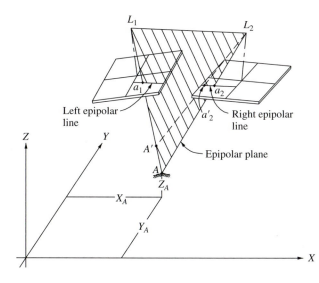

FIGURE 15-6
Epipolar geometry of a stereopair of photos.

By using the principle of epipolar geometry, searching for matching points is much more efficient. This is especially important when a large number of points must be matched, such as when a digital elevation model is created (see Sec. 15-7). Many softcopy systems perform epipolar resampling after relative orientation so that the rows of pixels in both images lie along epipolar lines. This resampling can increase the efficiency of pattern matching even further.

15-6 DIGITAL IMAGE MATCHING

A common task associated with the use of stereoplotters is to place the floating mark at a three-dimensional model position of an object point. This requires the ability to recognize similar image characteristics (texture, shapes, etc.) in small regions from the left and right images of a stereopair. The human visual system is able to perform this task with little conscious effort. When our eyes fixate on an object, the two images merge and the three-dimensional nature of the object is revealed. When one is working with digital images in a softcopy stereoplotter, placing of the floating mark can be performed either manually or by computer processing. Computer software which accomplishes this task uses techniques of digital image matching.

Digital image-matching techniques fall into three general categories: *area-based, feature-based,* and *hybrid methods.* Area-based methods perform the image match by a numerical comparison of digital numbers in small subarrays from each image. This approach is straightforward and commonly used in softcopy systems. Feature-based methods are more complicated and involve extraction of features, which are comprised of edges at different scales, with subsequent comparison based on feature characteristics such as size and shape. Feature-based image matching requires techniques from the realm of artificial intelligence in computer science. Hybrid methods involve some combination of the first two approaches. Typically, hybrid methods involve preprocessing of the left and right images to highlight features (edges) by methods which were introduced in Chap. 14. After the features have been located, they are matched by area-based methods. While all three approaches have particular advantages and disadvantages, this section focuses on area-based image-matching techniques.

Perhaps the simplest area-based digital image-matching method is a technique known as *normalized cross-correlation.* In this approach, a statistical comparison is computed from digital numbers taken from same-size subarrays in the left and right images. A correlation coefficient is computed by the following equation, using digital numbers from subarrays A and B.

$$c = \frac{\sum_{i=1}^{m} \sum_{j=1}^{n} [(A_{ij} - \overline{A})(B_{ij} - \overline{B})]}{\sqrt{\left[\sum_{i=1}^{m} \sum_{j=1}^{n} (A_{ij} - \overline{A})^2\right]\left[\sum_{i=1}^{m} \sum_{j=1}^{n} (B_{ij} - \overline{B})^2\right]}} \tag{15-1}$$

In Eq. (15-1), c is the correlation coefficient; m and n are the numbers of rows and columns, respectively, in the subarrays; A_{ij} is the digital number from subarray A at row i, column j; $\overline{A}$ is the average of all digital numbers in subarray A; B_{ij} is the digital number from subarray B at row i, column j; and $\overline{B}$ is the average of all digital numbers in subarray B. The correlation coefficient can range from -1 to $+1$, with $+1$ indicating perfect correlation (an exact match). A coefficient of -1 indicates negative correlation, which would occur if identical images from a photographic negative and positive were being compared. Coefficient values near zero indicate a nonmatch, and could result from a comparison of any two sets of random numbers. Due to factors such as image noise, perfect $(+1)$ correlation is extremely rare. Generally a threshold value, such as 0.7, is chosen and if the correlation coefficient exceeds that value, the subarrays are assumed to match.

Normalized cross-correlation is essentially the same operation as *linear regression* in statistics. Details of linear regression can be found in most elementary statistics texts, and only general concepts are discussed here. In linear regression, a set of ordered pairs (abscissas and ordinates) is statistically analyzed to determine how well the numbers correspond to a straight-line relationship. In the process, most-probable values are determined for the parameters (slope and intercept) of a best-fit line through the data points. For example, assume the following pair of 3×3 arrays of digital numbers are to be analyzed using linear regression.

$$A = \begin{bmatrix} 25 & 48 & 89 \\ 43 & 94 & 47 \\ 76 & 21 & 57 \end{bmatrix} \qquad B = \begin{bmatrix} 33 & 56 & 81 \\ 40 & 98 & 54 \\ 84 & 16 & 49 \end{bmatrix}$$

To compute the linear regression, a tabular solution can be used, as shown in Table 15-1. In this table, the abscissas and ordinates used for the regression are listed in the columns labeled a and b, respectively; a^2 and b^2 are their corresponding squares; and $a \times b$ are the products. (*Note:* In typical notation used for linear regression, x_i and y_i are used for abscissas and ordinates, respectively, and this notation is also shown in the table.)

TABLE 15-1
Tabular form for computing linear regression

$a(x_i)$	$b(y_i)$	$a^2(x_i^2)$	$b^2(y_i^2)$	$a \times b \; (x_i y_i)$
25	33	625	1089	825
48	56	2304	3136	2688
89	81	7921	6561	7209
43	40	1849	1600	1720
94	98	8836	9604	9212
47	54	2209	2916	2538
76	84	5776	7056	6384
21	16	441	256	336
57	49	3249	2401	2793
$\sum = 500$	$\sum = 511$	$\sum = 33{,}210$	$\sum = 34{,}619$	$\sum = 33{,}705$

In order to compute the regression, the following terms are computed.

$$S_x^2 = \sum(x_i - \bar{x})^2 = \sum x_i^2 - \frac{\left(\sum x_i\right)^2}{n} = 33{,}210 - \frac{500^2}{9} = 5432.2 \qquad (15\text{-}2)$$

$$S_y^2 = \sum(y_i - \bar{y})^2 = \sum y_i^2 - \frac{\left(\sum y_i\right)^2}{n} = 34{,}619 - \frac{511^2}{9} = 5605.6 \qquad (15\text{-}3)$$

$$S_{xy} = \sum[(x_i - \bar{x})(y_i - \bar{y})] = \sum(x_i y_i) - \frac{\left(\sum x_i\right)\left(\sum y_i\right)}{n}$$

$$= 33{,}705 - \frac{500 \times 511}{9} = 5316.1 \qquad (15\text{-}4)$$

$$\beta = \frac{S_{xy}}{S_x^2} = \frac{5316.1}{5432.2} = 0.979 \qquad (15\text{-}5)$$

$$\alpha = \frac{\sum y_i}{n} - \beta \frac{\sum x_i}{n} = \frac{511}{9} - 0.979\left(\frac{500}{9}\right) = 2.41 \qquad (15\text{-}6)$$

$$r = \frac{S_{xy}}{\sqrt{S_x^2 S_y^2}} = \frac{33{,}705}{\sqrt{33{,}210 \times 34{,}619}} = 0.963 \qquad (15\text{-}7)$$

In Eqs. (15-2) through (15-7), n is the number of data points (9 in this case), and the other terms are as indicated in the table. Parameter β in Eq. (15-5) is the slope of the regression line; parameter α in Eq. (15-6) is the y intercept of the regression line; and parameter r in Eq. (15-7) is the *sample correlation coefficient*. Note that r is the same as the normalized cross-correlation coefficient c from Eq. (15-1). Figure 15-7 shows a plot

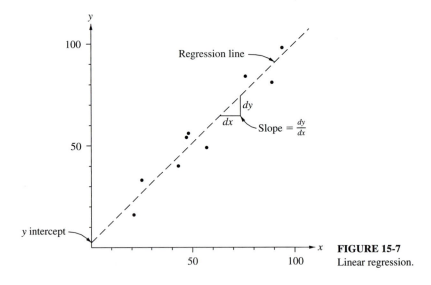

FIGURE 15-7
Linear regression.

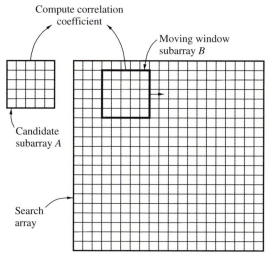

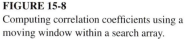

FIGURE 15-8
Computing correlation coefficients using a moving window within a search array.

of the nine data points, along with the regression line. In this figure, the nine data points lie nearly along the regression line, which is also indicated by the correlation coefficient r being nearly equal to 1.

Digital image matching by correlation can be performed in the following manner. A candidate subarray from the left photo is chosen, and a search will be performed for its corresponding subarray in the right image. Since the exact position of the image in the right image is not initially known, a search array is selected with dimensions much larger than those of the candidate subarray. A moving window approach is then used, comparing the candidate subarray from the left image with all possible window locations within the search array from the right image, as illustrated in Fig. 15-8. At each window location in the search array, the correlation coefficient is computed in a manner similar to moving window convolution (see Sec. 14-6), resulting in a correlation matrix C. After all coefficients have been calculated, the largest correlation value in C is tested to see if it is above the threshold. If it exceeds the threshold, the corresponding location within the search array is considered to be the match.

Example 15-1. The candidate array A is an ideal template for a fiducial cross, and the following search array S is a portion of a digital image containing a fiducial cross. Compute the position of the fiducial within the search array by correlation.

$$A = \begin{bmatrix} 0 & 0 & 50 & 0 & 0 \\ 0 & 0 & 50 & 0 & 0 \\ 50 & 50 & 50 & 50 & 50 \\ 0 & 0 & 50 & 0 & 0 \\ 0 & 0 & 50 & 0 & 0 \end{bmatrix} \quad S = \begin{bmatrix} 41 & 43 & 43 & 49 & 60 & 43 & 41 & 40 & 44 \\ 43 & 44 & 45 & 50 & 64 & 45 & 43 & 43 & 45 \\ 42 & 43 & 44 & 48 & 63 & 49 & 45 & 42 & 42 \\ 42 & 45 & 47 & 50 & 65 & 45 & 45 & 41 & 41 \\ 59 & 62 & 62 & 64 & 69 & 64 & 62 & 63 & 60 \\ 50 & 48 & 48 & 51 & 68 & 55 & 50 & 54 & 53 \\ 42 & 41 & 44 & 48 & 63 & 42 & 47 & 47 & 45 \\ 42 & 44 & 42 & 45 & 62 & 44 & 44 & 45 & 43 \\ 42 & 43 & 44 & 48 & 60 & 47 & 44 & 38 & 35 \end{bmatrix}$$

Solution. *Note:* The correlation coefficient at the first window position (with the upper left element of a 5 × 5 subarray at the 1, 1 position of the search array) will be calculated as an example.

1. Extract subarray B from the search array at position 1, 1.

$$B = \begin{bmatrix} 41 & 43 & 43 & 49 & 60 \\ 43 & 44 & 45 & 50 & 64 \\ 42 & 43 & 44 & 48 & 63 \\ 42 & 45 & 47 & 50 & 65 \\ 59 & 62 & 62 & 64 & 69 \end{bmatrix}$$

2. Compute the average digital number for subarrays A and B.

$$\overline{A} = \frac{0 + 0 + 50 + \cdots + 50 + 0 + 0}{25} = 18$$

$$\overline{B} = \frac{41 + 43 + 43 + \cdots + 62 + 64 + 69}{25} = 51.48$$

3. Compute the summation terms for the correlation coefficient.

$$\sum_{i=1}^{m}\sum_{j=1}^{n} [(A_{ij} - \overline{A})(B_{ij} - \overline{B})] = (0 - 18)(41 - 51.48) + (0 - 18)(43 - 51.48)$$

$$+ (50 - 18)(43 - 51.48) + \cdots + (50 - 18)$$

$$(62 - 51.48) + (0 - 18)(64 - 51.48)$$

$$+ (0 - 18)(69 - 51.48)$$

$$= -1316$$

$$\sum_{i=1}^{m}\sum_{j=1}^{n} (A_{ij} - \overline{A})^2 = (0 - 18)^2 + (0 - 18)^2 + (50 - 18)^2 + \cdots + (50 - 18)^2$$

$$+ (0 - 18)^2 + (0 - 18)^2$$

$$= 14,400$$

$$\sum_{i=1}^{m}\sum_{j=1}^{n} (B_{ij} - \overline{B})^2 = (41 - 51.48)^2 + (43 - 51.48)^2 + (43 - 51.48)^2$$

$$+ \cdots + (62 - 51.48)^2 + (64 - 51.48)^2 + (69 - 51.48)^2$$

$$= 2102.24$$

4. Compute the correlation coefficient.

$$c_{11} = \frac{\sum_{i=1}^{m}\sum_{j=1}^{n}[(A_{ij} - \overline{A})(B_{ij} - \overline{B})]}{\sqrt{\left[\sum_{i=1}^{m}\sum_{j=1}^{n}(A_{ij} - \overline{A})^2\right]\left[\sum_{i=1}^{m}\sum_{j=1}^{n}(B_{ij} - \overline{B})^2\right]}} = \frac{-1316}{\sqrt{2102.24 \times 14,400}} = -0.24$$

5. Compute the remaining coefficients in the same manner.

$$C = \begin{bmatrix} -0.24 & -0.09 & 0.35 & -0.19 & -0.19 \\ -0.24 & -0.16 & 0.32 & -0.21 & -0.32 \\ 0.25 & 0.37 & 0.94 & 0.29 & 0.27 \\ -0.08 & 0.02 & 0.50 & -0.06 & 0.03 \\ -0.28 & -0.23 & 0.27 & -0.18 & -0.22 \end{bmatrix}$$

6. Select the maximum correlation coefficient. The maximum value, 0.94, occurs at row 3, column 3 of the C array. This value was computed when the upper left element of the moving window was at that position (row $= 3$, column $= 3$) in the search array. Since the center of the cross in the template is 2 columns to the right and 2 rows down from the upper left corner, the center of the cross in the search array is at row $3 + 2 = 5$ and column $3 + 2 = 5$.

A second area-based digital image-matching method is the *least squares matching* technique. (The reader may wish to refer to App. B for general information on least squares.) Conceptually, least squares matching is closely related to the correlation method, with the added advantage of being able to obtain the match location to a fraction of a pixel. Different implementations of least squares matching have been devised, with the following form being commonly used.

$$A(x, y) = h_0 + h_1 B(x', y') \tag{15-8}$$

$$x' = a_0 + a_1 x + a_2 y \tag{15-9}$$

$$y' = b_0 + b_1 x + b_2 y \tag{15-10}$$

In Eq. (15-8), $A(x, y)$ is the digital number from the candidate subarray of the left image at location x, y; $B(x', y')$ is the digital number from a subarray in the search area of the right image at location x', y'; h_0 is the radiometric shift; and h_1 is the radiometric scale. Note that parameter h_0 is the same as the y intercept α of Eq. (15-6), and h_1 is the same as the slope β of Eq. (15-5). Equations (15-9) and (15-10) specify an affine relationship (see Sec. C-6) between the coordinates of the pixel on the left photo and the coordinates of the corresponding pixel on the right photo. Figure 15-9 illustrates the positions of subarrays A and B in the left and right images. In this figure, the x and y axes are the

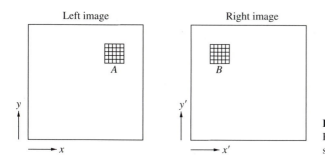

FIGURE 15-9
Positions of subarrays for least squares matching.

basis for coordinates on the left image, and the x' and y' axes are the basis for coordinates on the right image. Coordinates in both images are expressed in units of pixels.

Combining Eqs. (15-8), (15-9), and (15-10) and expressing the result in the form of a least squares observation equation gives the following:

$$f = h_0 + h_1 B(a_0 + a_1 x + a_2 y, b_0 + b_1 x + b_2 y) = A(x, y) + V_A \qquad (15\text{-}11)$$

In Eq. (15-11), f is the function relating the digital numbers from the two images, V_A is the residual, and the other variables are as previously defined. Equation (15-11) is nonlinear, and application of Taylor's series gives the following linear form.

$$f = f_0 + f'_{h_0}\, dh_0 + f'_{h_1}\, dh_1 + f'_{a_0}\, da_0 + f'_{a_1}\, da_1 + f'_{a_2}\, da_2 + f'_{b_0}\, db_0$$

$$+ f'_{b_1}\, db_1 + f'_{b_2}\, db_2 = A(x, y) + V_A \qquad (15\text{-}12)$$

where

$$f'_{h_0} = 1$$
$$f'_{h_1} = B(x', y')$$
$$f'_{a_0} = h_1 B'_x$$
$$f'_{a_1} = x h_1 B'_x$$
$$f'_{a_2} = y h_1 B'_x$$
$$f'_{b_0} = h_1 B'_y$$
$$f'_{b_1} = x h_1 B'_y$$
$$f'_{b_2} = y h_1 B'_y$$

and

$$B'_x = \frac{B(x' + 1, y') - B(x' - 1, y')}{2} \qquad (15\text{-}13)$$

$$B'_y = \frac{B(x', y' + 1) - B(x', y' - 1)}{2} \qquad (15\text{-}14)$$

Since the function f of Eq. (15-11) includes digital numbers from subarray B, partial derivative terms must be obtained using discrete values to estimate the slope of B in both the x and y directions. Equation (15-13) computes the estimate for slope in the x direction by taking the difference between the digital numbers of pixels to the right and left divided by 2, and Eq. (15-14) computes the estimate for slope in the y direction in a corresponding manner. Use of these discrete slope estimates, in conjunction with the chain rule from calculus, allows the partial derivatives (the f' terms) to be determined, as listed above.

Least squares matching is an iterative process which requires an accurate estimate for the position of B within the right image. Initial approximations must be obtained for the unknown parameters h_0, h_1, a_0, a_1, a_2, b_0, b_1, and b_2. Estimates for h_0 and h_1 can be obtained by linear regression as illustrated earlier in this section. If the coordinates of the lower left pixels of A and B are x_0, y_0 and x_0', y_0', respectively, the following initial approximations can be used for the affine parameters.

$$a_0 = x_0' - x_0 \qquad a_1 = 1 \qquad a_2 = 0$$

$$b_0 = y_0' - y_0 \qquad b_1 = 0 \qquad b_2 = 1$$

Each iteration of the solution involves forming the linearized equations, solving the equations by least squares to obtain corrections to the approximations, and adding the corrections to the approximations. At the beginning of an iteration, the pixels of subarray B (along with a 1-pixel-wide border around B which is needed for derivative estimates) are resampled (see App. E) from the right image. This is done by stepping through the pixels of subarray A, taking the x and y coordinates of each pixel, and transforming them to the right image x' and y' by using Eqs. (15-9) and (15-10). A corresponding digital number is then resampled from the right image at position x', y'. Once subarray B has been filled, the least squares equations can be formed and solved. The solution is then iterated until the corrections become negligible.

Some final comments are appropriate at this point. First, the estimated position of subarray B should be within a couple of pixels in order for the solution to converge properly. This can be achieved efficiently through the use of an image pyramid (see Sec. 14-7). Corresponding points can be matched at an upper level of the pyramid where the search area contains fewer pixels. Once the point is matched at a particular level of the pyramid, the position on the next-lower level will be known to within 2 pixels. By progressively matching from upper levels down to the bottom level, accurate position estimates can be obtained at each subsequent level. Another concern is the size of the subarrays to be matched. Generally, a subarray size of 20×20 to 30×30 gives satisfactory results. If the subarray is much smaller, the low redundancy can result in a weak solution. Larger subarrays can lead to problems due to terrain variations within the image area causing distortions that are not affine. Finally, the transformation equation for y' can be simplified by performing *epipolar resampling* on the images prior to matching. In epipolar resampling, the images are resampled so that the rows line up with epipolar lines. When this is done, Eqs. (15-9) and (15-10) can be simplified to

$$x' = a_0 + a_1 x \tag{15-15}$$

$$y' = b_0 \tag{15-16}$$

15-7 AUTOMATIC PRODUCTION OF DIGITAL ELEVATION MODELS

An important application of softcopy stereoplotters, which relies upon digital image matching, is the automatic production of digital elevation models. In this process, a set of image points is selected from the left image and subsequently matched with corresponding points in the right image. The set of selected points is typically arranged in a nominal grid pattern, which results in a nearly uniform regular grid DEM (see Sec. 13-6). In general, the resulting DEM will differ somewhat from a perfectly regular grid due to tilt and relief displacements which exist in the images.

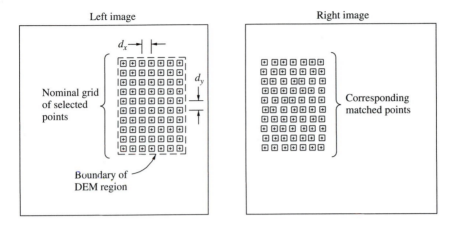

FIGURE 15-10
Positions of automatically matched DEM points in a stereopair.

Figure 15-10 is an illustration of the positions of DEM points within the overlap area of a stereopair of images. A boundary is chosen for the region in which the DEM points will be selected. The grid of selected points within the region on the left image is arranged at spacing of d_x and d_y in the x and y directions, respectively. Each of these points is matched with corresponding points from the right image, using image-matching techniques as discussed in the preceding section. Since the ground is generally not flat, the x parallax of DEM points will not be constant. As can be seen in Fig. 15-10, this varying x parallax causes the pattern of corresponding points from the right image to differ from a perfect grid arrangement.

Once the full set of DEM points has been matched and their object space coordinates calculated, their three-dimensional positions in the stereomodel can be represented by a set of floating marks superimposed upon the terrain. If desired, the plotter operator can then view these points in stereo, one at a time, and adjust the positions of points which do not appear to rest on the terrain. When the operator is satisfied with all vertical positions, the resulting DEM can be stored in a computer file. This DEM is then available for additional operations such as automatic contouring or the production of digital orthophotos, as discussed next.

15-8 AUTOMATIC PRODUCTION OF DIGITAL ORTHOPHOTOS

Another highly useful product which can be generated in a softcopy plotter is a digital orthophoto. As mentioned in Sec. 13-7, an orthophoto is a product which has the pictorial qualities of a photograph and the planimetric correctness of a map. Orthophotos are produced through the process of differential rectification, for which softcopy systems are particularly well suited. The essential inputs for the process of differential rectification are a DEM and a digital aerial photo having known exterior orientation parameters

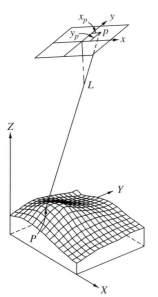

FIGURE 15-11
Collinearity relationship for a DEM point P
and its corresponding image p.

$(\omega, \phi, \kappa, X_L, Y_L,$ and $Z_L)$. It is also necessary to obtain the digital image coordinates (row and column) of the fiducials so that a transformation can be computed to relate photo coordinates to digital image coordinates. A systematic application of the collinearity equations (see Sec. 11-4) is then performed to produce the orthophoto.

Figure 15-11 illustrates the collinearity condition for a particular groundel point P in the DEM. The X and Y coordinates of point P are based upon the row and column within the DEM array, and its Z coordinate is stored in the DEM groundel array at that position. Given the X, Y, and Z coordinates of point P and the known exterior orientation parameters for the photograph, the collinearity equations [see Eqs. (11-1) and (11-2)] can be solved to determine photo coordinates x_p and y_p. These photo coordinates define the position where the image of groundel point P will be found. Since photo coordinates are related to the fiducial axis system, a transformation must be performed on these coordinates to obtain row and column coordinates in the digital image. The transformed row and column coordinates will generally not be whole numbers, so resampling is done within the scanned photo to obtain the digital number associated with groundel point P.

The process of creating the digital orthophoto requires repetitive application of the collinearity equations for all the points in the DEM array. Figure 15-12 gives a schematic illustration of the process. In this figure, two arrays are shown in vertical alignment, where each of the groundels of the DEM array corresponds one-to-one with a pixel of the orthophoto array. The orthophoto array is initially empty, and will be populated with digital numbers (shown in the figure as x's) as the process is carried out. At each step of the process, the X, Y, Z coordinates of the center point of a particular groundel of the DEM are substituted into the collinearity equations as discussed in the preceding paragraph. The resulting photo coordinates are then transformed to row and column coordinates of the digital aerial photo. Resampling is performed to obtain a

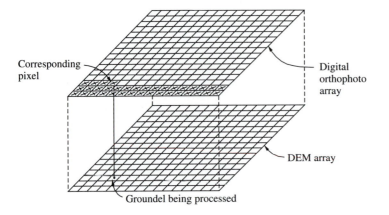

FIGURE 15-12
Schematic representation of digital orthophoto production process.

digital number, which is then placed into the corresponding pixel of the digital orthophoto. The process is complete when all pixels of the orthophoto have been populated with digital numbers.

Advances in digital imaging technology and softcopy plotters have made the production of digital orthophotos very economical. These highly useful images are fast becoming one of the most popular photogrammetric mapping products. They are often used as base maps for geographic information systems because of their accuracy and visual characteristics.

REFERENCES

Agouris, P., and T. Schenk: "Automated Aerotriangulation Using Multiple Image Multipoint Matching," *Photogrammetric Engineering and Remote Sensing,* vol. 62, no. 6, 1996, p. 703.

American Society for Photogrammetry and Remote Sensing: *Digital Photogrammetry, An Addendum to the Manual of Photogrammetry,* Bethesda, MD, 1996.

Bhattacharyya, G. K., and R. A. Johnson: *Statistical Concepts and Methods,* Wiley, New York, 1977.

Boulianne, M., and C. Nolette: "Virtual Reality Applied to User Interfaces for Digital Photogrammetric Workstations," *Photogrammetric Engineering and Remote Sensing,* vol. 65, no. 3, 1999, p. 277.

Graham, L. N., Jr., K. E. Ellison, Jr., and C. S. Riddell: "The Architecture of a Softcopy Photogrammetry System," *Photogrammetric Engineering and Remote Sensing,* vol. 63, no. 8, 1997, p. 1013.

Hannah, M. J.: "A System for Digital Stereo Matching," *Photogrammetric Engineering and Remote Sensing,* vol. 55, no. 12, 1989, p. 1765.

Heipke, C.: "A Global Approach for Least-Squares Image Matching and Surface Reconstruction in Object Space," *Photogrammetric Engineering and Remote Sensing,* vol. 58, no. 3, 1992, p. 317.

Helava, U. V.: "Object-Space Least-Squares Correlation," *Photogrammetric Engineering and Remote Sensing,* vol. 54, no. 6, 1988, p. 711.

Hohle, J.: "Experiences with the Production of Digital Orthophotos," *Photogrammetric Engineering and Remote Sensing,* vol. 62, no. 10, 1996, p. 1189.

Kersten, T., and S. Haering: "Automatic Interior Orientation of Digital Aerial Images," *Photogrammetric Engineering and Remote Sensing,* vol. 63, no. 8, 1997, p. 1007.

Kruphik, A.: "Using Theoretical Intensity Values as Unknowns in Multiple-Patch Least-Squares Matching," *Photogrammetric Engineering and Remote Sensing,* vol. 62, no. 10, 1996, p. 1151.

Li, M.: "Hierarchical Multipoint Matching," *Photogrammetric Engineering and Remote Sensing,* vol. 57, no. 8, 1991, p. 1039.

Norvelle, F. R.: "Stereo Correlation: Window Shaping and DEM Corrections," *Photogrammetric Engineering and Remote Sensing,* vol. 58, no. 1, 1992, p. 111.

Rosenholm, D.: "Multi-Point Matching Using the Least-Squares Technique for Evaluation of Three-Dimensional Models," *Photogrammetric Engineering and Remote Sensing,* vol. 53, no. 6, 1987, p. 621.

Toth, C. K., and A. Krupnik: "Concept, Implementation, and Results of an Automatic Aerotriangulation System," *Photogrammetric Engineering and Remote Sensing,* vol. 62, no. 6, 1996, p. 711.

Tseng, Y., J. Tzen, K. Tang, and S. Lin: "Image-to-Image Registration by Matching Area Features Using Fourier Descriptors and Neural Networks," *Photogrammetric Engineering and Remote Sensing,* vol. 63, no. 8, 1997, p. 975.

PROBLEMS

15-1. Describe the principle of the active polarizing screen method of stereoviewing for a softcopy plotter.

15-2. Describe the principle of the alternating LCD shutter display method of stereoviewing for a softcopy plotter.

15-3. Discuss the advantages of the fixed floating mark and moving images approach for stereoviewing over that of the moving floating mark and fixed images.

15-4. Briefly describe how automation can be applied to interior and relative orientation of a softcopy plotter.

15-5. Describe the concept of epipolar geometry as it applies to image matching.

15-6. Briefly describe the three categories of digital image-matching.

15-7. Describe how normalized cross-correlation relates to linear regression in image matching.

15-8. Given the following two subarrays from a pair of digital images, compute the normalized cross-correlation coefficient.

$$
A = \begin{bmatrix} 47 & 51 & 53 & 50 & 54 \\ 43 & 49 & 50 & 48 & 57 \\ 39 & 52 & 67 & 71 & 68 \\ 42 & 55 & 69 & 72 & 67 \\ 45 & 49 & 56 & 61 & 63 \end{bmatrix}
\qquad
B = \begin{bmatrix} 48 & 62 & 68 & 39 & 42 \\ 31 & 52 & 38 & 41 & 44 \\ 49 & 67 & 55 & 79 & 74 \\ 31 & 42 & 74 & 87 & 70 \\ 32 & 61 & 42 & 51 & 66 \end{bmatrix}
$$

15-9. Repeat Prob. 15-8, except that the following subarrays are to be used.

$$
A = \begin{bmatrix} 74 & 60 & 85 & 76 & 88 \\ 75 & 77 & 70 & 53 & 86 \\ 45 & 79 & 84 & 95 & 91 \\ 68 & 65 & 99 & 85 & 95 \\ 58 & 61 & 84 & 87 & 93 \end{bmatrix}
\qquad
B = \begin{bmatrix} 59 & 62 & 91 & 91 & 101 \\ 80 & 66 & 82 & 66 & 81 \\ 42 & 70 & 98 & 80 & 100 \\ 59 & 61 & 96 & 89 & 88 \\ 67 & 59 & 99 & 83 & 92 \end{bmatrix}
$$

15-10. Discuss the differences between digital image matching by normalized cross-correlation and least squares matching.

15-11. Explain how a DEM can be automatically generated by digital image matching with a stereopair of digital photos.

15-12. Describe the process of automatic generation of digital orthophotos. What key inputs are required?

CHAPTER
16

GROUND CONTROL FOR AERIAL PHOTOGRAMMETRY

16-1 INTRODUCTION

Photogrammetric control consists of any points whose positions are known in an object-space reference coordinate system and whose images can be positively identified in the photographs. In aerial photogrammetry the object space is the ground surface, and various reference ground coordinate systems are used to describe control point positions (see Chap. 5). Photogrammetric control, or *ground control* as it is commonly called in aerial photogrammetry, provides the means for orienting or relating aerial photographs to the ground. Almost every phase of photogrammetric work requires some ground control.

Photogrammetric control is generally classified as either *horizontal control* (the position of the point in object space is known with respect to a horizontal datum) or *vertical control* (the elevation of the point is known with respect to a vertical datum). Separate classifications of horizontal and vertical control have resulted primarily because of differences in horizontal and vertical reference datums, and because of differences in surveying techniques for establishing horizontal and vertical control. Also, horizontal and vertical control are considered separately in some photogrammetric processes. Often, however, both horizontal and vertical object-space positions of points are known, so that these points serve a dual control purpose.

Field surveying for photogrammetric control has historically been a two-step process, although now with the widespread use of GPS, this distinction is not as clear. The first step consists of establishing a network of *basic control* in the project area. This basic control consists of horizontal control monuments and benchmarks of vertical control which will serve as a reference framework for subsequent photo control surveys. The second step involves establishing object-space positions of *photo control* by means of surveys originating from the basic control network. Photo control points are the actual image points appearing in the photos that are used to control photogrammetric operations. The accuracy of basic control surveys is generally higher than that of subsequent photo control surveys. If GPS is used for the control surveying work, in some cases the intermediate step of establishing basic control may be bypassed and photo control established directly. This is discussed further in Sec. 16-8.

The two-step procedure of field surveying for photogrammetric control is illustrated in Fig. 16-1. In the figure a basic GPS control survey originates from existing control stations E_1 through E_4 and establishes a network of basic control points B_1 through B_6 in the project area. With these basic stations established, the second step of conducting subordinate surveys to locate photo control can occur. This is illustrated with the surveys that run between B_5 and B_6 and locate photo control points P_1 and P_2.

The establishment of good ground control is an extremely important element of almost any type of photogrammetric project. The accuracy of finished photogrammetric products can be no better than the ground control upon which they are based. Many maps and other items that have been carefully prepared in the office to exacting standards have failed to pass field inspection simply because the ground control was of poor quality. Because of its importance, the ground control phase of every photogrammetric project should be carefully planned and executed. The cost of establishing ground

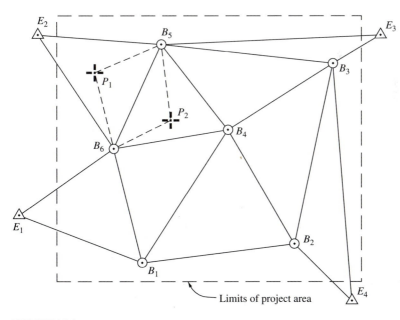

FIGURE 16-1
Field surveys for establishing photogrammetric control.

control for photogrammetric mapping is normally substantial, and depending upon conditions it can be expected to constitute between 10 and 50 percent of the total project cost.

16-2   SELECTING PHOTO CONTROL IMAGES

In general, images of acceptable photo control points must satisfy two requirements: (1) They must be sharp, well defined, and positively identified on all photos; and (2) they must lie in favorable locations in the photographs. (Reasons for this latter requirement are described in Sec. 16-3.) Control surveys for photogrammetry are normally conducted after the photography has been obtained. This ensures that the above two requirements can be met. Photo control images are selected after careful study of the photos. The study should include the use of a stereoscope to ensure a clear stereoscopic view of all points selected. This is important because many of the subsequent photogrammetric measurements will be made stereoscopically. A preliminary selection of photo control images may be made in the office, but the final selection should be made in the field with the photos in hand. This enables positive identification of objects to be made, and it also permits making a firsthand assessment of object point accessibility, terrain conditions, and surveying convenience.

Images for horizontal control have slightly different requirements than images for vertical control. Because their horizontal positions on the photographs must be precisely measured, images of horizontal control points must be very sharp and well defined horizontally. Some objects whose images are commonly satisfactory for horizontal control are intersections of sidewalks, intersections of roads, manhole covers, small lone bushes, isolated rocks, corners of buildings, fence corners, power poles, points on bridges, intersections of small trails or watercourses, etc. Care must be exercised to ensure that control points do not fall in shadowed areas on some photos.

Images for vertical control need not be as sharp and well defined horizontally. Points selected should, however, be well defined vertically. Best vertical control points are small, flat or slightly crowned areas. The small areas should have some natural features nearby, such as trees or rocks, which help to strengthen stereoscopic depth perception. Large, open areas such as the tops of grassy hills or open fields should be avoided, if possible, because of the difficulties they cause in stereoscopic depth perception. Intersections of roads and sidewalks, small patches of grass, small bare spots, etc., make excellent vertical control points.

The importance of exercising extreme caution in locating and marking objects in the field that correspond to selected photo images cannot be overemphasized. Mistakes in point identification are common and costly. A power pole, for example, may be located in the field, but it may not be the same pole whose image was identified on the photos. Mistakes such as this can be avoided by identifying enough other details in the immediate vicinity of each point so that verification is certain. A pocket stereoscope taken into the field can be invaluable in point identification, not only because it magnifies images but also because hills and valleys which aid in object verification can be seen both on the photos and on the ground.

16-3 NUMBER AND LOCATION OF PHOTO CONTROL

The required number of control points and their optimum location in the photos depend upon the use that will be made of them. For a very simple problem such as calculating the flying height of a photo which is assumed to be vertical (see Sec. 6-9), only the horizontal length of a line and the elevations of its endpoints are needed. A line of as great a length as possible should be chosen. For controlling mosaics (see Chap. 9), only a sparse network of horizontal control may be needed. The network should be uniformly distributed throughout the block of photos.

In solving the *space resection* problem for determining the position and orientation of a tilted photo (see Sec. 11-6), a minimum of three *XYZ* control points is required. The images of the control points should ideally form a large, nearly equilateral triangle. Although three control points are the required minimum for space resection, redundant control is recommended to increase the accuracy of the photogrammetric solution and to prevent mistakes from going undetected.

If photo control is being established for the purpose of orienting stereomodels in a plotting instrument for topographic map compilation, the absolute minimum amount of control needed in each stereomodel is three vertical and two horizontal control points. Again, the prudent photogrammetrist will utilize some amount of redundant control. As a practical minimum, each stereomodel oriented in a plotter should have three horizontal and four vertical control points. The horizontal points should be fairly widely spaced, and the vertical control points should be near the corners of the model. A satisfactory configuration is shown in Fig. 16-2. Some organizations require a fifth vertical control point in the center of each stereomodel.

If aerotriangulation (see Chap. 17) is planned to supplement photo control, then fewer ground-surveyed photo control points are needed. The amount of ground-surveyed

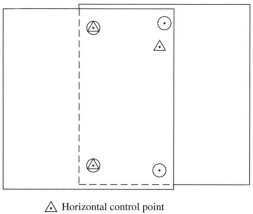

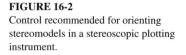
△ Horizontal control point
⊙ Vertical control point
⧌ Horizontal and vertical control point

FIGURE 16-2
Control recommended for orienting stereomodels in a stereoscopic plotting instrument.

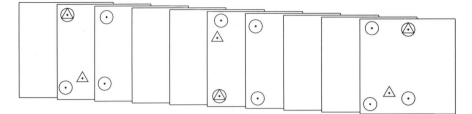

△ Horizontal control point
⊙ Vertical control point
⊘ Horizontal and vertical control point

FIGURE 16-3
Example control configuration for a strip of photographs.

photo control needed for aerotriangulation will vary, depending upon the size, shape, and nature of the area to be covered, the resulting accuracy required, and the procedures, instruments, and personnel to be used. In general, the more dense the ground-surveyed network of photo control, the better the resulting accuracy in the supplemental control determined by aerotriangulation. There is an optimum amount of ground-surveyed photo control, however, which affords maximum economic benefit from aerotriangulation and at the same time maintains a satisfactory standard of accuracy. On average, if aerotriangulation of a strip of photos is to be performed for the purpose of obtaining control for orienting stereomodels in a stereoplotter, a minimum of about two horizontal and three or four vertical ground-surveyed photo control points should appear in approximately every fifth stereomodel along the strip. This configuration is shown in Fig. 16-3. For aerotriangulation of blocks of photos, the ground-surveyed control should be systematically arranged throughout the block. Best control configurations consist of horizontal control along the periphery of the block with a uniform distribution of vertical control throughout the block. Figures 17-6 and 17-9a illustrate two examples. Experience generally dictates the best control configurations to use, and organizations involved in aerotriangulation normally develop their own standards which meet accuracy requirements for their particular combination of procedures, instruments, and personnel.

16-4   PLANNING THE CONTROL SURVEY

A great deal of planning should precede the photogrammetric control survey. A decision must be made early about the required accuracy of the survey. The type of equipment needed and the field techniques to be used must also be settled. These aspects and many more are interrelated and must be considered simultaneously to reach the best overall control survey plan. Previous experience in planning photogrammetric control surveys is a most valuable asset. A thorough understanding of surveying field techniques and a knowledge of instrument capabilities are also essential.

The accuracy required in any photogrammetric control survey will govern, to a large extent, the type of equipment and surveying techniques that can be used. Required accuracy of photo control depends primarily upon the accuracy required in the photogrammetric mapping or other project that it controls. This is not the only consideration, however, since accuracy may also depend upon whether the control will serve other purposes in addition to controlling the photogrammetric work. As an example, suppose a map is to be prepared by photogrammetric techniques for the purpose of planning and designing an urban transportation system. In that case the basic ground control should be accurate enough to also be used in describing property for right-of-way acquisition and to serve in precisely laying out the construction alignment, structures, etc. In this example the basic ground control should at least meet second-order standards, and it should be referenced to the state plane coordinate system (see Chap. 5). Basic horizontal control points and benchmarks should be permanently monumented and well described. If, on the other hand, a map is being prepared photogrammetrically for the purpose of forest inventory, not nearly as accurate a control survey is required; in fact, control taken directly from U.S. Geological Survey quadrangle maps is sometimes adequate.

The *national map accuracy standards* (NMAS) are commonly used to govern accuracy requirements of maps and therefore indirectly will also govern control surveying accuracy. If a map produced photogrammetrically is to be labeled as complying with these standards, it is required that at least 90 percent of the principal planimetric features be plotted to within $\frac{1}{30}$ in (0.8 mm) of their true positions for map scales of 1:20,000 or larger, and to within $\frac{1}{50}$ in (0.5 mm) for scales smaller than 1:20,000. On a map plotted at a scale of 1:600 (1 in/50 ft), this represents an allowable horizontal map error of 0.5 m (1.7 ft) on the ground. On a map at a scale of 1:24,000 (1 in/2000 ft), the allowable horizontal map error is 12 m (40 ft). If NMAS are to be met, horizontal photo control must be located to greater accuracy than the allowable horizontal map error. As a general rule of thumb, photo control should contain error no greater than about one-fourth to one-third the horizontal map accuracy tolerance. Some organizations require stricter tolerances than this. Of course, basic control must be more accurate than photo control.

NMAS also require that at least 90 percent of all points tested for elevation be correct to within one-half the contour interval. To meet this standard, a rule of thumb in topographic mapping states that elevations of vertical photo control points should be correct to within plus or minus about one-fifth of the contour interval; but as an additional safety factor, some agencies require that their accuracy be within one-tenth of the contour interval. According to this latter rule, a map being plotted with a contour interval of 1 m requires vertical photo control accurate to within ± 0.1 m. Again, the basic control must be more accurate than this.

A more current set of accuracy standards has been drafted by the Federal Geographic Data Committee.[1] Titled *Geospatial Positioning Accuracy Standards,* these standards are readily applicable to *digital maps* which are stored in a computer and

---

[1] The Web address of the Federal Geographic Data Committee is: http://www.fgdc.gov

manipulated with CAD software. Since digital map information may be displayed at any scale, accuracy standards for such products must be specified independently of plotting scale. The *Geospatial Positioning Accuracy Standards* address this by specifying accuracy in terms of ground distance at the 95 percent confidence level. As with NMAS, accuracy is expressed with separate horizontal and vertical components. The preferred test involves checking a set of at least 20 well-defined points by an independent source of higher accuracy. The standards state that this independent source "shall be of the highest accuracy feasible and practicable to evaluate the accuracy of the data set." Root mean square errors are computed and converted to the 95 percent confidence level by the appropriate multipliers. A digital planimetric map which passes at the 1-m level, for example would contain the statement "Tested 1-meter horizontal accuracy at 95% confidence level."

Other sets of accuracy standards have been established by various organizations such as the American Society for Photogrammetry and Remote Sensing, the Federal Highway Administration, and the American Society of Civil Engineers. These standards are all based on plotted scale of a map and as such are more appropriate for hard-copy maps than for digital maps.

In planning the control survey, maximum advantage should be taken of existing control in the area. The National Geodetic Survey has established numerous horizontal control monuments and vertical control benchmarks in its work of extending the national control network. The U.S. Geological Survey has also established a network of reliable horizontal and vertical control monuments in its topographic mapping operations. In certain localities, other agencies of the federal government such as the Tennessee Valley Authority, Army Corps of Engineers, and Bureau of Land Management have established control. Also, various state, county, and municipal agencies may have established control monuments. Caution should always be exercised in using existing control if it is of unknown accuracy.

16-5 TRADITIONAL FIELD SURVEY METHODS FOR ESTABLISHING HORIZONTAL CONTROL

Traditional instruments and techniques for field surveying are numerous and varied.[2] In this text only a very brief discussion of some basic methods is presented. The textbooks on surveying listed as references at the end of this chapter provide a much more thorough treatment of these subjects.

Horizontal control surveys, both for basic control and for photo control, may be conducted using any of the traditional field methods: *traversing, triangulation,* or *trilateration.* Of these methods, traversing is most common. Regardless of the method used, however, the survey must be connected to some existing reference control in the proximity of the project area. Existing control normally consists of a minimum of two (preferably three or more) intervisible points whose horizontal positions (e.g., state

---

[2] In this context, the Global Positioning System (GPS) is considered to be nontraditional. The use of GPS for photo control surveys is discussed in Secs. 16-7 and 16-8.

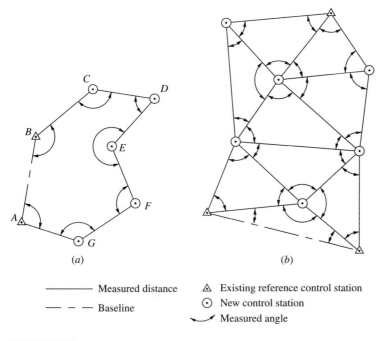

——————— Measured distance   ▵   Existing reference control station

— — — Baseline   ⊙   New control station

                                ⌣   Measured angle

FIGURE 16-4

(*a*) Example of a traverse network. (*b*) Example of a combined triangulation and trilateration network.

plane coordinates) are accurately known. The direction (e.g., azimuth from north) of the line connecting the two points is thus also normally known.

Traversing, as illustrated in Fig. 16-4*a*, consists of measuring horizontal angles and horizontal distances between consecutive stations of a closed network. The existing reference control stations are included in the network. Based on the existing control coordinates and reference direction, along with the newly measured horizontal angles and distances, coordinates of all new stations may be calculated trigonometrically in the rectangular coordinate system of the existing reference control. An adjustment is normally made to account for measurement errors, and the least squares method is best for this purpose.

Triangulation involves measurement of horizontal angles between intervisible stations of a network of triangular figures. Trilateration, on the other hand, involves measurement of horizontal distances in such a network. Often, triangulation and trilateration are combined in a given network, as shown in Fig. 16-4*b*. A least squares adjustment can be performed on the combined network to yield coordinates for the new stations. Using triangulation-combined trilateration enables highly accurate coordinates to be determined.

Total station instruments (see Fig. 16-5) are most commonly used for measuring the angles and distances for horizontal control surveys. These instruments combine

FIGURE 16-5
Total station instrument. (*Courtesy Leica Geosystems.*)

an automated angle-measuring component with an electronic distance-measuring component. The functions of these two components are controlled by an interfaced computer. A telescope with reticle is provided for sighting. As illustrated in Fig. 16-5, a total station instrument stands on a tripod. In measuring angles and distances, the instrument is first precisely centered over a point by means of an optical plumbing mechanism, and it is roughly leveled by means of a bull's-eye bubble. With most total station instruments, careful leveling is then achieved by centering the bubble of a precise level vial, but with some instruments this is done electronically. The angle-measuring component includes a graduated circle, which is placed in a horizontal plane during the leveling process.

Assume that the horizontal angle at station A, from B to G, of Fig. 16-4a will be measured. To begin the process, the total station instrument is carefully centered over station A. Then while looking through the instrument's telescope, the line of sight is directed precisely at station B. The reading of the graduated circle is then zeroed, and the telescope turned so that the line of sight is aimed precisely at station G. The arc between stations B and G is automatically resolved by the angle-measuring component, and its value displayed, and also stored, by the computer. The instrument's components for automatically measuring and displaying angular values are similar to those used in modern checkout machines at grocery and department stores. As the circle is turned in measuring an angle, a beam of light is interrupted by unique patterns of opaque bars with varying thickness and spacing. Based upon these light interruptions, the computer is able to resolve the angle. To eliminate mistakes and increase precision, the mean value of several repetitions of the angle measurement is taken. This angle-measuring process is repeated at each of the stations in the network.

Total station instruments measure distances electronically by employing the *phase shift* principle. To measure the distance from A to G of Fig. 16-4a, for example, the total station instrument is centered over station A, and a reflector is centered over station G. The instrument transmits an electromagnetic signal of precisely known wavelength to station G, where it is reflected and returned to station A. Components within the total station instrument compare the phase angles of the transmitted signal at station A, and the returned signal from G, and their difference is the so-called phase shift. But this measured phase shift would only indicate the length of the last partial wavelength in the distance from A to G and back to A, and the number of full wavelengths within that distance would be unknown. The number of full wavelengths in the double distance is resolved by transmitting additional frequencies of electromagnetic energy, each having a unique wavelength. The measuring operation is completely controlled by the total station's computer, and the distance is automatically displayed and stored.

If the line from the transmitter of the total station instrument to the reflector is not horizontal (the usual case), then its slope angle is also automatically measured by the total station instrument. The instrument's method of measuring slope angles is basically the same as that for measuring horizontal angles, except that a second graduated circle, which is oriented in a vertical plane, is employed. From measured slope distances and vertical angles, the interfaced computer determines horizontal distances.

For any horizontal control survey, an important task which must precede the taking of field measurements is establishing the network of stations in the project area whose positions are to be determined. In basic control surveys, the stations will normally be artificial monuments such as wooden stakes or iron rods driven into the ground. These are carefully referenced to permanent nearby features so that they can be recovered at a later date if lost. In photo control surveys, some of the stations will be artificial monuments and others will be the natural features selected for photo control points.

The specific procedures that should be used for any horizontal control survey will depend upon the conditions of the project. Existing topography or the presence of certain constructed features such as roads and railroads may render one procedure more economical than others. A fundamental limitation of triangulation, trilateration, and traversing is that clear lines of sight are required between points. This limitation generally necessitates that measurements be made in open areas, along clear paths (e.g., roads and railroads), or between hilltops. In the past, when control surveys were being performed in densely wooded areas, towers were used to raise the instruments above the treetops. Today, the use of towers is rare, since the measurements may be performed using GPS (see Secs. 16-7 and 16-8) which does not require clear lines of sight between points, although a clear view of the sky is required at each point.

The accuracy required of photo control is another major consideration in deciding upon equipment and procedures. If a high-accuracy standard for horizontal control is necessary, high standards of surveying will, of course, be required. This will generally require using the most precise instruments and taking the means of several repetitions of each distance and angle measurement. For lower-order work, instruments of lower precision may be used and the procedures relaxed somewhat. For the highest accuracy, GPS is the preferred method.

16-6 TRADITIONAL FIELD SURVEY METHODS FOR ESTABLISHING VERTICAL CONTROL

For vertical control surveys, *differential* leveling is the most common field procedure, and where highest accuracy is required, it is the method of choice. The basic equipment for differential leveling is a leveling instrument and a graduated rod. An automatic level, mounted on a tripod, and a graduated rod are shown in Fig. 16-6. The leveling instrument consists of a sighting telescope with reticle for reading the graduated rod and a means for orienting the telescope's line of sight in a horizontal plane. In using the instrument, rough leveling is accomplished first by centering a bull's-eye bubble. An *automatic compensator,* which is controlled by gravity, then precisely orients the line of sight in a horizontal plane.

The procedure of differential leveling is illustrated in Fig. 16-7. The vertical control survey originates from a *benchmark* (monument of known elevation) such as BM_X. The elevation of BM_X above datum is h_{BM_X}. To establish the elevation of new point A, the leveling instrument is set up between BM_X and point A, the line of sight is made horizontal, and readings R_1 and R_2 are taken on the graduated rod held vertically at BM_X and A, respectively. The elevation of point A is then equal to the elevation of BM_X, plus rod reading R_1, minus rod reading R_2.

To compensate for the earth's curvature, atmospheric refraction of the line of sight, and instrumental errors, lengths D_1 and D_2 of sights to BM_X and A should be approximately equal. Then the magnitude of error from these sources will be equal in readings R_1 and R_2, and since the error is added in R_1 and subtracted in R_2, it is eliminated. With the elevation of point A known, the elevation of point B may be established

FIGURE 16-6
Automatic leveling instrument, and level rod with metric graduations. (*Courtesy Leica Geosystems.*)

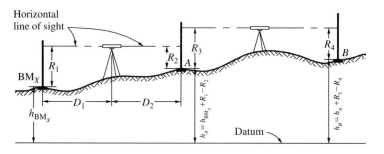

FIGURE 16-7
Differential or "spirit" leveling.

by setting the instrument between A and B and taking rod readings R_3 and R_4. The elevation of B is then equal to the elevation of A plus rod reading R_3, minus rod reading R_4. This process is continued until elevations have been determined for all desired points. The leveling circuit should terminate either on the initial benchmark or on some other benchmark so that adjustments can be made for errors that accumulated in the leveling process.

Another technique for determining differences in elevation is *trigonometric leveling*. It may be used where moderate accuracy is required and is especially well suited for rugged terrain. In trigonometric leveling, as illustrated in Fig. 16-8, vertical (zenith) angle Z and slope distance S are measured from the instrument at A to the reflector at B. As discussed in the preceding section, these measurements can be conveniently made with a total station instrument. The difference in elevation from the center of the instrument to the center of the reflector is $S \cos Z$. The elevation of point B is then equal to the elevation of point A, plus the instrument height i_h, plus $S \cos Z$, minus the reflector height r_h, as illustrated in Fig. 16-8. Compensation for errors due to earth curvature and atmospheric refraction is generally made by standard correction formulas found in many surveying texts. For highest accuracy in compensating for these errors, the measurements should be made in both directions and averaged.

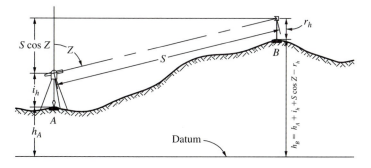

FIGURE 16-8
Trigonometric leveling.

16-7   FUNDAMENTALS OF THE GLOBAL POSITIONING SYSTEM

GPS was implemented by the U.S. Department of Defense to address the problem of navigation and positioning on a global basis. It consists of a network of 24 satellites orbiting the earth at an altitude of approximately 20,000 km, along with associated ground tracking and control stations. The satellites are arranged in six orbital planes and make a complete revolution every 12 hours (h). This arrangement guarantees that at least four satellites will be "visible" from anywhere on earth at any instant in time. The satellites are equipped with transmitters which can broadcast electromagnetic signals, as well as atomic clocks which establish a highly precise and accurate time basis.

Each satellite broadcasts digital messages or codes on two carrier frequencies: L1 at 1575.42 megahertz (MHz) and L2 at 1227.60 MHz. The codes are a pseudorandom series of bits (0s and 1s) which have been modulated onto the carrier frequencies. Two primary codes are broadcast: the C/A or *coarse acquisition* code, which is carried only by the L1 frequency, and the P or *precise* code, which is broadcast on both L1 and L2. In addition to these two primary codes, additional data are broadcast, including timing information and a satellite ephemeris.

The C/A code is broadcast at a *chip rate* of 1.023 MHz or 1.023×10^6 bits per second (bits/s). Given the speed of light of 2.9979246×10^8 m/s, each bit of the C/A code has an *equivalent length* of 293 m. The P code is broadcast at a frequency of 10.23 MHz, giving an equivalent length of 28.3 m/bit. The carrier waves are broadcast at much higher frequencies and therefore have wavelengths which are substantially shorter than the code bits. For the L1 frequency the wavelength is 0.190 m, and for L2 it is 0.244 m. The equivalent lengths of the code bits and the wavelengths of the carrier frequencies relate directly to the precision with which positions can be determined. Thus positions computed through *carrier-phase* processing can be determined much more precisely than those computed through *code-phase* processing.

The primary components of receivers used in conjunction with GPS are an antenna and a module which contains the receiver electronics, a clock, and a computer. Figure 16-9 shows a typical GPS receiver with the antenna set up on a tripod directly over a control monument. Virtually all GPS receivers are capable of measuring the C/A code phase while fewer can measure the P code phase. Only the more expensive *geodetic*-grade receivers can measure the carrier phase. In addition, some units can receive only the L1 frequency while *dual-frequency* receivers can receive both L1 and L2.

The Department of Defense operates a series of ground tracking and control stations around the globe having GPS receivers which continuously monitor the entire constellation of satellites. Based on observations from these stations, accurate orbit information, given with respect to the WGS84 datum, is determined for each satellite. It is periodically transmitted to the satellites for subsequent broadcast as part of the GPS signal. This orbit information comprises what is called the *predicted ephemeris*. It enables GPS users to compute estimated positions of satellites through extrapolation from previously computed orbit parameters. A more accurate *precise ephemeris,* which contains directly computed parameters (as opposed to predicted), may be obtained

FIGURE 16-9
GPS receiver with antenna mounted on a tripod.
(*Courtesy University of Florida.*)

from the National Geodetic Survey[3] several days after GPS observations have been obtained.

The fundamental mode of operation of GPS employs a single receiver and is called *point positioning*. In this mode, a receiver tracks the modulated code (C/A or P) from several satellites simultaneously. This information, the so-called code phase observable, is used to determine the time it takes for the signal to travel from the satellite to the receiver. The signal travel times from the various satellites are converted to distances known as *pseudoranges,* by multiplying by the speed of light. Given the positions of the satellites from the ephemeris, these distances are used in a three-dimensional spherical intersection to determine the coordinates of the receiver. Generally, at least four satellites are used which enable cancellation of the clock error of the receiver.

Errors inherent to point positioning are several. Most individually contribute an error of perhaps 1 to 5 m in the computed position of the receiver. These include errors due to ephemeris accuracy, timing accuracy, ionospheric and tropospheric interference, antenna construction, and multipath (or signal reflection). These errors are further amplified by a factor known as *PDOP* which is related to satellite geometry. The accumulation of all these errors can result in an error in receiver position of up to 25 m or more. In addition, to limit the usefulness of GPS to hostile forces, the Department of Defense generally implements the policy known as *selective availability,* or SA. Selective availability is an intentional degradation of timing and ephemeris information and when in effect, it can cause the point positioning error to increase to about 100 m.

A GPS method known as *differential positioning* can be used to determine locations of points with much greater accuracy than point positioning. The basic concept behind differential positioning is to employ two receivers which collect GPS signals

[3] Archival precise orbit information can be obtained at the following Web site: http://www.ngs.noaa.gov

simultaneously. One of the receivers is placed on a control point having known position, called a *base station;* and the other receiver, the *rover,* is placed at an unknown point. Since the position of the base station is known, errors due to satellite timing, ephemeris, and atmospheric interference can be calculated. Since both receivers collected data simultaneously, the errors calculated at the base station can be applied to the computed position of the rover. These errors, which would include those introduced by SA, are then absent in the computed coordinates of the unknown point.

For the ultimate in accuracy, *differential positioning* using carrier-phase observations is employed. In this approach, the phase changes of the L1 carrier wave (and perhaps L2) are measured at the receivers in order to determine their distances from the satellites. This is similar to the method employed by total station instruments in measuring distances electronically. The fundamental problem associated with this approach is that any particular cycle of the carrier wave is indistinguishable from the other cycles. The result is that there will be an unknown number of full cycles of the carrier wave between the satellite and the receiver. This *integer ambiguity* is a problem that must be overcome in the software used to process the information.

16-8   GROUND CONTROL SURVEYS BY GPS

Many modes of operation and data reduction techniques are available when performing ground control surveys by GPS. The method chosen will depend primarily upon accuracy requirements, but will also depend upon available equipment and processing software as well as the amount of time available to field crews. While an exhaustive list of GPS methods is not presented here, those that are commonly applied in photogrammetric control surveys are briefly discussed. The methods are broadly categorized into code-phase and carrier-phase techniques.

In general, single-receiver point positioning is far too inaccurate for useful photogrammetric control work. Therefore, differential techniques are used to establish control. If small-scale mapping is being performed, differential code-phase techniques may be employed. As mentioned in Sec. 16-7, differential GPS requires that one receiver be placed at a known base station. One or more roving receivers may then be employed to collect simultaneous observations at unknown points. Depending on the quality of receivers used and the distance between the base station and rover, errors of less than 5 m can generally be achieved. Currently the best-attainable accuracies of differential code-phase GPS are in the range of approximately 0.5 m.

Several carrier-phase methods of differential GPS are commonly used for establishing photogrammetric control. Of these, the *static* method is the most accurate. As the name implies, static GPS involves placing fixed receivers on points and collecting carrier-phase data for as long as an hour or more. After data have been collected, *baseline vectors* are computed between pairs of receivers which give ΔX, ΔY, and ΔZ components between corresponding points in a three-dimensional coordinate system. Generally, interconnected networks (see Fig. 16-1) of these vectors are created which are subsequently adjusted by least squares to obtain coordinates for the unknown points. Using the static technique, coordinates of unknown points can be determined with errors at the centimeter level.

If shorter observation times are desired, the method of *rapid static* observation may be employed. With this approach, station occupation times may be reduced to 10 min or even less, while still achieving centimeter-level accuracy. The rapid static method requires more expensive equipment and more sophisticated data processing techniques than static GPS, however. Generally, receivers capable of tracking both the L1 and L2 carrier frequencies as well as the C/A and P code signals are used. By including all four phase measurements in a highly redundant solution, accurate coordinates can be determined despite the reduced station occupation time.

Another technique known as *kinematic* positioning can be used for establishing ground control. With kinematic GPS, the rover receiver is in constant motion except for the brief period during which a station is occupied. The method requires a stationary receiver to be located at a known base station, and an initial start-up period during which the rover receiver is held stationary. This start-up period is required so that the processing software can calculate the unknown integer ambiguities. Once the ambiguities have been resolved, the rover receiver can be moved, and as long as it remains in contact with the satellite transmissions, the changes in integer ambiguities can be determined. The accuracy of kinematic GPS can be nearly as good as that of static, i.e., within a few centimeters. A major drawback associated with kinematic GPS surveys is that loss of signal cannot be tolerated. When the rover receiver "loses lock" on the satellite transmissions, which can happen if it goes under a tree, bridge, or other obstruction, then it must be returned to a previously surveyed position so that the integer ambiguities can be redetermined. This restricts the use of kinematic GPS to open areas. To reduce the inconvenience associated with loss of lock, the technique of "on-the-fly" (OTF) ambiguity resolution has been developed. This method of ambiguity resolution is performed by a software technique whereby many trial combinations of integer ambiguities for the different satellites are tested in order to determine the correct set. This essentially trial-and-error approach generally requires a period of uninterrupted data (i.e., no loss of lock) of as much as several minutes in order for the ambiguities to be correctly determined. Use of dual-frequency receivers which track C/A as well as P code provides redundancy which can greatly enhance the computational process of determining the integer ambiguities.

One final note concerning GPS surveys is warranted. While GPS is most often used to compute horizontal position, it is capable of determining vertical position (elevation) to nearly the same level of accuracy. The problem with the vertical position, however, is that it will be related to the ellipsoid, not the geoid or mean sea level (see Secs. 5-2 and 5-3). To relate the GPS-derived elevation (ellipsoid height) to the more conventional elevation (orthometric height), a geoid model is necessary (see Sec. 5-7).

16-9 ARTIFICIAL TARGETS FOR PHOTO IDENTIFIABLE CONTROL POINTS

In some areas such as prairies, forests, and deserts, natural points suitable for photogrammetric control may not exist. In these cases artificial points called *panel points* may be placed on the ground prior to taking the aerial photography. Their positions are then determined by field survey or in some cases by aerotriangulation. This procedure is called *premarking* or *paneling*.

Artificial targets provide the best possible photographic images, and therefore they are used for controlling the most precise photogrammetric work, whether or not natural points exist. Artificial targets are also used to mark section corners and boundary lines for photogrammetric cadastral work.

Besides their advantage of excellent image quality, their unique appearance makes misidentification of artificial targets unlikely. Disadvantages of artificial targets are that (1) extra work and expense are incurred in placing the targets, (2) the targets could be moved between the time of their placement and the time of photography, and (3) the targets may not appear in favorable locations on the photographs. To guard against the second disadvantage, the photography should be obtained as near as possible to the time of placing targets. To obtain target images in favorable positions on the photographs, the coverage of each photo can be planned in relation to target locations, and the positions of ground principal points can be specified on the flight plan.

A number of different types of artificial targets have been successfully used for photogrammetric control. The main elements in target design are good color contrast, a symmetric target that can be centered over the control point, and a target size that yields a satisfactory image on the resulting photographs. Contrast is best obtained using light-colored targets against a dark background or dark-colored targets against light backgrounds. The target shown in Fig. 16-10 provides good symmetry for centering over the control point. The center panel of the target should be centered over the control point, since this is the image point to which measurements will be taken. The legs help in identifying targets on the photos, and they also help in determining the exact center of the target should the image of the center panel be unclear. While the target shown in Fig. 16-10 is perhaps the ideal shape, circumstances may dictate use of other target shapes. Figure 16-11*a* shows a target which is often used where a smaller target is needed. The target of Fig. 16-11*b* is nearly as effective as that of Fig. 16-10, and it has the advantage of being more easily and quickly constructed. The target of Fig. 16-11*c* is less than optimal due to lack of biaxial symmetry; however, it may be needed in confined areas such as edges of highways.

Target sizes must be designed on the basis of intended photo scale so that the target images are the desired size on the photos. An image size of about 0.03 mm to about

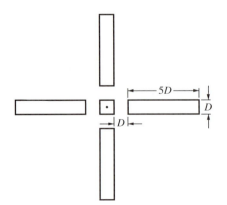

FIGURE 16-10
Artificial photogrammetric target.

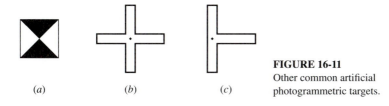

(a) (b) (c)

FIGURE 16-11
Other common artificial
photogrammetric targets.

0.10 mm for the sides of the central panel is generally ideal. As shown in Fig. 16-10, if the ground dimension of the central panel of the target is D, then the leg width should also be D, leg length should be $5D$, and the open space between the central panel and the leg should be D. Target sizes are readily calculated once the photo scale and optimum target image size are selected. If, for example, a central panel size of 0.05 mm is desired and photography at a scale of $1:12{,}000$ is planned, then D should be 0.6 m.

Materials used for targeting are quite variable. In some cases, satisfactory targets are obtained by simply painting white crosses on blacktop roads. In other cases, targets are painted on plywood, masonite, or heavy cloth, in which case they may be salvaged and reused. Satisfactory targets have also been made by placing stones against an earth background in the shape of a cross. The stones may be painted white for added contrast. Lime placed in the shape of a cross against a dark background has also produced satisfactory targets. Old tires painted white centered over the control points are also good for low-altitude, large-scale photography.

A procedure known as *post marking* can be performed if panel points are needed in an area after the photography has already been obtained. In this method, targets as described above are placed in the desired positions. Supplemental vertical photographs are taken of each target and its surrounding area with a small-format camera carried in a light aircraft flying at low altitude. Flying height can be calculated for the supplemental photography so that it has the same scale as the original photography. Locations of the targets can then be transferred stereoscopically from the supplemental photography to the original photography by using a point-transfer device. Zoom magnification of the individual viewing systems of the point-transfer device eliminates the need for having the scales of the supplemental and original photography equal.

16-10 INDEXING GROUND CONTROL

For each project it is advisable to prepare a set of paper prints for indexing the ground control. This set of photos should be examined carefully, and images of all control appearing on them should be identified from their field descriptions. When positive identification is made, the control point images should be lightly pricked with a pin to avoid the possibility of later misidentification mistakes; for example, if a control point lies at a particular sidewalk intersection corner, pricking that corner will offset the possibility of mistakenly using a different corner at some later time simply because the written description was misinterpreted. The pinprick should just penetrate the emulsion, and the pin should be held at right angles to the paper. Pricking should be done with the aid of a magnifying glass.

Control point images should be further marked by surrounding them with appropriate symbols, depending upon the type of control point. Triangles are commonly used to identify horizontal control points, while circles may be used for vertical control. A triangle inside a circle indicates both horizontal and vertical control on that point. Identifying numbers or names of control points should be written on the control photos beside the points. Some organizations have adopted numbering systems which correspond to the type of control point. As an example, vertical points may be given numbers from 1 to 999, horizontal points may be numbered from 1000 to 1999, and points of both vertical and horizontal control may be numbered from 2000 to 2999. Short, written descriptions should be placed on the backs of the photos near the pinpricks. A set of control index photos carefully prepared can be a valuable asset in performing subsequent photogrammetric operations.

REFERENCES

Abidin, H. Z.: "On-the-Fly Ambiguity Resolution," *GPS World,* vol. 5, no. 4, 1994, p. 40.

American Society of Photogrammetry: *Manual of Photogrammetry,* 4th ed., Falls Church, VA, 1980, chap. 8.

Anderson, J. M., and E. M. Mikhail: *Surveying: Theory and Practice,* WCB/McGraw-Hill, New York, 1998.

Bomford, G.: *Geodesy,* 4th ed., Clarendon Press, Oxford, 1980.

Leick, A.: *GPS Satellite Surveying,* 2d ed., Wiley, New York, 1995.

Thompson, M. M.: *Maps for America,* U.S. Geological Survey, Government Printing Office, Washington, 1987.

VanWijk, M. C.: "Test Areas and Targeting in the Hull Project," *Canadian Surveyor,* vol. 25, no. 5, 1971, p. 514.

Wolf, P. R., and R. C. Brinker: *Elementary Surveying,* 9th ed., HarperCollins, New York, 1993.

———and C. Ghilani: *Adjustment Computations: Statistics and Least Squares in Surveying and GIS,* Wiley, New York, 1997.

PROBLEMS

16-1. Explain the difference between basic control and photo control.

16-2. Describe the characteristics of good horizontal photo control points.

16-3. Describe the characteristics of good vertical photo control points.

16-4. State the national map accuracy standards for both horizontal positions and elevations.

16-5. Discuss the geospatial positioning accuracy standards, and contrast them with national map accuracy standards.

16-6. If a map is being prepared photogrammetrically to a scale of 500 ft/in, and photo control must be established to an accuracy 3 times greater than the allowable error for plotted points as specified by national map accuracy standards, how accurately on the ground must photo control points be located?

16-7. Repeat Prob. 16-6, except that map scale is $1:24,000$.

16-8. What are the photo dimensions of the square at the intersection of two sidewalks of 1.5-m width if photo scale is $1:30,000$?

16-9. What are the photographic dimensions in millimeters of a 0.75-m-diameter manhole cover if photo scale is $1:6000$?

16-10. Briefly describe three traditional field methods used in horizontal control surveys.

16-11. Give a brief description of two traditional field methods used in vertical control surveys.

16-12. Briefly describe the fundamental components of the Global Positioning System.

16-13. Explain the difference between carrier and code signals as it applies to GPS.

16-14. Discuss the errors which are inherent in single-receiver point positioning by GPS.

16-15. Briefly describe the principle of differential positioning by GPS. Why is this technique more accurate than point positioning?

16-16. Give a brief description of the concept of integer ambiguity as it applies to carrier-phase GPS.

16-17. Briefly discuss the principle of on-the-fly ambiguity resolution in kinematic GPS.

16-18. Discuss the advantages and disadvantages of using artificial targets as opposed to using natural targets for photo control.

16-19. What must be the ground dimension D (see Fig. 16-10) of artificial targets if their corresponding photo dimension is to be 0.05 mm on photos exposed from 3050 m above ground with a 152-mm-focal-length camera?

16-20. Repeat Prob. 16-19, except that the photo dimension is 0.10 mm, flying height above ground is 6000 ft, and camera focal length is 6.00 in.

16-21. Briefly discuss the importance of indexing ground control.

CHAPTER
17

AEROTRIANGULATION

17-1 INTRODUCTION

Aerotriangulation is the term most frequently applied to the process of determining the *X*, *Y*, and *Z* ground coordinates of individual points based on photo coordinate measurements. *Phototriangulation* is perhaps a more general term, however, because the procedure can be applied to terrestrial photos as well as aerial photos. The principles involved are extensions of the material presented in Chap. 11. In recent years, with improved photogrammetric equipment and techniques, accuracies to which ground coordinates can be determined by these procedures have become very high.

Aerotriangulation is used extensively for many purposes. One of the principal applications lies in extending or densifying ground control through strips and/or blocks of photos for use in subsequent photogrammetric operations. When used for this purpose, it is often called *bridging*, because in essence a "bridge" of intermediate control points is developed between field-surveyed control that exists in only a limited number of photos in a strip or block. Establishment of the needed control for compilation of topographic maps with stereoplotters is an excellent example to illustrate the value of aerotriangulation. In this application, as described in Chap. 12, the practical minimum number of control points necessary in each stereomodel is three horizontal and four vertical points. For large mapping projects, therefore, the number of control points needed is extensive, and the cost of establishing them can be extremely high if it is done exclusively by field survey methods. Much of this needed control is now routinely being established by aerotriangulation from only a sparse network of field-surveyed ground control and at a substantial cost savings. A more recent innovation involves the use of kinematic GPS in the aircraft to provide coordinates (indirectly) of the camera at the instant each photograph is exposed. In theory, this method of GPS control can eliminate the need for ground control entirely, although in practice a small amount of ground control is still used to strengthen the solution.

Besides having an economic advantage over field surveying, aerotriangulation has other benefits: (1) most of the work is done under laboratory conditions, thus minimizing delays and hardships due to adverse weather conditions; (2) access to much of the property within a project area is not required; (3) field surveying in difficult areas, such as marshes, extreme slopes, and hazardous rock formations, can be minimized; and (4) the accuracy of the field-surveyed control necessary for bridging is verified during the aerotriangulation process, and as a consequence, chances of finding erroneous control values after initiation of compilation are minimized and usually eliminated. This latter advantage is so meaningful that some organizations perform bridging even though adequate field-surveyed control exists for stereomodel control. It is for this reason also that some specifications for mapping projects require that aerotriangulation be used to establish photo control.

Apart from bridging for subsequent photogrammetric operations, aerotriangulation can be used in a variety of other applications in which precise ground coordinates are needed although most of these uses have been largely supplanted by GPS. In property surveying, aerotriangulation can be used to locate section corners and property corners or to locate evidence that will assist in finding these corners. In topographic mapping, aerotriangulation can be used to develop digital elevation models by computing X, Y, and Z ground coordinates of a systematic network of points in an area. Aerotriangulation has been used successfully for densifying geodetic control networks in areas surrounded by tall buildings where problems due to multipath cause a loss of accuracy in GPS surveys. Special applications include the precise determination of the relative positions of large machine parts during fabrication. It had been found especially useful in such industries as shipbuilding and aircraft manufacture. Many other applications of aerotriangulation are also being pursued.

Methods of performing aerotriangulation may be classified into one of three categories: *analog, semianalytical,* and *analytical.* Early analog procedures involved manual interior, relative, and absolute orientation of the successive models of long strips of photos using stereoscopic plotting instruments having several projectors. This created long strip models from which coordinates of pass points could be read directly. Later, universal stereoplotting instruments were developed which enabled this process to be accomplished with only two projectors. These procedures are now principally of historical interest, having given way to the other two methods.

Semianalytical aerotriangulation involves manual interior and relative orientation of stereomodels within a stereoplotter, followed by measurement of model coordinates. Absolute orientation is performed numerically—hence the term *semianalytical* aerotriangulation.

Analytical methods consist of photo coordinate measurement followed by numerical interior, relative, and absolute orientation from which ground coordinates are computed. Various specialized techniques have been developed within each of the three aerotriangulation categories. This chapter briefly describes some of these techniques. Part I covers semianalytical aerotriangulation, and Part II covers analytical methods. This discussion predominantly relates to bridging for subsequent photogrammetric operations because this is the principal use of aerotriangulation. Extension of these basic principles can readily be translated to the other areas of application, however.

17-2 PASS POINTS FOR AEROTRIANGULATION

Pass points for aerotriangulation are normally selected in the general photographic locations shown in Fig. 17-1*a*. The points may be images of natural, well-defined objects that appear in the required photo areas, but if such points are not available, pass points

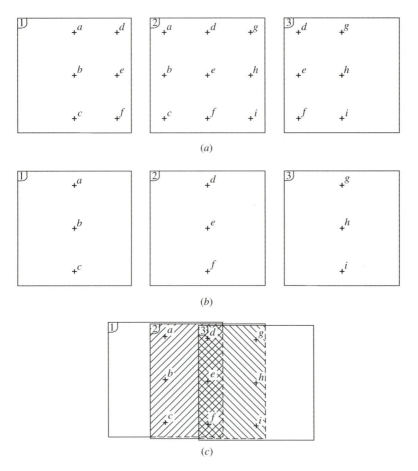

FIGURE 17-1
(*a*) Idealized pass point locations for aerotriangulation. (*b*) Pass point locations for stereoscopic measurement. (*c*) Locations of pass points in two adjacent stereomodels.

FIGURE 17-2
Wild PUG-4 stereoscopic
point-transfer instrument. (*Courtesy
LH Systems, LLC.*)

may be artificially marked by using a special stereoscopic point-marking device such as that shown in Fig. 17-2. Point-marking devices make small holes in the emulsion which become the pass points. Pass points can also be selected by automated procedures using digital photography.

Even though satisfactory natural points may exist in the required general locations on the photographs, many photogrammetrists prefer to mark pass points artificially for two reasons. First, a more discrete point is obtained so that more accurate measurements of its position can be obtained. Second, the likelihood of misidentifying pass points is greatly reduced.

The approach used in marking pass points depends upon the aerotriangulation method and whether the measurements will be made monoscopically or stereoscopically. With analytical aerotriangulation using a monocomparator (see Sec. 4-7), it is necessary to see the images of all pass points in each photograph. In artificially marking pass points for monocomparator measurement, some points, such as d, e, and f of Fig. 17-1a, appear on three successive photos. Each of these points may be located quite arbitrarily on one photograph, which is usually the center photo on which they appear; but once marked, they must be *carefully* transferred to their corresponding locations on the two adjacent plates. The Wild PUG-4 point-transfer instrument of Fig. 17-2 is a device designed specifically for marking corresponding pass points. With this instrument a stereopair of diapositives is placed on illuminated stage plates. The diapositives are viewed stereoscopically through a binocular viewing system which has a reference half mark superimposed within each of its optical paths. By means of slow-motion screws, the diapositives can be adjusted slightly in the x and y directions so that the floating mark, comprised of the two half marks, appears to rest exactly on the point to be marked. Tiny drills which are collinear with the half marks are then lowered, and holes are drilled into the emulsion to mark corresponding points. With any point-marking and point-transfer instrument, it must be recognized that, upon making the holes, the emulsion in those spots is destroyed. Thus no further refinement in the identification of the image position can be made, and any error in positioning is fixed. *This operation must therefore be done with extreme caution!*

When point measurements for aerotriangulation are made while viewing the photographs in stereo with either a stereoplotter or stereocomparator, only three pass points near the y axis of each photo are usually marked, as shown in Fig. 17-1b. Stereocomparators simultaneously measure photo coordinates of corresponding points on a stereopair of diapositives. These instruments have two separate measuring systems, one for each photo. During viewing through binoculars along optical paths, the positions of the diapositives are adjusted until a reference floating mark appears to rest exactly on the desired point. In this position measurements are recorded for both photos. Formerly, a number of companies manufactured stereocomparators; however, the more versatile analytical stereoplotter is now generally used for making stereo measurements.

When stereopairs of photos with pass points marked in the manner of Fig. 17-1b are oriented in a plotter, six points appear in each stereomodel as shown in Fig. 17-1c. The operator will see a slightly confusing scene. While the area surrounding the pass point will be seen properly in stereo, the drilled hole will appear in only the left or right ocular but not both. With practice, however, an operator can accurately set the floating mark on the image feature. Once the floating mark is properly placed, the X, Y, and Z model coordinates, or alternatively, the x and y photo coordinates of the left and right half marks, are then measured.

A recent innovation which is an outgrowth of softcopy photogrammetry is the automated selection of pass points by digital image matching, which is the essential step of a process commonly referred to as *automatic aerotriangulation*. In this method, points are selected in the overlap areas of digital images and are automatically matched between adjacent images, thus achieving pass point selection and photo coordinate measurement simultaneously. This method requires little operator intervention and is therefore a very economical process. An added benefit is that a large number of pass points can be generated with minimal effort, which adds redundancy and strengthens the aerotriangulation solution.

PART I
SEMIANALYTICAL AEROTRIANGULATION

17-3 GENERAL DESCRIPTION

Semianalytical aerotriangulation, often referred to as *independent model* aerotriangulation, is a partly analog and partly analytical procedure that emerged with the development of computers. It involves manual relative orientation in a stereoplotter of each stereomodel of a strip or block of photos. After the models have been measured, they are numerically adjusted to the ground system by either a sequential or a simultaneous method. In the sequential approach, contiguous models are joined analytically, one by one, to form a continuous strip model, and then absolute orientation is performed numerically to adjust the strip model to ground control. In the simultaneous approach, all models in a strip or block are joined and adjusted to ground control in a single step, much like the simultaneous transformation technique described in Sec. 9-9.

Regardless of whether the sequential or simultaneous method is employed, the process yields coordinates of the pass points in the ground system. Additionally, coordinates of the exposure stations can be determined in either process; however, their values are seldom required.

17-4 SEQUENTIAL CONSTRUCTION OF A STRIP MODEL FROM INDEPENDENT MODELS

In the sequential approach to semianalytical aerotriangulation, each stereopair of a strip is relatively oriented in a stereoplotter, the coordinate system of each model being independent of the others. When relative orientation is completed, model coordinates of all control points and pass points are read and recorded. This is done for each stereomodel in the strip. Figures 17-3*a* and *b* illustrate the first three relatively oriented stereomodels of a strip and show plan views of their respective independent coordinate systems. By means of pass points common to adjacent models, a three-dimensional conformal coordinate transformation (see Sec. C-7) is used to tie each successive model to the previous one. To gain needed geometric strength in the transformations, the coordinates of the perspective centers (model exposure stations) are also measured in each independent model and included as common points in the transformation. The right exposure station of model 1-2, O_2, for example, is the same point as the left exposure station of model 2-3. To transform model 2-3 to model 1-2, therefore, coordinates of common points d, e, f, and O_2 of model 2-3 are made to coincide with their corresponding model 1-2 coordinates. Once the parameters for this transformation have been computed, they are applied to the coordinates of points g, h, i, and O_3 in the system of model 2-3 to obtain their coordinates in the model 1-2 system. These points in turn become control for a transformation of the points of model 3-4. By applying successive coordinate transformations, a continuous strip of stereomodels may be formed, as illustrated in Fig. 17-3*c*. The entire strip model so constructed is in the coordinate system defined by model 1-2.

17-5 ADJUSTMENT OF A STRIP MODEL TO GROUND

After a strip model has been formed, it is numerically adjusted to the ground coordinate system using all available control points. If the strip is short, i.e., up to about four models, this adjustment may be done using a three-dimensional conformal coordinate transformation. This requires that a minimum of two horizontal control points and three vertical control points be present in the strip. More control than the minimum is desirable, however, as it adds stability and redundancy to the solution. As discussed later in this section, if the strip is long, a polynomial adjustment is preferred to

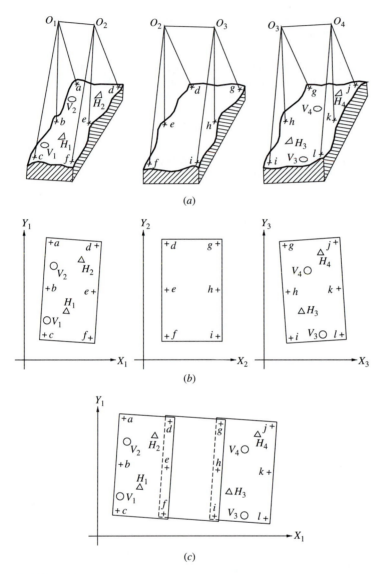

FIGURE 17-3
Independent model or semianalytical aerotriangulation. (*a*) Three adjacent relatively oriented stereomodels.
(*b*) Individual arbitrary coordinate systems of three adjacent stereomodels. (*c*) Continuous strip of
stereomodels formed by numerically joining the individual arbitrary coordinate systems into one system.

transform model coordinates to the ground coordinate system. In the short strip illustrated in Fig. 17-3*c*, horizontal control points H_1 through H_4 and vertical control points V_1 through V_4 would be used in a three-dimensional conformal coordinate transformation to compute the ground coordinates of pass points *a* through *l* and exposure stations O_1 through O_4.

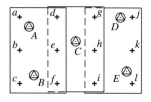

FIGURE 17-4
Configuration of pass points and control
for semianalytical aerotriangulation of
Example 17-1.

Example 17-1. Figure 17-4 illustrates a continuous strip of three stereomodels
with pass points a through l and ground control points A through E. Independent
model coordinates for points and exposure stations of each model are listed below
along with ground coordinates of control points A through E. Compute the ground
coordinates of the pass points and exposure stations by the sequential method of
semianalytical aerotriangulation. Use the three-dimensional conformal coordinate
transformation program provided.[1]

Model coordinates (mm)

	Model 1-2				Model 2-3				Model 3-4		
Point	x	y	z	Point	x	y	z	Point	x	y	z
a	−1.568	91.538	−0.243	d	−0.120	82.885	9.473	g	−0.869	104.502	−12.763
b	−3.995	−0.725	−1.124	e	−6.933	−5.499	10.627	h	0.623	9.411	−16.914
c	−2.453	−88.126	−3.378	f	−5.356	−84.482	9.860	i	3.259	−86.164	−18.112
d	92.922	92.610	−0.588	g	78.090	83.401	8.690	j	98.487	98.233	−12.522
e	81.781	−1.480	−1.881	h	80.939	2.593	7.601	k	100.807	3.273	−15.224
f	80.019	−85.903	−4.881	i	84.843	−78.510	8.957	l	98.299	−98.052	−16.039
A	23.295	65.291	−0.620	C	44.480	20.499	8.869	D	62.943	101.805	−12.672
B	34.886	−65.717	−3.919	O_2	0.000	0.000	152.292	E	83.271	−56.709	−15.386
O_1	0.000	0.000	152.292	O_3	85.079	−0.955	151.424	O_3	0.000	0.000	152.292
O_2	89.009	0.223	149.734					O_4	96.068	−1.067	152.499

Ground control coordinates (m)

Point	X	Y	Z
A	1444.632	2892.290	88.904
B	1525.490	2109.281	83.661
C	2133.597	2655.634	89.583
D	2705.480	3035.452	92.118
E	2811.033	2173.558	94.052

---

[1]The program used in the example is titled "3dconf" and can be downloaded at the following Web site:
http://www. surv.ufl.edu/wolfdewitt

Solution:

1. With an ASCII text editor, create the following data file with a ".dat" extension (see Example 11-3 for a description of the data file format):

```
d   -0.120   82.885    9.473  92.922   92.610   -0.588
e   -6.933   -5.499   10.627  81.781   -1.480   -1.881
f   -5.356  -84.482    9.860  80.019  -85.903   -4.881
02   0.000    0.000  152.292  89.009    0.223  149.734
#
g   78.090   83.401    8.690
h   80.939    2.593    7.601
i   84.843  -78.510    8.957
c   44.480   20.499    8.869
03  85.079   -0.955  151.424
#
```

2. Run the "3dconf" program to produce the following results. (Only the portion of the output which gives transformed points is shown.)

```
Transformed Points:
 Point      X          Y         Z      SDev.X   SDev.Y  SDev.Z
   g      176.501    89.791   -1.310   0.037    0.036   0.035
   h      176.019     3.397   -4.702   0.027    0.032   0.031
   i      176.645   -83.426   -5.488   0.036    0.036   0.036
   c      137.848    24.079   -2.920   0.024    0.025   0.024
   03     179.857    -4.535  148.932   0.040    0.038   0.038
```

The above output gives the coordinates of points g, h, i, C, and O_3 in the model 1-2 system.

3. With an ASCII text editor, create the following data file with a ".dat" extension:

```
g   -0.869  104.502  -12.763  176.501   89.791   -1.310
h    0.623    9.411  -16.914  176.019    3.397   -4.702
i    3.259  -86.164  -18.112  176.645  -83.426   -5.488
03   0.000    0.000  152.292  179.857   -4.535  148.932
#
j   98.487   98.233  -12.522
k  100.807    3.273  -15.224
1   98.299  -98.052  -16.039
D   62.943  101.805  -12.672
E   83.271  -56.709  -15.386
04  96.068   -1.067  152.499
#
```

4. Run the "3dconf" program to produce the following results. (Only the portion of the output which gives transformed points is shown.)

```
Transformed Points:
 Point      X          Y         Z      SDev.X   SDev.Y  SDev.Z
   j      266.567    82.280   -3.733   0.010    0.010   0.010
   k      266.877    -3.993   -5.860   0.008    0.009   0.009
   1      262.742   -95.950   -6.117   0.011    0.010   0.010
   D      234.369    86.171   -2.926   0.009    0.009   0.008
```

```
        E      249.870  -58.135   -5.288  0.009   0.009   0.008
        O4     267.032   -7.252  146.536  0.011   0.011   0.010
```

The output gives the coordinates of points j, k, l, D, E, and O_4 in the model 1-2 system.

5. With an ASCII text editor, create the following data file with a ".dat" extension:

```
A    23.295    65.291   -0.620  1444.632  2892.290  88.904
B    34.886   -65.717   -3.919  1525.490  2109.281  83.661
C   137.848    24.079   -2.920  2133.597  2655.634  89.583
D   234.369    86.171   -2.926  2705.480  3035.452  92.118
E   249.870   -58.135   -5.288  2811.033  2173.558  94.052
#
a    -1.568    91.538   -0.243
b    -3.995    -0.725   -1.124
c    -2.453   -88.126   -3.378
d    92.922    92.610   -0.588
e    81.781    -1.480   -1.881
f    80.019   -85.903   -4.881
O1    0.000     0.000  152.292
O2   89.009     0.223  149.734
g   176.501    89.791   -1.310
h   176.019     3.397   -4.702
i   176.645   -83.426   -5.488
O3  179.857    -4.535  148.932
j   266.567    82.280   -3.733
k   266.877    -3.993   -5.860
l   262.742   -95.950   -6.117
O4  267.032    -7.252  146.536
#
```

6. Run the "3dconf" program to produce the following results (only the portion of the output which gives transformed points is shown):

```
Transformed Points:
  Point      X         Y         Z       SDev.X  SDev.Y  SDev.Z
      a   1293.555  3047.090    86.231   0.062   0.062   0.080
      b   1287.111  2494.926    90.169   0.057   0.057   0.064
      c   1304.113  1972.055    85.755   0.065   0.065   0.084
      d   1858.691  3061.799    92.594   0.047   0.047   0.062
      e   1800.320  2497.897    93.466   0.040   0.040   0.042
      f   1797.369  1992.472    83.989   0.050   0.050   0.069
     O1   1296.858  2515.085  1008.088   0.081   0.106   0.080
     O2   1829.486  2524.009  1000.806   0.070   0.097   0.063
      g   2358.935  3052.234    96.113   0.046   0.046   0.060
      h   2363.847  2535.093    84.608   0.038   0.038   0.039
      i   2375.185  2015.747    88.834   0.048   0.048   0.067
     O3   2373.385  2503.479  1004.702   0.070   0.096   0.063
      j   2898.543  3015.005    90.524   0.059   0.059   0.073
      k   2908.066  2498.785    86.644   0.055   0.055   0.062
      l   2891.322  1948.364    94.128   0.064   0.064   0.087
     O4   2895.283  2494.672   998.523   0.079   0.103   0.078
```

   This completes the solution. Note that the output of step 6 contains the computed ground coordinates in meters for pass points a through l as well as the exposure stations O_1 through O_4. The computed standard deviations are also listed in meters.

   Due to the nature of sequential strip formation, random errors will accumulate along the strip. Often, this accumulated error will manifest itself in a systematic manner with the errors increasing in a nonlinear fashion. This effect, illustrated in Fig. 17-5, can be significant, particularly in long strips. Figure 17-5a shows a strip model comprised of seven contiguous stereomodels from a single flight line. Note from the figure that sufficient ground control exists in model 1-2 to absolutely orient it (and thereby the entire strip) to the ground system. The remaining control points (in models 4-5 and 7-8) can then be used as *checkpoints* to reveal accumulated errors along the strip. Figure 17-5b shows a plot of the discrepancies between model and ground coordinates for the checkpoints as a function of X coordinates along the strip. Except for the ground control in the first model, which was used to absolutely orient the strip, discrepancies exist between model positions of horizontal and vertical control points and their corresponding field-surveyed positions. Smooth curves are fit to the discrepancies as shown in the figure.

   If sufficient control is distributed along the length of the strip, a three-dimensional polynomial transformation can be used in lieu of a conformal transformation to perform

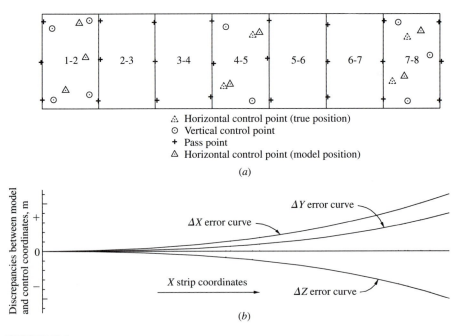

(a)

(b)

FIGURE 17-5
(a) Plan view of control extension of a seven-model strip. (b) Smooth curves indicating accumulation of errors in X, Y, and Z coordinates during control extension of a strip.

absolute orientation and thus obtain corrected coordinates for all pass points. This polynomial transformation yields higher accuracy through modeling of systematic errors along the strip. Most of the polynomials in use for adjusting strips formed by aerotriangulation are variations of the following third-order equations:

$$\overline{X} = a_0 + a_1X + a_2Y + a_3X^2 + a_4XY + a_5Y^2 + a_6X^3 + a_7X^2Y + a_8XY^2 + a_9Y^3$$

$$\overline{Y} = b_0 + b_1X + b_2Y + b_3X^2 + b_4XY + b_5Y^2 + b_6X^3 + b_7X^2Y + b_8XY^2 + b_9Y^3$$

$$\overline{Z} = c_0 + c_1X + c_2Y + c_3X^2 + c_4XY + c_5Y^2 + c_6X^3 + c_7X^2Y + c_8XY^2 + c_9Y^3$$

$$(17\text{-}1)$$

In Eqs. (17-1), $\overline{X}$, $\overline{Y}$, and $\overline{Z}$ are the transformed ground coordinates; X and Y are strip model coordinates; and the a's, b's, and c's are coefficients which define the shape of the polynomial error curves. The equations contain 30 unknown coefficients (a's, b's, and c's). Each three-dimensional control point enables the above three polynomial equations to be written, and thus 10 three-dimensional control points are required in the strip for an exact solution. When dealing with transformations involving polynomials, however, it is imperative to use redundant control which is well distributed throughout the strip. It is important that the control points occur at the periphery as well, since extrapolation from polynomials can result in excessive corrections. As illustrated by Fig. 17-5b, errors in X, Y, and Z are principally functions of the linear distance (X coordinate) of the point along the strip. However, the nature of error propagation along strips formed by aerotriangulation is such that discrepancies in X, Y, and Z coordinates are also each somewhat related to the Y positions of the points in the strip. Depending on the complexity of the distortion, certain terms may be eliminated from Eqs. (17-1) if they are found not to be significant. This serves to increase redundancy in the transformation which generally results in more accurate results. Further discussion of the application of polynomials in adjusting strip models to ground can be found in references cited at the end of this chapter.

17-6 INDEPENDENT MODEL AEROTRIANGULATION BY SIMULTANEOUS TRANSFORMATIONS

A more robust method for performing independent model aerotriangulation is to employ simultaneous transformations rather than sequential. This approach avoids the accumulation of errors inherent in the sequential strip-forming process and thus requires no polynomial error modeling. The method is similar to the simultaneous two-dimensional transformation discussed in Sec. 9-9. The basis of simultaneous independent model aerotriangulation is the three-dimensional conformal coordinate transformation (see Sec. C-7). The transformation equations [Eqs. (C-39)] are repeated here in a slightly modified form as Eqs. (17-2).

$$X = s(m_{11}x + m_{21}y + m_{31}z) + T_X$$

$$Y = s(m_{12}x + m_{22}y + m_{32}z) + T_Y \qquad (17\text{-}2)$$

$$Z = s(m_{13}x + m_{23}y + m_{33}z) + T_Z$$

In Eqs. (17-2), the terms are as described in Sec. C-7. It is important to note, however, that in the simultaneous transformation approach there will be as many individual sets of unknown parameters (s, m's, T_X, T_Y, and T_Z) as there are models. Figure 17-6 shows a small block of 6 contiguous models formed by two adjacent strips of 4 photos (3 models) each. The block contains 20 pass points (points 1 through 20) and 6 ground control points (A through F). Since the block contains 6 independent models, 6 sets of 7 parameters (total = 42) must be determined in order to transform the individual models to the ground coordinate system. The computation requires a least squares adjustment involving two sets of observation equations. The first set of observation equations is used for ground control points and has the same form as Eqs. (17-2), except that appropriate subscripts are included for the parameters and coordinates. Equations (17-3) give the general form for control point observations.

$$X_a = s_i(m_{11_i}x_{a_i} + m_{21_i}y_{a_i} + m_{31_i}z_{a_i}) + T_{X_i}$$
$$Y_a = s_i(m_{12_i}x_{a_i} + m_{22_i}y_{a_i} + m_{32_i}z_{a_i}) + T_{Y_i} \qquad (17\text{-}3)$$
$$Z_a = s_i(m_{13_i}x_{a_i} + m_{23_i}y_{a_i} + m_{33_i}z_{a_i}) + T_{Z_i}$$

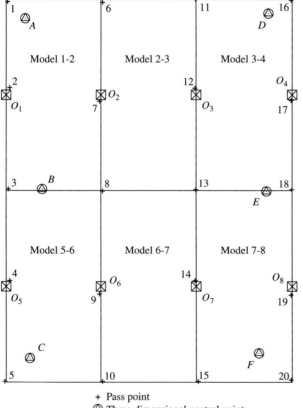

+ Pass point
⊖ Three-dimensional control point
⊠ Exposure station

FIGURE 17-6
Configuration of pass points, control, and exposure stations of a six-model block adjusted by method of simultaneous independent model aerotriangulation.

TABLE 17-1
Observed control points for the models of Figure 17-6

Point	Model i	Point	Model i
A	1-2	D	3-4
B	1-2	E	3-4
B	5-6	E	7-8
C	5-6	F	7-8

In these equations, the m's are the rotation matrix terms which relate model i to the ground system; s_i is the scale factor for model i; T_{X_i}, T_{Y_i}, and T_{Z_i} are the X, Y, and Z translations, respectively, for model i; x_{a_i}, y_{a_i}, and z_{a_i} are model coordinates for point a from model i; and X_a, Y_a, and Z_a are ground control coordinates for point a. Equations of this type are formed for every occurrence of a ground control point in each model. Table 17-1 lists all eight occurrences of control points in the individual models. Each entry in the table will have an associated set of equations of the form of Eqs. (17-3).

The second type of observation equations enforces the condition that tie points (pass points and common exposure stations) between adjacent models must coincide. The condition used is that the difference in transformed coordinates of the tie point from each adjacent model is equal to zero. Equations (17-4) express this condition for tie point a between models i and j.

$$s_i(m_{11_i}x_{a_i} + m_{21_i}y_{a_i} + m_{31_i}z_{a_i}) + T_{X_i} - [s_j(m_{11_j}x_{a_j} + m_{21_j}y_{a_j} + m_{31_j}z_{a_j}) + T_{X_j}] = 0$$

$$s_i(m_{12_i}x_{a_i} + m_{22_i}y_{a_i} + m_{32_i}z_{a_i}) + T_{Y_i} - [s_j(m_{12_j}x_{a_j} + m_{22_j}y_{a_j} + m_{32_j}z_{a_j}) + T_{Y_j}] = 0$$

$$s_i(m_{13_i}x_{a_i} + m_{23_i}y_{a_i} + m_{33_i}z_{a_i}) + T_{Z_i} - [s_j(m_{13_j}x_{a_j} + m_{23_j}y_{a_j} + m_{33_j}z_{a_j}) + T_{Z_j}] = 0$$
$$(17\text{-}4)$$

In these equations, the terms are as defined for Eqs. (17-3), except that the subscript j is used for parameters and coordinates of the adjacent model j. A set of equations of the form of Eqs. (17-4) is written for each occurrence of a tie point between adjacent models. Table 17-2 lists all the unique combinations of tie points between the models of Fig. 17-6. In this table, there are 28 unique tie point connections between adjacent models shown in the figure, including common exposure stations and control points B and E which can also serve as tie points. Note that tie point 6, which exists only in two models, contributes only one set of equations of the form of Eqs. (17-4), whereas tie point 8 contributes six sets of equations.

Each of the 28 tie point entries of Table 17-2 contributes 3 equations for a total of $28 \times 3 = 84$ equations. Each of the 8 control point entries of Table 17-1 also contributes 3 equations for a total of $8 \times 3 = 24$ equations. Thus the full set of observations comprises $84 + 24 = 108$ equations. Since there are only 42 unknown transformation parameters associated with the block, least squares is employed in the solution. After the transformation parameters have been computed, coordinates of pass points 1 through 20 can be obtained by applying Eqs. (17-2), using model coordinates and corresponding transformation parameters. Ground coordinates of pass points which

TABLE 17-2
Connections between models of Fig. 17-6 based on model coordinates of tie points

Point	Model i	Model j	Point	Model i	Model j	Point	Model i	Model j
3	1-2	5-6	10	5-6	6-7	15	6-7	7-8
6	1-2	2-3	11	2-3	3-4	18	3-4	7-8
7	1-2	2-3	12	2-3	3-4	B	1-2	5-6
8	1-2	2-3	13	2-3	3-4	E	3-4	7-8
8	1-2	5-6	13	2-3	6-7	O_2	1-2	2-3
8	1-2	6-7	13	2-3	7-8	O_3	2-3	3-4
8	2-3	5-6	13	3-4	6-7	O_6	5-6	6-7
8	2-3	6-7	13	3-4	7-8	O_7	6-7	7-8
8	5-6	6-7	13	6-7	7-8			
9	5-6	6-7	14	6-7	7-8			

served as tie points (i.e., those that appeared in adjacent models) can be computed from more than one model. For example, ground coordinates of point 6 of Fig. 17-6 can be computed by using coordinates and parameters of model 1-2. They can also be computed by using coordinates and parameters of model 2-3. In these cases, it is recommended that the ground coordinates be computed using each model, and the results averaged to obtain final pass point coordinates.

17-7 PERSPECTIVE CENTER COORDINATES

Since model perspective centers (exposure stations) are used in coordinate transformations associated with independent model aerotriangulation, their coordinates must be determined. The most convenient method is their direct readout in the model coordinate system, but this requires a stereoplotter specifically designed to obtain these coordinates. Optical projection plotters do not have this capability, nor do most mechanical projection plotters. Analytical plotters, however, compute these values as part of analytical relative orientation so they will be available for this type of plotter.

For plotters that do not have a direct capability of providing perspective center coordinates, either of two methods can be used. The first method, sometimes referred to as the *two-point* method, is illustrated in Fig. 17-7. In this method, discrete points from any diapositive are read monoscopically in the model coordinate system at the extreme high and low levels of the stereoplotter's Z range. The following two-point form of the equation for a straight line in three-dimensional space is then written for each point:

$$\frac{X_0 - X_{A_1}}{X_{A_2} - X_{A_1}} = \frac{Y_0 - Y_{A_1}}{Y_{A_2} - Y_{A_1}} = \frac{Z_0 - Z_{A_1}}{Z_{A_2} - Z_{A_1}} \tag{17-5}$$

In this equation, X_0, Y_0, and Z_0 are perspective center coordinates and X_{A_1}, Y_{A_1}, Z_{A_1}, X_{A_2}, Y_{A_2}, and Z_{A_2} are the model coordinates for point a read monoscopically at the two different Z levels, as shown in Fig. 17-7. A minimum of two points must be read at two Z levels to obtain a solution; but, as a practical minimum, four corner points should be

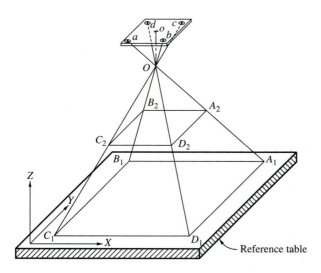
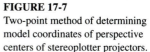

FIGURE 17-7
Two-point method of determining model coordinates of perspective centers of stereoplotter projectors.

read and the solution obtained by least squares. The advantage of this method is that no special equipment is needed.

A second method, commonly called the *grid plate* method, is illustrated in Fig. 17-8. A precisely manufactured grid plate is carefully centered in each projector, and X, Y, and Z model coordinates of a number of grid intersection points (preferably those in the corners of the model space) are read monoscopically in the model coordinate system. By treating the projector principal distance as analogous to the camera focal length, the coordinates of grid intersection points on the grid plate as analogous to photo coordinates, and model coordinates of grid intersections as ground coordinates, the space resection problem described in Sec. 11-6 is solved to obtain the X, Y, and Z

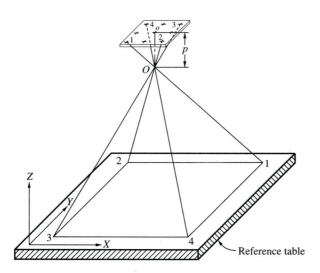
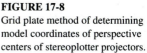

FIGURE 17-8
Grid plate method of determining model coordinates of perspective centers of stereoplotter projectors.

coordinates of the perspective center. A minimum of three grid intersection points must be read for a solution, but four corner points are recommended as a practical minimum. If four or more points are read, a least squares solution is obtained. Tests show that the accuracy with which perspective center coordinates can be determined increases by increasing the number of grid points read; however, no appreciable accuracy is gained by reading more than about 10 points.

The two methods described are both capable of yielding perspective center coordinates with good accuracy; however, the most precise method for the determination of perspective center coordinates involves the measurement of a stereomodel formed by relatively orienting a pair of precise grid plates that have been carefully centered in the projectors. Ideally, the resultant model should be a plane with a uniform grid size (the actual size will be dependent upon the model base set in the stereoplotter). After an operator performs relative orientation and measures X, Y, and Z model coordinates of a number of grid intersections, transformation constants as expressed by Eqs. (17-2) between model X, Y, and Z coordinates and an $X'Y'Z'$ "control" coordinate system are analytically determined. The control coordinate system in $X'Y'$ consists of coordinate values of the actual grid intersections, and Z' is zero for all points. The X, Y, and Z coordinates of the left perspective center are then determined by applying the transformation parameters according to Eqs. (17-2) to a point whose X', Y' coordinates correspond to the principal point of the left grid plate and whose Z' coordinate is equal to the principal distance set in the two projectors. Likewise, X, Y, and Z coordinates of the right perspective center are determined by applying the transformation parameters to a point whose X', Y' coordinates correspond to the principal point of the right grid plate and whose Z' coordinate is also equal to the principal distance set in the projector.

As previously mentioned, the stereomodel formed by the grid plates should be a uniform grid that exists in a plane. Discrepancies between the measured X, Y, and Z model coordinates of grid intersections and their X', Y', and Z' coordinates are an indication of any departure from the ideal condition. These discrepancies may be due to instrument maladjustment, scale differences between the X and Y projections in the stereoplotter, nonorthogonality between the X and Y axes of the stereoplotter coordinatograph, etc. These discrepancies (or errors) may be used to mathematically express stereomodel deformations through the use of observation equations such as

$$d_x = a_1X + a_2Y + a_3XY + a_4X^2 + a_5Y^2 + a_6X^2Y + a_7XY^2 + a_8X^2Y^2$$

$$d_y = b_1X + b_2Y + b_3XY + b_4X^2 + b_5Y^2 + b_6X^2Y + b_7XY^2 + b_8X^2Y^2 \qquad (17\text{-}6)$$

$$d_z = c_1X + c_2Y + c_3XY + c_4X^2 + c_5Y^2 + c_6X^2Y + c_7XY^2 + c_8X^2Y^2$$

In Eqs. (17-6), the d's are discrepancies in X, Y, and Z; and the a's, b's, and c's are coefficients which describe the model deformation. A set of three equations of the type of Eqs. (17-6) can be written for each grid model point read. If 8 points are read, a unique solution for the 24 unknown coefficients (a's, b's, and c's) can be made; if more than 8 are available, then least squares can be applied in their calculation. As with any transformation involving polynomials, it is advised to use several redundant points that are well distributed in the model. After the coefficients have been determined, Eqs. (17-6) can be used to correct for model deformations for every model point subsequently read in semianalytical aerotriangulation. Thus, this grid model procedure not only yields perspective center coordinates but also enables corrections to be made for systematic errors of the stereoplotter, thereby increasing the overall accuracy of the semianalytical aerotriangulation.

PART II
ANALYTICAL AEROTRIANGULATION

17-8 INTRODUCTION

The most elementary approaches to analytical aerotriangulation consist of the same basic steps as those of analog and semianalytical methods and include (1) relative orientation of each stereomodel, (2) connection of adjacent models to form continuous strips and/or blocks, and (3) simultaneous adjustment of the photos from the strips and/or blocks to field-surveyed ground control. What is different about analytical methods is that the basic input consists of precisely measured photo coordinates of control points and pass points. Relative orientation is then performed analytically based upon the measured coordinates and known camera constants. Finally, the entire block of photographs is adjusted simultaneously to the ground coordinate system.

Analytical aerotriangulation tends to be more accurate than analog or semianalytical methods, largely because analytical techniques can more effectively eliminate systematic errors such as film shrinkage, atmospheric refraction distortions, and camera lens distortions. In fact, X and Y coordinates of pass points can quite routinely be located analytically to an accuracy of within about $1/15,000$ of the flying height, and Z coordinates can be located to an accuracy of about $1/10,000$ of the flying height. With specialized equipment and procedures, planimetric accuracy of $1/350,000$ of the flying height and vertical accuracy of $1/180,000$ have been achieved. Another advantage of analytical methods is the freedom from the mechanical or optical limitations imposed by stereoplotters. Photography of any focal length, tilt, and flying height can be handled with the same efficiency. The calculations involved are rather complex; however, a number of suitable computer programs are available to perform analytical aerotriangulation. With the advances in computer power, these calculations have become routine.

Several different variations in analytical aerotriangulation techniques have evolved. Basically, however, all methods consist of writing condition equations that express the unknown elements of exterior orientation of each photo in terms of camera constants, measured photo coordinates, and ground coordinates. The equations are solved to determine the unknown orientation parameters, and either simultaneously or subsequently, coordinates of pass points are calculated. By far the most common condition equations used are the collinearity equations which are presented in Sec. 11-4 and derived in App. D. Analytical procedures have been developed which can simultaneously enforce collinearity conditions onto units which consist of hundreds of photographs.

17-9   SIMULTANEOUS BUNDLE ADJUSTMENT

The ultimate extension of the principles described in the preceding sections is to adjust all photogrammetric measurements to ground control values in a single solution known as a *bundle adjustment*. The process is so named because of the many light rays that pass through each lens position constituting a bundle of rays. The bundles from all photos are adjusted simultaneously so that corresponding light rays intersect at positions of the pass points and control points on the ground. The process is an extension of the principles of analytical photogrammetry presented in Chap. 11, applied to an unlimited number of overlapping photographs.

Figure 17-9*a* shows a small block in the same configuration as that of Fig. 17-6, consisting of two strips with four photos per strip. The photo block contains images of 20 pass points labeled 1 through 20 and 6 control points labeled *A* through *F*, for a total of 26 object points. Points 3, 8, 13, 18, *B*, and *E* also serve as *tie points* which connect the two adjacent strips. Figure 17-9*b* shows the individual photos in a nonoverlapping configuration. Note that photos 1, 4, 5, and 8 each contain images of 8 points; and photos 2, 3, 6, and 7 each contain images of 11 points, for a grand total of $4 \times 8 + 4 \times 11 = 76$ point images.

The unknown quantities to be obtained in a bundle adjustment consist of (1) the X, Y, and Z object space coordinates of all object points and (2) the exterior orientation parameters ($\omega, \phi, \kappa, X_L, Y_L$, and Z_L) of all photographs. The first group of unknowns (object space coordinates) is the necessary result of any aerotriangulation, analytical or otherwise. Exterior orientation parameters, however, are generally not of interest to the photogrammetrist, but they must be included in the mathematical model for consistency. In the photo block of Fig. 17-9*a* the number of unknown object coordinates is $26 \times 3 = 78$ (number of object points times the number of coordinates per point). The number of unknown exterior orientation parameters is $8 \times 6 = 48$ (number of photos times the number of exterior orientation parameters per photo). Therefore the total number of unknowns is $78 + 48 = 126$.

The measurements (observed quantities) associated with a bundle adjustment are (1) x and y photo coordinates of images of object points; (2) X, Y, and/or Z coordinates of ground control points; and (3) direct observations of the exterior orientation parameters ($\omega, \phi, \kappa, X_L, Y_L$, and Z_L) of the photographs. The first group of observations, photo coordinates, is the fundamental photogrammetric measurements made with a comparator or analytical plotter. For a proper bundle adjustment they need to be weighted according to the accuracy and precision with which they were measured. The next group of observations is coordinates of control points determined through field survey. Although ground control coordinates are indirectly determined quantities, they can be included as observations provided that proper weights are assigned. The final set of observations, exterior orientation parameters, has recently become important in bundle adjustments with the use of airborne GPS control as well as *inertial navigation systems* (INSs) which have the capability of measuring the angular attitude of a photograph.

Returning to the block of Fig. 17-9, the number of photo coordinate observations is $76 \times 2 = 152$ (number of imaged points times the number of photo coordinates per point), and the number of ground control observations is $6 \times 3 = 18$ (number of three-dimensional control points times the number of coordinates per point). If the exterior

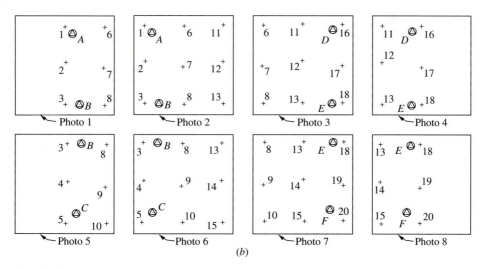

(a)

+ Pass point
⬭ Three-dimensional control point
⊠ Exposure station

(b)

FIGURE 17-9
(a) Block of photos in overlapped position. (b) Separated photos showing image points.

orientation parameters were measured, the number of additional observations would be $8 \times 6 = 48$ (number of photos times the number of exterior orientation parameters per photo). Thus, if all three types of observations are included, there will be a total of $152 + 18 + 48 = 218$ observations; but if only the first two types are included, there will be only $152 + 18 = 170$ observations. Regardless of whether exterior orientation parameters were observed, a least squares solution is possible since the number of observations is greater than the number of unknowns (126) in either case.

The observation equations which are the foundation of a bundle adjustment are the collinearity equations [see Eqs. (D-5) and (D-6)]. These equations are given below in a slightly modified form as Eqs. (17-7) and (17-8).

$$x_{ij} = x_o - f\left[\frac{m_{11_i}(X_j - X_{L_i}) + m_{12_i}(Y_j - Y_{L_i}) + m_{13_i}(Z_j - Z_{L_i})}{m_{31_i}(X_j - X_{L_i}) + m_{32_i}(Y_j - Y_{L_i}) + m_{33_i}(Z_j - Z_{L_i})}\right] \tag{17-7}$$

$$y_{ij} = y_o - f\left[\frac{m_{21_i}(X_j - X_{L_i}) + m_{22_i}(Y_j - Y_{L_i}) + m_{23_i}(Z_j - Z_{L_i})}{m_{31_i}(X_j - X_{L_i}) + m_{32_i}(Y_j - Y_{L_i}) + m_{33_i}(Z_j - Z_{L_i})}\right] \tag{17-8}$$

In these equations, x_{ij} and y_{ij} are the measured photo coordinates of the image of point j on photo i related to the fiducial axis system; x_o and y_o are the coordinates of the principal point in the fiducial axis system; f is the focal length (or more correctly, principal distance) of the camera; $m_{11_i}, m_{12_i}, \ldots, m_{33_i}$ are the rotation matrix terms for photo i; X_j, Y_j, and Z_j are the coordinates of point j in object space; and X_{L_i}, Y_{L_i}, and Z_{L_i} are the coordinates of the incident nodal point of the camera lens in object space. Since the collinearity equations are nonlinear, they are linearized by applying the first-order terms of Taylor's series at a set of initial approximations. After linearization (see Sec. D-4) the equations can be expressed in the following matrix form:

$$\dot{B}_{ij}\dot{\Delta}_i + \ddot{B}_{ij}\ddot{\Delta}_j = \varepsilon_{ij} + V_{ij} \tag{17-9}$$

where

$$\dot{B}_{ij} = \begin{bmatrix} b_{11_{ij}} & b_{12_{ij}} & b_{13_{ij}} & -b_{14_{ij}} & -b_{15_{ij}} & -b_{16_{ij}} \\ b_{21_{ij}} & b_{22_{ij}} & b_{23_{ij}} & -b_{24_{ij}} & -b_{25_{ij}} & -b_{26_{ij}} \end{bmatrix} \qquad \ddot{B}_{ij} = \begin{bmatrix} b_{14_{ij}} & b_{15_{ij}} & b_{16_{ij}} \\ b_{24_{ij}} & b_{25_{ij}} & b_{26_{ij}} \end{bmatrix}$$

$$\dot{\Delta}_i = \begin{bmatrix} d\omega_i \\ d\phi_i \\ d\kappa_i \\ dX_{L_i} \\ dY_{L_i} \\ dZ_{L_i} \end{bmatrix} \qquad \ddot{\Delta}_j = \begin{bmatrix} dX_j \\ dY_j \\ dZ_j \end{bmatrix} \qquad \varepsilon_{ij} = \begin{bmatrix} J_{ij} \\ K_{ij} \end{bmatrix} \qquad V_{ij} = \begin{bmatrix} v_{x_{ij}} \\ v_{y_{ij}} \end{bmatrix}$$

The above terms are defined as for Eqs. (D-11) and (D-12), except that subscripts i and j are used for the photo designation and point designation, respectively. (In App. D, the subscript A is used for the point designation, and no subscript is used for the photo designation.) Matrix $\dot{B}_{ij}$ contains the partial derivatives of the collinearity equations with respect to the exterior orientation parameters of photo i, evaluated at the initial approximations. Matrix $\ddot{B}_{ij}$ contains the partial derivatives of the collinearity equations with respect to the object space coordinates of point j, evaluated at the initial approximations. Matrix $\dot{\Delta}_i$ contains corrections for the initial approximations of the exterior orientation parameters for photo i, and matrix $\ddot{\Delta}_j$ contains corrections for the initial approximations of the object space coordinates of point j. Matrix ε_{ij} contains measured minus computed x and y photo coordinates for point j on photo i, and finally matrix V_{ij} contains residuals for the x and y photo coordinates.

Proper weights must be assigned to photo coordinate observations in order to be included in the bundle adjustment. Expressed in matrix form, the weights for x and y photo coordinate observations of point j on photo i are

$$W_{ij} = \sigma_o^2 \begin{bmatrix} \sigma_{x_{ij}}^2 & \sigma_{x_{ij}y_{ij}} \\ \sigma_{y_{ij}x_{ij}} & \sigma_{y_{ij}}^2 \end{bmatrix}^{-1} \tag{17-10}$$

where σ_o^2 is the reference variance; $\sigma_{x_{ij}}^2$ and $\sigma_{y_{ij}}^2$ are variances in x_{ij} and y_{ij}, respectively; and $\sigma_{x_{ij}y_{ij}} = \sigma_{y_{ij}x_{ij}}$ is the covariance of x_{ij} with y_{ij}. The reference variance is an arbitrary parameter which can be set equal to 1, and in many cases, the covariance in photo coordinates is equal to zero. In this case, the weight matrix for photo coordinates simplifies to

$$W_{ij} = \begin{bmatrix} \dfrac{1}{\sigma_{x_{ij}}^2} & 0 \\ 0 & \dfrac{1}{\sigma_{y_{ij}}^2} \end{bmatrix} \tag{17-11}$$

The next type of observation to be considered is ground control. Observation equations for ground control coordinates are

$$X_j = X_j^{00} + v_{X_j}$$
$$Y_j = Y_j^{00} + v_{Y_j} \tag{17-12}$$
$$Z_j = Z_j^{00} + v_{Z_j}$$

where X_j, Y_j, and Z_j are unknown coordinates of point j; X_j^{00}, Y_j^{00}, and Z_j^{00} are the measured coordinate values for point j; and v_{X_j}, v_{Y_j}, and v_{Z_j} are the coordinate residuals for point j.

Even though ground control observation equations are linear, in order to be consistent with the collinearity equations, they will also be approximated by the first-order terms of Taylor's series.

$$X_j^0 + dX_j = X_j^{00} + v_{X_j}$$
$$Y_j^0 + dY_j = Y_j^{00} + v_{Y_j} \tag{17-13}$$
$$Z_j^0 + dZ_j = Z_j^{00} + v_{Z_j}$$

In Eqs. (17-13), X_j^0, Y_j^0, and Z_j^0 are initial approximations for the coordinates of point j; dX_j, dY_j, and dZ_j are corrections to the approximations for the coordinates of point j; and the other terms are as previously defined.

Rearranging the terms of Eqs. (17-13) and expressing the result in matrix form gives

$$\ddot{\Delta}_j = \ddot{C}_j + \ddot{V}_j \tag{17-14}$$

where $\ddot{\Delta}_j$ is as previously defined and

$$\ddot{C}_j = \begin{bmatrix} X_j^{00} - X_j^{0} \\ Y_j^{00} - Y_j^{0} \\ Z_j^{00} - Z_j^{0} \end{bmatrix} \qquad \ddot{V}_j = \begin{bmatrix} v_{X_j} \\ v_{Y_j} \\ v_{Z_j} \end{bmatrix}$$

As with photo coordinate measurements, proper weights must be assigned to ground control coordinate observations in order to be included in the bundle adjustment. Expressed in matrix form, the weights for X, Y, and Z ground control coordinate observations of point j are

$$\ddot{W}_j = \sigma_o^2 \begin{bmatrix} \sigma_{X_j}^2 & \sigma_{X_jY_j} & \sigma_{X_jZ_j} \\ \sigma_{Y_jX_j} & \sigma_{Y_j}^2 & \sigma_{Y_jZ_j} \\ \sigma_{Z_jX_j} & \sigma_{Z_jY_j} & \sigma_{Z_j}^2 \end{bmatrix}^{-1} \tag{17-15}$$

where σ_o^2 is the reference variance; $\sigma_{X_j}^2$, $\sigma_{Y_j}^2$, and $\sigma_{Z_j}^2$ are the variances in X_j^{00}, Y_j^{00}, and Z_j^{00}, respectively; $\sigma_{X_jY_j} = \sigma_{Y_jX_j}$ is the covariance of X_j^{00} with Y_j^{00}; $\sigma_{X_jZ_j} = \sigma_{Z_jX_j}$ is the covariance of X_j^{00} with Z_j^{00}; and $\sigma_{Y_jZ_j} = \sigma_{Z_jY_j}$ is the covariance of Y_j^{00} with Z_j^{00}. As before, the reference variance can be arbitrarily set equal to 1; however, in general, since ground control coordinates are indirectly determined quantities, their covariances are not equal to zero.

The final type of observation consists of measurements of exterior orientation parameters. The form of their observation equations is similar to that of ground control and given as Eqs. (17-16).

$$\begin{aligned} \omega_i &= \omega_i^{00} + v_{\omega_i} \\ \phi_i &= \phi_i^{00} + v_{\phi_i} \\ \kappa_i &= \kappa_i^{00} + v_{\kappa_i} \\ X_{L_i} &= X_{L_i}^{00} + v_{X_{L_i}} \\ Y_{L_i} &= Y_{L_i}^{00} + v_{Y_{L_i}} \\ Z_{L_i} &= Z_{L_i}^{00} + v_{Z_{L_i}} \end{aligned} \tag{17-16}$$

The weight matrix for exterior orientation parameters has the following form:

$$\dot{W}_i = \begin{bmatrix} \sigma_{\omega_i}^2 & \sigma_{\omega_i\phi_i} & \sigma_{\omega_i\kappa_i} & \sigma_{\omega_iX_{L_i}} & \sigma_{\omega_iY_{L_i}} & \sigma_{\omega_iZ_{L_i}} \\ \sigma_{\phi_i\omega_i} & \sigma_{\phi_i}^2 & \sigma_{\phi_i\kappa_i} & \sigma_{\phi_iX_{L_i}} & \sigma_{\phi_iY_{L_i}} & \sigma_{\phi_iZ_{L_i}} \\ \sigma_{\kappa_i\omega_i} & \sigma_{\kappa_i\phi_i} & \sigma_{\kappa_i}^2 & \sigma_{\kappa_iX_{L_i}} & \sigma_{\kappa_iY_{L_i}} & \sigma_{\kappa_iZ_{L_i}} \\ \sigma_{X_{L_i}\omega_i} & \sigma_{X_{L_i}\phi_i} & \sigma_{X_{L_i}\kappa_i} & \sigma_{X_{L_i}}^2 & \sigma_{X_{L_i}Y_{L_i}} & \sigma_{X_{L_i}Z_{L_i}} \\ \sigma_{Y_{L_i}\omega_i} & \sigma_{Y_{L_i}\phi_i} & \sigma_{Y_{L_i}\kappa_i} & \sigma_{Y_{L_i}X_{L_i}} & \sigma_{Y_{L_i}}^2 & \sigma_{Y_{L_i}Z_{L_i}} \\ \sigma_{Z_{L_i}\omega_i} & \sigma_{Z_{L_i}\phi_i} & \sigma_{Z_{L_i}\kappa_i} & \sigma_{Z_{L_i}X_{L_i}} & \sigma_{Z_{L_i}Y_{L_i}} & \sigma_{Z_{L_i}}^2 \end{bmatrix}^{-1} \tag{17-17}$$

With the observation equations and weights defined as above, the full set of normal equations may be formed directly. In matrix form, the full normal equations are

$$N\Delta = K \qquad (17\text{-}18)$$

where

$$
N =
\begin{bmatrix}
\dot{N}_1 + \dot{W}_1 & {}_6 0^6 & {}_6 0^6 & \cdots & {}_6 0^6 & \overline{N}_{11} & \overline{N}_{12} & \overline{N}_{13} & \cdots & \overline{N}_{1n} \\
{}_6 0^6 & \dot{N}_2 + \dot{W}_2 & {}_6 0^6 & \cdots & {}_6 0^6 & \overline{N}_{21} & \overline{N}_{22} & \overline{N}_{23} & \cdots & \overline{N}_{2n} \\
{}_6 0^6 & {}_6 0^6 & \dot{N}_3 + \dot{W}_3 & \cdots & {}_6 0^6 & \overline{N}_{31} & \overline{N}_{32} & \overline{N}_{33} & \cdots & \overline{N}_{3n} \\
\vdots & & & & & & & & & \\
{}_6 0^6 & {}_6 0^6 & {}_6 0^6 & \cdots & \dot{N}_m + \dot{W}_m & \overline{N}_{m1} & \overline{N}_{m2} & \overline{N}_{m3} & \cdots & \overline{N}_{mn} \\
\overline{N}_{11}^T & \overline{N}_{21}^T & \overline{N}_{31}^T & \cdots & \overline{N}_{m1}^T & \ddot{N}_1 + \ddot{W}_1 & {}_3 0^3 & {}_3 0^3 & \cdots & {}_3 0^3 \\
\overline{N}_{12}^T & \overline{N}_{22}^T & \overline{N}_{32}^T & \cdots & \overline{N}_{m2}^T & {}_3 0^3 & \ddot{N}_2 + \ddot{W}_2 & {}_3 0^3 & \cdots & {}_3 0^3 \\
\overline{N}_{13}^T & \overline{N}_{23}^T & \overline{N}_{33}^T & \cdots & \overline{N}_{m3}^T & {}_3 0^3 & {}_3 0^3 & \ddot{N}_3 + \ddot{W}_3 & \cdots & {}_3 0^3 \\
\vdots & & & & & & & & & \\
\overline{N}_{1n}^T & \overline{N}_{2n}^T & \overline{N}_{3n}^T & \cdots & \overline{N}_{mn}^T & {}_3 0^3 & {}_3 0^3 & {}_3 0^3 & \cdots & \ddot{N}_n + \ddot{W}_n
\end{bmatrix}
$$

$$
\Delta =
\begin{bmatrix}
\dot{\Delta}_1 \\
\dot{\Delta}_2 \\
\dot{\Delta}_3 \\
\vdots \\
\dot{\Delta}_m \\
\ddot{\Delta}_1 \\
\ddot{\Delta}_2 \\
\ddot{\Delta}_3 \\
\vdots \\
\ddot{\Delta}_n
\end{bmatrix}
\qquad
K =
\begin{bmatrix}
\dot{K}_1 + \dot{W}_1 \dot{C}_1 \\
\dot{K}_2 + \dot{W}_1 \dot{C}_2 \\
\dot{K}_3 + \dot{W}_1 \dot{C}_3 \\
\vdots \\
\dot{K}_m + \dot{W}_m \dot{C}_m \\
\ddot{K}_1 + \ddot{W}_1 \ddot{C}_1 \\
\ddot{K}_2 + \ddot{W}_2 \ddot{C}_2 \\
\ddot{K}_3 + \ddot{W}_3 \ddot{C}_3 \\
\vdots \\
\ddot{K}_n + \ddot{W}_n \ddot{C}_n
\end{bmatrix}
$$

The submatrices in the above forms are

$$\dot{N}_i = \sum_{j=1}^{n} \dot{B}_{ij}^T W_{ij} \dot{B}_{ij} \qquad \overline{N}_{ij} = \dot{B}_{ij}^T W_{ij} \ddot{B}_{ij} \qquad \ddot{N}_j = \sum_{i=1}^{m} \ddot{B}_{ij}^T W_{ij} \ddot{B}_{ij}$$

$$\dot{K}_i = \sum_{j=1}^{n} \dot{B}_{ij}^T W_{ij} \varepsilon_{ij} \qquad \ddot{K}_j = \sum_{i=1}^{m} \ddot{B}_{ij}^T W_{ij} \varepsilon_{ij}$$

In the above expressions, m is the number of photos, n is the number of points, i is the photo subscript, and j is the point subscript. Note that if point j does not appear on photo i, the corresponding submatrix will be a zero matrix. Note also that the $\dot{W}_i$ contributions to the N matrix and the $\dot{W}_i \dot{C}_i$ contributions to the K matrix are made only when observations for exterior orientation parameters exist; and the $\ddot{W}_j$ contributions to the N matrix and the $\ddot{W}_j \ddot{C}_j$ contributions to the K matrix are made only for ground control point observations.

While the normal equations are being formed, it is recommended that the estimate for the standard deviation of unit weight be calculated (see Sec. B-10). Assuming the initial approximations are reasonable, matrices ε_{ij}, $\dot{C}_i$, and $\ddot{C}_j$ are good estimates of the negatives of the residuals. Therefore, the estimate of the standard deviation of unit weight can be computed by

$$S_0 = \sqrt{\frac{\sum\limits_{j=1}^{n}\sum\limits_{i=1}^{m} \varepsilon_{ij}^T W_{ij} \varepsilon_{ij} + \sum\limits_{i=1}^{m} \dot{C}_i^T \dot{W}_i \dot{C}_i + \sum\limits_{j=1}^{n} \ddot{C}_j^T \ddot{W}_j \ddot{C}_j}{\text{n.o.} - \text{n.u.}}} \tag{17-19}$$

In Eq. (17-19), n.o. is the number of observations and n.u. is the number of unknowns in the solution.

After the normal equations have been formed, they are solved for the unknowns Δ, which are corrections to the initial approximations for exterior orientation parameters and object space coordinates. The corrections are then added to the approximations, and the procedure is repeated until the estimated standard deviation of unit weight converges. At that point, the variance-covariance matrix for the unknowns can be computed by

$$\Sigma_{\Delta\Delta} = S_0^2 \, N^{-1} \tag{17-20}$$

Computed standard deviations for the unknowns can then be obtained by taking the square root of the diagonal elements of the $\Sigma_{\Delta\Delta}$ matrix.

17-10 INITIAL APPROXIMATIONS FOR THE BUNDLE ADJUSTMENT

In Sec. 17-9, the equations involved in a bundle adjustment were presented. Since the equations are nonlinear, Taylor's series was used to linearize the equations; therefore, initial approximations are required for the unknowns. For the solution to converge properly, the initial approximations must be reasonably close to the correct values. Several methods may be used to obtain initial approximations; however, preliminary strip adjustments are most commonly employed.

The first step is to perform analytical relative orientation (see Sec. 11-10) for each stereopair in the block. The photo coordinate residuals should be inspected at this point as an initial check on the measurements. Next, the relatively oriented models are connected to form strips by the method presented in Sec. 17-4. Residuals from this step can also provide a quality check on the photo coordinate measurements and point identification. After all the strip models have been formed and validated, each strip is individually adjusted to ground control points located within each strip. This adjustment to ground control can be performed either by a three-dimensional conformal coordinate transformation or by a three-dimensional polynomial transformation (see Sec. 17-5). Residuals from this step provide a check on the ground control coordinates as well as point identification.

At this point, ground coordinates will have been calculated for all points in the photo block. An additional check can be performed to validate the identification of tie points between strips. If the identification of tie points is consistent, their coordinates as determined in adjacent strips should agree within a small tolerance. Assuming everything is consistent at this point, the resulting ground coordinates can be used as initial approximations for the bundle adjustment.

Approximations for the exterior orientation parameters can also be obtained directly from the strip adjustment if the adjustment is performed using a three-dimensional conformal coordinate transformation. In that case, since perspective centers (camera stations) are included when adjacent models are connected, their object space coordinates will be available after the final adjustment to ground control. Assuming vertical photography, zeros can be used as approximations for ω and ϕ. Approximations for κ can be obtained directly from the final three-dimensional conformal coordinate transformation to ground control, which contains a compatible κ angle. If a polynomial strip adjustment is performed, the perspective centers are not included in the adjustment. In that case, after the polynomial adjustment is completed, the space resection problem can be solved for each photo (see Sec. 11-6). In these calculations, the ground coordinates obtained for the pass points in the polynomial adjustment are used as control coordinates.

17-11 BUNDLE ADJUSTMENT WITH AIRBORNE GPS CONTROL

As mentioned in Sec. 17-1, kinematic GPS observations can be taken aboard the aircraft as the photography is being acquired to determine coordinates for exposure stations. Use of GPS in the aircraft to control a bundle adjustment of a block of photographs is termed *airborne GPS control*. By including coordinates of the exposure stations in the adjustment, the amount of ground control can be greatly reduced.

Figure 17-10 illustrates the geometric relationship between a camera and GPS antenna on an aircraft. In this figure, x, y, and z represent the standard three-dimensional coordinate system of a mapping camera; and x_A, y_A, and z_A represent the coordinates of

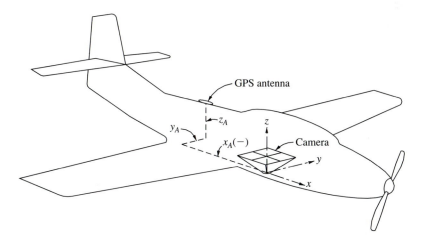

FIGURE 17-10
Configuration of camera and GPS antenna for airborne GPS control.

the GPS antenna relative to the camera axes. The x axis of the camera is parallel to the longitudinal axis of the aircraft, the z axis is vertical, and the y axis is perpendicular to the x and z axes. Since object space coordinates obtained by GPS pertain to the phase center of the antenna but the exposure station is defined as the incident nodal point of the camera lens, the GPS coordinates of the antenna must be translated to the camera lens. To properly compute the translations, it is necessary to know the angular orientation of the camera with respect to the object space coordinate system. Determining the correct angular orientation is complicated by the use of a gimbaled camera mount which allows relative rotations between the camera and the aircraft frame.

If the camera in its mount was fixed, the angular orientation parameters of the camera (ω, ϕ, and κ) would translate directly to angular orientation of the camera-to-antenna vector. However, differential rotations ($\Delta\omega$, $\Delta\phi$, and $\Delta\kappa$) between the camera and the aircraft frame must also be taken into account in order to determine the angular attitude of the camera-to-antenna vector in object space. Note that even in a so-called fixed mount there will generally be a crab adjustment ($\Delta\kappa$) to ensure proper photographic coverage (see Sec. 3-6). Some camera mounts such as the Leica PAV30 shown in Fig. 3-8 have the capability of measuring the differential rotations, and they can be recorded by a computer.

Assuming each differential rotation of the camera with respect to the mount is positive counterclockwise when viewed from the positive axis toward the origin, the following equations specify the rotation of the camera-to-antenna vector with respect to object space.

$$\omega' = \omega - \Delta\omega$$

$$\phi' = \phi - \Delta\phi \qquad (17\text{-}21)$$

$$\kappa' = \kappa - \Delta\kappa$$

In Eqs. (17-21), ω, ϕ, and κ are conventional angular exterior orientation parameters of the camera with respect to the object space coordinate system; $\Delta\omega$, $\Delta\phi$, and $\Delta\kappa$ are the differential rotations of the camera with respect to the mount; and ω', ϕ', and κ' are the rotations of the camera-to-antenna vector with respect to object space coordinates.

Once ω', ϕ', and κ' have been determined, a rotation matrix, M', can be computed. [See Eqs. (C-33). Note that ω', ϕ', and κ' are used in place of ω, ϕ, and κ in Eqs. (C-33).] After M' has been computed, the coordinates of the camera lens can be computed by the following:

$$\begin{bmatrix} X_L \\ Y_L \\ Z_L \end{bmatrix} = \begin{bmatrix} X_{\text{GPS}} \\ Y_{\text{GPS}} \\ Z_{\text{GPS}} \end{bmatrix} - M'^T \begin{bmatrix} x_A \\ y_A \\ z_A \end{bmatrix} \qquad (17\text{-}22)$$

In Eqs. (17-22), X_L, Y_L, and Z_L are the object space coordinates of the camera station; X_{GPS}, Y_{GPS}, and Z_{GPS} are the object space coordinates of the GPS antenna; M' is the rotation matrix previously defined; and x_A, y_A, and z_A are the camera-to-antenna vector components as illustrated in Fig. 17-10.

When a camera mount is used which does not provide for measurement of the differential rotations $\Delta\omega$, $\Delta\phi$, and $\Delta\kappa$, they are assumed to be equal to zero, resulting in

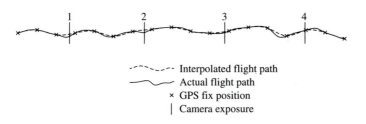

--------- Interpolated flight path
〰〰〰 Actual flight path
× GPS fix position
| Camera exposure

FIGURE 17-11
Interpolation between GPS fixes for position at time of exposure.

errors in the computed position of the camera lens. This error can be minimized by mounting the GPS antenna vertically above the camera in the aircraft, which effectively eliminates the error due to unaccounted crab adjustment $\Delta\kappa$. As long as the differential rotations $\Delta\omega$ and $\Delta\phi$ are small (less than a couple of degrees) and the antenna-to-camera vector is short (less than 2 m), the lens positional error will be less than 10 cm. One last comment must be made concerning the translation of GPS antenna coordinates to the lens. Since the values of ω, ϕ, and κ are required to compute the translation, the antenna offset correction must be included within the iterative loop of the analytical bundle adjustment.

A second consideration regarding airborne GPS positioning is the synchronization (or lack thereof) of GPS fixes with the photographic exposures. The GPS receiver will be recording data at uniform time intervals called *epochs,* which may be on the order of 1 s each. The camera shutter, on the other hand, operates asynchronously with respect to the GPS fixes, as shown in Fig. 17-11. The result is that the position of the GPS antenna must first be interpolated from adjacent fixes before the coordinates can be translated to the lens. Depending upon atmospheric turbulence and the epoch recording rate of the GPS receiver, the error due to this interpolation can be quite severe. In Fig. 17-11, interpolation for exposures 1 and 2 results in sizable errors from the actual positions, whereas interpolation for exposure 4 is nearly perfect.

A final consideration regarding airborne GPS positioning is the problem of *loss of lock* on the GPS satellites, especially during banked turns. When a GPS receiver operating in the kinematic mode loses lock on too many satellites, the integer ambiguities must be redetermined (see Sec. 16-8). Since returning to a previously surveyed point is generally out of the question, on-the-fly (OTF) techniques are used to calculate the correct integer ambiguities. With high-quality, dual-frequency, P-code receivers, OTF techniques are often successful in correctly redetermining the integer ambiguities. In some cases, however, an integer ambiguity solution may be obtained which is slightly incorrect. This results in an approximately linear drift in position along the flight line, which causes exposure station coordinate errors to deteriorate. This problem can be detected by using a small number of ground control points at the edges of the photo block. Inclusion of additional parameters in the adjustment corresponding to the linear drift enables a correction to be applied which eliminates this source of error. Often *cross strips* are flown at the ends of the regular block strips, as shown in Fig. 17-12. The cross strips contain ground control points at each end which allow drift due to incorrect OTF integer

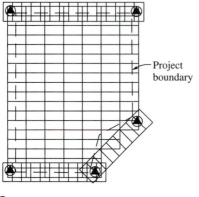

— Project
boundary

FIGURE 17-12
Configuration of flight strips for airborne
GPS control.

⬤ Three-dimensional control point

ambiguities to be detected and corrected. The corrected cross strips in turn serve to provide endpoint coordinates for the remainder of the strips in the block, thus enabling drift corrections to be made for those strips as well.

Two additional precautions regarding airborne GPS should be noted. First, it is recommended that a bundle adjustment with analytical self-calibration (see Sec. 11-12) be employed when airborne GPS control is used. Often, due to inadequate modeling of atmospheric refraction distortion, strict enforcement of the calibrated principal distance (focal length) of the camera will cause distortions and excessive residuals in photo coordinates. Use of analytical self-calibration will essentially eliminate that effect. Second, it is essential that appropriate object space coordinate systems be employed in data reduction. GPS coordinates in a geocentric coordinate system should be converted to local vertical coordinates for the adjustment (see Secs. 5-5 and F-4). After aerotriangulation is completed, the local vertical coordinates can be converted to whatever system is desired. Elevations relative to the ellipsoid can be converted to orthometric elevations by using an appropriate geoid model.

17-12 AEROTRIANGULATION WITH SATELLITE IMAGES

For certain applications with low accuracy requirements, aerotriangulation from satellite images may be suitable. For example, for small-scale topographic mapping over mountainous regions, panchromatic images from the French *Système Pour d'Observation de la Terre* (SPOT) satellite may be used. The SPOT satellite uses a linear array sensor, the geometry of which was discussed in Sec. 3-14.2. Stereopairs of SPOT images can be acquired for a region by using the off-axis pointing capability of the satellite. Photogrammetric analysis of the resulting images can be performed through the use of modified collinearity equations.

As mentioned in Sec. 3-14.2, linear array sensors capture images having a line perspective geometry as opposed to the point perspective of a frame camera. The effect of this is that each scan line of the scene has its own set of exterior orientation parameters

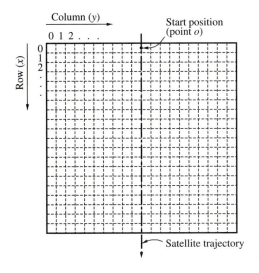

FIGURE 17-13
Illustration of linear array sensor image.

as well as its own principal point in the center of the line. Since the satellite is highly stable during acquisition of the image, the exterior orientation parameters can be assumed to vary in a systematic fashion. Figure 17-13 illustrates an image from a linear array sensor. In this figure, the start position (point o) is the projection of the center of row 0 on the ground. At this point, the satellite sensor has a particular set of exterior orientation parameters ω_o, ϕ_o, κ_o, X_{L_o}, Y_{L_o}, and Z_{L_o}. These parameters can be assumed to vary systematically as a function of the x coordinate (row in which the image appears). Various functional relationships have been tested for modeling these systematic variations, and the following have been found to consistently yield satisfactory results:

$$\omega_x = \omega_o + a_1 x$$

$$\phi_x = \phi_o + a_2 x$$

$$\kappa_x = \kappa_o + a_3 x$$

$$X_{L_x} = X_{L_o} + a_4 x \qquad (17\text{-}23)$$

$$Y_{L_x} = Y_{L_o} + a_5 x$$

$$Z_{L_x} = Z_{L_o} + a_6 x + a_7 x^2$$

In Eqs. (17-23), x is the row number of some image position; ω_x, ϕ_x, κ_x, X_{L_x}, Y_{L_x}, and Z_{L_x} are the exterior orientation parameters of the sensor when row x was acquired; ω_o, ϕ_o, κ_o, X_{L_o}, Y_{L_o}, and Z_{L_o} are the exterior orientation parameters of the sensor at the start position; and a_1 through a_7 are coefficients which describe the systematic variations of the exterior orientation parameters as the image is acquired. Note that according to Eqs. (17-23) the variation in Z_L is second order, whereas the other variations are linear (first-order). This is due to the curved orbital path of the satellite and is based on an assumption that a local vertical coordinate system (see Sec. 5-5) is being used. Depending upon the accuracy requirements and measurement precision, the coefficient of the second-order term a_7 may often be assumed to be equal to zero.

Given the variation of exterior orientation parameters described above, the collinearity equations which describe linear array sensor geometry for any image point a are

$$0 = -f\left[\frac{m_{11_x}(X_A - X_{L_x}) + m_{12_x}(Y_A - Y_{L_x}) + m_{13_x}(Z_A - Z_{L_x})}{m_{31_x}(X_A - X_{L_x}) + m_{32_x}(Y_A - Y_{L_x}) + m_{33_x}(Z_A - Z_{L_x})}\right] \tag{17-24}$$

$$y_a = y_o - f\left[\frac{m_{21_x}(X_A - X_{L_x}) + m_{22_x}(Y_A - Y_{L_x}) + m_{23_x}(Z_A - Z_{L_x})}{m_{31_x}(X_A - X_{L_x}) + m_{32_x}(Y_A - Y_{L_x}) + m_{33_x}(Z_A - Z_{L_x})}\right] \tag{17-25}$$

In Eqs. (17-24) and (17-25), y_a is the y coordinate (column number) of the image of point A; y_o is the y coordinate of the principal (middle) point of the row containing the image; f is the sensor focal length; m_{11_x} through m_{33_x} are the rotation matrix terms [see Eqs. (C-33)] for the sensor attitude when row x_a was acquired; X_{L_x}, Y_{L_x}, and Z_{L_x} are the coordinates of the sensor when row x_a was acquired; and X_A, Y_A, and Z_A are the object space coordinates of point A. Note that the exterior orientation terms and hence the rotation matrix terms are functions of the form of Eqs. (17-23). It is also important to note that the units of the image coordinates and the focal length must be the same. For example, the first three SPOT sensor systems had focal lengths of 1082 mm and, when operating in the panchromatic mode, pixel dimensions of 0.013 mm in their focal planes.[2] Therefore, if standard row and column image coordinates (in terms of pixels) are used, the focal length is expressed as 1082 mm/0.013 mm/pixel = 83,200 pixels.

Implementation of aerotriangulation using linear array sensor images is more complicated than that of frame camera images. Difficulties can arise due to correlation between exterior orientation parameters, as well as other factors. The reader may consult references listed at the end of this chapter for more details.

17-13 EFFICIENT COMPUTATIONAL STRATEGIES FOR AEROTRIANGULATION

Section 17-9 presented the mathematical form for least squares adjustment of photogrammetric blocks. As presented, however, the equations are somewhat inefficient for computational purposes. Methods are available for reducing the matrix storage requirements and solution time for large blocks of photographs. One approach is briefly presented here.

The first step is to partition the full normal equations [Eq. (17-18)] so that the exterior orientation terms and the object space coordinate terms are separated, giving

$$\begin{bmatrix} \dot{N} & \overline{N} \\ \overline{N}^T & \ddot{N} \end{bmatrix} \begin{bmatrix} \dot{\Delta} \\ \ddot{\Delta} \end{bmatrix} = \begin{bmatrix} \dot{K} \\ \ddot{K} \end{bmatrix} \tag{17-26}$$

---

[2] SPOT 1, 2, and 3 sensors could be operated in either a panchromatic or multispectral mode. In panchromatic mode, pixel dimensions were 0.013 mm, and ground resolution was 10 m. In multispectral mode, pixel dimensions were 0.026 mm, and ground resolution was 20 m.

In Eq. (17-26), $\dot{N}$ is the block-diagonal submatrix from the upper left portion of N having dimensions $6m \times 6m$, where m is the number of photos in the block; $\ddot{N}$ is the block-diagonal submatrix from the lower right portion of N having dimensions $3n \times 3n$, where n is the number of object points in the block; $\overline{N}$ is the submatrix from the upper right portion of N having dimensions $6m \times 3n$ and $\overline{N}^T$ is its transpose; $\dot{\Delta}$ is the submatrix from the upper portion of Δ having dimensions of $6m \times 1$, consisting of the correction terms for the exterior orientation parameters for all photos; $\ddot{\Delta}$ is the submatrix from the lower portion of Δ having dimensions of $3n \times 1$, consisting of the correction terms for the object space coordinates for all points; $\dot{K}$ is the submatrix from the upper portion of K having dimensions $6m \times 1$; and $\ddot{K}$ is the submatrix from the lower portion of K having dimensions $3n \times 1$.

A block-diagonal matrix consists of nonzero submatrices along the main diagonal and zeros everywhere else. This kind of matrix has the property that its inverse is also block-diagonal, where the submatrices are inverses of the corresponding submatrices of the original matrix. As such, the inverse of a block-diagonal matrix is much easier to compute than the inverse of a general, nonzero matrix. With this in mind, Eq. (17-26) can be rearranged to a form which can be solved more efficiently. First, Eq. (17-26) is separated into two separate matrix equations.

$$\dot{N}\dot{\Delta} + \overline{N}\ddot{\Delta} = \dot{K} \tag{17-27}$$

$$\overline{N}^T\dot{\Delta} + \ddot{N}\ddot{\Delta} = \ddot{K} \tag{17-28}$$

Equation (17-28) is then rearranged to solve for $\ddot{\Delta}$.

$$\ddot{\Delta} = \ddot{N}^{-1}(\ddot{K} - \overline{N}^T\dot{\Delta}) \tag{17-29}$$

Next the right side of Eq. (17-29) is substituted for $\ddot{\Delta}$ in Eq. (17-27).

$$\dot{N}\dot{\Delta} + \overline{N}\ddot{N}^{-1}(\ddot{K} - \overline{N}^T\dot{\Delta}) = \dot{K} \tag{17-30}$$

Rearranging Eq. (17-30) to collect the $\dot{\Delta}$ terms gives

$$(\dot{N} - \overline{N}\ddot{N}^{-1}\overline{N}^T)\dot{\Delta} = (\dot{K} - \overline{N}\ddot{N}^{-1}\ddot{K}) \tag{17-31}$$

Matrix Eq. (17-31) is referred to as the *reduced normal equations*. These equations are solved for $\dot{\Delta}$, which can then be substituted into Eq. (17-29) to compute $\ddot{\Delta}$. This approach is more efficient since the largest system of equations which must be solved has only $6m$ unknowns, as opposed to $6m + 3n$ unknowns in the full normal equations. This efficiency is made possible by the block-diagonal structure of the $\ddot{N}$ matrix.

An additional enhancement to the solution can be made to increase computational efficiency even further. This enhancement exploits the fact that the coefficient matrix of the reduced normal equations is *sparse;* i.e., it has a large number of elements that are zero. Special computational techniques and data storage methods are available which take advantage of sparsity, reducing both computational time and data storage requirements. Details concerning these special computational techniques may be found in references listed at the end of this chapter.

Figure 17-14 shows a small block with three strips of nine photos each, having end lap and side lap equal to 60 and 30 percent, respectively. The outlines of photo coverage

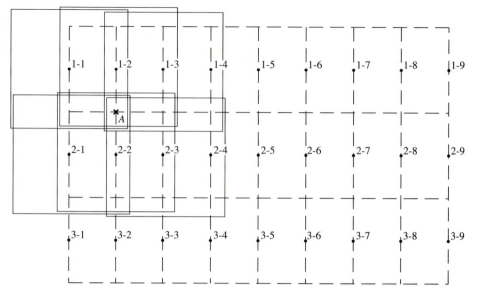

FIGURE 17-14
Configuration of a photo block having three strips of nine photos each.

for only the first three photos in strips 1 and 2 are shown in the figure, and the remainder are represented as neat models (see Sec. 18-7). In Fig. 17-14, the image of a representative pass point A exists on photos 1-1, 1-2, 1-3, 2-1, 2-2, and 2-3. This pass point causes "connections" between each possible pair of photos from the set of six on which it is imaged. Connections for the entire block are illustrated in Fig. 17-15. This figure shows a *graph* which indicates the connections (shown as lines or arcs) caused by shared pass points over the entire block.

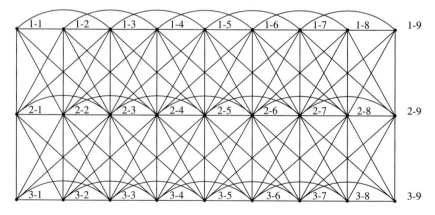

FIGURE 17-15
Graph showing connections between photos caused by shared pass points.

TABLE 17-3
Down-strip and cross-strip ordering for the photos of Fig. 17-14

Position	Down-strip order	Cross-strip order	Position	Down-strip order	Cross-strip order
1	1-1	1-1	15	2-6	3-5
2	1-2	2-1	16	2-7	1-6
3	1-3	3-1	17	2-8	2-6
4	1-4	1-2	18	2-9	3-6
5	1-5	2-2	19	3-1	1-7
6	1-6	3-2	20	3-2	2-7
7	1-7	1-3	21	3-3	3-7
8	1-8	2-3	22	3-4	1-8
9	1-9	3-3	23	3-5	2-8
10	2-1	1-4	24	3-6	3-8
11	2-2	2-4	25	3-7	1-9
12	2-3	3-4	26	3-8	2-9
13	2-4	1-5	27	3-9	3-9
14	2-5	2-5			

These connections cause nonzero submatrices to appear at corresponding locations in the reduced normal equations. The positions where these nonzero submatrices appear depend upon the order in which the photo parameters appear in the reduced normal equation matrix. Two ordering strategies, known as *down-strip* and *cross-strip,* are commonly employed. In the down-strip ordering, the photo parameters are arranged by strips, so that the nine photos from strip 1 appear first, followed by the nine photos of strip 2, and the nine photos from strip 3. With cross-strip ordering, the photo parameters are arranged so that the first photo of strip 1 appears first, followed by the first photos of strips 2 and 3; then the second photos of strips 1, 2, and 3; and so on. These two photo orders are listed in Table 17-3. As will be demonstrated, cross-strip ordering leads to a more efficient solution than down-strip ordering in this case.

Figure 17-16 shows a schematic representation of the reduced normal equations when down-strip ordering is employed. Notice from the figure that the nonzero elements tend to cluster in a band about the main diagonal of the matrix. The width of the band from the diagonal to the farthest off-diagonal nonzero element is the *bandwidth* of the matrix. The bandwidth of the matrix shown in Fig. 17-16 is $6 \times 12 = 72$. With cross-strip ordering of the photos, the reduced normal equation matrix shown in Fig. 17-17 results. Here, the bandwidth is $6 \times 8 = 48$, which is substantially smaller than that for down-strip ordering. The narrower the bandwidth, the faster the solution and the less storage required.

Solution time for non-banded reduced normal equations is proportional to the number of unknowns ($6m$) raised to the third power. For the example, with 27 photos, the time is proportional to $(6 \times 27)^3 = 4.2 \times 10^6$. For banded equations, the solution time is proportional to the bandwidth squared, times the number of unknowns. For the example with down-strip number, the time is proportional to $72^2 \times (6 \times 27) = 8.4 \times 10^5$, which is 5 times faster than the nonbanded case. With cross-strip numbering, the time is proportional to $48^2 \times (6 \times 27) = 3.7 \times 10^5$, which is more than 11 times faster than the nonbanded case!

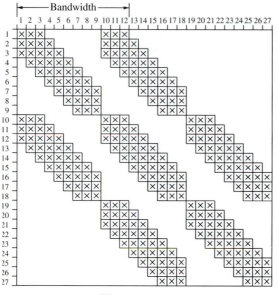

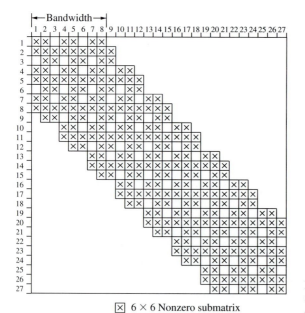

☒ 6 × 6 Nonzero submatrix

FIGURE 17-16
Structure of the reduced normal
equations using down-strip ordering.

☒ 6 × 6 Nonzero submatrix

FIGURE 17-17
Structure of the reduced normal
equations using cross-strip ordering.

Down-strip and cross-strip ordering generally apply only to regular, rectangular photo blocks. In cases where photo blocks cover irregular areas, other more complicated approaches should be used to achieve a minimal bandwidth. Details of these other approaches can be found in references which follow.

REFERENCES

Ackermann, F., and H. Schade: "Application of GPS for Aerial Triangulation," *Photogrammetric Engineering and Remote Sensing,* vol. 59, no. 11, 1993, p. 1625.

American Society of Photogrammetry: *Manual of Photogrammetry,* 4th ed., Bethesda, MD, 1980, chap. 2.

Brown, D. C.: "New Developments in Photogeodesy," *Photogrammetric Engineering and Remote Sensing,* vol. 60, no. 7, 1994, p. 877.

Curry, S., and K. Schuckman: "Practical Considerations for the Use of Airborne GPS for Photogrammetry," *Photogrammetric Engineering and Remote Sensing,* vol. 59, no. 11, 1993, p. 1611.

Duff, I. S., A. M. Erisman, and J. K. Reid: *Direct Methods for Sparse Matrices,* Oxford University Press, New York, 1990.

Ebadi, H., and M. A. Chapman: "GPS-Controlled Strip Triangulation Using Geometric Constraints of Man-Made Structures," *Photogrammetric Engineering and Remote Sensing,* vol. 64, no. 4, 1998, p. 329.

El-Hakim, S. F., and H. Ziemann: "A Step-by-Step Strategy for Gross-Error Detection," *Photogrammetric Engineering and Remote Sensing,* vol. 50, no. 6, 1984, p. 713.

Erio, G.: "Three-Dimensional Transformations of Independent Models," *Photogrammetric Engineering and Remote Sensing,* vol. 41, no. 9, 1975, p. 1117.

George, A., and J. W. H. Liu: *Computer Solution of Large Sparse Positive-Definite Systems,* Prentice-Hall, Englewood Cliffs, NJ, 1981.

Goad, C. C., and M. Yang: "A New Approach to Precision Airborne GPS Positioning for Photogrammetry," *Photogrammetric Engineering and Remote Sensing,* vol. 63, no. 9, 1997, p. 1067.

Gruen, A., M. Cocard, and H.-G. Kahle: "Photogrammetry and Kinematic GPS: Results of a High Accuracy Test," *Photogrammetric Engineering and Remote Sensing,* vol. 59, no. 11, 1993, p. 1643.

Jacobsen, K.: "Experiences in GPS Photogrammetry," *Photogrammetric Engineering and Remote Sensing,* vol. 59, no. 11, 1993, p. 1651.

Kubik, K., D. Merchant, and T. Schenk: "Robust Estimation in Photogrammetry," *Photogrammetric Engineering and Remote Sensing,* vol. 53, no. 2, 1987, p. 167.

Novak, K.: "Rectification of Digital Imagery," *Photogrammetric Engineering and Remote Sensing,* vol. 58, no. 3, 1992, p. 339.

Schut, G. H.: "Development of Programs for Strip and Block Adjustment at the National Research Council of Canada," *Photogrammetric Engineering,* vol. 30, no. 2, 1964, p. 283.

Schwarz, K. P., M. A. Chapman, M. W. Cannon, and P. Gong: "An Integrated INS/GPS Approach to the Georeferencing of Remotely Sensed Data," *Photogrammetric Engineering and Remote Sensing,* vol. 59, no. 11, 1993, p. 1667.

Theodossiou, E. I., and I. J. Dowman: "Heighting Accuracy of SPOT," *Photogrammetric Engineering and Remote Sensing,* vol. 56, no. 12, 1990, p. 1643.

Toth, C. K., and A. Krupnik: "Concept, Implementation, and Results of an Automatic Aerotriangulation System," *Photogrammetric Engineering and Remote Sensing,* vol. 62, no. 6, 1996, p. 711.

Westin, T.: "Precision Rectification of SPOT Imagery," *Photogrammetric Engineering and Remote Sensing,* vol. 56, no. 2, 1990, p. 247.

Wolf, P. R.: "Independent Model Triangulation," *Photogrammetric Engineering,* vol. 36, no. 12, 1970, p. 1262.

PROBLEMS

17-1. Discuss the advantages of aerotriangulation over field surveys.

17-2. List the three categories of aerotriangulation. Which categories are currently used?

17-3. Discuss the approaches for marking pass points for monoscopic measurement versus stereoscopic measurement.

17-4. Briefly describe independent model aerotriangulation by the sequential method.

17-5. Why is a three-dimensional polynomial transformation often used for transforming a sequentially constructed strip model to the ground coordinate system?

17-6. A continuous strip of three stereomodels has pass points a through l and ground control points A through D. Independent model coordinates for points and exposure stations of each model are listed below along with ground coordinates of control points. Compute the ground coordinates of the pass points and exposure stations by the sequential method of semianalytical aerotriangulation. Use the three-dimensional conformal coordinate transformation program provided (see Example 17-1).

Model coordinates (mm)

	Model 1-2				Model 2-3				Model 3-4		
Point	x	y	z	Point	x	y	z	Point	x	y	z
a	12.872	99.960	−1.308	d	1.652	82.499	3.641	g	−4.384	95.574	−11.554
b	0.969	1.586	−1.203	e	5.950	−3.959	4.158	h	−1.578	2.610	−14.159
c	−8.682	−96.182	−7.983	f	−2.634	−90.656	2.838	i	−1.816	−94.498	−11.206
d	95.626	91.726	1.727	g	87.684	82.045	2.312	j	103.388	96.458	−10.641
e	99.370	0.273	−1.181	h	90.551	−2.573	1.567	k	92.207	−3.266	−8.861
f	89.504	−91.341	−6.273	i	90.719	−90.849	6.096	l	91.996	−102.817	−11.437
A	41.352	73.058	−1.046	O_2	0.000	0.000	152.673	C	74.595	79.408	−11.492
B	57.398	−84.026	−4.765	O_3	93.605	−1.916	153.414	D	85.973	−111.232	−8.786
O_1	0.000	0.000	152.673					O_3	0.000	0.000	152.673
O_2	90.920	−1.423	155.945					O_4	93.613	−6.377	155.413

Ground control coordinates (m)

Point	X	Y	Z
A	12,613.434	17,302.493	265.517
B	12,676.810	16,850.242	260.045
C	13,252.875	17,331.091	258.497
D	13,302.036	16,801.338	263.223

17-7. Briefly describe how the method of independent model aerotriangulation by simultaneous transformations differs from the sequential approach.

17-8. Compute model coordinates of one of the perspective centers of a stereomodel, using the two-point equation [Eq. (17-5)]. Model coordinates of two points A and B measured at two different elevations in the model are as follows:

	X and Y at $Z = 50.00$ mm		X and Y at $Z = 280.00$ mm	
Point	X, mm	Y, mm	X, mm	Y, mm
A	591.81	785.71	594.30	656.30
B	902.47	505.10	762.31	504.55

17-9. Repeat Prob. 17-8, except that the model coordinates of points A and B are as follows:

Point	X and Y at $Z = 100.00$ mm		X and Y at $Z = 330.00$ mm	
	X, mm	Y, mm	X, mm	Y, mm
A	420.53	217.89	409.83	347.48
B	652.77	787.13	535.43	655.30

17-10. Discuss the advantages of analytical methods of photogrammetric control extension as opposed to semianalytical methods.

17-11. Briefly describe the unknowns and measurements associated with a bundle adjustment of a block of photographs.

17-12. Describe how coordinates of the GPS antenna are related to the exposure station when airborne GPS control is used to control photography.

17-13. Briefly discuss the problem associated with the lack of synchronization of GPS fixes with camera exposures in airborne GPS control.

17-14. What is the purpose of cross strips at the ends of photo blocks when airborne GPS is used to control photography?

17-15. Briefly explain how a line perspective image differs from a point perspective image.

17-16. Briefly discuss the characteristic of the $\ddot{N}$ submatrix that makes the method of reduced normal equations more efficient for a bundle adjustment than solving the full normal equations.

17-17. Discuss the difference between down-strip and cross-strip numbering as they apply to the bandwidth of the reduced normal equations of a bundle adjustment.

CHAPTER
18

PROJECT
PLANNING

18-1 INTRODUCTION

Successful execution of any photogrammetric project requires that thorough planning be done prior to proceeding with the work. Planning, more than any other area of photogrammetric practice, must be performed by knowledgeable and experienced persons who are familiar with all aspects of the subject.

The first and most important decision to be made in the planning process concerns the selection of the products that will be prepared. In addition to selecting the products, their scales and accuracies must be fixed. These decisions can only be made if the planner thoroughly understands what the client's needs are, so that the best overall products can be developed to meet those needs. The client will also naturally be concerned with the anticipated costs of the items as well as the proposed schedule for their delivery. Therefore, successful planning will probably require several meetings with the client prior to commencing the work, and depending upon the nature and magnitude of the project, continued meetings may be needed as production progresses.

A variety of products may be developed in a given photogrammetric project, including prints of aerial photos, photo indexes, photomaps, mosaics, orthophotos, planimetric and topographic maps, digital maps for GIS databases and other purposes, cross sections, digital elevation models, cadastral maps, and others. In addition to the wide variation in products that could be developed for a given project, there are normally other major considerations that will have definite bearing on procedures, costs, and scheduling. These include the location of the project; its size, shape, topography, and vegetation cover; the availability of existing ground control; etc. Thus, every project presents unique problems to be considered in the planning stages.

Assuming that the photogrammetric products, and their scales and accuracies, have been agreed upon with the client, the balance of the work of project planning can generally be summarized in the following categories:

1. Planning the aerial photography[1]
2. Planning the ground control
3. Selecting instruments and procedures necessary to achieve the desired results
4. Estimating costs and delivery schedules

When planning has been completed for these categories, the photogrammetrist will normally prepare a detailed proposal which outlines plans, specifications, an estimate of costs, and delivery schedules for the project. The proposal often forms the basis of an agreement or contract for the performance of the work.

Of the above four categories, this chapter concentrates primarily on planning of the aerial photography. Planning of the ground control has been discussed in detail in Chap. 16, and instrument and procedure selection has been discussed in earlier chapters where the various photogrammetric products and instruments for producing them have been described. A brief discussion of the subjects of cost estimating and scheduling is given in Sec. 18-12.

18-2 IMPORTANCE OF FLIGHT PLANNING

Because the ultimate success of any photogrammetric project probably depends more upon good-quality photography than on any other aspect, planning the aerial photography, also called *flight planning,* is of major concern. If the photography is to satisfactorily serve its intended purposes, the photographic mission must be carefully planned and faithfully executed according to the "flight plan." A flight plan generally consists of two items: a *flight map,* which shows where the photos are to be taken; and *specifications,* which outline how to take them, including specific requirements such as camera and film requirements, scale, flying height, end lap, side lap, and tilt and crab tolerances. A flight plan which gives optimum specifications for a project can be prepared only after careful consideration of all the many variables which influence aerial photography.

An aerial photographic mission is an expensive operation involving two or more crewpersons and high-priced aircraft and equipment. In addition, periods of time that are acceptable for aerial photography are quite limited in many areas by weather and ground cover conditions, which are related to seasons of the year. Failure to obtain satisfactory photography on a flight mission not only necessitates costly reflights, but

---

[1] Although a significant number of projects involve terrestrial or close-range photogrammetry, in this chapter aerial photogrammetry is assumed.

also in all probability will cause long and expensive delays on the project for which the photos were ordered. For these reasons flight planning is one of the most important operations in the overall photogrammetric project.

18-3 PHOTOGRAPHIC END LAP AND SIDE LAP

Before discussing the many aspects which enter into consideration in planning an aerial photographic mission, it will be helpful to redefine the terms *end lap* and *side lap*. As discussed in Sec. 1-4, vertical aerial photographic coverage of an area is normally taken as a series of overlapping flight strips. As illustrated in Fig. 18-1, *end lap* is the overlapping of successive photos along a flight strip. Figure 18-2 illustrates *side lap,* or the overlap of adjacent flight strips.

In Fig. 18-1, G represents the dimension of the square of ground covered by a single vertical photograph (assuming level ground and a square camera focal-plane format), and B is the air base or distance between exposure stations of a stereopair. The amount of end lap of a stereopair is commonly given in percent. Expressed in terms of G and B, it is

$$\text{PE} = \frac{G - B}{G} \times 100 \tag{18-1}$$

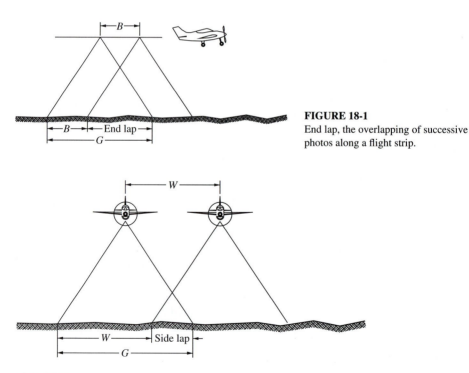

FIGURE 18-1
End lap, the overlapping of successive photos along a flight strip.

FIGURE 18-2
Side lap, the overlapping of adjacent flight strips.

In Eq. (18-1), PE is percent end lap. If stereoscopic coverage of an area is required, the absolute minimum end lap is 50 percent. However, to prevent gaps from occurring in the stereoscopic coverage due to crab, tilt, flying height variations, and terrain variations, end laps greater than 50 percent are used. Also, if the photos are to be used for photogrammetric control extension, images of some points must appear on three successive photographs—a condition requiring greater than 50 percent end lap. For these reasons aerial photography for mapping purposes is normally taken with about 60 percent end lap.

Crab, as explained in Sec. 3-6, exists when the edges of the photos in the *x* direction are not parallel with the direction of flight. It causes a reduction in stereoscopic coverage, as was indicated in Fig. 3-7. Figures 18-3 through 18-5 illustrate reductions in end lap causing loss of stereoscopic coverage due to tilt, flying height variation, and relief variations, respectively.

Side lap is required in aerial photography to prevent gaps from occurring between flight strips as a result of drift, crab, tilt, flying height variation, and terrain variations. *Drift* is the term applied to a failure of the pilot to fly along planned flight lines. It is often caused by strong winds, but can also result from a lack of definite features and objects shown on the flight map which can also be identified from the air to guide the pilot during photography. Excessive drift is the most common cause for gaps in photo coverage; when this occurs, reflights are necessary.

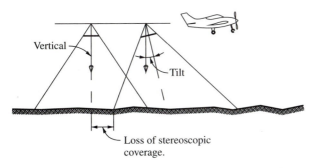

FIGURE 18-3
Failure to achieve stereoscopic coverage due to tilt.

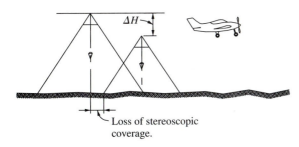

FIGURE 18-4
Failure to achieve stereoscopic
coverage due to flying height variations.

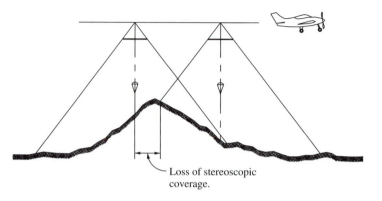

Loss of stereoscopic
coverage.

FIGURE 18-5
Failure to achieve stereoscopic coverage due to terrain variations.

In Fig. 18-2, again G represents the dimension of the square of ground coverage of a single vertical photograph, and W is the spacing between adjacent flight lines. An expression for PS *(percent side lap)* in terms of G and W

$$\text{PS} = \frac{G - W}{G} \times 100 \tag{18-2}$$

Mapping photography is normally taken with a side lap of about 30 percent. Besides helping to prevent gaps in coverage, another advantage realized from using this large a percentage is the elimination of the need to use the extreme edges of the photography, where the imagery is of poorer quality. Photography for orthophoto or mosaic work is sometimes taken with greater than 30 percent side lap since this reduces the sizes of the central portions of the photographs that must be used, thereby lessening distortions of images due to tilt and relief. In certain cases, such as for very precise photogrammetric control extension, the aerial photos may be taken with 60 percent side lap as well as 60 percent end lap to increase the redundancy in the bundle adjustment (see Chap. 17).

Example 18-1. The air base of a stereopair of vertical photos is 1400 m, and flying height above average ground is 2440 m. The camera has a focal length of 152.4 mm and a 23-cm format. What is the percent end lap?

Solution

1. Average photo scale $= \dfrac{f}{H'_{\text{avg}}} = \dfrac{152.4 \text{ mm}}{2440 \text{ m} \times 1000 \text{ mm/m}}$

   $= 1 : 16{,}000 \ (1 \text{ in} = 1330 \text{ ft})$

2. Ground coverage dimension $G = \dfrac{23 \text{ cm} \times 16{,}000}{100 \text{ cm/m}} = 3680 \text{ m} \ (12{,}100 \text{ ft})$

3. Percent end lap, by Eq. (18-1), is

$$PE = \frac{3680 \text{ m} - 1400 \text{ m}}{3680 \text{ m}} \times 100 = 62\%$$

Example 18-2. In Example 18-1, assume that the spacing between adjacent flight strips is 2500 m. What is the percent side lap?

Solution. By Eq. (18-2),

$$PS = \frac{3680 \text{ m} - 2500 \text{ m}}{3680 \text{ m}} \times 100 = 32\%$$

18-4 PURPOSE OF THE PHOTOGRAPHY

In planning aerial photographic missions, the first and foremost consideration is the purpose for which the photography is being taken. Only with the purpose defined can optimum equipment and procedures be selected. In general, aerial photographs are desired which have either good *metric* qualities or high *pictorial* qualities. Photos having good metric qualities are needed for topographic mapping or other purposes where precise quantitative photogrammetric measurements are required. High pictorial qualities are required for qualitative analysis, such as for photographic interpretation or for constructing orthophotos, photomaps, and aerial mosaics.

Photographs of good metric quality are obtained by using calibrated cameras and films having fine-grained, high-resolution emulsions. For topographic mapping and other quantitative operations, photography is preferably taken with a wide- or super-wide-angle (short-focal-length) camera so that a large *base-height* ratio (B/H') is obtained. The B/H' ratio, as described in Sec. 7-8, is the ratio of the air base of a pair of overlapping photographs to the average flying height above ground. The larger the B/H' ratio, the greater the intersection angles (parallactic angles) between intersecting light rays to common points. In Figs. 18-6a and b, for example, the air bases are equal, but the focal length and flying height in Fig. 18-6a are one-half those in Fig. 18-6b. The photographic scales are therefore equal for both cases, but the B/H' ratio of Fig. 18-6a is double that of Fig. 18-6b, and parallactic angle ϕ_1 to point A in Fig. 18-6a is nearly double the corresponding angle ϕ_2 in Fig. 18-6b.

It can be shown that the errors in computed positions and elevations of points in a stereopair increase with increasing flying heights, and decrease with increasing x parallax. Large B/H' ratios denote low flying heights and large x parallaxes, conditions favorable to higher accuracy. The photos of Fig. 18-6a are therefore superior to those of Fig. 18-6b for mapping or quantitative analyses.

Photography of high pictorial quality does not necessarily require a calibrated camera, but the camera must have a good-quality lens. In many cases, films having fast, large-grained emulsions produce desirable effects. For some photo interpretation work, normal color films are useful. For other special applications, black-and-white infrared or color infrared films are desirable. Special effects can also be obtained by using filters in

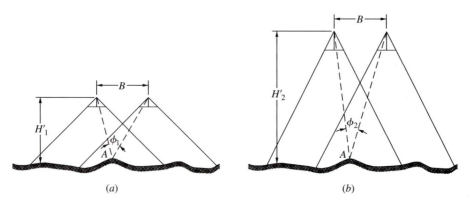

FIGURE 18-6
Parallactic angles increase with increasing B/H' ratios.

combination with various types of films. Timber types, for example, can be delineated quite effectively by using a red filter in combination with black-and-white infrared film.

For mosaic work, relief displacements, tilt displacements, and scale variations produce objectionable degradations of pictorial quality. These may be minimized, however, by increasing flying height, thereby decreasing the B/H' ratio. Increased flying height, of course, reduces photo scale; but compensation for this can be achieved by using a longer-focal-length camera. The photo of Fig. 18-7a was exposed at one-half the flying height of the photo of Fig. 18-7b. The scales of the two photos are equal, however,

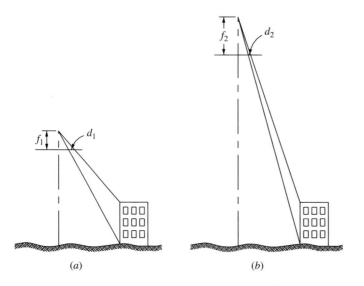

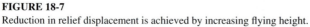

FIGURE 18-7
Reduction in relief displacement is achieved by increasing flying height.

because focal length f_2 of Fig. 18-7b is double f_1 of Fig. 18-7a. The photo of Fig. 18-7b is more desirable for mosaic construction because its scale variations and image distortions due to relief, tilt, and flying height variations are much less than those of the photo of Fig. 18-7a. On Fig. 18-7a, for example, relief displacement d_1 is more than double the corresponding relief displacement d_2 of Fig. 18-7b.

18-5 PHOTO SCALE

Average photographic scale is one of the most important variables that must be selected in planning aerial photography. It is generally fixed within certain limits by specific project requirements. For topographic mapping, photo scale is usually dictated by the map's required scale and/or horizontal and vertical accuracy. On the other hand, in photo interpretation applications, or for orthophoto or mosaic preparation, ensuring that the smallest ground objects of importance can be resolved on the photos may be the governing consideration in selecting the photo scale.

> **Example 18-3.** A particular project involves the use of aerial photography to study the centerline markings of highways. The actual width of painted centerlines on the highways is 100 mm (4 in). A high-resolution film (80 line pairs per millimeter) will be used. What minimum photo scale is required?
>
> **Solution.** With a resolution of 80 line pairs per millimeter (see Sec. 3-13), the smallest objects that could reasonably be resolved with that film would be $\frac{1}{80}$ mm = 0.0125 mm in size. Thus the minimum scale required would be roughly
>
> $$S_{\min} = \frac{0.0125 \text{ mm}}{100 \text{ mm}} = 1/8000 \text{ or } 1:8000 \ (1 \text{ in} = 670 \text{ ft})$$

In topographic mapping, the enlargement ratio from photo scale to the scale of the plotted map must be considered. With older direct optical projection stereoplotters, where maps were drawn in real time as the operator viewed and traced the stereomodel, the enlargement ratio was fixed within narrow limits, with 5 being most common (see Chap. 12, Part I). Today these types of instruments are seldom used in practice. Mechanical projection stereoplotters (see Chap. 12, Part II) have enlargement ratio capabilities which range from less than 1 up to 10 or more, depending upon the particular instrument. However, the number of these instruments in use today is also declining, and those being used are seldom employed for direct map compilation anymore. Rather they are equipped with digitizers, and topographic information is digitized and stored in files. From these digital files, maps are plotted using computers. Today, analytical plotters and softcopy systems account for the majority of topographic mapping, and for topographic mapping both of these systems are normally used to produce digital files.

In using computers to plot maps from digitized files, virtually any enlargement ratio is possible. It is important to note, however, that the digitized information contains errors, and these errors are magnified by whatever enlargement ratio is used. Thus to ensure the integrity and accuracy of a map compiled from digital data, the enlargement ratios that are actually utilized must be held within reasonable limits. To ensure that their plotted maps meet required accuracy standards, many organizations will not enlarge more than about five or six times over photo scale. Higher enlargement ratios are sometimes used but this should be done with caution, and generally only when experience with previous projects has shown through field checks that satisfactory accuracies have been obtained.

Example 18-4. A map must be compiled at a scale of 1 : 6000. If an enlargement ratio of 5 will be used in producing this map, what is the required photo scale?

Solution. Photo scale is one-fifth as large as map scale. Therefore,

$$S_{\text{photo}} = \frac{1}{6000} \times \frac{1}{5} = 1 : 30,000 \ (1 \text{ in} = 2500 \text{ ft})$$

Selection of optimum map scale depends upon the purpose of the map. It should be carefully planned, because compilation at a larger scale than necessary is uneconomical, and compilation at too small a scale reduces the usefulness of the map or may even render it unsatisfactory. The horizontal accuracy (accuracy to which planimetric positions of points can be measured from a map) depends directly upon the map's scale. Assume, for example, that map positions of planimetric features can be plotted correctly to within $\frac{1}{30}$ in, a condition necessary to meet National Map Accuracy Standards (see Sec. 16-4) for large-scale maps.[2] Now if a particular cadastral map requires that points be accurate to within ± 2.0 ft, required map scale is 1 in = 60 ft (1 : 720). Then for that situation, if an enlargement ratio of 5 is employed, photo scale is fixed at 5 × 60 = 300 ft/in (1 : 3600). In another example, if points only need to be accurate to within ± 20 ft on a topographic map for preliminary planning, and again assuming the same accuracy of $\frac{1}{30}$ in, then a scale of 1 in = 600 ft (1 : 7200) is all that would be required, and minimum photo scale would be 5 × 7200, or 1 : 36,000.

As previously noted, vertical mapping accuracy is also an important factor to be considered in planning aerial photography. In past photogrammetric practice, contours which portrayed topographic relief were compiled directly from stereomodels, and thus the guidelines and standards for specifying vertical accuracy in topographic mapping were commonly given in terms of contour interval. As noted earlier in this text, however, for modern photogrammetry projects digital elevation models (DEMs) are now generally compiled rather than contours. From the DEMs, triangulated irregular network (TIN) models are constructed using computers, and then contours, cross sections, and profiles can be generated automatically from the TIN models. But even though different procedures are now employed to compile topographic information, contours are still the end product that is often used in representing topographic relief. Thus guidelines and standards that are based on contour interval are still appropriate for quantifying vertical mapping accuracy. And as will be discussed later in this section, as vertical

---

[2] The complete set of National Map Accuracy Standards is available at the following Web site: http://rmmcweb.cr.usgs.gov/public/nmpstds/nmas647.html.

mapping accuracy requirements increase (contour interval decreases), flying height must decrease and photo scale must increase.

As with planimetric accuracy, the contour interval to be selected for a particular mapping project depends upon the intended use of the maps. Assume, for example, that elevations can be interpolated correctly from a map to within one-half the contour interval, a condition required for meeting National Map Accuracy Standards. If elevations must be interpolated to within ±0.5 ft on a highway design map, then a 1-ft contour interval is necessary. If elevations must be interpolated to ±10 ft on a map prepared for studying the volume of water impounded in the reservoir of a large dam, then a 20-ft contour interval is all that is required.

The recommended contour interval depends on not only the use to be made of the map but also the type of terrain. If the map is being prepared for planning a sewer system for a city such as Las Vegas, Nevada, which lies on relatively flat terrain, perhaps a 1-ft contour interval would be required. On the other hand, if a topographic map of San Francisco is being prepared for the same purpose, because of the large range of relief in that city, perhaps a 5- or 10-ft contour interval would be used.

In planning a topographic mapping project, contour interval and map scale must be selected so that they are compatible. As map scale decreases, contour interval must increase; otherwise, the contours would become too congested on the map. In large-scale mapping of average types of terrain, the scale and contour interval relationships shown in Table 18-1 generally provide satisfactory compatibility.

Relative accuracy capabilities in photogrammetric mapping, whether planimetric or vertical, depend upon many variables, but the most important is flying height above ground. Others include the quality of the stereoplotting instrument that is used and the experience and ability of its operator, the camera and its calibration, the quality of the photography, the density and accuracy of the ground control, and the nature of the terrain and its ground cover. A rule of thumb for quantifying vertical accuracy capability, based on contour interval, employs a term called the *C factor*. The *C* factor is the ratio of the flying height above ground of the photography (H') to the contour interval that can be reliably plotted using that photography, or in equation form

$$C \text{ factor} = \frac{H'}{\text{CI}} \tag{18-3}$$

TABLE 18-1
Compatible map scales and contour intervals for average terrain

English system		Metric system	
Map scale	**Contour interval**	**Map scale**	**Contour interval**
1 in/50 ft	1 ft	1:500	0.5 m
1 in/100 ft	2 ft	1:1000	1 m
1 in/200 ft	5 ft	1:2000	2 m
1 in/500 ft	10 ft	1:5000	5 m
1 in/1000 ft	20 ft	1:10,000	10 m

The units of H' and CI (contour interval) of Eq. (18-3) are the same. The C factors that are employed by photogrammetric mapping organizations are based upon their experiences, and this experience will include field checks of map accuracies achieved on a variety of previous projects. To ensure that their maps meet required accuracy standards, many organizations use a C factor of from about 1200 to 1500. Other organizations may push the value somewhat higher, but this must be done with extreme caution.

> **Example 18-5.** A topographic map having a scale of 200 ft/in with 5-ft contour interval is to be compiled from contact-printed diapositives using a stereoplotter having a nominal 6-in (152-mm) principal distance. Determine the required flying height for the photography if the maximum values to be employed for the C factor and enlargement ratio are 1500 and 5, respectively.

Solution

1. Considering contour interval and C factor:

$$H' = 1500(5 \text{ ft}) = 7500 \text{ ft}$$

2. Considering map scale and enlargement ratio:

$$\text{Photo scale} = \frac{1 \text{ in}}{200 \text{ ft}} \times \frac{1}{5} = \frac{1 \text{ in}}{1000 \text{ ft}} = \frac{f}{H'}$$

Thus, $\qquad\qquad H' = 6 \text{ in} \times 1000 \text{ ft/in} = 6000 \text{ ft}$

In this instance, the lower of the two flying heights (6000 ft) must be selected, and therefore the enlargement ratio from photo scale to map scale governs over contour interval.

In some topographic mapping projects, particularly where there is little relief, it is impractical to show elevations using contours because few, if any, may result. In these situations, a grid of spot elevations can be read throughout the area to depict the relief. A rule of thumb relating accuracy of spot elevations and flying height is that the ratio of flying height above ground to the accuracy with which spot elevations can be reliably read is approximately 5000. Thus if spot elevations are needed to an accuracy of $\pm\frac{1}{2}$m, then the flying height above ground necessary to achieve those results would be in the range of about $\frac{1}{2}$ (5000), or roughly 2500 m. Again, in addition to flying height, the actual accuracies that can be achieved also relate to the stereoplotting instrument, the operator's ability, the quality of the aerial photography, the density and accuracy of the ground control, and other factors.

18-6   FLYING HEIGHT

Once the camera focal length and required average photo scale have been selected, required flying height above average ground is automatically fixed in accordance with scale [see Eq. (6-3)].

Example 18-6. Aerial photography having an average scale of 1:6000 is required to be taken with a 152.4-mm-focal-length camera over terrain whose average elevation is 425 m above mean sea level. What is required flying height above mean sea level?

Solution. By Eq. (6-3)

$$S = \frac{f}{H - h_{avg}}$$

Thus

$$\frac{1}{6000} = \frac{152.4 \text{ mm}}{H - 425 \text{ m}} \left(\frac{1 \text{ m}}{1000 \text{ mm}} \right)$$

$$H = 6000(0.1524 \text{ m}) + 425 \text{ m} = 1340 \text{ m above MSL}$$

Flying heights above average ground may vary from a hundred meters or so in the case of large-scale helicopter photography, to several hundred kilometers if satellites are used to carry the camera. Flying heights used in taking photos for topographic mapping normally vary between about 500 and 10,000 m. If one portion of the project area lies at a substantially higher or lower elevation than another part, two different flying heights may be necessary to maintain uniform photo scale.

Ground coverage per photo for high-altitude photography is greater than that for low-altitude photography (see Tables 18-2 and 18-3). Fewer high-altitude photos are therefore required to cover a given area. Very high-altitude coverage is more expensive to obtain than low-altitude photography because of the special equipment required. Some of the problems encountered at high flying heights are the decreasing available oxygen, decreasing pressure, and extreme cold. When flying heights exceed about 3000 m, an oxygen supply system is necessary for the flight crew. At altitudes above 10,000 m, pure oxygen under pressure is required. Also, the cabin must be pressurized, and heaters are required to protect the crew against the cold. Most aerial photography is taken by using single- or twin-engine aircraft. Supercharged single-engine aircraft can reach altitudes of about 6 km, and supercharged twin-engine aircraft are capable of approaching 10 km. Higher altitudes require turbocharged or jet aircraft.

During photography the pilot maintains proper flying height by means of an altimeter or GPS receiver. Since altimeters give elevations above mean sea level, the proper reading is the sum of average ground elevation and required flying height above ground necessary to achieve proper photo scale. Altimeters are barometric instruments, and consequently their readings are affected by varying atmospheric pressure. They must be checked daily and adjusted to base airport air pressure. GPS receivers, while essentially unaffected by barometric pressure, give elevations relative to the ellipsoid (see Sec. 16-8); therefore a geoid model is required to relate these elevations to mean sea level.

18-7 GROUND COVERAGE

Once average photographic scale and camera format dimensions have been selected, the ground surface area covered by a single photograph may be readily calculated. In addition, if end lap and side lap are known, the ground area covered by the stereoscopic *neat model* can be determined. The neat model, as illustrated in Fig. 18-8, is the stereoscopic area between adjacent principal points and extending out sideways in both directions to the middle of the side lap. The neat model has a width of B and a breadth of W. Its coverage is important since it represents the approximate mapping area of each stereopair.

Example 18-7. Aerial photography is to be taken from a flying height of 6000 ft above average ground with a camera having a 6-in (152.4-mm) focal length and a 9-in (23-cm) format. End lap will be 60 percent, and side lap will be 30 percent. What is the ground area covered by a single photograph and by the stereoscopic neat model?

Solution

1. By Eq. (6-1),

$$S = \frac{6 \text{ in}}{6000 \text{ ft}} = \frac{1 \text{ in}}{1000 \text{ ft}} \text{ or } 1:12{,}000$$

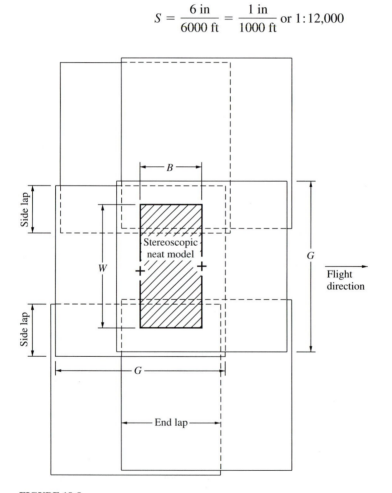

FIGURE 18-8
The area covered by a stereoscopic neat model.

2. The dimension G of the square ground area covered by a single photo is

$$G = 1000 \text{ ft/in} \times 9 \text{ in} = 9000 \text{ ft}$$

3. The area in acres covered on the ground by a single photo is

$$A = \frac{(9000 \text{ ft})^2}{43,560 \text{ ft}^2/\text{acre}} = 1900 \text{ acres}$$

4. At 60 percent end lap, B is $0.4G$; and at 30 percent side lap, W is $0.7G$. Therefore the dimensions of the rectangular stereoscopic neat model are

$$B = 0.4(9000 \text{ ft}) = 3600 \text{ ft}$$

$$W = 0.7(9000 \text{ ft}) = 6300 \text{ ft}$$

And the area of the neat model is

$$\frac{3600 \text{ ft} \times 6300 \text{ ft}}{43,560 \text{ ft}^2/\text{acre}} = 520 \text{ acres}$$

Tables 18-2 and 18-3 give ground dimensions and ground areas covered by a single photo and by the stereoscopic neat model for various commonly used photo scales. Table 18-2, given in the English system, is based on a 6-in-focal-length camera having a 9-in-square format. Table 18-3, given in the metric system, is based on a 152-mm-focal-length camera having a 23-cm-square format. End lap and side lap of 60 percent and 30 percent, respectively, are assumed for both tables.

TABLE 18-2
English system dimensions and areas of single photos and neat models for a 6-in-focal-length camera and various commonly used photo scales

Photo scale	Photo scale, in/ft	Flying height, ft	G, ft	W (0.7G), ft	B (0.4G), ft	Acres per photo	Acres per neat model
1:1,800	1/150	900	1,400	940	540	42	12
1:2,400	1/200	1,200	1,800	1,300	720	74	21
1:3,000	1/250	1,500	2,200	1,600	900	120	33
1:3,600	1/300	1,800	2,700	1,900	1,100	170	47
1:4,200	1/350	2,100	3,200	2,200	1,300	230	64
1:4,800	1/400	2,400	3,600	2,500	1,400	300	83
1:5,400	1/450	2,700	4,000	2,800	1,600	380	100
1:6,000	1/500	3,000	4,500	3,200	1,800	460	130
1:6,600	1/550	3,300	5,000	3,500	2,000	560	160
1:7,200	1/600	3,600	5,400	3,800	2,200	670	190
1:7,800	1/650	3,900	5,800	4,100	2,300	790	220
1:8,400	1/700	4,200	6,300	4,400	2,500	910	260
1:9,000	1/750	4,500	6,800	4,700	2,700	1,000	290

(continued)

TABLE 18-2 (*continued*)

Photo scale	Photo scale, in/ft	Flying height, ft	G, ft	W (0.7G), ft	B (0.4G), ft	Acres per photo	Acres per neat model
1:9,600	1/800	4,800	7,200	5,000	2,900	1,200	330
1:10,800	1/900	5,400	8,100	5,700	3,200	1,500	420
1:12,000	1/1,000	6,000	9,000	6,300	3,600	1,900	520
1:15,000	1/1,250	7,500	11,000	7,900	4,500	2,900	810
1:18,000	1/1,500	9,000	14,000	9,400	5,400	6.5*	1.8*
1:24,000	1/2,000	12,000	18,000	13,000	7,200	12*	3.3*
1:30,000	1/2,500	15,000	22,000	16,000	9,000	18*	5.1*
1:40,000	1/3,333	20,000	30,000	21,000	12,000	32*	9.0*
1:50,000	1/4,167	25,000	38,000	26,000	15,000	50*	14*
1:60,000	1/5,000	30,000	45,000	32,000	18,000	73*	20*

*Square miles.

TABLE 18-3
Metric system dimensions and areas of single photos and neat models for a 152-mm-focal-length camera and various commonly used photo scales

Photo scale	Photo scale, in/ft	Flying height, m	G, m	W (0.7G), m	B (0.4G), m	Hectares per photo	Hectares per neat model
1,000	83	152	230	160	92	5.3	1.5
2,000	167	304	460	320	180	21	5.9
3,000	250	456	690	480	280	48	13
4,000	333	608	920	640	370	85	24
5,000	417	760	1,200	800	460	130	37
6,000	500	912	1,400	970	550	190	53
7,000	583	1,060	1,600	1,100	640	260	73
8,000	667	1,220	1,800	1,300	740	340	95
9,000	750	1,370	2,100	1,400	830	430	120
10,000	833	1,520	2,300	1,600	920	530	150
12,000	1,000	1,820	2,800	1,900	1,100	760	210
15,000	1,250	2,280	3,400	2,400	1,400	1,200	330
18,000	1,500	2,740	4,100	2,900	1,700	1,700	480
20,000	1,667	3,040	4,600	3,200	1,800	2,100	590
25,000	2,083	3,800	5,800	4,000	2,300	3,300	930
30,000	2,500	4,560	6,900	4,800	2,800	4,800	1,300
35,000	2,917	5,320	8,000	5,600	3,200	6,500	1,800
40,000	3,333	6,080	9,200	6,400	3,700	8,500	2,400
45,000	3,750	6,840	10,000	7,200	4,100	11,000	3,000
50,000	4,167	7,600	12,000	8,000	4,600	13,000	3,700
55,000	4,583	8,360	13,000	8,900	5,100	16,000	4,500
60,000	5,000	9,120	14,000	9,700	5,500	19,000	5,300

18-8   WEATHER CONDITIONS

The weather, which in most locations is uncertain for any given day, is a very important consideration in aerial photography. In most cases, an ideal day for aerial photography is one that is free from clouds, although if the sky is less than 10 percent cloud-covered, the day may be considered satisfactory. If clouds of greater than 10 percent coverage are present but are so high that they are above the planned flying height, this may still be objectionable since large cloud shadows will be cast on the ground, obscuring features. The number of satisfactory cloudless days varies with time of year and locality. There are certain situations in which overcast weather can be favorable for aerial photography. This is true, for example, when large-scale photos are being taken for topographic mapping over built-up areas, forests, steep canyons, or other features which would cast troublesome shadows on clear, sunny days.

A particular day can be cloudless and still be unsuitable for aerial photography due to atmospheric haze, smog, dust, smoke, high winds, or air turbulence. Atmospheric haze scatters almost entirely in the blue portion of the spectrum, and it can therefore be effectively eliminated from the photographs by using a yellow filter in front of the camera lens. Smog, dust, and smoke scatter throughout the entire spectrum and cannot be filtered out satisfactorily. Best days for photographing over industrial areas which are susceptible to smog, dust, and smoke occur after heavy rains or during moving cold fronts which clear the air. Windy, turbulent days can create excessive image motion and cause difficulties in keeping the camera oriented for vertical photography, in staying on planned flight lines, and in maintaining constant flying heights.

The decision to fly or not to fly is one that must be made daily. The flight crew should be capable of interpreting weather conditions and of making sound decisions as to when satisfactory photography can be obtained. If possible, the flight crew should be based near the project so that they can observe the weather firsthand and quickly take advantage of satisfactory conditions.

18-9   SEASON OF THE YEAR

The season of the year is a limiting factor in aerial photography because it affects ground cover conditions and the sun's altitude. If photography is being taken for topographic mapping, the photos should be taken when the deciduous trees are bare, so that the ground is not obscured by leaves. In many places this occurs twice a year for short periods in the late fall and in early spring. Oak trees tend to hold many of their leaves until spring, when the buds swell and cause the leaves to fall. In areas with heavy oak cover, therefore, the most satisfactory period for aerial photography is that very short period in the spring between budding and leafing out. Sometimes aerial photography is taken for special forestry interpretation purposes, in which case it may be desirable for the trees to be in full leaf. Normally aerial photography is not taken when the ground is snow-covered. Heavy snow not only obscures the ground but also causes difficulties in interpretation and in stereoviewing. Occasionally, however, a light snow cover can be helpful by making the ground surface more readily identifiable in tree-covered areas.

Another factor to be considered in planning aerial photography is the sun's altitude. Low sun angles produce long shadows, which can be objectionable because they obscure detail. Generally a sun angle of about 30° is the minimum acceptable for aerial photography. During the winter months of November through February, the sun never reaches a 30° altitude in some northern parts of the United States due to the sun's southerly declination. Aerial photography should therefore be avoided in those areas during these months, if possible. Often snow cover will prevent photography during these periods anyway. For the other months, photography should be exposed during the middle portion of the day after the sun rises above 30° and before it falls below that altitude. For certain purposes, shadows may be desirable, since they aid in identifying objects. Shadows of trees, for example, help to identify the species. Shadows may also be helpful in locating photo-identifiable features such as fenceposts and power poles to serve as photo control points.

18-10   FLIGHT MAP

A flight map, as shown in Fig. 18-9, gives the project boundaries and flight lines the pilot must fly to obtain the desired coverage. The flight map is prepared on some existing map which shows the project area. United States Geological Survey quadrangle maps are frequently used. The flight map may also be prepared on small-scale photographs of the area, if they are available. In executing the planned photographic mission, the pilot finds two or more features on each flight line which can be identified both on the flight map and on the ground. The aircraft is flown so that lines of flight pass over the ground points. Alternatively, an airborne GPS receiver can be employed to guide the aircraft along predefined flight lines.

Rectangular project areas are most conveniently covered with flight lines oriented north and south or east and west. As illustrated in Fig. 18-9, this is desirable because the pilot can take advantage of section lines and roads running in the cardinal directions and fly parallel to them.

If the project area is irregular in shape or if it is long and narrow and skewed to cardinal directions, it may not be economical to fly north and south or east and west. In planning coverage for such irregular areas, it may be most economical to align flight lines parallel to project boundaries as nearly as possible. Flight planning templates are useful for determining the best and most economical photographic coverage for mapping, especially for small areas. These templates, which show blocks of neat models, are prepared on transparent plastic sheets at scales that correspond to the scales of the base maps upon which the flight plan is prepared. The templates are then simply superimposed on the map over the project area and oriented in the position which yields best coverage with the fewest neat models. Such a template is shown in Fig. 18-10. The crosses represent exposure stations, and these may be individually marked on the flight map. This template method of flight planning is exceptionally useful in planning exposure station locations when artificial targets are used (see Sec. 16-9).

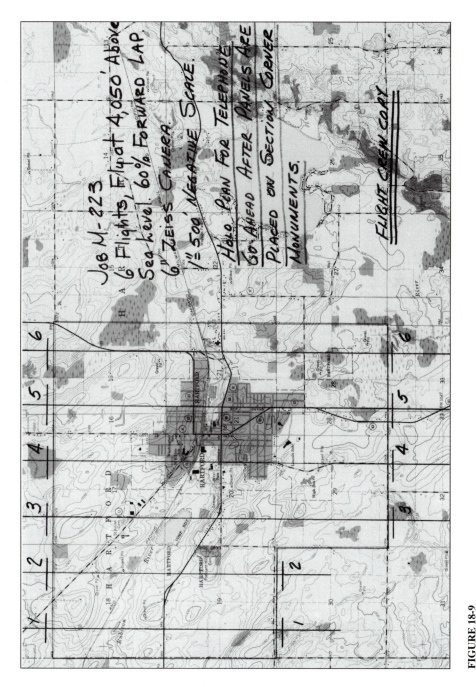

FIGURE 18-9
Example of a flight plan. (*Courtesy Ayres Associates, Inc.*)

421

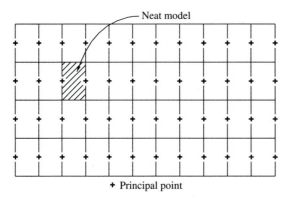

+ Principal point

FIGURE 18-10.
Transparent template of neat models used for planning aerial photography.

Once the camera focal length, photo scale, end lap, and side lap have been selected, the flight map can be prepared. The following example illustrates flight map preparation for a rectangular project area.

Example 18-8. A project area is 10 mi (16 km) long in the east-west direction and 6.5 mi (10.5 km) wide in the north-south direction (see Fig. 18-11). It is to be covered with vertical aerial photography having a scale of 1:12,000. End lap and side lap are to be 60 and 30 percent, respectively. A 6-in- (152.4-mm-) focal-length camera with a 9-in- (23-cm-) square format is to be used. Prepare the flight map on a base map whose scale is 1:24,000, and compute the total number of photographs necessary for the project.

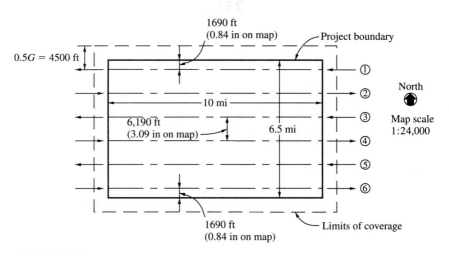

FIGURE 18-11
Project area for Example 18-9.

Solution

1. Fly east-west to reduce the number of flight lines.

2. Dimension of square ground coverage per photograph [photo scale = 1 : 12,000 (1 in/1000 ft)] is

$$G = 9 \text{ in} \times 1000 \text{ ft/in} = 9000 \text{ ft (2800 m)}$$

(*Note:* These values can also be obtained from Tables 18-2 and 18-3.)

3. Lateral advance per strip (at 30 percent side lap) is

$$W = 0.7G = (0.7)(9000 \text{ ft}) = 6300 \text{ ft (1900 m)}$$

(*Note:* These values can also be obtained from Tables 18-2 and 18-3.)

4. Number of flight lines. (Align the first and last lines with 0.3G (side-lap dimension) coverage outside the north and south project boundary lines, as shown in Fig. 18-11. This ensures lateral coverage outside of the project area.)

Distance of first and last flight lines inside their respective north and south project boundaries (see Fig. 18-11) is

$$0.5G - 0.3G = 0.2G = 0.2(9000) = 1800 \text{ ft (550 m)}$$

Number of spaces between flight lines:

$$\frac{6.5 \text{ mi} \times 5280 \text{ ft/mi} - 2 \times 1800 \text{ ft}}{6300 \text{ ft}} = 4.9 \text{ (round up to 5)}$$

Number of flight lines = number of spaces + 1 = 6

5. Adjust the percent side lap and flight line spacing.

Adjusted percent side lap for integral number of flight lines (include portion extended outside north and south boundaries):

$$2\left(0.5 - \frac{PS}{100}\right)G + (\text{no. spaces})\left(1 - \frac{PS}{100}\right)G = \text{total width}$$

$$2\left(0.5 - \frac{PS}{100}\right)9000 \text{ ft} + 5\left(1 - \frac{PS}{100}\right)9000 \text{ ft} = 6.5 \text{ mi} \times 5280 \text{ ft/mi}$$

$$2\left(0.5 - \frac{PS}{100}\right) + 5\left(1 - \frac{PS}{100}\right) = 3.813$$

$$PS = 31.2\% \quad \text{(Note slight increase.)}$$

Adjusted spacing W_a between flight lines for integral number of flight lines:

$$W_a = \left(1 - \frac{31.2}{100}\right)G = 6190 \text{ ft (1890 m)}$$

6. Linear advance per photo (air base at 60 percent end lap):

$$B = 0.4G = (0.4)(9000 \text{ ft}) = 3600 \text{ ft } (1100 \text{ m})$$

(*Note:* These values can also be obtained from Tables 18-2 and 18-3.)

7. Number of photos per strip (take two extra photos beyond the project boundary at both ends of each strip to ensure complete stereoscopic coverage):

$$\text{No. photos per strip} = \frac{10 \text{ mi} \times 5280 \text{ ft/mi}}{3600 \text{ ft}} + 1 + 2 + 2 = 19.7 \text{ (use 20)}$$

8. Total number of photos:

$$20 \text{ photos/strip} \times 6 \text{ strips} = 120$$

9. Spacing of flight lines on the map:

$$\text{Map scale} = 1 : 24,000 \ (1 \text{ in} = 2000 \text{ ft})$$

$$W_M = \frac{6190 \text{ ft per strip}}{2000 \text{ ft/in}} = 3.09 \text{ in } (78.6 \text{ mm})$$

Draw the flight lines at 3.09-in spacing on the map, with the first and last lines $[(0.5 - 31.2/100)9000 \text{ ft}]/2000 \text{ ft/in} = 0.84$ in inside the project boundaries.

Computer programs are now available for preparing flight plans. Fig. 18-12 illustrates a flight plan for a highway corridor being prepared with the aid of a computer. Design variables including camera focal length, photo scale, end lap, and side lap are input to the computer. Base maps upon which the flight maps will be prepared can either be scanned into the computer from existing hard copies, or they can be downloaded from

FIGURE 18-12
Flight plan prepared on a computer. (*Courtesy University of Florida*)

existing databases of topographic maps. Coordinates of points that delineate the boundaries of the area to be photographed must be input, and the coordinates of at least two control points must also be entered and their locations identified within the map area. The computer can then make all the same calculations that were demonstrated in Example 18-8 and prepare flight maps with superimposed flight lines. In the most modern aerial photography and navigation systems, after designing the flight map, the computer determines the coordinates of the ends of the flight lines. Then the aircraft's navigation system, aided by an onboard GPS receiver, automatically guides the aircraft along the desired flight lines at the required altitude and exposes the photographs according to the given percent end lap.

18-11 SPECIFICATIONS

Most flight plans include a set of detailed specifications which outline the materials, equipment, and procedures to be used on the project. These specifications include requirements and tolerances pertaining to photographic scale (including camera focal length and flying height), end lap, side lap, tilt, crab, and photographic quality. The following is a sample set of detailed specifications for aerial photography (courtesy Ayres Associates, Inc.).

1. *General.* The engineer shall perform the necessary flying and photography to provide photographic coverage of an area approximately 8 square miles in extent shown on the sketch map attached hereto as exhibit A. The engineer may sublet this phase of the work to a qualified and experienced aerial photographic firm. The city, however, retains the right to approve or reject any or all such firms which the engineer may wish to engage.

2. *Scale.* Flight height above average ground shall be such that the negatives will have an average scale of 1 in = 500 ft (1 : 6000). Negatives having a departure from the specified scale by more than 5 percent because of tilt or abrupt changes in flying altitude must be corrected. The photographs shall be suitable for the compilation of the topographic maps specified herein, and the mapping flight height shall not vary from 3000 ft above mean terrain by more than 5 percent.

3. *End lap and side lap.* End lap shall be sufficient to provide full stereoscopic coverage of the area to be mapped. End lap shall average 63 percent, plus or minus 5 percent. End lap of less than 58 percent or more than 68 percent in one or more negatives shall be cause for rejection of the negatives in which such deficiency or excess occurs; unless within a stereoscopic pair, end lap exceeding 68 percent is necessary in areas of low elevation to attain the minimum 58 percent end lap in adjacent areas of high elevation. Wherever there is a change in direction of the flight lines, vertical photography on the beginning of a forward section shall endlap the photography of a back section by 100 percent. Any negatives having side lap of less than 20 percent or more than 55 percent may be rejected.

4. *Tilt.* Negatives made with the optical axis of the aerial camera in a vertical position are desired. Tilt of any negative by more than 5°, an average tilt of more than 1° for

the entire project, or tilt between any two successive negatives exceeding 4° may be cause of rejection.

5. *Crab.* Crab in excess of 3° may be cause of rejection of the flight line of negatives or portions thereof in which such crab occurs.

6. *Quality.* The photographs shall be clear and sharp in detail and of uniform average density. They shall be free from clouds, cloud shadows, light streaks, static marks, or other blemishes which would interfere with their intended use. All photography shall be taken when the area to be mapped is free of snow, before foliation, and at such time as to ensure a minimum solar angle of 30°, except upon written authorization to the contrary by the city.

7. *Camera.* For topographic and contour mapping, photographs shall be exposed with a distortion-free 6-in- (152-mm-) focal-length precision aerial mapping camera equipped with a between-the-lens element shutter to produce negatives 9 in × 9 in (23 cm × 23 cm). The engineer shall furnish the city with a precision camera calibration report for the camera to be used.

8. *Contact prints.* The contact prints from the vertical negatives shall be printed on double-weight semimatte paper of suitable contrast.

9. *Photo index.* Photo indices shall be prepared by directly photographing, on safety base film at a convenient scale, the assembly of contact prints from all indexed and evaluated prints used. The photo index shall carry a suitable title, scale, and north point.

10. *Ownership of negatives.* All negatives shall become the property of the city and shall be delivered to the city upon completion of this contract, or may be stored indefinitely in the film library of the engineer at no added charge.

18-12 COST ESTIMATING AND SCHEDULING

Cost estimating is an area of critical concern in the operation of any photogrammetric business, because if projects are let to contract that are underestimated, devastating financial results can ensue. In general, the items that must be considered in a cost analysis include material, labor, and overhead. In addition, a reasonable allowance for profit must be included.

Material costs are directly related to the quantity of each photogrammetric product to be prepared, and the procedures for calculating these quantities are quite straightforward. Example 18-8 illustrated the method of estimating the total number of aerial photos needed to cover a project area. Similar procedures are used to estimate other materials, and thus these costs can usually be estimated with fair accuracy. Overhead costs, which consist of salaries of administrative personnel, office and laboratory rental, electricity, water, heat, telephone, miscellaneous office supplies, etc., are also rather straightforward to determine. Labor costs, which generally constitute the major expense on photogrammetric projects, are considerably more difficult to estimate accurately, and this presents the greatest challenge to the estimator.

Although estimates can be made of the number of hours needed to perform each task involved in photogrammetric procedures, there are many unforeseen circumstances that can cause the actual time expended to deviate significantly from the estimates. The most realistic approach to estimating labor, therefore, is to rely on past experiences with projects of a similar nature. Obviously, it then becomes very important to keep detailed records of the actual costs incurred on the individual items of all projects. Because of significant variations of project complexities and rapidly changing cost factors, past records alone cannot be relied upon completely, and a good deal of intuition and subjective judgment is also necessary. This can normally be obtained only through years of experience.

In estimating, it is easy to omit small items, but enough of these over time can accumulate to cause a significant loss of revenue. Therefore, care must be exercised to prevent these omissions, and the use of checklists is a good way to handle this problem.

Once the total number of labor hours has been estimated for each phase of a project, schedules for completion of the various operations can be planned on the basis of the number of instruments and personnel available to do the work. In addition to these factors, however, another important consideration is the amount of other work in progress and its status in relation to required completion dates. To arrive at realistic schedules, additional time in excess of that actually needed to perform the work must be added to account for uncontrollable circumstances. As an example, schedules for aerial photography and ground control surveys must account for possible delays due to inclement weather.

Every reasonable attempt should be made to accommodate clients with stringent scheduling needs. In some cases, to meet critical new scheduling requirements and still adhere to delivery dates already agreed upon, it may be necessary to consider hiring additional staff and running more than one work shift. Of course, the possibility of purchasing additional equipment also exists, but this should be done with caution and only when anticipated quantities of continued future work can justify the expenditures.

REFERENCES

American Society of Photogrammetry: *Manual of Photogrammetry,* 4th ed., Falls Church, VA, 1980, chap. 7.

Graham, L. C.: "Flight Planning for Stereo Radar Mapping," *Photogrammetric Engineering and Remote Sensing,* vol. 41, no. 9, 1975, p. 1131.

Hobbie, D.: "Orthophoto Project Planning," *Photogrammetric Engineering,* vol. 40, no. 8, 1974, p. 967.

Lafferty, M. E.: "Accuracy/Costs with Analytics," *Photogrammetric Engineering,* vol. 39, no. 5, 1973, p. 507.

Moffitt, F. H.: "Photogrammetric Mapping Standards," *Photogrammetric Engineering and Remote Sensing,* vol. 45, no. 12, 1979, p. 1637.

Paterson, G. L.: "Photogrammetric Costing," *Photogrammetric Engineering,* vol. 37, no. 12, 1971, p. 1267.

Ulliman, J. J.: "Cost of Aerial Photography," *Photogrammetric Engineering and Remote Sensing,* vol. 41, no. 4, 1975, p. 491.

U.S. Army Corps of Engineers: *Photogrammetric Mapping,* EM1110-1-1000, Hyattsville, MD, 1993.

Walker, P. M., and D. T. Trexler: "Low Sun-Angle Photography," *Photogrammetric Engineering and Remote Sensing,* vol. 43, no. 4, 1977, p. 493.

Wood, G.: "Photo and Flight Requirements for Orthophotography," *Photogrammetric Engineering,* vol. 38, no. 12, 1972, p. 1190.

PROBLEMS

18-1. The air base of a stereopair of vertical photos is 1200 m, and the flying height above average ground is 2000 m. If the camera has a 152-mm focal length and a 23-cm square format, what is the percent end lap?

18-2. Repeat Prob. 18-1, except that the air base is 850 m and flying height above average ground is 1300 m.

18-3. Repeat Prob. 18-1, except that the air base is 4200 ft and flying height above average ground is 6500 ft.

18-4. For Prob. 18-1, if adjacent flight lines are spaced at 2250 m, what is the percent side lap?

18-5. For Prob. 18-2, if adjacent flight lines are spaced at 1360 m, what is the percent side lap?

18-6. For Prob. 18-3, if adjacent flight lines are spaced at 6800 ft, what is the percent side lap?

18-7. An average photo scale of 1:20,000 is required of vertical photos. What air base is required to achieve 60 percent end lap if the camera has a 23-cm square format?

18-8. Repeat Prob. 18-7, except that required photo scale is 1:7500 and average end lap must be 55 percent.

18-9. Vertical photographs are to be exposed from 2300 m above average ground. If a B/H' ratio of 0.65 is required, what should be the length of the air base? What will the the percent end lap be for these photos if the camera focal length is 152 mm and the format is 23 cm square?

18-10. Repeat Prob. 18-9, except that the photos were exposed from 4900 ft above ground and the required B/H' ratio is 0.55.

18-11. What is the B/H' ratio for vertical photography exposed with 55 percent end lap using a camera having a 152-mm focal length and a 23-cm square format?

18-12. Repeat Prob. 18-11, except that end lap is 60 percent and camera focal length is 210 mm.

18-13. A project requires counting the number of people on a beach by using aerial photography. Assuming a 1-ft-diameter circle as a reasonable size for a person when viewed from above, and assuming a film with 80 line pairs per millimeter resolution will be used, what photo scale will be required? If a 152-mm-focal-length camera will be used, what is the required flying height above the beach?

18-14. A map with a scale of 1:9000 is to be compiled from vertical aerial photographs. The enlargement ratio from photo scale to map scale will be 5, and a 152-mm-focal-length camera will be used. What should be the flying height above average ground for the photography?

18-15. Repeat Prob. 18-14, except that a map with a scale of 1:6000 will be compiled and an enlargement ratio of 7.5 will be applied.

18-16. A C factor of 1500 will be applied in compiling a map having a contour interval of 4 m. What maximum flying height is acceptable, and what is corresponding photo scale if the camera has a 152-mm focal length?

18-17. Repeat Prob. 18-16 except that the contour interval is 5 ft and a C factor of 1200 will be applied.

18-18. An engineering design map is to be compiled from aerial photography. The map is to have a scale of 1:2400 and a 2-m contour interval. The enlargement ratio from photo scale to map scale is 5, and the C factor is 1500. If the camera focal length is 152 mm, what is the required flying height above average ground, based upon required map scale? Based upon contour interval? Which condition controls flying height?

18-19. Repeat Prob. 18-18, except that map scale is 500 ft/in, the contour interval is 10 ft, and the C factor and enlargement ratio to be applied are 1500 and 6, respectively.

18-20. Vertical aerial photographs are taken from a flying height of 3500 m above average ground using a camera with a 210-mm-focal-length lens and a 23-cm square format. End lap is 60 percent at average terrain elevation. How many acres of ground are covered in a single photograph? In the neat model? (Assume 15 percent side lap.)

18-21. For Prob. 18-20, if low, average, and high terrain is 550, 650, and 775 m, respectively, above datum, what is the percent end lap at low terrain? At high terrain? What is the percent side lap at low terrain? At high terrain?

18-22. A rectangular area 10 mi in the north-south direction by 40 mi in the east-west direction is to be covered with aerial photography having a scale of 1:6000. End lap and side lap are to be 60 and 30 percent, respectively. A camera having a 23-cm square format is to be used. Compute the total number of photographs in the project, assuming that the flight strips are flown in an east-west direction and that the coverage of the first and last flight lines is 75 percent within the project boundary. Also add two photos at the ends of each strip to ensure complete coverage.

18-23. If a flight map is to be prepared for Prob. 18-22 on a base map having a scale of 1:24,000, what should be the spacing of flight lines on the map? What is the map distance between successive exposures along a flight line?

18-24. A transparent template of neat models, similar to that shown in Fig. 18-10, is to be prepared to overlay on a map having a scale of 1:12,000. What should be the dimensions of neat models on the template if the camera format is 23 cm square, photo scale is 1:5000, end lap is 60 percent, and side lap is 30 percent?

18-25. Repeat Prob. 18-24, except that map scale is 1:100,000 and photo scale is 1:24,000.

18-26. A rectangular project area 3 km in the north-south direction and 4 km in the east-west direction is to be photographed at a scale of 1:3000. End lap and side lap are to be 60 and 30 percent, respectively, and the camera format is 23 cm square. Compute the total number of photographs needed to cover this area, assuming that flight lines will run east-west and that the first and last lines will be flown so that the adjusted side lap will extend outside the project boundaries. Add two photos at the ends of each strip to ensure complete coverage. Prepare a flight map showing the flight lines, assuming the base map has a scale of 1 : 24,000.

CHAPTER
19

TERRESTRIAL AND CLOSE-RANGE PHOTOGRAMMETRY

19-1 INTRODUCTION

Terrestrial photogrammetry is an important branch of the science of photogrammetry. It deals with photographs taken with cameras located on the surface of the earth. The cameras may be handheld, mounted on tripods, or suspended from towers or other specially designed mounts. The term *close-range photogrammetry* is generally used for terrestrial photographs having object distances up to about 300 m. With terrestrial photography the cameras are usually accessible, so that direct measurements can be made to obtain exposure station positions, similar to airborne GPS control with aerial photography. With some terrestrial cameras, angular orientation can also be measured or set to fixed values, so that all elements of exterior orientation of a terrestrial photo are commonly known and need not be calculated. These known exterior orientation parameters are a source of control for terrestrial photos, replacing in whole or in part the necessity for locating control points in the object space.

Terrestrial photography may be *static* (photos of stationary objects) or *dynamic* (photos of moving objects). For static photography, slow, fine-grained, high-resolution films may be used and the pictures taken with long exposure times. Stereopairs can be obtained by using a single camera and making exposures at both ends of a baseline. In taking dynamic terrestrial photos, fast films and rapid shutter speeds are necessary. If stereopairs of dynamic occurrences are required, two cameras located at the ends of a baseline must make simultaneous exposures.

19-2 APPLICATIONS OF TERRESTRIAL AND CLOSE-RANGE PHOTOGRAMMETRY

Historically the science of photogrammetry had its beginning with terrestrial photography, and topographic mapping was among its early applications. Terrestrial photos were found especially useful for mapping rugged terrain which was difficult to map by conventional field-surveying methods. Although it was known that topographic mapping could be done more conveniently using aerial photos, no practical method was available for taking aerial photographs until the airplane was invented. Following the invention of the airplane, emphasis in topographic mapping shifted from terrestrial to aerial methods. Terrestrial photogrammetry is still used in topographic mapping, but its application is usually limited to small areas and special situations such as deep gorges or rugged mountains that are difficult to map from aerial photography. Other topographic applications of terrestrial photogrammetry are in mapping construction sites, areas of excavation, borrow pits, material stockpiles, etc.

Through the years terrestrial photogrammetry has continued to gain prominence in numerous diversified nontopographic applications. Examples of nontopographic applications occur in such areas as aircraft manufacture, shipbuilding, telecommunications, robotics, forestry, archaeology, anthropology, architecture, geology, engineering, mining, nuclear industry, criminology, oceanography, medicine, dentistry, and many more. In the field of medicine, X-ray photogrammetry has been utilized advantageously for measuring sizes and shapes of body parts, recording tumor growth, studying the development of fetuses, locating foreign objects within the body, etc.

Figure 19-1 illustrates an application of close-range photogrammetry to rocket design. In this figure, close-range photographs are being taken of a hemispherical fuel tank

FIGURE 19-1
Application of close-range photogrammetry to the determination of the dimensions of a rocket fuel tank. (*Courtesy Geodetic Services, Inc.*)

FIGURE 19-2
Application of close-range
photogrammetry for determining the size
and shape of a floodgate. Note the
artificial targets (white dots) placed on
the structure. (*Courtesy Geodetic
Services, Inc.*)

for the Ariane 5 rocket. Coordinates of points on the tank were computed to verify the
dimensional integrity of the rocket system. Figure 19-2 demonstrates the use of close-
range photogrammetry to determine the dimensions of a floodgate for a dry dock at a
shipyard. In this application, the size and shape of the floodgate's lower portion were determined so that a replacement could be fabricated with the correct dimensions. Figure
19-3 illustrates the use of close-range photogrammetry to determine the shape of a military vehicle. This information was used to create a CAD model of the vehicle which
would be used for radar and infrared recognition applications.

Terrestrial photogrammetry has been used to great advantage as a reliable means
of investigating traffic accidents. Photos that provide all information necessary to
reconstruct the accident may be rapidly obtained. Time-consuming sketches and ground
measurements, which all too often are erroneous, are not needed, and normal traffic flow

FIGURE 19-3
Application of close-range
photogrammetry to determine a CAD
model of a military vehicle. (*Courtesy
Geodetic Services, Inc.*)

can be restored more quickly. Terrestrial photogrammetry has been widely practiced in accident investigation for many years in several European countries.

Terrestrial photogrammetry has become a very useful tool in many areas of scientific and engineering research for several reasons. One reason is that it makes possible measurements of objects which are inaccessible for direct measurement. Also, measurements can be obtained without actually touching delicate objects. In some experiments, such as measurements of water waves and currents, physical contact during measurement would upset the experiment and render it inaccurate. Cameras which freeze the action at a particular instant of time make possible measurements of dynamic events such as deflections of beams under impact loads. Because of the many advantages and conveniences offered by terrestrial photogrammetry, its importance in the future seems assured.

19-3   TERRESTRIAL CAMERAS

A variety of cameras are used in terrestrial photography. All fall into one of two general classifications: *metric* or *nonmetric.* The term *metric camera,* as used here, includes those cameras manufactured expressly for photogrammetric applications. They have fiducial marks built into their focal planes, or have calibrated CCD arrays, which enable accurate recovery of their principal points. Metric cameras are stably constructed and completely calibrated before use. Their calibration values for focal length, principal-point coordinates, and lens distortions can be applied with confidence over long periods. In the case of analytical self-calibration (see Sec. 11-12), the calibration is performed in conjunction with each project in which the cameras are used.

The CRC-1 (close-range camera) of Fig. 19-4 is a terrestrial camera which is used primarily for industrial applications. This camera uses 240-mm by 40-m roll film which

FIGURE 19-4
The CRC-1 close-range camera. (*Courtesy Geodetic Services, Inc.*)

is held against an ultraflat platen by vacuum pressure. Four illuminated corner fiducials and a system of 25 rear-projected reseau marks are imaged on each photo to provide compensation for film deformation. The CRC-1 can use interchangeable lenses of 240- and 120-mm focal lengths. Accuracies surpassing 1 part in 200,000 of the object size are routinely achieved with this camera.

The INCA (*In*telligent Digital *Ca*mera) of Fig. 19-5 is a digital terrestrial camera which is also used primarily for industrial applications. This camera uses a CCD array having either a 2029 × 2044 or 3060 × 2036 pixel format, depending on the configuration, with each pixel having an 8-bit gray level. The camera's angular field of view can range from approximately 35° to 80° through use of interchangeable lenses. Images are recorded on a solid-state memory card which can hold up to 80 standard images (800 when image compression is used). Typical accuracies of 1 part in 100,000 of the object size are achieved with this camera.

Phototheodolites and *stereometric cameras* are two special types of terrestrial camera systems in the metric classification. A phototheodolite (see Fig. 1-1) is an instrument that incorporates a metric camera with a surveyor's theodolite. With this instrument, precise establishment of the direction of the optical axis can be made. A stereometric camera system consists of two identical metric cameras which are mounted at the ends of a bar of known length. The optical axes of the cameras are oriented perpendicular to the bar and parallel with each other. The length of the bar provides a known baseline length between the cameras, which is important for controlling scale.

Nonmetric cameras are manufactured for amateur or professional photography where pictorial quality is important but geometric accuracy requirements are generally not considered paramount. These cameras do not contain fiducial marks, but they can be modified to include them. Nonmetric cameras can be calibrated and used with satisfactory results for many terrestrial photogrammetric applications. Lack of film-flattening

FIGURE 19-5
The INCA digital close-range camera. (*Courtesy Geodetic Services, Inc.*)

mechanisms and less stable construction render nonmetric cameras less accurate than metric cameras. Examples of nonmetric cameras are shown in Figs. 2-11 and 3-4.

19-4 HORIZONTAL AND OBLIQUE TERRESTRIAL PHOTOS

Terrestrial photos may be classified as being either *horizontal* or *oblique,* a classification that stems from the orientation of the camera axis at the time of the photography. Horizontal terrestrial photos are obtained if the camera axis is horizontal when the exposure is made. In that orientation, the plane of the photo is vertical, and if the camera is of the metric type, the *x* and *y* photo axes (defined by the fiducial lines) would be horizontal and vertical lines respectively. If a camera is available that has leveling capabilities, horizontal terrestrial photos can be obtained. It is generally not practical to expose horizontal photos because it is usually necessary to incline the camera axis up or down somewhat in order to center the object of interest in the field of view. Furthermore, no significant benefits result from taking horizontal photos.

Oblique terrestrial photos, in which the camera axis is inclined either up or down at the time of exposure, represent the more general case in terrestrial photogrammetry. Figure 19-6 illustrates an oblique terrestrial photo taken with the camera axis inclined at an angle θ from horizontal. In this case the axis is inclined downward, and θ is called a *depression* angle. If the camera axis were inclined upward, θ would be an *elevation* angle. The angle of inclination of the camera axis is an important variable for certain elementary methods of determining object space positions of points whose images appear on overlapping terrestrial photos. With some terrestrial cameras the angle of inclination can be set or measured so that it becomes a known quantity. If it is unknown for a particular photo or photos, methods are available for computing its value.

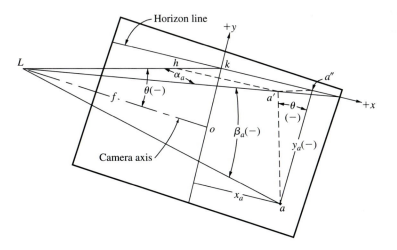

FIGURE 19-6
Horizontal and vertical angles from measurements on an oblique terrestrial photo.

The equations and methods of analyzing terrestrial photos for the balance of this chapter assume oblique terrestrial photos. These equations and procedures are general, however, and are therefore also applicable to the special case of horizontal photos.

19-5   DETERMINING THE ANGLE OF INCLINATION OF THE CAMERA AXIS

One elementary method of determining the angle of inclination of the camera axis of a terrestrial photo relies on the following two fundamental principles of *perspective geometry:* (1) horizontal parallel lines intersect at a vanishing point on the horizon and (2) vertical parallel lines intersect at the nadir (or zenith). If a photograph contains images of linear features which are horizontal or vertical, the horizon or nadir can be located through graphical construction. Figure 19-7 illustrates an oblique terrestrial photo of two buildings at a street corner. In this figure, dashed lines are extended from the tops and bottoms of the windows (horizontal parallel lines) to their intersections at vanishing points v_1 and v_2, which when connected define the horizon. Also shown are dashed lines extending from the vertical building edges to their intersection at the nadir point n. The line from n through the principal point o intersects the horizon at a right angle at point k.

A reference photo coordinate system can be established for the oblique photo of Fig. 19-7 from the preceding graphical construction. Figure 19-8a shows the x and y axes of the oblique photo coordinate system with its origin at point k. Note that the x axis coincides with the horizon. Figure 19-8b shows a profile view of the principal plane Lkn. The optical axis Lo is inclined downward at a depression angle θ. To conform with usual sign convention, depression angles are considered negative in algebraic sign. If the camera axis is inclined upward, θ is considered positive. Angle θ can be determined by either of two approaches. The first approach requires that the horizon be located by vanishing points. Then line ko is drawn at a right angle to the horizon through point o (the principal point of the photo) to define the y axis. Distance ko is then measured, and angle θ is computed from the following (see Fig. 19-8b):

$$\theta = \tan^{-1}\left(\frac{y_o}{f}\right) \tag{19-1}$$

In Eq. (19-1), θ is the depression (or elevation) angle, y_o is the y coordinate of the principal point in the oblique photo coordinate system, and f is the camera focal length. In this equation, the correct algebraic sign must be applied to y_o so that angle θ, in turn, will have the correct algebraic sign. It is also necessary to use an appropriate value for the focal length f. If the graphical analysis is being made on an enlarged photographic print, the focal length of the camera must be correspondingly enlarged.

A second approach to determining angle θ is based on the location of the nadir. After the nadir has been determined, distance on is measured and the tilt angle t computed by the following.

$$t = \tan^{-1}\left(\frac{on}{f}\right) \tag{19-2}$$

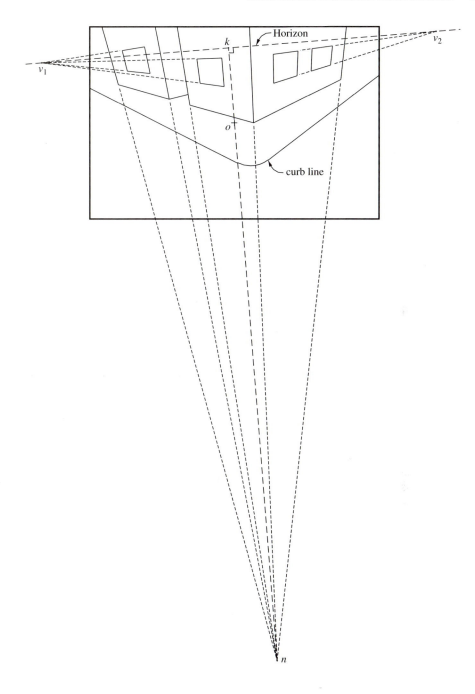

FIGURE 19-7
Location of horizon and nadir on an oblique photograph.

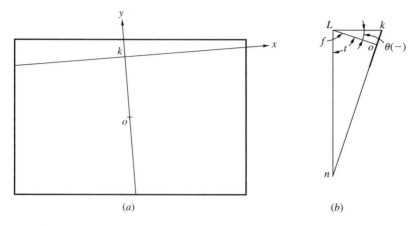

FIGURE 19-8
(*a*) Oblique photo coordinate axis system. (*b*) Side view of principal plane showing depression angle θ and tilt angle *t*.

Once the tilt angle has been determined, angle θ can be computed from

$$\theta = t - 90° \tag{19-3}$$

If a photograph is taken in which the optical axis is pointed upward from the horizon, angle θ is an elevation angle, and vertical parallel lines will intersect at the zenith. In this case, the distance *oz* from the principal point to the zenith point is measured, and angle θ can be computed by

$$\theta = 90° - \tan^{-1}\left(\frac{oz}{f}\right) \tag{19-4}$$

As discussed in Sec. 19-11, analytical methods can also be used for determining the angle of inclination of the camera axes for terrestrial photos, and this method provides the highest level of accuracy.

19-6   COMPUTING HORIZONTAL AND VERTICAL ANGLES FROM OBLIQUE TERRESTRIAL PHOTOS

Once angle θ has been determined, horizontal and vertical angles can be computed for points whose images appear in the photograph. Refer again to Fig. 19-6 which shows the geometry of an oblique terrestrial photograph. In this figure, *L* is the exposure station, and *Lo* is *f*, the camera focal length; *Lk* is a horizontal line intersecting the photo at *k*. The *x* axis coincides with the horizon line, and the *y* axis passes through the principal point *o*, perpendicular to the *x* axis at point *k*. Point *a* is an image point, and *aa'* is a vertical line, with *a'* being located in the horizontal plane of the exposure station and horizon line. Horizontal angle α_a between the vertical plane, (*La'a*), containing image point *a* and the vertical plane, *Lko*, containing the camera axis is

$$\alpha_a = \tan^{-1}\left(\frac{ha'}{Lk - hk}\right) = \tan^{-1}\left(\frac{x_a}{f \sec\theta - y_a \sin\theta}\right) \tag{19-5}$$

In Eq. (19-5) note that correct algebraic signs must be applied to x_a, y_a, and θ. Algebraic signs of α angles are positive if they are clockwise from the optical axis and negative if they are counterclockwise.

After the horizontal angle α_a has been determined, vertical angle β_a to image point a can be calculated from the following equation:

$$\beta_a = \tan^{-1}\left(\frac{aa'}{La'}\right) = \tan^{-1}\left[\frac{aa'}{(Lk - hk)\sec \alpha_a}\right] = \tan^{-1}\left[\frac{y_a \cos \theta}{(f \sec \theta - y_a \sin \theta)\sec \alpha_a}\right]$$

$$(19\text{-}6)$$

The algebraic signs of β angles are automatically obtained from the signs of the y coordinates used in Eq. (19-6).

> **Example 19-1.** An oblique terrestrial photo was exposed with the camera axis depressed at an angle $\theta = -14.3°$. The camera focal length was 60.00 mm. Compute the horizontal and vertical angles to an object point A whose image has photo coordinates $x_a = 27.4$ mm and $y_a = -46.2$ mm, measured with respect to the oblique photo coordinate axes.
>
> *Solution.* From Eq. (19-5),
>
> $$\alpha_a = \tan^{-1}\left[\frac{27.4}{60.00 \sec(-14.3°) - (-46.2)\sin(-14.3°)}\right] = 28.5°$$
>
> From Eq. (19-6),
>
> $$\beta_a = \tan^{-1}\left[\frac{-46.2\cos(-14.3°)}{[60.00\sec(-14.3°) - (-46.2)\sin(-14.3°)](\sec 28.5°)}\right] = -37.9°$$

19-7 EXPOSURE STATION LOCATION AND CAMERA AXIS DIRECTION

Sometimes the location of the exposure station and camera axis direction are unknown for a terrestrial photo and must be determined. A simple and convenient method for locating the horizontal position of the exposure station and the direction of the optical axis is *three-point resection*. It may be done either graphically or numerically; but to obtain the solution, angle θ must be known and images of at least three horizontal control points must appear in the photo. By using Eq. (19-5), horizontal angles between the camera axis and rays to the three control points are calculated. In the graphical three-point resection procedure, the three control points are plotted to scale on a base map according to their ground coordinates. A transparent template containing the three rays and the camera axis is prepared based on the calculated horizontal angles. The template is placed on the base map and adjusted in position and rotation until the three rays simultaneously pass through their respective plotted control points, as shown in Fig. 19-9. The exposure station position and direction of the optical axis are then marked on the map by pinpricking through the template. Numerical techniques of three-point resection are discussed in most textbooks on surveying and are not discussed here.

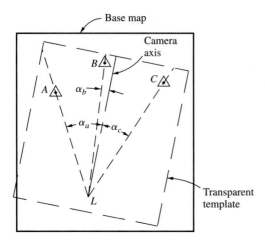

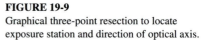

FIGURE 19-9

Graphical three-point resection to locate exposure station and direction of optical axis.

The elevation of the exposure station is the height of the camera lens above datum. If the elevation of the occupied station is known, the elevation of the camera lens is usually determined by measuring the vertical distance from the ground point up to the camera lens and adding it to the elevation of the ground point. If the exposure station elevation is unknown, it may be determined from a vertical control point, provided that the horizontal position of the camera and direction of its optical axis are known.

Assume that the position and elevation of point A in Fig. 19-10 are known. Vertical angle β_a to control point A is calculated from Eq. (19-6). With horizontal distance LA_h known, the camera station elevation may be calculated from

$$\text{Elev } L = \text{elev } A - LA_h \tan \beta_a \tag{19-7}$$

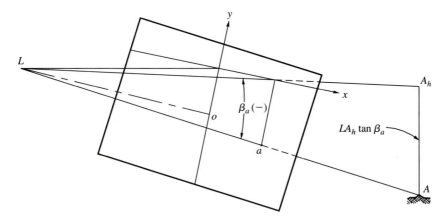

FIGURE 19-10

Determining elevation of camera station of terrestrial photo by using one vertical control point.

In Eq. (19-7), it is important to use the proper algebraic sign with the vertical angle β_a. If more than one control point is available, the average of exposure station elevations determined from all control points is adopted.

19-8 LOCATING POINTS BY INTERSECTION FROM TWO OR MORE OBLIQUE TERRESTRIAL PHOTOS

If images of an object point appear in two or more oblique photos, the position and elevation of the point can readily be determined, provided the camera positions and directions of the optical axes are known. Figure 19-11a illustrates two oblique photos

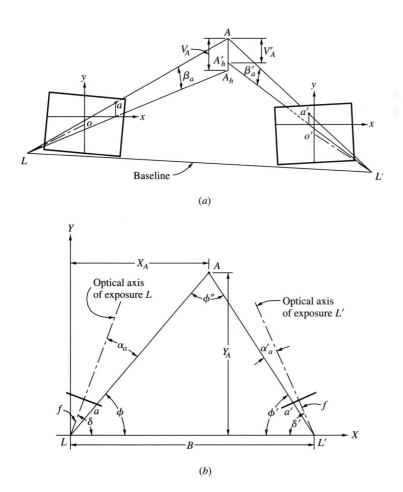

(a)

(b)

FIGURE 19-11
(a) Locating points by intersection from two oblique terrestrial photos. (b) Plan view of intersection from two oblique terrestrial photos.

taken from exposure stations L and L'. Figure 19-11b is a plan view of the situation. Images of object point A appear at a and a' on the two photos. Assume that angles δ and δ' of Fig. 19-11b have been measured with respect to the baseline, so that relative directions of the optical axes of the two exposure stations are known. (Alternatively, angles δ and δ' could have been determined by three-point resections of both photos, as discussed in Sec. 19-7.) Assume also that the horizontal length of the baseline has been measured and that the camera station elevations are known. An arbitrary XY object space coordinate system is adopted with its origin at exposure station L and the X axis in line with the baseline. It is required to determine the X and Y coordinates and elevation of point A.

This problem may be solved analytically or graphically. In the analytical solution, angles α_a, α_a', β_a, and β_a' are calculated from Eqs. (19-5) and (19-6). The angles ϕ, ϕ', and ϕ'' of Fig. 19-11b are calculated as follows:

$$\phi = \delta - \alpha_a$$
$$\phi' = \delta' + \alpha_a' \tag{19-8}$$
$$\phi'' = 180° - \phi - \phi'$$

By applying the law of sines, distance LA in Fig. 19-11b may be calculated as

$$LA = \frac{B \sin \phi'}{\sin \phi''} \tag{19-9}$$

Also, by the law of sines,

$$L'A = \frac{B \sin \phi}{\sin \phi''} \tag{19-10}$$

Coordinates X_A and Y_A may then be calculated as

$$X_A = LA \cos \phi$$
$$Y_A = LA \sin \phi \tag{19-11}$$

A check on these coordinates may be obtained as follows:

$$X_A = B - L'A \cos \phi'$$
$$Y_A = L'A \sin \phi' \tag{19-12}$$

With reference to Fig. 19-11a, the elevation of point A is determined as

$$\text{Elev } A = \text{elev } L + V_A \tag{19-13}$$

where $V_A = LA_h \tan \beta_a$, and LA_h is equal to horizontal length LA found in Eq. (19-9). A check may also be obtained on the elevation of A as follows:

$$\text{Elev } A = \text{elev } L' + V_A' \tag{19-14}$$

where $V_A' = L'A_h' \tan \beta_a'$, and $L'A_h'$ is equal to horizontal length $L'A$ found in Eq. (19-10).

If images of an object point appear on more than two photos, additional check values on position and elevation may be obtained and the averages of several solutions can be adopted.

Example 19-2. As illustrated in Figs. 19-11*a* and *b*, two oblique photos were taken with a terrestrial camera. From analysis of oblique photos, horizontal and vertical angles were determined as $\alpha_a = 15.6°$, $\alpha_a' = -6.2°$, $\beta_a = 13.5°$, and $\beta_a' = 16.2°$. Angles δ and δ' were determined from three-point resections as 69.5° and 66.2°, respectively. The horizontal baseline length was determined to be 75.0 m, and the exposure station elevations were 263.0 and 260.7 m above mean sea level for the left and right exposures, respectively. Calculate the *X* and *Y* coordinates of point *A* in a rectangular coordinate system with its origin at *L* and with the *X* axis in the plane of the baseline, as shown in Fig. 19-11*b*. Also compute the elevation of *A*.

Solution. By Eqs. (19-8), angles ϕ, ϕ', and ϕ'' are

$$\phi = 69.5° - 15.6° = 53.9°$$

$$\phi' = 66.2° - 6.2° = 60.0°$$

$$\phi'' = 180° - 53.9° - 60.0° = 66.1°$$

Horizontal distances *LA* and *L'A* by Eqs. (19-9) and (19-10) are

$$LA = \frac{75.0 \sin 60.0°}{\sin 66.1°} = 71.0 \text{ m}$$

$$L'A = \frac{75.0 \sin 53.9°}{\sin 66.1°} = 66.3 \text{ m}$$

The X_A and Y_A coordinates of point *A* by Eqs. (19-11) are

$$X_A = 71.0 \cos 53.9° = 41.9 \text{ m}$$

$$Y_A = 71.0 \sin 53.9° = 57.4 \text{ m}$$

The elevation of point *A* by Eq. (19-13) is

$$\text{Elev } A = 263.0 + 71.0 \tan 13.5° = 280.0 \text{ m above mean sea level}$$

By applying Eqs. (19-12) and (19-14), checks may be obtained for the X_A and Y_A coordinates and for the elevation of *A* as follows:

$$X_A = 75.0 - 66.3 \cos 60.0° = 41.9 \text{ m} \quad \text{(Check!)}$$

$$Y_A = 66.3 \sin 60.0° = 57.4 \text{ m} \quad \text{(Check!)}$$

$$\text{Elev } A = 260.7 + 66.3 \tan 16.2° = 280.0 \text{ m} \quad \text{(Check!)}$$

(Note that in general, the check on the elevation will not be exact because the values are obtained from independent measurements.)

19-9 CONTROL FOR TERRESTRIAL PHOTOGRAMMETRY

In terrestrial photogrammetry, the object space is often relatively close to the camera; in fact, in many cases the photographs are taken in laboratories. Object distances vary from a few centimeters up to 300 m or more, and the objects photographed vary in size from articles as small as human teeth and smaller, to very large buildings or ships. In any case, if accurate maps are to be made of photographed objects, control will be required.

In terrestrial photogrammetry there are basically four different methods of establishing control: (1) imposing the control on the camera by measuring its position and orientation with respect to a coordinate system or with respect to the photographed object, (2) locating control points in the object space in a manner similar to locating control for aerial photography, (3) combining camera control and object space control points, and (4) using a free-network adjustment with scale control only.

In the first method, no control points need appear in the object space. Rather, the position and orientation of the camera or cameras are measured with respect to the object itself. If a planar object is being photographed from a single camera station, control requirements may be satisfied by measuring the distance from the camera to the plane surface and orienting the camera optical axis perpendicular to the surface. Perpendicular orientation can be accomplished by mounting a plane-surfaced mirror parallel to the object plane and then moving the camera about until the reflection of the camera lens occupies the center of the field of view. If the camera focal length is known, a complete planimetric survey of the object can then be made.

If stereopairs of photos are taken, the control survey can consist of measuring the horizontal distance and difference in elevation between the two camera stations and also determining the orientations of the camera optical axis for each photo. Phototheodolites enable a complete determination of camera orientation and direction of optical axis. Stereometric cameras automatically provide control by virtue of their known baseline length and parallel optical axes. In exposing stereopairs with less elaborate cameras, horizontal orientation can be enforced by using level vials, and parallel orientation of the camera axes can be accomplished by reflection from parallel mirrors.

In the second method of controlling terrestrial photos, points should be selected in the object space which provide sharp and distinct images in favorable locations in the photographs. Their positions in the object space should then be carefully measured. If no satisfactory natural points can be found in the object space, artificial targets may be required. Targets should be designed so that their images appear sharp and distinct in the photos. White crosses on black cards may prove satisfactory. If the object space is small and the control points are close together, measurements for locating the targets may be made directly by means of graduated scales. If the object space is quite large or if the control points are inaccessible for direct measurement, triangulation with precise theodolites set up at the ends of a carefully measured baseline may be necessary. In some cases a premeasured grid pattern may be placed in the object space and photographed along with the object, thereby affording control.

If the object being photographed is stationary, control points may be located on the object. Corners of window frames, for example, may be used if a building is being photographed. If a dynamic event is being photographed at increments of time, for example, photographing beam deflections under various loads, then targets may have to be mounted on some stationary framework apart from the object. By means of surveyor's levels, targets may be set at equal elevations, thereby providing a horizontal line in the object space. Vertical lines may be easily established by hanging plumb bobs in the object space and attaching targets to the string.

The third method of controlling terrestrial photography is a combination of the first two methods. This third approach is generally regarded as prudent because it provides redundancy in the control, which prevents mistakes from going undetected and also enables increased accuracy to be obtained.

The fourth control method uses an arbitrary coordinate system, with the scale of the model being defined through one or more known lengths that appear in the object space. The known lengths may be based upon distance measurements made between target points on the object. In other cases, *scale bars* may be used. A scale bar is a long item such as a metal rod which has a known length. By placing one or more scale bars in proximity to the object, photo coordinate measurements can be made on the images of the ends of the bar, thus including the scale bar in the three-dimensional model of the object space. By constraining the distances between these endpoints to their known values, the scale of the overall object will be established. The arbitrary coordinate system can be established by setting the position and angular attitude of one of the exposures to some nominal set of values (say, $\omega = \phi = \kappa = X_L = Y_L = Z_L = 0$). The remaining exposures and object points will then be determined relative to this set of defined values. After all coordinates have been obtained, a three-dimensional conformal coordinate transformation (see Sec. C-7) can be used to relate the arbitrary system to any desired frame of reference.

19-10 PLANNING FOR CLOSE-RANGE PHOTOGRAMMETRY

In some cases, close-range photogrammetric analyses are done with existing amateur photography. For example, a police officer may have taken photographs of the scene of a vehicle accident, and subsequent photogrammetric analysis may be needed to determine the positions of the vehicles, tire marks, and other evidence. Such photography may not be ideally exposed, well focused, or generally in an optimal geometric configuration for photogrammetric analysis. On the other hand, under controlled situations (e.g., industrial applications), preliminary planning can be performed so as to control such factors as type of camera, lighting, and camera orientation.

Three main considerations for pictorial quality are resolution, depth of field, and exposure. Whether one is using digital or film cameras, resolution is important in that all points of interest must be clearly visible on the resulting image (see Secs. 3-13 and 3-14). A preliminary assessment should be made to ensure that the resolution is sufficient to adequately capture the smallest necessary details. Depth of field (see Sec. 2-3) is particularly important for ensuring proper focus. In close-range photogrammetry, since object depth is typically of significant size relative to the distance from the camera, a large f-stop setting may be required to ensure that the entire object is in focus. Proper exposure is necessary for the image points being measured to have sufficient contrast and definition. Particular attention should be paid to avoiding shadows on the object, especially when a flash is used for illumination. In some cases, special retro-reflective targets may be attached to points of interest prior to exposing the photographs. This allows the photographer to underexpose the background, with the targets remaining clear and bright.

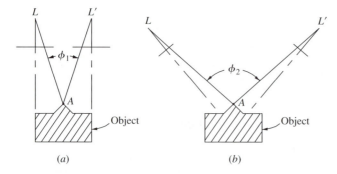

FIGURE 19-12
(*a*) Close-range stereo coverage of an object with parallel camera axes.
(*b*) Close-range stereo coverage with convergent photography.

There are also some physical constraints which must be considered in planning close-range photography. For example, object points must be visible from at least two (preferably more) photographs. A number of different camera locations may need to be considered in order to meet this constraint. Another constraint concerns the physical space around the object. In many applications, objects may be enclosed in confined spaces which makes effective determination of camera positions more difficult.

An important geometric consideration for close-range photography is the angular orientation of the camera exposure stations. Accuracy of the analytical solution depends, to a large extent, upon the angles of intersection between rays of light. The highest overall accuracy will be achieved when angles of intersection are near 90°. Figure 19-12*a* illustrates stereo photographic coverage of an object where the camera axes are parallel. In this figure, the parallactic angle ϕ_1 to object point A is approximately 35°. In Fig. 19-12*b*, stereo coverage of the same object is obtained from convergent photos. In this figure, the corresponding parallactic angle ϕ_2 is approximately 95°. Since ϕ_2 is closer to 90° than ϕ_1, the overall accuracy of the computed coordinates of point A will be higher in the configuration of Fig. 19-12*b*. Notice also that the stereoscopic coverage in Fig. 19-12*a* is only approximately 50 percent of the field of view, while, the stereoscopic coverage of Fig. 19-12*b* is 100 percent of the field of view. This enables the full format of the camera to be used, resulting in greater efficiency and higher effective image resolution.

19-11 ANALYTICAL SOLUTION

For the highest level of accuracy in close-range photogrammetry, a fully analytical solution is preferred. By applying the principles of analytical photogrammetry as described in Chap. 11 and Secs. 17-8 and 17-9, precisely measured photo coordinates of images can be used to directly compute X, Y, and Z coordinates in object space. The foundation of the analytical solution is the collinearity condition which gives rise to the collinearity equations [see Eqs. (11-1) and (11-2)]. These equations can be directly applied to terrestrial as well as aerial photographs.

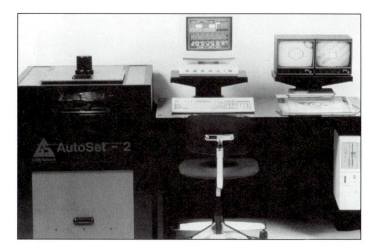

FIGURE 19-13
The AutoSet-2 automated comparator. (*Courtesy Geodetic Services, Inc.*)

In the preferred analytical method, the self-calibration approach (see Sec. 11-12) is used. This gives a calibration of the camera under the actual conditions (temperature, humidity, etc.) which existed when the photographs were taken. Certain geometric requirements must be met in order to effectively perform analytical self-calibration. First, numerous redundant photographs from multiple locations are required, with sufficient *roll diversity*. Roll diversity is a condition in which the photographs have angular attitudes that differ greatly from each other. Another requirement is that many well-distributed image points be measured over the entire format. This is important for accurate determination of lens distortion parameters.

Accurate measurement of photo coordinates is necessary to ensure accurate results from the analytical solution. High-precision comparators are generally used for film-based photographs. The AutoSet-2 system of Fig. 19-13, for example, is a precision comparator which can automatically measure targeted points to an accuracy of 0.5 μm. Digital camera systems, on the other hand, rely upon image-matching techniques (see Sec. 15-6) to obtain accurate photo coordinates. In any case, it is essential to properly identify object points as they appear on the different photos. Mislabeled points will result in an inaccurate analytical solution or, in some cases, will cause the solution to fail completely.

REFERENCES

American Society of Photogrammetry: *Manual of Photogrammetry*, 4th ed., Bethesda, MD, 1980, chap. 16.
———: *Handbook of Non-Topographic Photogrammetry,* 2d ed., Bethesda, MD, 1989.
Brown, D. C.: "New Developments in Photogeodesy," *Photogrammetric Engineering and Remote Sensing,* vol. 60, no. 7, 1994, p. 877.
———: "Close-Range Camera Calibration," *Photogrammetric Engineering,* vol. 37, no. 8, 1971, p. 855.

Fraser, C. S.: "Photogrammetric Measurement to One Part in a Million," *Photogrammetric Engineering and Remote Sensing,* vol. 58, no. 3, 1992, p. 305.

————: "Microwave Antenna Measurement," *Photogrammetric Engineering and Remote Sensing,* vol. 52, no. 10, 1986, p. 1627.

————: "Photogrammetric Measurement of Thermal Deformation of a Large Process Compressor," *Photogrammetric Engineering and Remote Sensing,* vol. 51, no. 10, 1985, p. 1569.

————: "Network Design Considerations for Non-Topographic Photogrammetry," *Photogrammetric Engineering and Remote Sensing,* vol. 50, no. 8, 1984, p. 1115.

————: "Optimization of Precision in Close-Range Photogrammetry," *Photogrammetric Engineering and Remote Sensing,* vol. 48, no. 4, 1982, p. 561.

Fryer, J. G., and D. C. Brown: "Lens Distortion for Close-Range Photogrammetry," *Photogrammetric Engineering and Remote Sensing,* vol. 52, no. 1, 1986, p. 51.

Ghosh, S. K.: "Photogrammetry for Police Use: Experience in Japan," *Photogrammetric Engineering and Remote Sensing,* vol. 46, no. 3, 1980, p. 329.

Karras, G. E., and E. Petsa: "DEM Matching and Detection of Deformation in Close-Range Photogrammetry without Control," *Photogrammetric Engineering and Remote Sensing,* vol. 59, no. 9, 1993, p. 1419.

Kenefick, J. F., M. S. Gyer, and B. F. Harp: "Analytical Self-Calibration," *Photogrammetric Engineering,* vol. 38, no. 11, 1972, p. 1117.

Lichti, D. D., and M. A. Chapman: "Constrained FEM Self-Calibration," *Photogrammetric Engineering and Remote Sensing,* vol. 63, no. 9, 1997, p. 1111.

Okamoto, A.: "The Model Construction Problem Using the Collinearity Condition," *Photogrammetric Engineering and Remote Sensing,* vol. 50, no. 6, 1984, p. 705.

Wong, K. W.: "Machine Vision, Robot Vision, Computer Vision, and Close-Range Photogrammetry," *Photogrammetric Engineering and Remote Sensing,* vol. 58, no. 8, 1992, p. 1197.

————: "Mathematical Formulation and Digital Analysis in Close-Range Photogrammetry," *Photogrammetric Engineering and Remote Sensing,* vol. 41, no. 11, 1975, p. 1355.

PROBLEMS

19-1. Discuss some of the uses of terrestrial or close-range photogrammetry.

19-2. Describe the differences between metric and nonmetric terrestrial cameras.

19-3. A horizontal terrestrial photo ($\theta = 0°$) was exposed with a phototheodolite having a focal length of 60.00 mm. Find the horizontal angle *ALB* at the exposure station subtended by points *A* and *B* if corresponding images *a* and *b* have photo coordinates of $x_a = 32.45$ mm, $y_a = -17.69$ mm, $x_b = -22.24$ mm, and $y_b = 29.73$ mm.

19-4. For the data of Prob. 19-3, calculate the vertical angles from the exposure station to points *A* and *B*.

19-5. Repeat Prob. 19-3, except that the camera focal length is 190.04 mm and the measured photo coordinates are $x_a = -79.28$ mm, $y_a = 39.84$ mm, $x_b = 45.00$ mm, and $y_b = 21.92$ mm.

19-6. Calculate the vertical angles for points *A* and *B* of Prob. 19-5.

19-7. A stereopair of oblique terrestrial photos was exposed as illustrated in Fig. 19-11. The camera had a 164.96-mm focal length, and the camera axis was inclined upward from horizontal at an angle of 6.30° for both photos. Horizontal angles δ and δ' measured from the baseline were 82.42° and 76.70°, respectively. The horizontal length of the baseline was 26.13 m, and the elevations of camera stations *L* and *L'* were 31.98 and 31.57 m, respectively. Calculate *X* and *Y* ground coordinates of point *A* if oblique photo coordinates of image *a* on the left and right photos were $x_a = -1.61$ mm, $y_a = 44.38$ mm, $x_a' = -63.57$ mm, and $y_a' = 41.45$ mm. Assume the origin of ground coordinates to be at camera station *L* and that the *X* axis coincides with the baseline, as illustrated in Fig. 19-11*b*.

19-8. Calculate the elevation of point *A* of Prob. 19-7.

19-9. On the overlapping pair of terrestrial photos of Prob. 19-7, a second point *B* has oblique photo coordinates on the left and right photos of $x_b = 63.42$ mm, $y_b = 40.72$ mm, $x_b' = 2.78$ mm, and $y_b' = 44.58$ mm. Calculate the horizontal length of line *AB*.

19-10. An oblique terrestrial photo was exposed with the camera axis depressed at an angle of 12.60°. The camera focal length was 194.95 mm. Calculate the horizontal and vertical angles between the rays from the camera station to object points A and B if their images have oblique photo coordinates of $x_a = -85.72$ mm, $y_a = -24.18$ mm, $x_b = 51.88$ mm, and $y_b = -89.85$ mm.

19-11. Repeat Prob. 19-10, except that the depression angle was 10.50°, the camera focal length was 90.01 mm, and the oblique photo coordinates were $x_a = 1.70$ mm, $y_a = -56.63$ mm, $x_b = 57.22$ mm, and $y_b = 24.18$ mm.

19-12. Discuss four basic approaches in establishing control for terrestrial or close-range photogrammetry.

19-13. Name and discuss three considerations which affect pictorial quality that should be considered in planning close-range photography.

19-14. Discuss how accuracy can be improved through the use on convergent photos versus stereo photographs with parallel camera axes.

19-15. Describe two geometric requirements for photography when analytical self-calibration is used for computing object coordinates in close-range photogrammetry.

CHAPTER
20

INTRODUCTION TO GEOGRAPHIC INFORMATION SYSTEMS[1]

20-1 INTRODUCTION

A *geographic information system* (GIS) can be briefly defined as an assemblage of hardware, software, data, and organizational structure for collecting, storing, manipulating, and analyzing "spatially referenced" data. In a more detailed definition,[2] a GIS is described as

> "any information management system which can: 1. Collect, store, and retrieve information based on its spatial location; 2. Identify locations within a targeted environment which meet specific criteria; 3. Explore relationships among data sets within that environment; 4. Analyze the related data spatially as an aid to making decisions about that environment; 5. Facilitate selecting and passing data to application-specific analytical models capable of assessing the impact of alternatives on the chosen environment; and 6. Display the selected environment both graphically and numerically either before or after analysis."

Other definitions can be found, but the important element which is common to all is that in a GIS, decisions are made based on spatial analyses performed on information

---

[1] Adapted from the 9th edition of *Elementary Surveying* by Paul R. Wolf and Russell C. Brinker, with permission from Prentice-Hall.

[2] F. Hanigan, "GIS by Any Other Name Is Still . . . ," *The GIS Forum,* vol. 1, 1988, p. 6.

that is referenced in a common geographical system. Any of the object space coordinate systems described in Chap. 5 could be employed as the geographic reference system, but the state plane and UTM coordinate systems described in Sec. 5-6 are most often used. In any GIS, the accuracy of the spatial analyses and hence the validity of decisions reached as a result of those analyses depend directly on the quality of the spatially related information in its database. As discussed in Sec. 20-2, photogrammetry is used extensively in collecting the data and information within those databases, and thus it plays a critical role in GIS activity.

The term *geographic information system* is a relatively new one, first appearing in published literature circa 1970. But although the term is new, many of its concepts have been in existence for centuries. The *map overlay* concept, for example, which is one of the important tools used in GIS spatial analysis, was used by French cartographer Louis-Alexandre Berthier during the American Revolution. He prepared a series of maps and overlaid them to analyze changes in troop positions over time. Another example that illustrates an early use of the overlay concept occurred in 1854, during the great cholera epidemic. To demonstrate the relationship between two important variables, Dr. John Snow overlaid a map of London showing locations where cholera deaths had occurred with another map that positioned the wells in that city. These early examples illustrate fundamentals that still comprise the basis of our modern GISs, that is, making decisions based on the simultaneous analysis of data of differing types that are all spatially located in a common geographic frame of reference. The full capabilities and benefits of our modern GISs could not be realized, of course, until the arrival of the computer age.

A generalized concept of how data of different types, also called *layers,* are collected and overlaid in a GIS is illustrated in Fig. 20-1. In that figure, maps *A* through *G* represent some of the different layers of spatially related information that can be digitally recorded and incorporated into a GIS. These include (*A*) parcels of different land ownership, (*B*) zoning, (*C*) floodplains, (*D*) wetlands, (*E*) land cover, and (*F*) soil types. Map *G* is the geodetic reference framework which consists of the network of survey control points that exist in the area. Note that these control points are included within each of the other layers, thereby providing the means for spatially locating all the data in a common reference system. This enables information from two or more different layers to be integrated. Also, maps can be created from the merged data and overlaid if desired. In Fig. 20-1, for example, map *H* is the composite of all layers.

A GIS incorporates software for manipulating spatial data with database management software. This combination enables the simultaneous storage, retrieval, overlay, and display of many different spatially related data sets in the manner illustrated by Fig. 20-1. Furthermore, sophisticated GIS software has been developed which can analyze and query the data sets that result from these different overlays and display combinations, a procedure which provides answers to questions that have previously been impossible to obtain. As a result, GISs have become extremely important in many areas of activity, including management, planning, design, impact assessment, regulation and enforcement, predictive modeling, and many others.

The importance and value of geographic information systems can perhaps best be illustrated by discussing a specific application. Consider, for example, a GIS for flood forecasting on a large regional basis, such as within a major river basin. Important

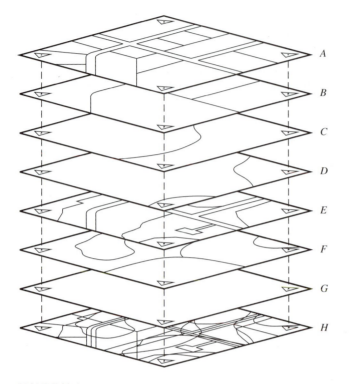

FIGURE 20-1
Concept of layers in a geographic information system. Map layers are
(*A*) parcels; (*B*) zoning; (*C*) floodplains; (*D*) wetlands; (*E*) land cover;
(*F*) soils; (*G*) reference framework; and (*H*) composite overlay.

spatially related information needed for this GIS database would include the basin's
topography; soil types; land cover; existing stream network within the basin, together
with the stream-gauging records of each; locations and sizes of existing bridges, culverts, and other drainage structures; data on existing dams and the water impoundment
capacities of their associated reservoirs; and records of past rainfall intensity and duration. Given these and other data sets, together with a model to estimate runoff, the GIS
software could perform a variety of analyses under varied conditions. Assessing the
magnitudes of floods and their associated danger created by rainfalls of high intensity
over long durations in specific areas of the basin is one type of analysis that could be
performed. If the database included information on population distribution; property
ownership; names, addresses, and telephone numbers of residents; dwelling locations;
and the transportation network in the basin; then the GIS could identify any residents
who might be endangered by a particular flooding condition. Furthermore, locations and
heights of emergency levees could be specified, and if necessary, endangered areas
could be evacuated. And should evacuation be required, the GIS could even identify the
best and safest escape routes. Another type of analysis could involve research on the

mitigation of flooding and damage resulting from the construction of permanent dams and other flood detention structures of varying sizes at varying locations.

Many other examples could be given to illustrate the potential values and benefits of geographic information systems. They are very powerful tools, and their use will increase significantly in the future.

20-2 THE IMPORTANCE OF PHOTOGRAMMETRY IN GENERATING GIS DATABASES

Data that are entered into a GIS can come from many different sources and may be of varying quality. To support a given GIS, some new information may be gathered specifically for its database. More than likely, however, part of the data can be obtained from existing sources such as maps, engineering plans, aerial photos, satellite images, and other documents and files which were produced for other purposes.

Photogrammetry plays a very important role in the collection of information for most GIS databases. This is true whether the data are being newly acquired or are being compiled from existing sources. Photogrammetric products frequently used as sources of information for GIS databases are of three types: *images, maps,* and *digital data files.* Products in the image category include aerial photos, photomaps, aerial mosaics, orthophotos, and satellite images. By using instruments and procedures described in Sec. 20-6, entire images or selected data from images can be digitized and entered into the database. Images are particularly useful as map substitutes, and they can be employed as convenient background reference frameworks for certain GIS applications.

Maps are scaled diagrams consisting of lines and symbols. They can be prepared photogrammetrically by using the elementary techniques described in Chap. 9, or they can be compiled by using stereoscopic plotting instruments, as discussed in Chap. 13. These graphic products can also be digitized for database entry using methods described in Sec. 20-6.

Photogrammetric digital data files consist of digital elevation models (DEMs), digital line graphs (DLGs), cross sections and profiles, and other digital files. DEMs are digital records of terrain elevations which are commonly given at regularly spaced horizontal intervals. They contain only elevation information and give no data on the locations of planimetric features. DLGs are digital representations of the points, lines, and areas that define natural and cultural features. They are limited to planimetric representation only, however, and give no elevation information. Profiles are a record of elevations taken at specified locations along a reference line, and cross sections are elevations taken transverse to a reference line at specified intervals. Other digital files may also be recorded photogrammetrically. These may include comprehensive topographic coverage (a digital record of all planimetric and elevation information) of an area, or they may include only one or more specific types or layers of information. These digital files will contain not only the positional information for each point or feature (X, Y, Z coordinates in the reference system), but also will include one or more associated codes that identify the type of point or feature, and give certain other attribute information as well.

Building the database is one of the most expensive and challenging aspects of developing a GIS. In fact, it has been estimated that this aspect may represent from about 60 to 80 percent of the total cost of implementing a GIS.

20-3 SPATIAL DATA

Two basic data classifications are used in GISs: *spatial* and *nonspatial.* Spatial data, sometimes interchangeably called *graphical data,* consist in general of natural and cultural features that can be shown with lines or symbols on maps, or that can be seen as images on photographs. In a GIS these data must be represented, and spatially located, in digital form, by using a combination of fundamental elements called *simple spatial objects.* The formats used in this representation are either *vector* or *raster.* The *relative spatial relationships* of the simple spatial objects are given by their *topology.*

Simple spatial objects, formats, and topology, are described in the following subsections.

20-3.1 Simple Spatial Objects

The simple spatial objects most commonly used in locating data in a frame of reference are illustrated in Fig. 20-2 and described as follows:

1. *Points* define single geometric positions. They are used to locate features such as houses, wells, or mines (see Fig. 20-2a). Spatial locations of points are given by their coordinates, commonly in state plane or UTM systems (see Chap. 5).
2. *Lines and strings* are obtained by connecting points. A line connects two points, and a string is a sequence of two or more connected lines. Lines and strings are used to represent and locate property lines, roads, streams, fences, etc. (see Fig. 20-2b).
3. *Interior areas* consist of the continuous space within three or more connected lines or strings that form a closed loop (see Fig. 20-2c). Interior areas are used to

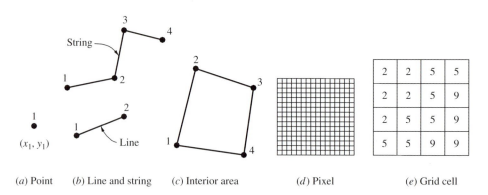

(a) Point   (b) Line and string   (c) Interior area   (d) Pixel   (e) Grid cell

FIGURE 20-2
Simple spatial objects used to represent data digitally in a GIS.

represent and locate, for example, the limits of government jurisdictions, parcels of land ownership, different types of land cover, or large buildings.

4. *Pixels,* as discussed in Sec. 2-13, are usually tiny squares that represent the smallest elements into which a digital image is divided (see Fig. 20-2*d*). Continuous arrays of pixels, arranged in rows and columns, are used to enter data from aerial photos, satellite images, orthophotos, etc. The distributions of colors or tones throughout the image are specified by assigning a numerical value to each pixel. Pixel size can be varied and is specified either in terms of image or object scale. At image scale, pixel size may be specified directly (e.g., 0.025 × 0.025 mm) or as a number of pixels per unit distance (e.g., 100 dots per inch, where a dot corresponds to a pixel). At object scale, pixel size is usually expressed directly by dimension (e.g., 10 m pixel size).

5. *Grid cells* are single elements, usually square, within a continuous geographic variable. Similarly to pixels, their sizes can be varied, with smaller cells yielding improved resolution. Grid cells may be used to represent terrain slopes, soil types, land cover, water table depths, land values, population density, etc. The distribution of a given data type within an area is indicated by assigning a numerical value to each cell, for example, showing soil types in an area using the number 2 to represent sand, 5 for loam, and 9 for clay, as illustrated in Fig. 20-2*e*.

20-3.2   Vector and Raster Formats

The simple spatial objects described in Sec. 20-3.1 give rise to two different formats, *vector* and *raster,* for storing and manipulating spatial data in a GIS. When data are depicted in the vector format, a combination of points, lines, strings, and interior areas is used. The raster format uses pixels and grid cells.

In the vector format, points are used to specify locations of objects such as survey control monuments, utility poles, and manholes; lines and strings depict linear features such as roads, transmission lines, and boundaries; and interior areas show regions having common characteristics, for example, areas of uniform land cover or common ownership. An example illustrating the vector format is given with Fig. 20-3 and Table 20-1.

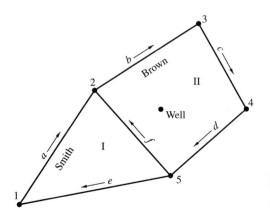

FIGURE 20-3
Vector representation of a simple graphical record.

TABLE 20-1
Vector representation of Fig. 20-3

(a)		(b)		(c)	
Point identifier	Coordinates	Line identifier	Points	Area identifier	Lines
1	$(X, Y)_1$	a	1, 2	I	a, f, e
2	$(X, Y)_2$	b	2, 3	II	b, c, d, f
3	$(X, Y)_3$	c	3, 4		
4	$(X, Y)_4$	d	4, 5		
5	$(X, Y)_5$	e	5, 1		
Well	$(X, Y)_{well}$	f	5, 2		

Figure 20-3 shows two adjacent land parcels, one designated parcel I, owned by Smith, and the other identified as parcel II, owned by Brown. As shown, the configuration consists of points, lines, and areas.

Vector representation of the data can be achieved by creating a set of tables that list these points, lines, and areas (see Table 20-1). Data within the tables are linked by using *identifiers* and are related spatially through the coordinates of points. As illustrated in column (*a*) of Table 20-1, all points in the area are indicated by an identifier. Similarly each line is described by its endpoints, as shown in column (*b*) of Table 20-1, and the endpoint coordinates locate the various lines spatially. Areas in Fig. 20-3 are defined by the lines that enclose them, as shown in column (*c*) of Table 20-1. As before, coordinates of endpoints locate the areas and enable the determination of their locations and magnitudes.

Other data types can also be represented in vector formats. Consider, for example, the simple case illustrated with the land cover map shown in Fig. 20-4*a*. In that figure, areas of different land cover (forest, marsh, etc.) are shown with standard topographic symbols. A vector representation of this region is shown in Fig. 20-4*b*. Here lines and strings locate boundaries of regions having a common type of land cover. Note that the stream consists of the string connecting points 1 through 11. By means of tables similar to Table 20-1 the data of this figure can be entered into a GIS using the vector format.

As an alternative to the vector approach, data can be depicted in the raster format using pixels or grid cells. Each equal-size pixel or cell is uniquely located by its row and column numbers, and it is labeled with a numerical value or *code* that corresponds to the properties of the specific area it covers. In the raster format, a point would be indicated with a single grid cell, a line would be depicted as a sequence of adjacent grid cells having the same code, and an area having common properties would be shown as a group of identically coded contiguous cells. It should be readily apparent, therefore, that in general the raster method yields a coarser level of accuracy or definition of points, lines, and areas than the vector method.

In the raster format, the size selected for the individual pixels or cells defines the *resolution*, or precision with which data are represented. The smaller the elements, the higher

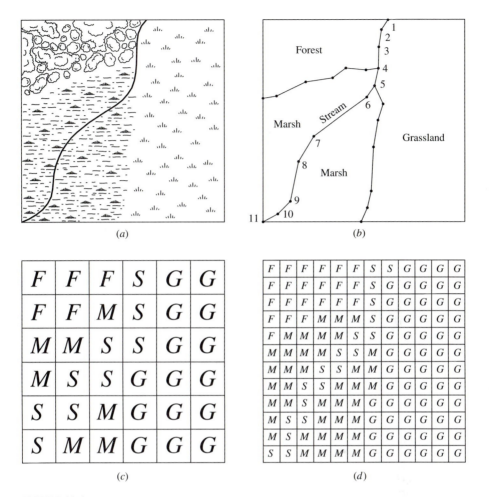

FIGURE 20-4
Land cover maps of a region. (*a*) Region using standard topographic symbols. (*b*) Vector representation of same region. (*c*) Raster representation of region using a coarse-resolution grid cell. (*d*) Raster representation using a finer-resolution grid cell.

the resolution. Examples illustrating raster representation, and the degradation of resolution with increasing grid size, are given in Figs. 20-4*c* and *d*. In these figures the land cover data from Fig. 20-4*a* have been entered as two different raster data sets. Each cell has been assigned a value representing one of the land cover classes, that is, *F* for forest, *G* for grassland, *M* for marsh, and *S* for stream. Figure 20-4*c* depicts the area with a relatively large-resolution grid; and, as shown, it yields a coarse representation of the original points, lines, and areas. With a finer-resolution grid, such as in Fig. 20-4*d*, the points, lines, and areas are rendered with greater precision. Note, however, that as grid resolution increases, so does the volume of data (number of grid cells) required to enter the data.

Despite the fact that raster formats yield coarser resolution of spatial features than the vector format, the raster format is still often used in GISs. One reason is that many data are available in raster format. Examples include digital aerial photos, orthophotos, and satellite images. Another reason for the popularity of the raster format is the ease with which it enables collection, storage, and manipulation of data using computers. Furthermore, various refinements of raster images are readily made using available image-processing software (see Chap. 14). Finally, for many data sets such as wetlands and soil types, the boundary locations are rather vague and the use of the raster format does not significantly degrade the inherent accuracy of the data.

20-3.3   Topology

Topology is a branch of mathematics which can be used for describing how spatial objects are related to one another. The unique sizes, dimensions, and shapes of the individual objects are not addressed by topology. Rather, only their *relative relationships* are specified.

In discussing topology, it is necessary to first define *nodes, chains,* and *polygons.* These are some additional simple, spatial objects which are commonly used for specifying the topological relationships of information entered into GIS databases. Nodes define the beginnings and endings of chains, or identify the junctions of intersecting chains. Chains are similar to lines (or strings) and are used to define the limits of certain areas or delineate specific boundaries. Polygons are closed loops similar to areas and are defined by a series of connected chains. In topology, sometimes single nodes exist within polygons for labeling purposes.

In GISs the most important topological relationships are

1. *Connectivity*—identifying which chains are connected at which nodes.
2. *Direction*—defining a "from node" and a "to node" of a chain.
3. *Adjacency*—indicating which polygons are on the left and which are on the right side of a chain.
4. *Nestedness*—specifying what simple spatial objects are within a polygon. They could be nodes, chains, or other smaller polygons.

The topological relationships just described are illustrated and described by example with reference to Fig. 20-3. In the figure, for example, through connectivity, it is established that nodes 2 and 3 are connected to form the chain labeled *b*. Connectivity would also indicate that at node 2, chains *a*, *b*, and *f* are connected. Topological relationships are normally listed in tables and stored within the database of a GIS. Part (*a*) of Table 20-2 summarizes all the connectivity relationships of Fig. 20-3.

Directions of chains are also indicated topologically in Fig. 20-3. For example, chain *b* proceeds from node 2 to node 3. Directions can be very important in a GIS for establishing such things as which way a river flows, or the direction traffic moves on one-way streets. Often a consistent direction convention is followed within a GIS, for example, always proceeding clockwise around polygons. Part (*b*) of Table 20-2 summarizes the directions of all chains within Fig. 20-3.

TABLE 20-2
Topological relationships of elements in the graphical record of Fig. 20-3

(a) Connectivity		(b) Direction			(c) Adjacency			(d) Nestedness	
Nodes	Chain	Chain	From node	To node	Chain	Left polygon	Right polygon	Polygon	Nested node
1-2	a	a	1	2	a	0	I	II	Well
2-3	b	b	2	3	b	0	II		
3-4	c	c	3	4	c	0	II		
4-5	d	d	4	5	d	0	II		
5-1	e	e	5	1	e	0	I		
2-5	f	f	5	2	f	I	II		

The topology of Fig. 20-3 would also describe, through adjacency, that Smith and Brown share a common boundary, which is chain f from node 5 to node 2, and that Smith is on the left side of the chain and Brown is on the right. Obviously the chain's direction must be stated before left or right positions can be declared. Part (c) of Table 20-2 lists the adjacency relationships of Fig. 20-3. Note that a zero has been used in the table to designate regions outside the polygons and beyond the area of interest.

Nestedness establishes that the well of Fig. 20-3 is contained within Brown's polygon. Part (d) of Table 20-2 lists that topological information.

The relationships expressed through the identifiers for points, lines, and areas of Table 20-1, and the topology in Table 20-2, conceptually yield a "map" (in the eyes through which a computer sees)! With these types of information available to the computer, the analysis and query processes of a GIS are made possible.

20-4 NONSPATIAL DATA

Nonspatial data, also often called *attribute data,* describe geographic regions or define characteristics of spatial features within geographic regions. Nonspatial data are usually alphanumeric and provide information such as color, texture, quantity, quality, and value of features. Smith and Brown, as the property owners of parcels I and II of Fig. 20-3, and the land cover classifications of forest, marsh, grassland, and stream in Fig. 20-4 are examples of attribute data. Other examples could include the addresses of the owners of the land parcels, their types of zoning, the dates the properties were purchased, and their assessed values; or data regarding a particular highway, including its route number, pavement type, number of lanes, lane widths, and year of last resurfacing. Nonspatial data are often derived from documents such as plans, reports, files, and tables.

In general, spatial data will have related nonspatial attributes, and thus some form of linkage must be established between these two different types of information. Usually this is achieved with *a common identifier* which is stored with both the graphical and the nongraphical data. Identifiers such as a unique parcel identification number, a grid cell label, or the specific mile point along a particular highway may be used.

20-5 DATA FORMAT CONVERSIONS

In manipulating information within a GIS database, it is often necessary to either integrate vector and raster data, or convert from one form to the other. Integration of the two types of data, i.e., using both types simultaneously, has become possible due to recent hardware and software improvements. It is usually accomplished by displaying vector data overlaid on a raster image background, as illustrated in Fig. 20-5. In that figure, vector data representing the property boundaries that exist within the different subdivisions (areas) are overlaid on an orthophoto of the same area. This graphic was developed as part of a population growth and distribution study being conducted for a municipality. The combination of vector and raster data is useful to provide a frame of reference and to assist GIS operators in interpreting displayed data.

Sometimes it is necessary or desirable to convert raster data to vector format or vice versa. Procedures for accomplishing these conversions are described in subsections that follow.

20-5.1 Vector-to-Raster Conversion

Vector-to-raster conversion is also known as *coding* and can be accomplished in several ways, three of which are illustrated in Fig. 20-6. Figure 20-6*a* is an overlay of the vector representation of Fig. 20-4*b* with a coarse raster of grid cells. In one conversion

FIGURE 20-5
Vector data overlaid on a raster image background. (*Courtesy University of Florida.*)

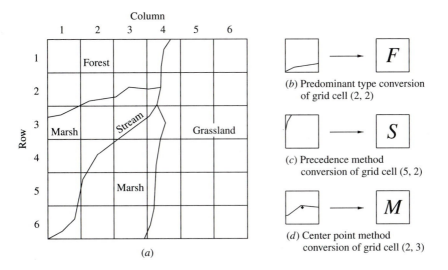

(a)

FIGURE 20-6
Methods for converting data from vector to raster format. (*a*) Vector representation of
Fig. 20-4*b* overlaid on a coarse grid. (*b*), (*c*), (*d*) Vector-to-raster conversion by predominant
type, precedence method, and center-point method, respectively.

method, called *predominant type coding,* each grid cell is assigned the value corresponding to the predominant characteristic of the area it covers. For example, the cell located in row 2, column 2 of Fig. 20-6*a* overlaps two polygons, one of forest (type *F*) and
one of marsh (type *M*). As shown in Fig. 20-6*b*, since the largest portion of this cell lies
in forest, the cell is assigned the value *F*—the predominant type.

In another coding method, called *precedence coding,* each category in the vector
data is ranked according to its importance or "precedence" with respect to the other categories. In other words, each cell is assigned the value of the highest-ranked category
present in the corresponding area of the vector data. A common example involves water.
While a stream channel may cover only a small portion of a cell area, it is arguably the
most important feature in that area. Also, it is important to avoid breaking up the stream.
Thus for the cell in row 5, column 2 of Fig. 20-6*a*, which is illustrated in Fig. 20-6*c*,
water would be given the highest precedence, and the cell coded *S*, even though most of
the cell is covered by marsh.

Center-point coding is a third technique for converting from vector to raster data.
Here a cell is simply assigned the category value at the vector location corresponding to
its center point. An example is shown in Fig. 20-6*d*, which represents the cell in row 2,
column 3 of Fig. 20-6*a*. Here, since marsh exists at the cell's center, the entire cell is
designated as category *M*. Note that the grid cell of row 3, column 4 would be classified
by predominant type as grassland, by precedence as stream, and by center point as
marsh. This illustrates how different conversion processes can yield different classifications for the same data.

The precision of these vector-to-raster conversions depends on the size of the grid used. Obviously, using a raster of large cells would result in a relatively inaccurate representation of the original vector data. On the other hand, a fine-resolution grid can very closely represent the vector data, but would require a large amount of computer memory. Thus the choice of grid resolution becomes a tradeoff between computing efficiency and spatial precision.

20-5.2 Raster-to-Vector Conversion

Raster-to-vector conversions are more vaguely defined than vector-to-raster ones. The procedure involves extracting lines from raster data which represent linear features such as roads, streams, or boundaries of common data types. Whereas the approach is basically a simple one, and consists of identifying the pixels or cells through which vector lines pass, the resulting jagged or "staircase" type of outline is not indicative of the true lines. One raster-to-vector conversion example is illustrated in Fig. 20-7a, which shows the cells of Fig. 20-4d identifying the stream. Once these cells have been selected, a problem that remains concerns how to fit a line to these jagged forms. One solution consists in simply connecting adjacent cell centers (see Fig. 20-7b) with line segments. Note, however, that the resulting line (see Fig. 20-7c) does not agree very well with the original stream line of Fig. 20-4a. This example illustrates that some type of line smoothing is usually necessary to properly represent the gently curved boundaries that normally occur in nature. However, fitting smooth lines to the jagged cell boundaries is a complicated mathematical problem which does not necessarily have a unique solution. The decision ultimately becomes a choice between accuracy of representation and cost of computation.

No matter which conversion is performed, errors are introduced during the process, and some information from the original data is lost. Use of smaller grid cells improves the results. Nevertheless, as illustrated by the example of Fig. 20-7, if a data set is converted from vector to raster and then back to vector (or vice versa), the final data set will not likely match the original. Thus it is important for GIS operators to be

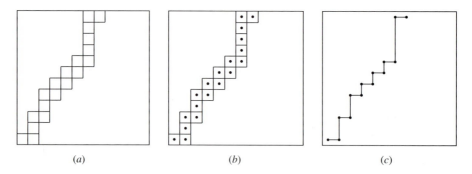

(a)   (b)   (c)

FIGURE 20-7
Conversion of data from raster to vector format. (a) Cells of Fig. 20-4d that identify the stream line.
(b) Cell centers. (c) Vector representation of stream line recreated by connecting adjacent cell centers.

aware of how their data have been manipulated, and what can be expected if conversion is performed.

20-6 CREATING GIS DATABASES

Before the database for a specific GIS is developed, several important factors should be considered and some critical decisions made. Included are decisions about the types of data that need to be obtained, the optimum formats for gathering these data, the reference coordinate system that will be most convenient and practical for spatially relating all data, and the necessary accuracy of each data type. Considerations for future updating of the database should also be addressed. Once these decisions and others are made, it is necessary to locate sources for the required data. As noted earlier, it may be possible, depending upon circumstances, to utilize some data from existing sources. In that event, a significant cost savings can result. In many situations, however, it is necessary to collect new data aimed directly at meeting the needs of the specific GIS.

The subsections that follow describe some methods available for generating the computer files needed to construct a GIS database. Procedures for digitizing existing sources of information are presented, as well as those for collecting and digitizing new information.

20-6.1 Digitizing Existing Graphical Materials

If sources such as maps, orthophotos, construction plans, or other graphical documents already exist and will meet the needs of the GIS database, often these can be conveniently and economically converted to digital files using a tablet digitizer (see Fig. 4-6). Many GIS software packages provide programs that directly support this procedure. As described in Sec. 4-6, a tablet digitizer contains an electronic grid and an attached cursor. Movement of the cursor across the grid creates an electronic signal unique to the cursor's position. At any desired point, the coordinates can be relayed to the computer for storage by pressing the Record button on the cursor. Data identifiers or attribute codes can be associated with each point through the computer's keyboard, or by pressing additional buttons on the cursor.

The digitizing process begins by securing the source document to the tablet digitizer. If the document is a map, the next step is to register its reference coordinate system to that of the digitizer. This is accomplished by digitizing a series of reference (tick) marks on the map for which geographic coordinates, such as state plane or UTM, are precisely known. With both reference and digitizer coordinates known for these tick marks, a two-dimensional coordinate transformation (see App. C) can be computed. This determines the parameters of scale change, rotation, and translation necessary to transform digitized coordinates to the reference geographic coordinate system. As a result of this step, any map features can be digitized, whereupon their coordinates are automatically transformed to the selected system of reference and entered directly into the database.

Both planimetric features and contours can be digitized from topographic maps. Planimetric features are recorded by digitizing the individual points, lines, or areas that portray them. As described in Sec. 20-3.2, this process creates data in a vector format.

Elevation data can be recorded as a digital elevation model by digitizing critical points along contours. The computer can then derive *triangulated irregular networks* (TINs) from these data. As described in Sec. 13-6, a TIN model is a network of adjoining triangles which are constructed by connecting points in a nonuniform DEM. Rigorous criteria are followed in selecting the specific points used to form each triangle, and when properly selected, the resulting overall TIN model provides a three-dimensional representation of the terrain in digital form. Commonly, two characteristics of TIN model triangles are assumed: (1) Each triangular surface represents a plane, and (2) their sides are of constant slope. GIS software can query a TIN model to obtain various types of information including point elevations, profiles, cross sections, slopes, aspects (slope directions), and contours having any specified contour interval.

20-6.2   Digitizing from Aerial Photos with Stereoplotters

Information from aerial photographs can also be entered directly into a GIS database in digital form by using the elementary procedures described in Chap. 9, or by using an optical or mechanical stereoplotter with digital encoders, an analytical plotter, or a softcopy plotter (see Chap. 13). In the digitizing procedures described in Chap. 13, both planimetric features and elevations can be recorded, and very high accuracy can be achieved. The data are registered to the selected reference coordinate system and vertical datum through the process of *absolute orientation* of the plotting instrument. The digitizing procedure will vary somewhat depending upon the particular instrument used. In general, however, in recording planimetric features, an operator views the oriented stereomodel, points to objects of interest, enters any necessary feature identifiers or codes, and depresses a key or foot pedal to transfer the information to a file in an interfaced computer. To digitize elevations, a DEM can be read directly from the three-dimensional stereomodel and stored in a computer file. Then, as discussed in the preceding section, TIN models can be developed from the DEM.

For each data file included within a GIS database, information regarding its source, the instruments and procedures used to generate it, and its accuracy should be included so GIS users can assess its quality. This information is commonly called *metadata* (i.e., data about data). Resultant accuracies will depend mainly on the quality of the source materials, which of course is directly related to the manner in which the data were originally collected. Accuracies will also be affected by errors in control points used for registration or orientation, differential shrinkage and expansion of the materials upon which the graphical products are printed, pointing errors, and imperfections in the digitizing equipment.

Data sets generated by digitizing will usually need to be checked carefully to ensure that all desired features have been included. Also the data must be corrected or "cleaned" before being used in a GIS. In this process, undesired points and line portions must be removed, and "unclosed" polygons, which result from imprecise pointing when returning to a polygon's starting node, must be closed. Finally, thin polygons or "slivers," created by lines being inadvertently digitized twice but not exactly in the same location, must be eliminated. This editing process can be performed by the operator, or with a program that can find and remove certain features that fall within a set of user-defined tolerances. Digitizing and editing require a relatively large investment in time and cost.

20-6.3 Scanning

Scanners, as described in Sec. 4-8, are instruments that can automatically convert graphical documents to a digital file in raster format. Their principal advantages are that the tedious work of manual digitizing is eliminated and that the process of converting graphical documents to digital form is greatly accelerated. Precise scanners for measuring photo coordinates were described in Sec. 4-8 and illustrated in Figs. 4-9 and 4-10. While these types of scanners could be used to scan small documents for entry into GIS databases, for scanning maps and other sizable documents, large-format scanners are more commonly used.

Scanners accomplish their objective by measuring the amount of light reflected from a document and encoding the information in a pixel array. This is possible because different areas of a document will reflect light in proportion to their tones, from a maximum for white through the various shades of gray to a minimum for black. The data are stored directly in the interfaced computer, and they can be viewed on a screen, edited, and manipulated. Documents such as subdivision plats, topographic maps, engineering drawings and plans, aerial photos, and orthophotos can be digitized using scanners. Because scanners will record everything, including blemishes, stains, and creases, editing is an important and necessary step in preparing the scanned files for entry into a GIS database. As noted, the output files are in raster format, but if necessary, they can be converted to vector form using techniques described in Sec. 20-5.2.

The geometric quality of any raster file obtained from scanning depends somewhat on the instrument's accuracy, but pixel size (or resolution) is generally a more significant factor. A smaller pixel size will normally yield superior precision, but there are certain tradeoffs that must be considered. Whereas a large pixel size will produce a coarse representation of the original document, it will require less scanning time and less computer storage. Conversely, a fine resolution, which generates a precise depiction of the original, requires more scanning time and greater computer storage. An additional problem is that at very fine resolution, the scanner will record a great deal of "noise," i.e., small impurities such as blemishes or specks of dirt.

20-6.4 Keyboard Entry and Coordinate Geometry

Data can be entered into a GIS file directly by using the keyboard on a computer. Often data input by this method are nonspatial, such as map annotations or numerical or tabular data. To facilitate the keying in of data, an intermediate file having a simple format is sometimes created. This file is then converted to a GIS-compatible format using special software.

Data for use in a GIS can also be generated directly from field survey data using coordinate geometry software. This procedure is particularly convenient when a total station instrument with interfaced data collector is available (see Sec. 16-5). The field data can be downloaded from the data collector into a computer, processed with the coordinate geometry software, and entered directly into the GIS database. Data generated in this manner are generally very accurate.

20-6.5 Existing Digital Data

Sometimes data are already available in a suitable digital format, so conversion is not necessary. Organizations located in the vicinity of the area covered by the new GIS may be able to provide such digital data. In addition, the federal government has produced digital data files of a wide variety of types. Examples are the digital line graphs (DLGs) and digital elevation models that are available from the U.S. Geological Survey.[3] Often existing digital data must undergo conversion of file structures and formats to be usable with specific GIS software. Because of variations in the way data are represented by different software, it is possible that information can be lost or that spurious data can creep in during this process.

20-7 GIS ANALYTICAL FUNCTIONS

Most GISs are equipped with a menu of basic analytical functions which enable data to be manipulated, spatially analyzed, and queried. These functions, when accompanied with appropriate databases, provide GISs with powerful capabilities for supplying information that can so significantly aid in planning, management, and decision making.

    The specific functions available within the software of any particular GIS system will vary, but some of the more common and useful spatial analysis and computational functions are briefly described in the subsections that follow.

20-7.1 Buffering

This spatial analysis function involves the creation of new polygons that are geographically related to nodes, chains, or existing polygons. *Point buffering,* also known as *radius searching,* is illustrated in Fig. 20-8a. It creates a circular buffer zone of radius R around a specific node. Information about the new zone can then be assembled, and analyses made of the new data. A simple example illustrates its value. Assume that a polluted well has just been discovered. With appropriate databases available, all dwellings within a critical radius of the well can be located; the names, addresses, and telephone numbers of all individuals living within the buffer zone tabulated; and the people quickly notified of the possibility that the water could be polluted.

    Line buffering, illustrated in Fig. 20-8b, creates new polygons along established lines such as roads, railroads, or streams. To illustrate the use of line buffering, assume that to preserve the natural streambank and prevent erosion, a zoning commission has

---

[3]Digital line graphs and digital elevation models can be obtained online at the following Web site: *http://edcwww.cr.usgs.gov.* These products can also be ordered from U.S. Geological Survey, EROS Data Center, 47914 252d St., Sioux Falls, SD 57198.

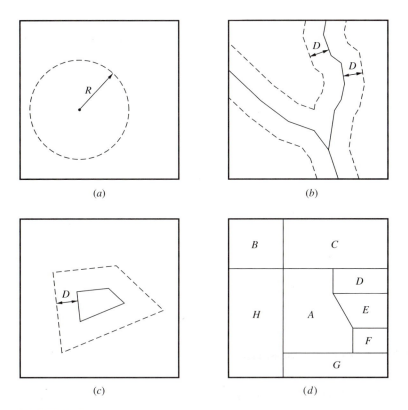

FIGURE 20-8
GIS spatial analysis functions. (*a*) Point buffering. (*b*) Line buffering. (*c*) Polygon buffering. (*d*) Adjacency analysis.

established the construction setback distance from a certain stream at *D*. The line buffering function can quickly identify the restricted areas. *Polygon buffering,* illustrated in Fig. 20-8*c*, creates a new polygon around an existing polygon. An example of its use could occur in identifying those landowners whose property lies within a certain distance *D* of the proposed site of a new industrial facility. Many other examples could be given to illustrate the value of buffering, but those cited above illustrate the value of this function for rapidly extracting critical information to support management and decision making in certain situations.

20-7.2 Adjacency and Connectivity

If the topological relationships discussed in Sec. 20-3.3 have been entered into a GIS database, certain analyses regarding relative positioning of features can be performed. *Adjacency* and *connectivity* are two such important spatial analysis functions which often assist significantly in management and decision making. An example of adjacency is illustrated in Fig. 20-8*d*. It relates to a zoning change requested by the owner of parcel *A*. Before taking action on the request, the jurisdiction's zoning administrators are required to notify the owners of all adjacent properties (in this case *B* through *H*). If the GIS database includes the parcel descriptions with topology and other appropriate attributes, an adjacency analysis can identify the properties that abut parcel *A* and can provide the names and addresses of the owners.

Connectivity involves analyses of the intersections of linear features. The need to repair a city water main can serve as an example to illustrate its value. Suppose that these repairs are scheduled to occur between the hours of 1:00 p.m. and 4:00 p.m. on a certain date. If infrastructure data are stored within the city's GIS database, all customers connected to this line whose water service will be interrupted by the repairs can be identified, and their names and addresses tabulated. The GIS can even print a letter and address labels, to facilitate a mailing announcing details of the planned interruption to all affected customers. Many similar examples could be given to illustrate benefits that can result from adjacency and connectivity.

20-7.3 Overlay

The overlay capability is one of the most widely used spatial analysis functions of a GIS. As indicated in Fig. 20-1, GIS information is usually divided into layers, with each containing data in a single category of closely related features. Nonspatial data or "attributes" are often associated with each category. The individual layers are spatially registered to each other through a common reference network or coordinate system. Any number of layers can be entered into a GIS database, and could include municipal boundaries, public land survey system, private ownership boundaries, soils, topography, land cover, hydrology, zoning, transportation networks, utilities, and many others. Having these various data sets available in spatially related layers makes the overlay function possible. Its employment in a GIS can be compared to using a collection of Mylar overlays in traditional mapping. However, much greater efficiency and flexibility are possible when operating in the computer environment of a GIS; and not only can graphical data be overlaid, but attribute information can be combined as well.

In Sec. 20-1, two examples were presented which illustrated some early applications of the overlay concept. Many other examples can be cited in which this function has been used to advantage in a modern GIS. Consider, for example, that the land in a particular area that is suitable for development must be identified. To perform an indepth analysis of this situation, the evaluation would normally have to consider numerous variables within the area, including the topography (slope and aspect of the terrain), soil type, land cover, land ownership, and others. Certain combinations of these variables could make land unsuitable for development. Figure 20-9 illustrates a simplified case of evaluating land suitability from the analysis of only two variables, slope and soil type. Figure 20-9*a* shows polygons within which the average slope is either 5 or 10 percent. Figure 20-9*b* classifies the soils in the area as *E* (erodible) or *S* (stable). The composite of the two data sets after a *polygon-on-polygon* overlay is shown in Fig. 20-9*c*. It identifies polygon I, which combines 5 percent slopes with low-erodibility class *S* soils. Since this combination does not present potential erosion problems, the area within polygon I is suitable for development, while areas II, III, and IV are not.

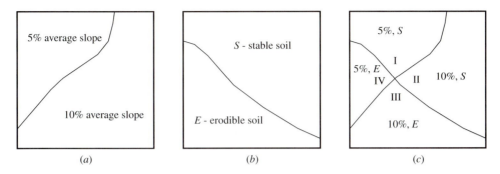

FIGURE 20-9
Example of GIS overlay function used to evaluate land suitability. (*a*) Polygons of differing slopes.
(*b*) Varying soil types in an area. (*c*) Overlay of (*a*) and (*b*), identifying polygon I as an area combining
lower slopes with stable soils which would be suitable for development.

Another GIS overlay function is a *point-in-polygon* overlay. This function re-
solves which point features are located within which polygons after layers are combined. As an example, a GIS operator may want to know which wells are located in an
area of highly permeable soil, in order to predict possible well contamination. A similar
overlay process, *line in polygon,* identifies specific linear features within polygons of interest. An example of its application would be the identification of all bituminous roads,
paved more than 10 years ago, in townships whose roadway maintenance budgets are
less than $250,000. Information of this nature would be valuable to support decisions
concerning the allocation of state resources for local roadway maintenance.

The GIS functions just described can be used individually, as has been illustrated
with the above examples, or they can be employed in combination. The flood prediction
example given in Sec. 20-1 can be extended to illustrate the simultaneous application of
line buffering, adjacency, and overlay. Suppose that an impending flood is predicted to
crest at a stage of 3 m above the river's normal elevation. To give an early warning to
people who may be endangered by this flood, it is first necessary to identify the lands
that lie at or below the expected flood stage. This can be done by using line buffering,
where the thread of the river is the line of reference. The width of the buffer zone, however, is variable and is determined by combining elevation data in the TIN model with
the buffering process. The adjacency and overlay functions are then used to determine
which landowners are within the flood zone. From this information the names, addresses, and telephone numbers of affected property owners can be tabulated, and the
people notified.

From the previous discussion, it is apparent that the simultaneous application of
several GIS functions has the capability of yielding very dramatic results.

20-7.4   Other GIS Functions

In addition to the spatial analysis functions described in preceding subsections, many
other functions are available with most GIS software. Some of these include the

capability of computing (1) the number of times a particular kind of point occurs within a certain polygon; (2) the distances between selected points, or from a selected line to a point; (3) locations of polygon centroids; (4) areas within polygons; and (5) volumes within polygons where depth or other conditions are specified.

A variety of different mapping functions may also be performed using GISs. These include (1) changing the reference coordinate system, (2) rotating the reference grid, (3) performing map scale changes, and (4) changing the contour interval used to represent elevations. Also, single layers of information or any combination of layers can be mapped.

Most GISs are also capable of performing several different digital terrain analysis functions. Some of these (1) calculate profiles along designated reference lines, (2) determine cross sections at specified points along a reference line; (3) generate perspective views where the viewpoint can be varied; (4) analyze visibility to determine what can or cannot be seen from a given vantage point; (5) compute slopes and aspects; and (6) make sun intensity analyses.

Output from GISs can be provided in graphical form as maps, charts, or diagrams; in numerical form as statistical tabulations; or in other files that result from computations and manipulations of the geographic data. These materials can be supplied in either hard-copy or softcopy form.

20-8 GIS APPLICATIONS

As stated earlier, and as indicated by the brief examples given in the preceding sections of this chapter, the areas of GIS applications are widespread. In Chap. 21, several additional examples are described to illustrate GIS applications. Further evidence of the diversity of GIS applications can be seen by reviewing the bibliographies at the ends of this chapter and Chap. 21. GIS technology is being used worldwide, at all levels of government, in business and industry, by public utilities, and in private engineering offices. Some of the more common areas of application occur in (1) land-use planning; (2) natural resource management; (3) environmental impact assessment; (4) census, population distribution, and related demographic analyses; (5) agriculture; (6) analysis of the geographic distributions of disasters such as hurricanes, floods, and fires; (7) routing of buses or trucks in a fleet; (8) tax mapping and mapping for other special purposes; (9) infrastructure and utility mapping and management; and (10) urban and regional planning.

As the use of GIS technology expands, there will be a growing need for trained individuals who understand the fundamentals of these systems. Users should be aware of the manner in which information is recorded, stored, managed, retrieved, analyzed, and displayed using a GIS. System users should also have a fundamental understanding of each of the GIS functions, including their bases for operation, their limits, and their capabilities. *Perhaps of greatest importance, users must realize that information obtained from a GIS can be no better than the quality of the data from which it was derived.* This is of particular importance to those engaged in photogrammetric activities related to the creation of the GIS databases.

REFERENCES

American Congress on Surveying and Mapping: "Implementing the Spatial Data Transfer Standard," *Cartography and Geographic Information Systems,* vol. 19, no. 5, 1992, p. 277.

Antennucei, J., K. Brown, P. Croswell, J. Kevany and H. Archer: *Geographic Information Systems,* Van Nostrand Reinhold, New York, 1991.

Aronoff, S.: *Geographic Information Systems: A Management Perspective,* WDL Publications, Ottawa, Canada, 1989.

Carter, J. R.: "Perspectives on Sharing Data in Geographic Information Systems," *Photogrammetric Engineering and Remote Sensing,* vol. 58, no. 11, 1992, p. 1557.

Croswell, P.: "GIS Software Overview," *Point of Beginning,* vol. 23, no. 2, 1997, p. 34.

Dangermond, J.: "Where Is GIS Technology Going?" *Professional Surveyor,* vol. 12, no. 2, 1992, p. 30.

Epstein, E., and T. Duchesneau.: "Use and Value of a Geodetic Reference System," *Journal of the Urban and Regional Information Systems Association,* vol. 2, no. 1, 1990, p. 11.

Falconer, A.: "Geographic Information Technology Fulfills Need for Timely Data," *GIS World,* vol. 5, no. 6, 1992, p. 37.

Fotheringham, S., and P. Rogerson: *Spatial Analysis and GIS,* Taylor & Francis Publishers, Bristol, PA, 1994.

Frank, A. U., M. J. Egenhofer, and W. Kuhn: "A Perspective on GIS Technology in the Nineties," *Photogrammetric Engineering and Remote Sensing,* vol. 57, no. 11, 1991, p. 1431.

Fuller, G.: "A Vision for a Global Geospatial Information Network (GGIN)," *Photogrammetric Engineering and Remote Sensing,* vol. 65, no. 5, 1999, p. 525.

Gibson, D.: "Conversion Is Out, Measurement Is in—Are We Beginning the Surveying and Mapping Era of GIS?" *Surveying and Land Information Systems,* vol. 59, no. 1, 1999, p. 69.

Goldstein, H.: "Mapping Convergence: GIS Joins the Enterprise," *Civil Engineering,* vol. 67, no. 6, 1997, p. 36.

Goodchild, M., and S. Gopal: *Accuracy of Spatial Databases,* Taylor & Francis Publishers, Bristol, PA, 1989.

Hanigan, F.: "GIS by Any Other Name Is Still . . . ," *The GIS Forum,* no. 1, 1988, p. 6.

Herring, J.: "The Open GIS Model," *Photogrammetric Engineering and Remote Sensing,* vol. 65, no. 5, 1999, p. 585.

Holmes, D. D.: "Automated Data Capture for Geographic Information Systems," *Surveying and Land Information Systems,* vol. 51, no. 2, 1991, p. 87.

Hubl, E.: "The Evolution of a GIS," *Point of Beginning,* vol. 23, no. 2, 1997, p. 26.

Huxhold, W.: *An Introduction to Urban Geographic Information Systems,* Oxford University Press, New York, 1991.

Kennedy, M.: *The Global Positioning System and GIS,* Ann Arbor Press, Chelsea, MI, 1996.

Korte, G.: "Maps on the Web," *Point of Beginning,* vol. 24, no. 4, 1999, p. 42.

————: "Across the States," *Point of Beginning,* vol. 24, no. 2, 1998, p. 34.

Lembo, J., and P. Hopkins: "The Use of Adjustment Computations in Geographic Information Systems for Improving the Positional Accuracy of Vector Data," *Surveying and Land Information Systems,* vol. 58, no. 4, 1998, p. 195.

Light, D.: "The National Aerial Photography Program as a Geographic Information System Resource," *Photogrammetric Engineering and Remote Sensing,* vol. 59, no. 1, 1993, p. 61.

Maguire, D., M. Goodchild, and D. Rhind: *Geographical Information Systems: Principles and Applications,* Wiley, New York, 1991.

McKee, L.: "Open GIS Specification Conformance—What Does It Mean?" *Geo Info Systems,* vol. 8, no. 11, 1998, p. 36.

National Institute of Standards and Technology: *Federal Information Processing Standard Publication 173: Spatial Data Transfer Standard,* U.S. Department of Commerce, 1992.

Onsrud, H., and D. Cook: *Geographic and Land Information Systems for Practicing Surveyors,* American Congress on Surveying and Mapping, Bethesda, MD, 1990.

Plewe, B.: "So You Want to Build an Online GIS?" *GIS World,* vol. 10, no. 11, 1997, p. 58.

Pollock, R. J., and J. D. McLaughlin: "Data Base Management System Technology and Geographic Information Systems," *ASCE Journal of Surveying Engineering,* vol. 116, no. 1, 1990, p. 9.

Quattrochi, D., and M. Goodchild: *Scaling of Remote Sensing Data for GIS,* Lewis Publishers, Boca Raton, FL, 1997.

Ripple, W.: *Fundamentals of GIS: A Compendium,* American Congress on Surveying and Mapping, Bethesda, MD, 1990.

Sionecker, E. T., and M. J. Hewitt: "Evaluating Locational Point Accuracy in a GIS Environment," *Geo Info Systems,* vol. 1, no. 6, 1991, p. 36.

PROBLEMS

20-1. List the fundamental components of a GIS.

20-2. Discuss the importance of the reference coordinate system layer in a GIS.

20-3. Name and describe the different simple spatial objects used for representing graphical data in digital form. Which objects are used in vector format representations?

20-4. Discuss the differences between vector and raster formats for storing data in a GIS. Cite the advantages and disadvantages of each.

20-5. How many pixels are required to convert the following documents to raster form for the conditions given?
(*a*) A 36-in square map scanned at 200 dpi
(*b*) An orthophoto of a 3 × 3 km area with a 10 m pixel size
(*c*) A 23 cm square aerial photo scanned at a 25 μm pixel size

20-6. Explain how data can be converted from (*a*) vector to raster format and (*b*) raster to vector format.

20-7. For what types of data is the raster format best suited?

20-8. Discuss the compromising relationships between pixel size and resolution in raster data representation.

20-9. Discuss the advantages and disadvantages of using tablet digitizers for converting maps and other graphical data to digital form.

20-10. What are the advantages and disadvantages of using scanners for converting graphical data to digital form?

20-11. Define the term *topology,* and discuss its importance in geographic information systems.

20-12. Compile a list of linear features for which the topological relationship of direction would be important.

20-13. What is connectivity in GIS terminology? Cite a GIS example application that illustrates its value.

20-14. Explain the concept of adjacency in GIS spatial analysis. Give an example illustrating its beneficial application.

20-15. If data are being represented in vector format, what simple spatial objects would be associated with each of the following topological properties?
(*a*) Connectivity   (*b*) Direction
(*c*) Adjacency   (*d*) Nestedness

20-16. What is the actual ground dimension of a pixel for the following conditions?
(*a*) A 1:20,000 scale hard-copy orthophoto scanned at a 50 μm pixel size
(*b*) An aerial photo having a scale of 1 in = 1:12,000 scanned at a 15 μm pixel size
(*c*) A 1:6000 map scanned at 100 dpi

20-17. Describe the following GIS functions: (*a*) point buffering, (*b*) line buffering, and (*c*) polygon buffering.

20-18. Figure 20-9 demonstrates one application which illustrates the value of the GIS overlay function. Cite two other examples where this function would be beneficial.

20-19. Identify possible sources in your state for locating the following data sets or GIS layers of information:
(*a*) Land ownership parcels   (*b*) Zoning
(*c*) Floodplains and wetlands   (*d*) Soil types
(*e*) Land cover

20-20. Compile a list of data layers and attributes that would likely be included in a GIS for

    (*a*) Selecting the optimum corridor for constructing a new rapid-transit system to connect two major cities

    (*b*) Selecting the optimum location for a new airport in a large metropolitan area

    (*c*) Routing a nationwide fleet of trucks

    (*d*) Selecting the fastest routes for reaching locations of fires from various fire stations in a large city

20-21. In Sec. 20-7.3, a flood warning example is given to illustrate the value of simultaneously applying more than one GIS analytical function. Describe another example.

20-22. Consult the literature on GISs and, based on your research, discuss an example that illustrates the application of a GIS in

    (*a*) Natural resource management   (*b*) Agriculture

    (*c*) Engineering   (*d*) Forestry

    (*e*) Demography   (*f*) Surveying

CHAPTER
21

PHOTOGRAMMETRIC APPLICATIONS IN GIS

21-1 INTRODUCTION

As noted in the preceding chapter, and in earlier sections of this book, photogrammetry and remote sensing play extremely important roles in the development and implementation of geographic information systems. In one very useful application, aerial images are frequently employed by GIS operators as background frames of reference for performing spatial analyses. The images may be obtained from either aerial cameras or satellite systems, and in general they will have undergone some form of georeferencing such as conversion to digital mosaics (see Chap. 9) or orthophotos (see Chap. 15) prior to being used in this manner. But perhaps the most important contribution to geographic information systems made by photogrammetry and remote sensing is their use in generating layers of information for databases.

Information for specific GIS databases can be compiled directly from stereomodels created from aerial photographs by photogrammetric restitution instruments such as analytical plotters or softcopy plotters (see Chaps. 13 and 15). As examples, layers of information such as roads, hydrography, land use, and many others can be coded and digitized directly from stereomodels, and entered into GIS databases. Digital line graphs (DLGs) and digital elevation models (DEMs) are two additional products that are frequently compiled directly from stereomodels. The former gives planimetric positions of objects, while the latter provides elevation data. Both are essential for many GIS applications. The digital orthophoto (see Chap. 15) is another photogrammetric product that has become indispensable in GIS work. Digital orthophotos can be automatically and very economically compiled. They are especially valuable because they are in image (raster) form, and GIS operators can analyze them visually. In addition they are planimetrically accurate, and thus two-dimensional features can be digitized directly from them. Because of these characteristics, they are extremely convenient for use as background reference frameworks for GIS applications. Roads, railroads, wetland boundaries, and other planimetric features that appear in raster form on a digital orthophoto, for example, can readily be converted to vector form so that they are amenable to various GIS spatial analyses.

Layers of information for geographic information systems are often compiled by simultaneously analyzing photogrammetric products in conjunction with other documents. As an example, a land ownership layer may be conveniently developed by making a visual analysis of digital orthophotos while simultaneously reading and interpreting parcel descriptions, dimensions, and other information given in deeds, plat books, and tax documents. Generating database information by photogrammetry is almost always faster and more cost-effective than doing so by any other means, and the process enables high orders of accuracy to be achieved.

In sections that follow, a variety of applications are presented in which geographic information systems were used as a means of solving problems. The contributions made by photogrammetry and remote sensing in these applications are described.

21-2   LAW ENFORCEMENT AND REGULATION

Geographic information systems have had many useful applications in law enforcement and regulation. A specific law enforcement example selected for illustration in this book involves a drug dealing arrest case. The litigation that followed this arrest benefitted from using the GIS buffer function (see Sec. 20-7), superimposed upon a digital orthophoto. In this example, illustrated in Fig. 21-1, the drug arrest was made in a city which had enacted a law extending the sentences for persons convicted of such conduct if the incident occurred within 1000 ft of any school grounds. The arresting officer was able to pinpoint the location of the incident on the digital orthophoto. Then by establishing a 1000-ft buffer around that point, as illustrated in the figure, all areas within the critical zone were identified. Note that part of the school grounds was indeed within the buffer. Because the orthophoto had been carefully georeferenced during its production, all features shown upon it were planimetrically correct, and the graphical display was accurate.

Prior to using this GIS procedure, law enforcement officers had to spend many hours making difficult measurements between houses, buildings, and trees in attempting to determine the proximity of arrests to schools. Maps of this nature are now routinely used in courts as exhibits in a variety of litigations.

21-3   HAZARDOUS WASTE MANAGEMENT

Figure 21-2 illustrates the application of a geographic information system in hazardous waste management. The figure identifies the location of a facility within a city that could potentially experience a hazardous materials spill. In this particular situation it was

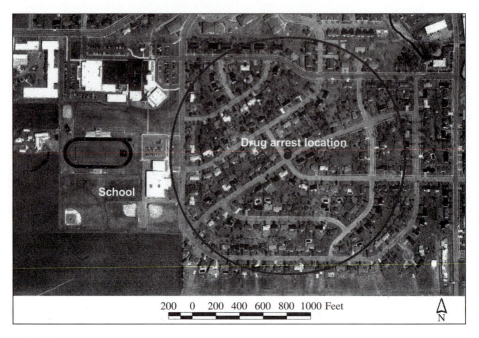

FIGURE 21-1
Application of the GIS buffer function in law enforcement. (*Courtesy Land Information Office, Dane County, Wisconsin.*)

anticipated that if a toxic spill were to occur, a hazardous plume could spread over an area having a radius of 0.7 mi from the site. To assist emergency management officials in preparing a disaster response plan for such an event, a GIS was developed.

With this GIS, a digital orthophoto was used as the frame of reference for spatial analyses. One layer of information within the GIS database identified and located institutions and special facilities within the city that were either highly populated or that would be particularly sensitive to the toxic plume. If a toxic spill were to occur, these areas might need immediate evacuation from the endangered area, and/or special assistance may be required to affect evacuation. Another database layer consisted of streets. Both the street and institution layers were developed by digitizing directly from stereomodels, using an analytical plotter. Supplemental information for these layers was obtained from other documents and through field evaluation. Information such as the identification of one-way streets, average daily traffic data, locations of stoplights and stop signs, and current construction activity would logically be included. By constructing a buffer of 0.7-mi radius around the potential toxic spill site, as shown in Fig. 21-2, the critical area affected was identified, and through a GIS analysis, emergency management officials could develop response plans.

A GIS such as that described above could be readily updated with time as institutions, activities, and conditions within the city change. Furthermore, with the addition of other layers of information the GIS could be useful for many other applications. As

Hazardous Materials Plan, Well Number 3

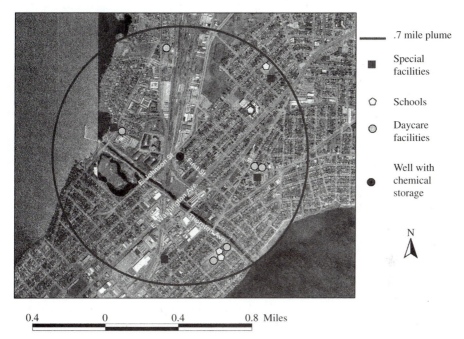

——— .7 mile plume

■ Special facilities

⬠ Schools

◎ Daycare facilities

● Well with chemical storage

N

0.4   0   0.4   0.8 Miles

FIGURE 21-2
Buffer function superimposed on digital orthophoto in hazardous waste management application.
(*Courtesy Land Information Office, Dane County, Wisconsin.*)

examples, it might be applied in plotting the most expedient routes for firetrucks to reach their destinations, or for planning the most efficient school bus routes. There are many other potential uses.

21-4 WATER QUALITY MANAGEMENT

Geographic information systems have been successfully applied to the solution of many problems in water quality management. A project of this type selected for discussion in this text involves a GIS that was developed for the Sugar River watershed in south-central Wisconsin (see Fig. 21-3). The objectives of the GIS were to provide a cost-effective means for assisting in preventing soil erosion and to improve and preserve the water quality in the Sugar River and its tributaries, as well as in Belle Lake which exists at the lower end of the watershed. The GIS was developed by the Dane County Land Conservation Department, with funding from the Wisconsin Department of Natural Resources, the U.S. Environmental Protection Agency, and Dane County.

In the Sugar River watershed, excessive amounts of sediment and phosphorus, generated by both agricultural and urban land uses, enter the surface water. These conditions

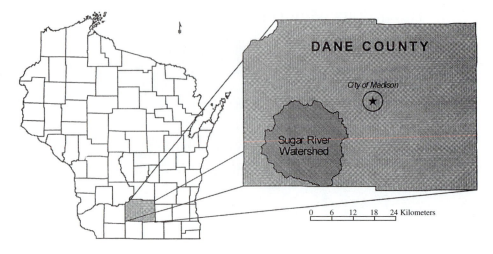

FIGURE 21-3
Location of Sugar River water quality management GIS project. (*Courtesy Dane County Land Conservation Department, M.S. Robinson and A. Roa-Espinosa.*)

are detrimental to the aquatic and wildlife habitat within the basin, and they also impact negatively upon the aesthetic qualities and recreational use of the water. To address this problem, a GIS was developed to provide the information necessary for a computer modeling program to identify areas of high sediment and phosphorus delivery within the watershed. Improved management practices could then be implemented in these areas.

The major layers of information developed for the GIS database included topography, hydrography, soils, and land use/land cover. Photogrammetry played a major role in the development of each of these layers. The elevation layer (see Fig. 21-4) was based upon a digital elevation model that was produced photogrammetrically. The DEM, which consisted of a 10-m grid with accuracies to within ±0.3 m, provided slope information that was critical in the computer modeling. The hydrography layer, which consisted of the rivers and streams within the watershed, was digitized from orthophotos. By combining the elevation and hydrography layers, small individual hydrologic drainage areas that were needed for the computer modeling were identified. There were approximately 500 individual areas in the watershed, as shown in Fig. 21-5 (see page 480), and they averaged about 1 km^2 in size. The soils layer was digitized from NHAP aerial photos.

The land use/land cover layer consisted of the combination of many individual layers. It included categories of cropland, grassland, pasture, woodland, wetland, water, roads, residential areas, industrial areas, and commercial areas (see Fig. 21-6 on page 481). Although other sources such as zoning maps and parcel maps were used to compile this information, most of these data were obtained from orthophotos. Individual farm tracts were included within this layer, and these tracts were further subdivided into separate fields. This was important because their differing crop rotations and land uses have an important bearing on their rates of sediment and phosphorus delivery.

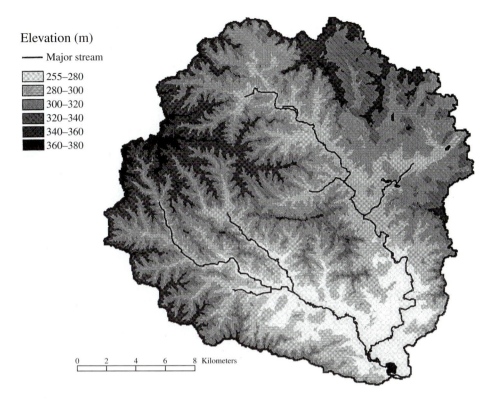

Elevation (m)

—— Major stream

255–280
280–300
300–320
320–340
340–360
360–380

0   2   4   6   8 Kilometers

FIGURE 21-4
Elevation model prepared from DEM of Sugar River watershed. (*Courtesy Dane County Land Conservation Department, M.S. Robinson and A. Roa-Espinosa.*)

The previous information was used by the computer program to model the sediment and phosphorus delivery rates across the watershed. Figure 21-7 (see page 482) is a graphic which shows the sediment yield of each of the hydrologic units in the watershed. A similar graphic was developed for phosphorus yields. Areas with high yields could be examined in the field and targeted for improved field and management practices. These practices could include such measures as practicing conservation tillage, using soil binders, establishing buffer strips, constructing terraces, or installing water and sediment control basins.

21-5 WILDLIFE MANAGEMENT

An example illustrating a GIS application in wildlife management involves a portion of the Lower Virgin River in southwestern United States. The study area, shown in Fig. 21-8 (page 483), includes the entire riparian corridor of the river from the Virgin River Gorge near Littlefield in northwest Arizona, to the Lake Mead National Recreation Area in southeast Nevada. This involves a total of 12,349 acres of riparian

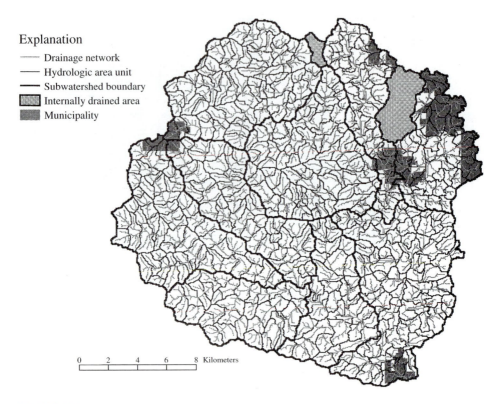

Explanation

‐‐‐‐ Drainage network
——— Hydrologic area unit
——— Subwatershed boundary
▨ Internally drained area
▧ Municipality

0   2   4   6   8 Kilometers

FIGURE 21-5
Hydrologic area units within the Sugar River watershed obtained by combining elevation and hydrography layers. (*Courtesy Dane County Land Conservation Department, M.S. Robinson and A. Roa-Espinosa.*)

vegetation and open water/river channel over a length of approximately 52 mi of river. The study was conducted by the Southern Nevada Water Authority. Its objectives were to delineate the vegetation communities and other habitat types (open water, etc.) within the riparian corridor of the river, determine how the different vegetation species are used by wildlife, and predict how the distribution and composition of vegetation and wildlife change with varying flow levels in the river. Evaluations of the status of the federally endangered southwestern willow flycatcher and the federally threatened desert tortoise were matters of special interest in this project.

To achieve the project's goals, a GIS was developed which covered the area. The base map, or frame of reference for performing GIS analyses, was prepared from color infrared aerial photos. These photos, taken at a scale of 1:24,000, were scanned, rectified, and joined digitally to form a mosaic of the study area. A total of 38 photos were involved in the mosaic construction. Control for the digital mosaic process was obtained from USGS digital line graphs.

In this study, one of the critical layers of information compiled was vegetation. To assist in preparing this layer, portions of the base map in selected areas were

Explanation

☐ Cropland
▨ Grassland / pasture
▨ Wetland / water
▨ Woodland
■ Urban / road

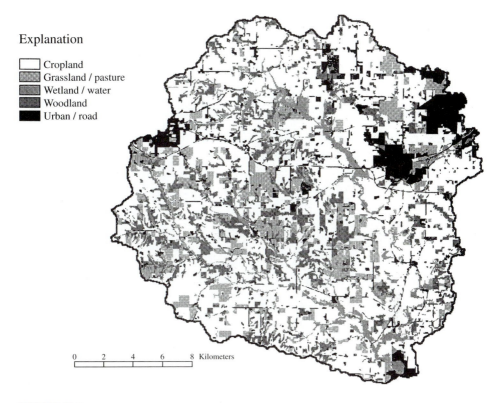

0   2   4   6   8 Kilometers

FIGURE 21-6
Land use/land cover layer of the Sugar River GIS database. (*Courtesy Dane County Land Conservation Department, M.S. Robinson and A. Roa-Espinosa.*)

enlarged to 1 : 6000 scale. This increased the visual resolution which facilitated their use for fieldwork. These large-scale photomaps were used during field surveys to identify and delineate the locations of vegetation classifications in several key areas of the river corridor. Based upon these "ground truth" samples, the remaining images in the corridor were interpreted and classified, using the heads-up digitizing process (see Sec. 9-5). A hard copy of this preliminary vegetation map was then taken into the field, and the delineations of vegetation types were verified and modified as necessary. GPS was also used in some cases to check visual analyses and to locate certain features that were not visible on the base map, such as special habitat areas and nesting grounds. Tablet digitizing was then used to incorporate these modifications into the vegetation layer of the database. Next the resulting vegetation map was utilized to select representative areas for evaluating wildlife and their use of the resources within the riparian corridor. A total of 11 vegetation classifications were included on the map. Figure 21-9 (page 484) shows the portion of the vegetation map at the location where the Virgin River corridor enters Lake Mead. Note that only five classifications existed within this section of the corridor.

Explanation

- ·········· Drainage network
- —— Hydrologic area unit
- —— Subwatershed boundary
- ▓▓ Internally drained area

Sediment Yield (MT/year)

- ▓ 0–450
- ▓ 450–1200
- ▓ 1200–2100
- ▓ 2100–3400
- ▓ 3400–6800

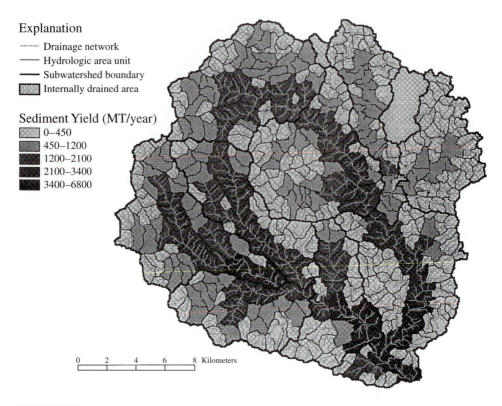

0   2   4   6   8 Kilometers

FIGURE 21-7
Sediment yield within the Sugar River watershed obtained through GIS analysis. (*Courtesy Dane County Land Conservation Department, M.S. Robinson and A. Roa-Espinosa.*)

The vegetation/habitat layer was used as a guide in selecting areas for conducting wildlife surveys in the field. The areas were chosen so that each of the vegetation classifications and habitats were represented within the survey. In addition to vegetation types, the topography, floodplain characteristics, and stream gradient were also considered in selecting the wildlife survey areas. Accessibility was a consideration as well. Sampling transects which included small mammal live traps, pitfall traps, and bird observation points were placed at selected stations within the survey areas to sample the wildlife populations. GPS was used to control the placement and mapping of the transect lines for data capture within the GIS model. To date, a total of 170 different species of birds, 34 different species of mammals, 13 species of reptiles, and 4 species of amphibians have been found to utilize the Lower Virgin River riparian corridor.

This project has been meeting all its objectives. A great deal of valuable data have been gathered to further existing knowledge of the wildlife in this southwest habitat, and to aid in understanding the dependence these wild creatures have upon the different vegetation communities and other habitat types that exist in the river corridor. The study of this 52-mi-long riparian corridor would have been a daunting task if performed by ground methods alone. However, as discussed, photogrammetry has contributed significantly in

NEVADA ARIZONA

Littlefield, AZ

Mesquite, NV

Virgin Mountains

Lake Mead National Recreation Area

6   0   6   12 Miles

Lower Virgin River Riparian Corridor

FIGURE 21-8
Study area of Virgin River wildlife management GIS project. (*Courtesy Southern Nevada Water Authority.*)

the conduct of this study. Not only has photogrammetry enabled the work to be completed more rapidly and efficiently, but also it has permitted greater amounts of data to be collected, analyzed, and managed, thereby improving the overall end results. This method of study has the added advantage that the information collected can conveniently be shared with other agencies having overlapping or common interests and goals.

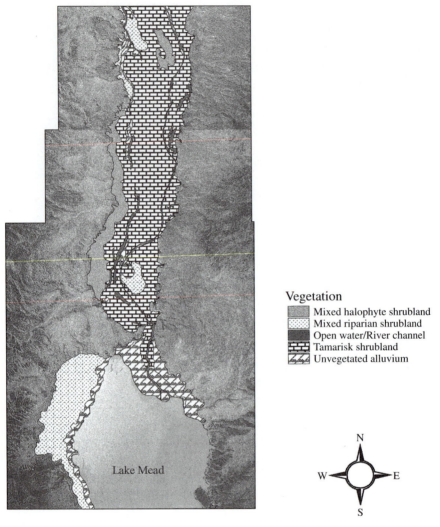

Vegetation
- Mixed halophyte shrubland
- Mixed riparian shrubland
- Open water/River channel
- Tamarisk shrubland
- Unvegetated alluvium

500   0   500   1000   1500 Feet

FIGURE 21-9
Vegetation layer of Virgin River wildlife management GIS project. (*Courtesy Southern Nevada Water Authority.*)

21-6 ENVIRONMENTAL RESTORATION

The Kissimmee River, located in south Florida, plays a vital role in the Everglades ecosystem. In the 1960s this river was rechanneled to improve navigation and flood control. The subsequent impact of this project, however, had a detrimental effect on the Florida Everglades. In 1994, a cooperative effort was launched to restore the Kissimmee River basin to its natural state by diverting the channel back to its original course. To

FIGURE 21-10
TIN model of a portion of the Kissimmee River and its floodplain with shaded relief, spot elevations, and breaklines. (*Courtesy 3001, Inc.*)

facilitate the restoration effort, a GIS was established for a 65,000-hectare (160,000-acre) region surrounding the river. The GIS will be used as part of a 25-year study to evaluate changes in biodiversity as the region slowly returns to its former condition.

Aerial photography was flown over the region using airborne GPS control. The resulting photography was subsequently scanned and used to produce digital orthophotos as well as DEMs. Figure 21-10 shows a three-dimensional view of the DEM in a small portion of the channel where a plug was placed to divert the river course. These photogrammetrically derived data were combined with other information to form the GIS. A land use/land cover determination was performed over the region to establish baseline information for plant species prior to restoration. In addition, animal species were inventoried at various locations determined by GPS and included in the GIS. As the restoration progresses, temporal changes in plant and animal species will continue to be cataloged to determine the effectiveness of the restoration effort.

21-7 LAND DEVELOPMENT

The coastline of Florida is susceptible to erosion due to natural disasters such as hurricanes. The Federal Emergency Management Agency (FEMA) sponsored a study to assess the impact of erosion and flooding on existing and future land development along the coast. This study included a comparison of flood zones as established by FEMA with flood zones that were accurately determined through field surveys. In addition, locations of critical infrastructure facilities were determined by GPS. The information will be used to determine susceptibility of properties to flooding and erosion, and to provide estimates of property values for insurance purposes.

A digital orthophoto was used as a base map for the spatial information. On this base, various layers of information were overlaid, including parcel boundaries,

FIGURE 21-11
GIS for coastal erosion and flooding study. The figure shows a digital orthophoto with overlaid flood zones and property boundaries. (*Courtesy 3001, Inc.*)

infrastructure, FEMA flood zones, and newly surveyed flood zones. Figure 21-11 illustrates a portion of the GIS for Brevard County, Florida. This figure portrays the digital orthophoto background upon which flood zones and parcel boundaries are shown. From this view, the user can then "point and click" on a particular parcel, and the GIS will retrieve specific information relating to that property. Figure 21-12 shows a table which gives the parcel-specific information along with a street-level view of the structure located on the property. This table includes flood zone and erosion information as well as an indication of the building type, property elevation, horizontal position (latitude and longitude), and ocean frontage. Additional spaces in the table are reserved for age of the building, floor area, and assessed value of the property. The GIS contains information for both developed and undeveloped properties and can be used as a guide for future land development in the region.

21-8   TRANSPORTATION

The Roadway Characteristics Inventory (RCI) is a GIS database which contains information about signs, pavement, drainage structures, bridges, etc., along highways in Florida. First introduced as a pilot project, the RCI was developed to facilitate pavement maintenance, accident cataloging, replacement or repair of damaged structures (guardrails, signs, fences, etc.), and other items related to highway issues. The GIS replaced manual inventory methods in which facilities were located by driving along the highways and noting the positions of inventory features from odometer readings (dead

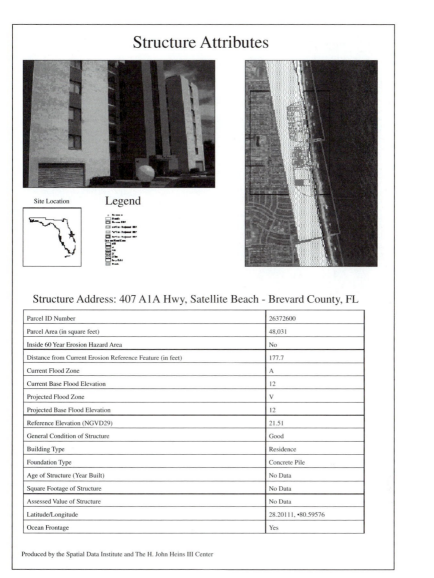

Structure Attributes

Site Location   Legend

Structure Address: 407 A1A Hwy, Satellite Beach - Brevard County, FL

Parcel ID Number	26372600
Parcel Area (in square feet)	48,031
Inside 60 Year Erosion Hazard Area	No
Distance from Current Erosion Reference Feature (in feet)	177.7
Current Flood Zone	A
Current Base Flood Elevation	12
Projected Flood Zone	V
Projected Base Flood Elevation	12
Reference Elevation (NGVD29)	21.51
General Condition of Structure	Good
Building Type	Residence
Foundation Type	Concrete Pile
Age of Structure (Year Built)	No Data
Square Footage of Structure	No Data
Assessed Value of Structure	No Data
Latitude/Longitude	28.20111, •80.59576
Ocean Frontage	Yes

Produced by the Spatial Data Institute and The H. John Heins III Center

FIGURE 21-12
Parcel-specific information for a selected property from Fig. 21-11. (*Courtesy 3001, Inc.*)

reckoning). The GIS will be used initially for planning purposes; however, the spatial accuracy is good enough to support other purposes such as preliminary design.

Aerial photography was acquired along the highway at a flying height of 600 m above terrain. This photography was used to compile planimetric features such as road centerlines and edges of pavement. Digital orthophotos were also produced, at a resolution of 10 cm, to provide a spatially accurate visual base map. To complement the photogrammetric data, field inventories were conducted to locate and catalog the various

FIGURE 21-13
Selection from the Roadway Characteristics Inventory GIS showing relevant information for a specific feature. (*Courtesy 3001, Inc.*)

highway features. Field locations were performed using differential, code-phase GPS techniques which obtained positions to submeter accuracy. Figure 21-13 shows a portion of the GIS along a stretch of highway. When a user selects a feature by pointing and clicking, a table pops up, giving detailed information about the particular item. In Fig. 21-13, the box culvert was selected, giving a table showing its position, diameter, type of material, and other information.

The plan calls for 5-year updates to be performed to keep the information in the GIS current. Updates will consist of revisiting previously surveyed features to report on their current condition, as well as obtaining information for new features. Due to the high accuracy of the GIS database and its completeness, it is anticipated that the RCI will provide for better decision making now and in the future.

21-9 HYDROGRAPHY

The Atchafalaya River Hydrographic Survey Book was produced by the U.S. Army Corps of Engineers as a part of the Regional Engineering and Environmental Geographic Information System (REEGIS). The information will be used as an aid in flood control, navigation, and resource management as well as to provide an online database for public access. The GIS incorporates many hydrographic features including river and levee elevations, buoys, and navigational hazards, as well as topographic details of the surrounding terrain. Figure 21-14 shows a map sheet produced from the GIS data.

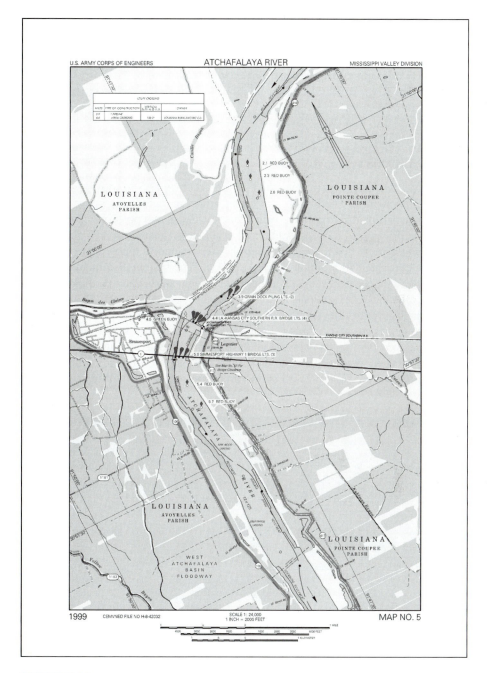

FIGURE 21-14
Map from the Atchafalaya River Book GIS. (*Courtesy 3001, Inc.*)

Topographic information was produced from aerial photography by stereoplotter compilation. In addition, digital orthophotos were generated and included as an overlay. Hydrographic surveys were included to provide water depth information as an aid to navigation. Hydrographic features were linked to an information database, to support user queries. The River Book GIS has been published on the Internet to provide public access to the information.

21-10 MULTIPURPOSE LAND INFORMATION SYSTEM

The city of Sulphur, Louisiana, established a high-accuracy GIS to provide spatial information for a variety of needs. Users of the information include the police, fire, and public works departments. To achieve high accuracy, the spatial positioning of the GIS was based on GPS control surveys. Additional data layers, created through photogrammetric techniques, were tied to the GPS control. These include a topographic map, digital orthophotography, and a digital elevation model. The system also includes parcel boundaries based on tax assessor data, utility information, stormwater facilities, fire hydrants, and crime incident locations.

This multipurpose GIS supports a variety of uses. For example, if the public works department needs to perform maintenance on a section of sewer pipe, the GIS user can point and click on the pipe and all customers connected to the sewer line will be listed. Notices can then be mailed to these customers to inform them of the impending service interruption. Figure 21-15 shows a portion of the GIS with parcels, build-

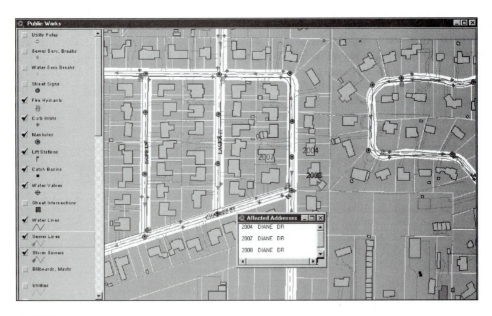

FIGURE 21-15
Sanitary sewer facility map from the Sulphur, Louisiana, GIS showing addresses affected by pipe maintenance. (*Courtesy 3001, Inc.*)

ings, streets, stormwater facilities, water lines, and sanitary sewer lines. A section of sewer pipe was selected, and the table lists the affected addresses. This illustrates just one of the many uses of this multipurpose GIS.

21-11 SUMMARY

The examples presented in this chapter are only a very small sample of the many ways in which GISs are being applied in problem solving. And as indicated by these examples, photogrammetry is being widely used to provide the spatially accurate data layers needed to enable GISs to function. In particular, the examples verify that digital orthophotos are a very commonly used photogrammetric product in GISs. Georeferenced satellite imagery is also frequently used in applications where lower image resolution is suitable. By virtue of their positional accuracies, areawide coverage capabilities, and cost effectiveness, photogrammetric products will continue to be used as data layers in many GIS applications. The references which follow cite numerous other GIS applications where photogrammetry and remote sensing play vital roles.

REFERENCES

American Society for Photogrammetry and Remote Sensing: "Special Issue: Geographic Information Systems," *Photogrammetric Engineering and Remote Sensing,* vol. 63, no. 10, 1997.

————: "Special Issue: Remote Sensing and GIS for Hazards," *Photogrammetric Engineering and Remote Sensing,* vol. 64, no. 10, 1998.

————: "GIS Special Issue: U.S./Mexico Border Region," *Photogrammetric Engineering and Remote Sensing,* vol. 64, no. 11, 1998.

Cadwallader, W., and R. Riethmueller: "Geospatial Technology in a Developing Country," *Geo Info Systems,* vol. 9, no. 1, 1999, p. 22.

Carter, G.: "Estimation of Nonpoint Source Phosphorous and Nitrogen Loads in Five Watersheds in New Jersey's Atlantic Coastal Drainage Basin," *Surveying and Land Information Systems,* vol. 58, no. 3, 1998, p. 167.

Decker, D., and R. Seekins: "Creating a Statewide Digital Base Map: The Texas Orthoimagery Program," *Surveying and Land Information Systems,* vol. 57, no. 1, 1997, p. 23.

Donnelly, J.: "Geographic Information Systems in Map Making," *Bulletin, American Congress on Surveying and Mapping,* no. 134, 1991, p. 30.

Dooley, D.: "Linear Transformation: Delaware Department of Transportation Uses GIS to Fill in the Gaps," *Geo Info Systems,* vol. 8, no. 5, 1998, p. 24.

Douglas, W.: *Environmental GIS: Applications to Industrial Facilities,* Lewis Publishers, Boca Raton, FL, 1995.

Duhaime, R. J., P. V. August, and W. R. Wright: "Automated Vegetation Mapping Using Digital Orthophotography," *Photogrammetric Engineering and Remote Sensing,* vol. 63, no. 11, 1997, p. 1295.

Faber, B. G., W. W. Wallace, and G. E. Johnson: "Active Response GIS: For Resource Management Spatial Decision Support Systems," *Photogrammetric Engineering and Remote Sensing,* vol. 64, no. 1, 1998, p. 7.

Fry, W., and J. Dozzi: "A New Look at Old Man River," *Civil Engineering,* vol. 67, no. 6, 1997, p. 49.

Gilbrook, M. J.: GIS Paves the Way, *Civil Engineering,* vol. 69, no. 11, 1999, p. 34.

Goldstein, H.: "Mapping Convergence: GIS Joins the Enterprise," *Civil Engineering,* vol. 67, no. 6, 1997, p. 36.

Greenfield, J.: "Consistent Property Line Analysis for Land Surveying and GIS/LIS," *Surveying and Land Information Systems,* vol. 57, no. 2, 1997, p. 69.

Greenwood, D., and D. Hathaway: "Assessing Opal's Impact," *Civil Engineering,* vol. 66, no. 1, 1996, p. 40.

Hendricksen, C., and L. Hall: "GIS Measures Water Use in the Arid West," *Geo Info Systems,* vol. 2, no. 7, 1992, p. 63.

Hinton, C.: "North Carolina City Saves Time, Lives, and Money with Award-Winning GIS," *Geo Info Systems,* vol. 7, no. 9, 1997, p. 35.

Hodgson, M., and R. Palm: "Attitude toward Disaster: A GIS Design for Analyzing Human Response to Earthquake Hazards," *Geo Info Systems,* vol. 2, no. 7, 1992, p. 40.

Kautz, R.: "Satellite Imagery and GIS Help Protect Wildlife Habitat in Florida," *Geo Info Systems,* vol. 2, no. 1, 1992, p. 37.

Koch, M., and F. El-Baz: "Identifying the Effects of the Gulf War on the Geomorphic Features of Kuwait by Remote Sensing and GIS," *Photogrammetric Engineering and Remote Sensing,* vol. 64, no. 7, 1998, p. 739.

Lang, L.: "Use of GIS, GPS and Remote Sensing Spreads to California's Winegrowers," *Modern Agriculture,* vol. 1, no. 2, 1997, p. 12.

Lembo, A. J., C. Powers, and E. S. Gorin: "The Use of Innovative Data Collection Techniques in Support of Enterprise Wide GIS Development," *Photogrammetric Engineering and Remote Sensing,* vol. 64, no. 9, 1998, p. 861.

Lindquist, R.: "Illinois Cleans Up: Using GIS for Landfill Siting," *Geo Info Systems,* vol. 1, no. 2, 1991, p. 30.

Loukes, D., and J. McLaughlin: "GIS and Transportation: Canadian Perspective," *ASCE Journal of Surveying Engineering,* vol. 117, no. 3, 1991, p. 123.

Lyon, J., and J. McCarthy: *Wetland and Environmental Applications of GIS,* Lewis Publishers, Boca Raton, FL, 1995.

Mettel, C.: "GIS and Satellite Images Provide Precise Calculations of Probable Maximum Flood," *Geo Info Systems,* vol. 2, no. 6, 1992, p. 44.

Oppmann, R., and N. von Meyer: "Oakland County, Michigan, Uses Enterprise GIS to Maintain Its High Quality of Life," *Geo Info Systems,* vol. 9, no. 7, 1999, p. 28.

Padgett, D.: "Assessing the Safety of Transportation Routes for Hazardous Materials," *Geo Info Systems,* vol. 2, no. 2, 1992, p. 46.

Pottle, D.: "Land and Sky," *Point of Beginning,* vol. 23, no. 2, 1997, p. 46.

Quattrochi, D., and J. Luvall: "Urban Sprawl and Urban Pall: Assessing the Impacts of Atlanta's Growth on Meterology and Air Quality Using Remote Sensing and GIS," *Geo Info Systems,* vol. 9, no. 5, 1999, p. 26.

Ripple, W.: *Geographic Information Systems for Resource Management,* American Congress on Surveying and Mapping, Bethesda, MD, 1986.

Voss, D.: "Integrating Pavement Management and GIS," *Geo Info Systems,* vol. 1, no. 9, 1991, p. 18.

Welch, R., M. Madden, and R. F. Doren: "Mapping the Everglades," *Photogrammetric Engineering and Remote Sensing,* vol. 65, no. 2, 1999, p. 163.

Wilkins, D., and J. Beem: "Flight Plan for the Future," *Point of Beginning,* vol. 24, no. 7, 1999, p. 22.

PROBLEMS

21-1. Contact the utility service provider in your local area. Describe how it is using GIS to support its operation. Note where photogrammetry is being used to provide data for the GIS.

21-2. Repeat Prob. 21-1, except contact your local tax assessor's office.

21-3. List 10 layers of information which would be useful in a GIS for fire rescue and emergency medical assistance.

21-4. List 10 layers of information which would be useful in a GIS for landfill site selection.

21-5. Give five examples of GIS applications in which digital orthophotos would be useful.

21-6. Cite five examples of GIS applications in which digital elevation models would be useful.

21-7. Give five examples of GIS applications in which topographic data produced by stereocompilation would be useful.

APPENDIX

A

UNITS, ERRORS, SIGNIFICANT FIGURES, AND ERROR PROPAGATION

A-1 UNITS

The solution of photogrammetric problems generally requires some type of length, angle, or area measurement. Length measurements can be based on either the metric system of meters, centimeters, millimeters, micrometers, etc., or the English system of feet and inches.

While the metric system is preferred for linear units, the English system is still widely used in the United States. Conversion between the two systems is frequently necessary, and it can be complicated by the fact that two common conversion standards exist. Prior to 1959, the accepted metric-English conversion was that 1 m is exactly equal to 39.37 in, or 12 m equals 39.37 ft. This is the basis of the so-called *U.S. survey foot*. In 1959 that conversion was officially changed to 1 m is exactly equal to 2.54 cm or 0.3048 m equals 1 ft, which is the *international standard foot*. The difference between these definitions is approximately 1 part in 500,000 and is therefore negligible in most circumstances. Use of a particular definition is essentially a matter of accepted standards, although in some cases legislative statute dictates the use of a specific conversion. In a practical sense, the only situation of concern occurs in converting geographic coordinates whose values have in excess of six significant figures. In this text, the international standard conversion will be used unless the U.S. survey standard is specifically indicated.

The following list of length, angle, and area unit equivalents should be helpful to students using this book:

1. Length equivalents
   a. *Metric*
      1 meter (m) = defined standard
      1 micrometer (μm) = 0.000001 m*
      1 millimeter (mm) = 0.001 m*
      1 centimeter (cm) = 0.01 m*
      1 kilometer (km) = 1000 m*

   b. *English*
      12 inches (in) = 1 foot (ft)*
      3 ft = 1 yard (yd)*
      5280 ft = 1 mile (mi)*

   c. *Metric-English (international standard)*
      2.54 cm = 1 in*
      1 m = 3.2808399 ft
      0.3048 m = 1 ft*
      1 km = 0.6213712 mi

   d. *Metric-English (U.S. survey foot)*
      1 m = 39.37 in*
      1 m = 3.2808333 ft
      0.30480061 m = 1 ft
      1 km = 0.6213699 mi

2. Angle equivalents
   π = 3.141592654
   1 circle = 2π radians (2π rad) = 360 degrees (360°) = 400 grads or gons (400^g)*
   1° = 60 minutes (60′)*
   1′ = 60 seconds (60″)*
   1 rad = 57.29578° = 57°17′44.8″ = 206,264.8″
   1° = 0.01745329 rad = 1.111111^g
   1^g = 0.9° = 54′ = 3240″*
   1^g = 0.01570796^r

3. Area equivalents
   a. *Metric*
      1 hectare (ha) = 10,000 m²*
      1 km² = 100 ha*

   b. *English*
      1 acre = 43,560 ft²*
      640 acres = 1 mi²*

---

*The asterisk denotes an exact conversion factor; other factors are correct only to the digits shown.

  c. *Metric-English (International Standard)*
     4046.856 m^2 = 1 acre
     1 ha = 2.471054 acres
     0.4046856 ha = 1 acre

  d. *Metric-English (U.S. survey foot)*
     4046.873 m^2 = 1 acre
     1 ha = 2.471044 acres
     0.4046873 ha = 1 acre

A-2   ERRORS

In the processes of measuring any quantity, factors such as human limitations, instrumental imperfections, and instabilities in nature render the measured values inexact.
Due to these factors, no matter how carefully a measurement is performed, it will always contain some error. Photogrammetry is a science which frequently requires measurements, and therefore an understanding of errors—including how they occur and
how they are treated in computation—is important. Before learning about errors, it is
helpful to have an understanding of the concepts of *accuracy* and *precision.*

   Accuracy can be defined as the degree of conformity to the true value. A value
which is very close to the true value has high accuracy, and a value that is far from true
has low accuracy. Since the true value for a continuous physical quantity is never
known, accuracy is likewise never known; therefore, it can only be estimated. An acceptable method for assessing accuracy is by checking against an independent, higher-
accuracy standard.

   Precision, on the other hand, is the degree of refinement of a quantity. The level
of precision can be assessed by making repeated measurements and checking the consistency of the values. If the values are very close to each other, the measurements have
high precision; and if the values vary widely, the measurements have low precision.

   An *error* is defined as the difference between a particular value and the true or
correct value. Whenever a measurement is made of a continuous physical quantity (such
as distance), the result will contain some amount of error. Errors can be categorized as
random errors, systematic errors, and *mistakes.* Mistakes or *blunders* are gross errors
caused by carelessness or negligence. Common examples of blunders are point misidentification, a transcription error in recording a value, and misreading of a scale. Large
blunders can generally be avoided or detected and eliminated through careful procedures and subsequent quality control checks. Small blunders can be particularly troublesome when they are large enough to render the results unsatisfactory, yet small
enough that they are undistinguishable from acceptable random errors. Blunders can be
prevented by exercising care and remaining alert while measurements are being taken.
If they occur, they can usually be detected by careful checking, and then eliminated.

   Systematic error in a measurement is an error which follows some mathematical
or physical law. If the conditions causing the error are measured and properly modeled,
a correction can be calculated and the systematic error eliminated. Systematic errors
will remain constant in magnitude and algebraic sign if the conditions producing them

remain the same. Because their algebraic sign tends to remain the same, systematic errors accumulate, and consequently they are often referred to as *cumulative* errors. Examples of systematic errors in photogrammetry are shrinkage or expansion of photographs, camera lens distortions, and atmospheric refraction distortions.

After blunders and systematic errors have been eliminated, the errors that remain are called random or *accidental* errors. Random errors are generally small, but they can never be avoided entirely in measurements. They do not follow physical laws as systematic errors do, and therefore they must be dealt with according to the mathematical laws of probability. Random errors are as likely to be positive as negative; hence they tend to compensate each other and consequently are often called *compensating* errors. Random errors occur in photogrammetry, for example, in estimating between least graduations of a scale, or in indexing a scale.

A-3   SIGNIFICANT FIGURES

When real numbers are used to represent measured or computed values, it is important to convey the value's precision as well as the number itself. One approach for indicating precision involves enforcement of the rules of *significant figures*. The technically proper method for representing precision involves *statistics,* although significant figures generally yield satisfactory results and are substantially easier to use.

By definition, measured values have a number of significant figures equal to the number of digits that are *certain,* plus one *estimated digit.* As an example, suppose a distance of 24.37 mm is measured with a scale whose smallest graduations are $\frac{1}{10}$ mm. The first three digits are certain, and the fourth (7) is estimated on the scale between the 0.3- and 0.4-mm graduations; thus it is also significant. It is important to record *all* significant figures in measurement. Rounding off or failing to record the last estimated digit (even when it is a zero) is a waste of the extra time spent in obtaining that added precision. Conversely, if more than the actual number of significant figures is recorded, a precision is implied which, in fact, does not exist.

While nonzero digits are always significant, zeros in a recorded value may or may not be significant, depending upon the circumstances. The following rules apply:

1. Zeros between nonzero digits are significant.

   Examples:
   1001   four significant figures
   200.013   six significant figures
   4.05   three significant figures

2. Zeros to the left of the first nonzero digit serve only to position the decimal point and are not significant.

   Examples:
   0.003   one significant figure
   0.057   two significant figures
   0.00281   three significant figures

3. Zeros to the right of nonzero digits which are also to the right of a decimal point are significant.

Examples:

0.10	two significant figures
7.50	three significant figures
483.000	six significant figures

4. Zeros to the right of the rightmost nonzero digit but to the left of an implied decimal point are not significant unless specified by placing a bar over the rightmost significant zero or by moving the decimal point to the left and expressing the number in scientific notation. In this book, scientific notation will be used only on rare occasions since it is more difficult to visualize.

Examples:

$380 = 3.8 \times 10^2$	two significant figures
$38\bar{0} = 3.80 \times 10^2$	three significant figures
$160\bar{0} = 1.600 \times 10^3$	four significant figures
$30\bar{0}0 = 3.00 \times 10^3$	three significant figures

In computations using measured values, it is important that answers be given to a number of significant figures consistent with the number of significant figures in the data used to compute them. Specifying an answer to less than the proper number of significant figures does not take advantage of the precision achieved in measuring the quantities, and giving the answer to more significant figures than are justified is misleading because it implies precision that does not exist.

In adding and subtracting, the calculations are performed without regard to significant figures, but the answer is rounded off to the column dictated by the least-precise value. The least-precise value in a set of numbers that are being added or subtracted is that value whose rightmost significant figure is farthest to the left relative to the decimal point.

Examples:

```
 4.735
 0.67
24.
```
29.405   Answer = 29 (rounded to the ones column governed by 24)

```
 1130
-83.073
```
1046.927   Answer = 1050 (rounded to the tens column governed by 1130)

In multiplication and division, the number of significant figures in answers is equal to the least number of significant figures of any data used in the calculation.

Examples:

$1738 \times 24 = 41,712$	Answer = 42,000 (two significant figures governed by 24)
$648.1 \times 0.0523 = 33.89563$	Answer = 33.9 (three significant figures governed by 0.0523)
$\dfrac{23.985}{13} = 1.845$	Answer = 1.8 (two significant figures governed by 13)

In multiplying or dividing by exact constants (or taking reciprocals), the constants do not govern the number of significant figures in the answer.

Examples:

$$\frac{1}{209.37} = 0.004776233$$

Answer = 0.0047762 (five significant figures governed by 209.37)

Convert 15.73 ft to inches.
15.73 ft × 12 in/ft = 188.76 in

Answer = 188.8 in (four significant figures governed by 15.73, not 12)

In mixed calculations, significant figures must be determined by taking into account the order in which the calculations are performed.

Examples:

$$\frac{(104.3 - 101.7)(538)}{104.3} = \frac{2.6(538)}{104.3} = 13.411$$

Answer = 13 (two significant figures governed by 2.6, the result of the subtraction which was performed first)

Solve for i:

$$\frac{1}{152.417} = \frac{1}{30,000} + \frac{1}{i}$$

$$0.0065609\overline{4}8 = 0.00003\overline{3}333 + \frac{1}{i}$$

$$0.0065609\overline{4}8 - 0.00003\overline{3}333 = 0.006527615 = \frac{1}{i}$$

$i = 153.195$ Answer = 153 (three significant figures governed by 0.0065\overline{2}7615, the result of the subtraction).

Note the use of the overbar which serves as a mnemonic to indicate the rightmost significant figure. This avoids rounding of intermediate results, which can contaminate the answer in some cases.

Trigonometric functions involving angles are complicated with regard to assessing the proper number of significant figures. A good method for determining significant figures is to vary the rightmost significant figure in the value by one unit and compute the function a second time, noting the amount of change from the original computed result. This can be readily seen by subtracting the two values and noting the leftmost significant figure in the difference.

Examples:

tan(7.50°) = 0.1316525

tan(7.51°) = 0.1318301

Difference = 0.0001776

Therefore the answer = 0.1317 (round off in the ten-thousandths place, the position of the first significant figure in the difference).

$$\cos(0.05°) = 0.999999619$$

$$\cos(0.06°) = 0.999999452$$

$$\text{Difference} = 0.000000168$$

Therefore the answer = 0.9999996 (round off in the ten-millionth place, the position of the first significant figure in the difference).

$$\tan^{-1}(0.457) = 24.5604°$$

$$\tan^{-1}(0.458) = 24.6078°$$

$$\text{Difference} = 0.0474°$$

Therefore the answer = 24.56° (round off in the hundredths place, the position of the first significant figure in the difference).

In this text the term *nominal* is frequently used to imply nearness to a given value, e.g., a nominal 6-in-focal-length camera. Nominal in this context may be assumed to imply two additional significant figures (a nominal 6-in-focal-length camera lens therefore means a focal length of 6.00 in).

In making computations it is important to carry out intermediate values to more than the required number of significant figures so as not to contaminate the final result. This is easily achieved by storing intermediate values in a calculator's memory storage registers and recalling them as needed. If this is not feasible, carrying one additional digit in intermediate computations should give adequate results.

A-4   ERROR PROPAGATION

Often it is necessary to estimate the error in a computed quantity which is indirectly determined from measurements that contain error. This can be accomplished through statistical error propagation. Full error propagation assumes that errors in the variables of the equations are *correlated;* i.e., error in one variable is dependent upon error in other variables. Full error propagation is generally not necessary when computing a value as a function of measurements. In its simplest form, error propagation assumes that errors in the measurements are *independent*. Assume that the computed value F, as shown in Eq. (A-1), is a function of n independent variables $x_1, x_2, \ldots, x_n$, which have standard errors $\sigma_1, \sigma_2, \ldots, \sigma_n$, respectively.

$$F = f(x_1, x_2, \ldots, x_n) \tag{A-1}$$

The standard error of the computed value is given by Eq. (A-2).

$$\sigma_F = \pm \sqrt{\left(\frac{\partial F}{\partial x_1}\right)^2 \sigma_1^2 + \left(\frac{\partial F}{\partial x_2}\right)^2 \sigma_2^2 + \cdots + \left(\frac{\partial F}{\partial x_n}\right)^2 \sigma_n^2} \tag{A-2}$$

Example A-1. Compute the value of h and its standard error, based on the following relief displacement equation:

$$h = \frac{dH}{r}$$

where
$$d = 11.3 \text{ mm} \qquad \sigma_d = \pm 0.15 \text{ mm}$$
$$H = 912 \text{ m} \qquad \sigma_H = \pm 3 \text{ m}$$
$$r = 97.2 \text{ mm} \qquad \sigma_r = \pm 0.3 \text{ mm}$$

Solution. First, compute h.

$$h = \frac{11.3 \text{ mm} \times 912 \text{ m}}{97.2 \text{ mm}} = 106.02 \text{ m} \qquad (\text{Answer} = 106 \text{ m, based on significant figures})$$

Next, apply Eq. (A-2).

$$\sigma_h = \pm \sqrt{\left(\frac{\partial h}{\partial d}\right)^2 \sigma_d^2 + \left(\frac{\partial h}{\partial H}\right)^2 \sigma_H^2 + \left(\frac{\partial h}{\partial r}\right)^2 \sigma_r^2}$$

$$= \pm \sqrt{\left(\frac{H}{r}\right)^2 \sigma_d^2 + \left(\frac{d}{r}\right)^2 \sigma_H^2 + \left(\frac{-d \cdot H}{r^2}\right)^2 \sigma_r^2}$$

$$= \pm \sqrt{\left(\frac{912 \text{ m}}{97.2 \text{ mm}}\right)^2 (0.15 \text{ mm})^2 + \left(\frac{11.3 \text{ mm}}{97.2 \text{ mm}}\right)^2 (3 \text{ m})^2 + \left[\frac{-(11.3 \text{ mm})(912 \text{ m})}{(97.2 \text{ mm})^2}\right]^2 (0.3 \text{ mm})^2}$$

$$= \pm \sqrt{1.98 \text{ m}^2 + 0.12 \text{ m}^2 + 0.11 \text{ m}^2} = \pm 1.5 \text{ m}$$

REFERENCES

Anderson, J. M., and E. M. Mikhail: *Surveying: Theory and Practice,* McGraw-Hill, New York, 1998.
Brinker, R. C., and R. Minnick: *The Surveying Handbook,* Chapman & Hall, New York, 1995, chap. 3.
Wolf, P. R., and R. C. Brinker: *Elementary Surveying,* HarperCollins, New York, 1994.
Wolf, P. R., and C. G. Ghilani: *Adjustment Computations, Statistics and Least Squares in Surveying and GIS,* John Wiley & Sons, Inc., New York, 1997.

PROBLEMS

A-1. Convert the following lengths to inches:
  (a) 90.33 mm
  (b) 9.00 mm
  (c) 22.5 cm
  (d) 0.1527 m

A-2. Convert the following lengths to millimeters:
  (a) 6.00 in
  (b) 3.741 in
  (c) 0.17 ft
  (d) 1.22 ft

A-3. Make the following length conversions:
  (a) Express 459.2 m in feet.
  (b) Express 100.00 ft in meters.
  (c) Express 61.5 km in miles.
  (d) Express 87 mi in kilometers.

A-4. Convert the following angles to degrees, minutes, and seconds.
  (a) 24.907°
  (b) 0.106^g
  (c) −12.509^g
  (d) 0.024936^{rad}

A-5. Convert the following angles to grads:
  (a) 71°06′51″
  (b) 0°16′37″
  (c) 10.55°
  (d) 0.9015^{rad}

A-6. Convert the following angles to radians:
  (a) 40.25°
  (b) 5°50′22″
  (c) 11.405^g
  (d) 0.0027^g

A-7. Make the following area conversions.
  (a) Express 1.8917 in square feet.
  (b) Express 9527 ft^2 in hectares.
  (c) Express 175.3 m^2 in square feet.
  (d) Express 21,450 ft^2 in square meters.

A-8. How many significant figures are in the following numbers?
  (a) 4000
  (b) 159.0
  (c) 0.001
  (d) 27,400
  (e) 30$\overline{0}$0
  (f) 24°40′10″

A-9. Express the answers to the following problems to the correct number of significant figures.
  (a) 129 + 3.60 + 0.3 + 20.175
  (b) 3.9 × 508 ÷ 95.3
  (c) The reciprocal of 152.35
  (d) 161.5 × (135.0 − 133.7) ÷ 135.0
  (e) (104.2)^2
  (f) $\sqrt{27.15}$
  (g) The average of 43.29, 43.31, 43.31, and 43.32
  (h) sin(89.7°)

A-10. Compute the estimated error in h_A based on the following parallax equation:

$$h_A = h_C + \frac{(p_a - p_c)(H - h_C)}{p_a}$$

where

$$
\begin{array}{ll}
h_C = 57.31 \text{ m} & \sigma_{h_C} = \pm 0.02 \text{ m} \\
p_a = 88.04 \text{ mm} & \sigma_{p_a} = \pm 0.05 \text{ mm} \\
p_c = 89.10 \text{ mm} & \sigma_{p_c} = \pm 0.05 \text{ mm} \\
H = 918.2 \text{ m} & \sigma_H = \pm 1.0 \text{ m}
\end{array}
$$

INTRODUCTION TO LEAST SQUARES ADJUSTMENT

B-1 INTRODUCTION

As discussed in Sec. A-2, all measurements contain error. With appropriate care, blunders can be avoided; and if an appropriate mathematical model is used, compensation for systematic errors can be provided. No matter how much care is taken, however, random errors will still remain. The theoretically correct method of treating these errors is known as *least squares adjustment*. Least squares is by no means a new method. Karl Gauss, a German mathematician, used the method as early as the latter part of the 18th century. Until the invention of computers, however, it was employed rather sparingly because of the lengthy calculations involved. In order to introduce least squares, some fundamental background concepts will be presented first.

B-2 DEFINITIONS

The following definitions of terms necessarily must precede a discussion of least squares:

Observations. Directly observed (or measured) quantities which contain random errors.

True value. The theoretically correct or exact value of a quantity. From measurements, however, the true value can never be determined, because no matter how much care is exercised in measurement, small random errors will still always be present.

Error. The difference between any measured quantity and the true value for that quantity. Since the true value of a measured quantity can never be determined, errors are likewise indeterminate, and hence they are strictly theoretical quantities. Errors may be estimated by comparing measured or computed values with those obtained by an independent method known to have higher accuracy. For example, error in a ground distance obtained through scaling from a vertical photograph can be estimated by comparing it with a ground-surveyed value obtained with an electronic distance-measuring device.

Most probable value. That value for a measured or indirectly determined quantity which, based upon the observations, has the highest probability. The most probable value (MPV) is determined through least squares adjustment, which is based on the mathematical laws of probability. The most probable value for a quantity that has been *directly* and *independently* measured several times with observations of equal weight is simply the mean, or

$$MPV = \frac{\sum x}{m} \tag{B-1}$$

In Eq. (B-1), $\sum x$ is the sum of the individual measurements, and m is the number of observations. Methods for calculating most probable values of quantities determined through *indirect* observations, which may or may not be equally weighted, are described in later sections of this appendix.

Residual. The difference between any measured quantity and the most probable value for that quantity. It is the value which is dealt with in adjustment computations, since errors are indeterminate. The term *error* is frequently used when *residual* is in fact meant, and although they are very similar, there is this theoretical distinction.

Degrees of freedom. The number of redundant observations (those in excess of the number actually needed to calculate the unknowns). Redundant observations reveal discrepancies in observed values and make possible the practice of least squares adjustment for obtaining most probable values.

Weight. The relative worth of an observation compared to any other observation. Measurements may be weighted in adjustment computations according to their precisions. A very precisely measured value logically should be weighted heavily in an adjustment so that the correction it receives is smaller than that received by a less precise measurement. If the same equipment and procedures are used on a group of measurements, each observation is given an equal weight. Weights are discussed further in Sec. B-6.

Standard deviation. A quantity used to express the precision of a group of measurements. Standard deviation is sometimes called the *root mean square error,* although that designation is somewhat inaccurate. It may also be called the *68 percent error,* since according to the theory of probability, 68 percent of the observations in a group should have residuals smaller than the standard deviation. An expression for the standard deviation of a quantity for which a number of direct, equally weighted observations have been made is

$$S = \pm \sqrt{\frac{\sum v^2}{r}} \tag{B-2}$$

In Eq. (B-2), S is the standard deviation, Σv^2 is the sum of the squares of the residuals, and r is the number of degrees of freedom. In the case of m repeated measurements of the same unknown quantity (a common occurrence in photogrammetry), the first measurement establishes a value for the unknown, and all additional measurements, $m - 1$ in number, are redundant. The units of standard deviation are the same as those of the original measurements.

Example B-1. The 10 values listed in column (a) below were obtained in measuring a photographic distance with a glass scale. Each value was measured using the same instrument and procedures; thus equal weights are assumed. What are the most probable value and the standard deviation of the group of measurements?

(a) Measured values, mm	(b) Residuals, mm	(c) Squared residuals, mm^2
105.27	−0.005	0.000025
105.26	−0.015	0.000225
105.29	0.015	0.000225
105.29	0.015	0.000225
105.30	0.025	0.000625
105.27	−0.005	0.000025
105.26	−0.015	0.000225
105.28	0.005	0.000025
105.28	0.005	0.000025
105.25	−0.025	0.000625
$\Sigma = 1052.75$	$\Sigma = 0.000$	$\Sigma = 0.002250$

1. By Eq. (B-1), the most probable value for a quantity that has been directly and independently measured several times is

$$\text{MPV} = \frac{1052.75}{10} = 105.275 \text{ mm}$$

2. Residuals listed in column (b) above are obtained by subtracting the MPV from each measurement. The squared residuals are listed in column (c).
3. By Eq. (B-2), standard deviation is

$$S = \pm \sqrt{\frac{0.002250}{10 - 1}} = \pm 0.016 \text{ mm}$$

B-3 HISTOGRAMS

A *histogram* is a graphical representation of the distribution of a group of measurements or of the residuals for a group of measurements. It illustrates in an "easily digestible" form the nature of occurrence of random errors. A histogram is simply a bar graph of the sizes of measured values, or the sizes of residuals, as abscissas versus their frequency of occurrence as ordinates. An example histogram of residuals for 50 measurements of a photo distance is shown in Fig. B-1.

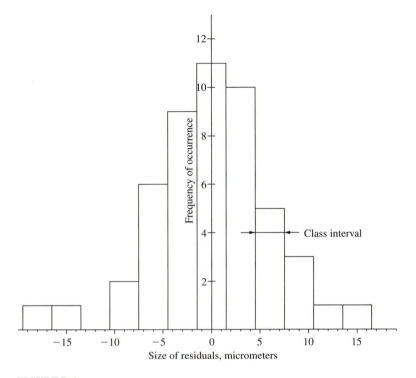

FIGURE B-1
Histogram for 50 measurements of a photographic distance.

Histograms graphically display the following information about a particular group of measurements:

1. Whether the measurements or residuals are symmetrically distributed about some central value.
2. The total spread or dispersion in the measured values or in the residuals.
3. The precision of the measured values. (A tall, narrow histogram represents good precision, while a short, wide one represents poor precision.)

Histograms of varying shape will be obtained by different personnel and for variations in equipment used. Many comparator measurements of a photographic distance, for example, would likely produce a very narrow histogram with a high ordinate at the center. A histogram of the same distance measured the same number of times with an engineer's scale, and plotted at the same scale, would be wider, with a much lower ordinate value at the center.

When a histogram of residuals is plotted, the residuals are calculated and classified into groups or *classes* according to their sizes. The range of residual size in each class is called the *class interval*. The width of bars on a histogram is equal to the class interval, and for the example of Fig. B-1 it is 3 μm. It is important to select a class interval that portrays the distribution of residuals adequately. The value selected will depend upon the total number of measurements in the group. Usually a class interval that produces from 8 to 15 bars (classes) on the graph is ideal. The number of residuals within each class (frequency of residuals) is then counted and plotted on the ordinate scale versus the residual size for the class on the abscissa scale.

B-4   NORMAL DISTRIBUTION OF RANDOM ERRORS

In the remaining discussion it is assumed that all error distributions are *normal*. This is a good assumption in photogrammetry since most distributions are in fact normal or very nearly normal.

A normal distribution, or *gaussian distribution* as it is often called, is one in which the graph has the typical bell-shaped curve shown in Fig. B-2. It is symmetric about the ordinate for a residual of zero. For a very large group of measurements, this curve may be obtained in much the same way as a histogram except that the sizes of residuals are plotted on the abscissa versus their *relative frequencies* of occurrence on the ordinate. The relative frequency of residuals occurring within an interval Δv is simply the ratio of the number of residuals within that interval to the total number of residuals. If the number of measurements is infinitely large and if the size of the interval is taken as infinitesimally small, the resulting curve becomes smooth and continuous, as shown in Fig. B-2.

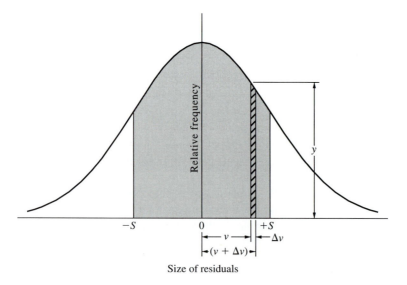

FIGURE B-2
Normal distribution curve.

The equation of the normal distribution curve is

$$y = \frac{1}{\sqrt{2\pi}\sigma} e^{-\frac{(x-\mu)^2}{2\sigma^2}}$$ (B-3)

where y = ordinate of normal distribution curve, equal to relative frequency of occurrence of residuals between the size of v and $v + \Delta v$

   σ = standard deviation

   e = base of natural logarithms

   μ = mean of distribution ($\mu = 0$ in Fig. B-2)

   x = abscissa

The derivation of Eq. (B-3) is beyond the scope of this text, but it can be found in references listed at the end of this appendix. In Fig. B-2 the probability of a residual occurring between the limits of v and $v + \Delta v$ is equal to the crosshatched area under the curve between those limits. It is the product of the ordinate y and the interval Δv. For any single measurement of a group of measurements, the probability that its residual occurs between any two abscissas on the curve (such as between $-S$ and $+S$ of Fig. B-2) is equal to the area under the normal distribution curve between those abscissas. Since for a group of measurements all residuals must fall somewhere on the abscissa scale of the normal distribution curve, the total area under the curve represents total probability and is therefore equal to 1.0. The area under the curve between any two abscissas may be found by integrating Eq. (B-3) between those two abscissa limits. The integration is beyond the scope of this text, but it is pertinent to point out that the area between $-S$ and $+S$ (shown as shaded in Fig. B-2) is 68 percent of the total area under the curve. Hence S is called the 68 percent error, as previously mentioned.

B-5 FUNDAMENTAL CONDITION OF LEAST SQUARES

For a group of equally weighted observations, the fundamental condition which is enforced in least squares adjustment is that *the sum of the squares of the residuals is minimized*. This condition, which has been developed from the equation for the normal distribution curve, provides most probable values for the adjusted quantities. Suppose a group of m equally weighted measurements were taken having residuals $v_1, v_2, v_3, \ldots, v_m$. Then, in equation form, the fundamental condition of least squares is expressed as

$$\sum_{i=1}^{m} v_i^2 = v_1^2 + v_2^2 + v_3^2 + \cdots + v_m^2 = \text{minimum}$$ (B-4)

Some basic assumptions which underlie least squares theory are that the number of observations being adjusted is large and that the frequency distribution of the errors is *normal*. Although these basic assumptions are not always met, least squares adjustment still provides the most rigorous error treatment available, and hence it has become very popular and important in many areas of modern photogrammetry. Besides yielding most probable values for the unknowns, least squares adjustment enables precisions of adjusted quantities to be determined; and it often reveals the presence of large errors and mistakes so that steps can be taken to eliminate them.

B-6   WEIGHTED OBSERVATIONS

Weights of individual observed values may be assigned according to a priori estimates, or they may be obtained from the standard deviations of the observations, if available. An equation expressing the relation between standard deviation and weight is

$$w_i = \frac{1}{S_i^2} \tag{B-5}$$

In Eq. (B-5), w_i is the weight of the ith observed quantity, and S_i^2 is the square of the standard deviation, or *variance* of that observation. Equation (B-5) implies that *weight is inversely proportional to variance*. If measured values are to be weighted in least squares adjustment, then the fundamental condition to be enforced is that the *sum of the weights times their corresponding squared residuals is minimized,* or, in equation form,

$$\sum_{i=1}^{m} w_i v_i^2 = w_1 v_1^2 + w_2 v_2^2 + w_3 v_3^2 + \cdots + w_m v_m^2 = \text{minimum} \tag{B-6}$$

B-7   APPLYING LEAST SQUARES

In the "observation equation" method of least squares adjustment, *observation equations* are written which relate measured values to their residual errors and the unknown parameters. One observation equation is written for each measurement. For a unique solution, the number of equations must equal the number of unknowns. If redundant observations are made, then more observation equations can be written than are needed for a unique solution, and most probable values of the unknown parameters can be determined by the method of least squares. For a group of equally weighted observations, an equation for each residual error is obtained from each observation. The residuals are squared and added to obtain the function $\sum_{i=1}^{m} v_i^2$. To minimize the function, partial derivatives are taken with respect to each unknown variable and set equal to zero. This yields a set of equations called *normal equations* which are equal in number to the number of unknowns. The normal equations are solved to obtain the most probable values for the unknowns.

> **Example B-2.** As an elementary example illustrating the method of least squares adjustment by the observation equation method, consider the following three equally weighted measurements taken between points A, B, and C of Fig. B-3:
>
> $$x + y = 3.0$$
> $$x = 1.5$$
> $$y = 1.4$$
>
> These three equations relate the two unknowns x and y to the observations. Values for x and y could be obtained from any two of these equations; therefore, the remaining equation is redundant. Notice, however, that the values obtained for x

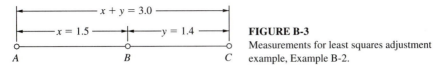

FIGURE B-3
Measurements for least squares adjustment
example, Example B-2.

and y will differ, depending upon which two equations are solved. It is therefore apparent that the measurements contain errors. The equations may be rewritten as observation equations by including residual errors as follows:

$$x + y = 3.0 + v_1$$

$$x = 1.5 + v_2$$

$$y = 1.4 + v_3$$

To arrive at the least squares solution, the observation equations are rearranged to obtain expressions for the residuals; these are squared and added to form the function $\sum_{i=1}^{m} v_i^2$ as follows:

$$\sum_{i=1}^{m} v_i^2 = (x + y - 3.0)^2 + (x - 1.5)^2 + (y - 1.4)^2$$

The above function is minimized, enforcing the condition of least squares, by taking partial derivatives with respect to each unknown and setting them equal to zero. This yields the following two equations:

$$\frac{\partial \sum v^2}{\partial x} = 0 = 2(x + y - 3.0) + 2(x - 1.5)$$

$$\frac{\partial \sum v^2}{\partial y} = 0 = 2(x + y - 3.0) + 2(y - 1.4)$$

The above equations are the normal equations, and expressed in reduced form they are

$$2x + y = 4.5$$

$$x + 2y = 4.4$$

Solving the reduced normal equations simultaneously yields $x = 1.533$ and $y = 1.433$. According to the theory of least squares, these values have the highest probability. Having the most probable values for the unknowns, the residuals can be calculated by substitution back into the original observation equations, or

$$v_1 = 1.533 + 1.433 - 3.000 = -0.033$$

$$v_2 = 1.533 - 1.500 = 0.033$$

$$v_3 = 1.433 - 1.400 = 0.033$$

The above example is indeed simple, but it serves to illustrate the method of least squares without complicating the mathematics. Least squares adjustment of large systems of observations is performed in the same manner.

B-8 SYSTEMATIC FORMULATION OF NORMAL EQUATIONS

In large systems of observations, it is helpful to utilize systematic procedures to formulate normal equations. Consider the following system of m linear observation equations of equal weight containing n unknowns:

$$
\begin{aligned}
a_{11} X_1 + a_{12} X_2 + a_{13} X_3 + \cdots + a_{1n} X_n - L_1 &= v_1 \\
a_{21} X_1 + a_{22} X_2 + a_{23} X_3 + \cdots + a_{2n} X_n - L_2 &= v_2 \\
a_{31} X_1 + a_{32} X_2 + a_{33} X_3 + \cdots + a_{3n} X_n - L_3 &= v_3 \\
\cdots\cdots\cdots\cdots\cdots\cdots\cdots\cdots\cdots\cdots\cdots\cdots \\
a_{m1} X_1 + a_{m2} X_2 + a_{m3} X_3 + \cdots + a_{mn} X_n - L_m &= v_m
\end{aligned} \tag{B-7}
$$

In Eqs. (B-7) the a_{ij}'s are coefficients of the unknown X_j's; the L_i's are the observations; and the v_i's are the residuals. By squaring the residuals and summing them, the function Σv^2 is formed. Taking partial derivatives of Σv^2 with respect to each unknown X_j yields n normal equations. After reducing and factoring the normal equations, the following generalized system for expressing normal equations results:

$$
\begin{aligned}
\sum_{i=1}^{m}(a_{i1}a_{i1})X_1 + \sum_{i=1}^{m}(a_{i1}a_{i2})X_2 + \sum_{i=1}^{m}(a_{i1}a_{i3})X_3 + \cdots + \sum_{i=1}^{m}(a_{i1}a_{in})X_n &= \sum_{i=1}^{m}(a_{i1}L_i) \\
\sum_{i=1}^{m}(a_{i2}a_{i1})X_1 + \sum_{i=1}^{m}(a_{i2}a_{i2})X_2 + \sum_{i=1}^{m}(a_{i2}a_{i3})X_3 + \cdots + \sum_{i=1}^{m}(a_{i2}a_{in})X_n &= \sum_{i=1}^{m}(a_{i2}L_i) \\
\sum_{i=1}^{m}(a_{i3}a_{i1})X_1 + \sum_{i=1}^{m}(a_{i3}a_{i2})X_2 + \sum_{i=1}^{m}(a_{i3}a_{i3})X_3 + \cdots + \sum_{i=1}^{m}(a_{i3}a_{in})X_n &= \sum_{i=1}^{m}(a_{i3}L_i) \\
\cdots\cdots\cdots\cdots\cdots\cdots\cdots\cdots\cdots\cdots\cdots\cdots \\
\sum_{i=1}^{m}(a_{in}a_{i1})X_1 + \sum_{i=1}^{m}(a_{in}a_{i2})X_2 + \sum_{i=1}^{m}(a_{in}a_{i3})X_3 + \cdots + \sum_{i=1}^{m}(a_{in}a_{in})X_n &= \sum_{i=1}^{m}(a_{in}L_i)
\end{aligned} \tag{B-8}
$$

It may be similarly shown that normal equations may be systematically formed from weighted observation equations in the following manner:

$$
\begin{aligned}
\sum_{i=1}^{m}(w_i a_{i1}a_{i1})X_1 + \sum_{i=1}^{m}(w_i a_{i1}a_{i2})X_2 + \sum_{i=1}^{m}(w_i a_{i1}a_{i3})X_3 + \cdots + \sum_{i=1}^{m}(w_i a_{i1}a_{in})X_n &= \sum_{i=1}^{m}(w_i a_{i1}L_i) \\
\sum_{i=1}^{m}(w_i a_{i2}a_{i1})X_1 + \sum_{i=1}^{m}(w_i a_{i2}a_{i2})X_2 + \sum_{i=1}^{m}(w_i a_{i2}a_{i3})X_3 + \cdots + \sum_{i=1}^{m}(w_i a_{i2}a_{in})X_n &= \sum_{i=1}^{m}(w_i a_{i2}L_i) \\
\sum_{i=1}^{m}(w_i a_{i3}a_{i1})X_1 + \sum_{i=1}^{m}(w_i a_{i3}a_{i2})X_2 + \sum_{i=1}^{m}(w_i a_{i3}a_{i3})X_3 + \cdots + \sum_{i=1}^{m}(w_i a_{i3}a_{in})X_n &= \sum_{i=1}^{m}(w_i a_{i3}L_i) \\
\cdots\cdots\cdots\cdots\cdots\cdots\cdots\cdots\cdots\cdots\cdots\cdots \\
\sum_{i=1}^{m}(w_i a_{in}a_{i1})X_1 + \sum_{i=1}^{m}(w_i a_{in}a_{i2})X_2 + \sum_{i=1}^{m}(w_i a_{in}a_{i3})X_3 + \cdots + \sum_{i=1}^{m}(w_i a_{in}a_{in})X_n &= \sum_{i=1}^{m}(w_i a_{in}L_i)
\end{aligned} \tag{B-9}
$$

In Eqs. (B-9) the terms are as described previously, except that the w_i's are the relative weights of the individual observations.

The formulation of normal equations from observation equations may be further systematized by handling the systems of Eqs. (B-8) or (B-9) in a tabular manner.

Example B-3. Using the tabular method, form the normal equations for Example B-2.

Solution

Equation number (i)	a_{i1}	a_{i2}	L_i	$a_{i1}a_{i1}$	$a_{i1}a_{i2}$	$a_{i2}a_{i2}$	$a_{i1}L_i$	$a_{i2}L_i$
1	1	1	3.0	1	1	1	3.0	3.0
2	1	0	1.5	1	0	0	1.5	0.0
3	0	1	1.4	0	0	1	0.0	1.4
				$\Sigma = 2$	$\Sigma = 1$	$\Sigma = 2$	$\Sigma = 4.5$	$\Sigma = 4.4$

For this example, normal equations are formed by satisfying Eqs. (B-8) as follows (note that because of the commutative property of multiplication, $a_{i1}a_{i2} = a_{i2}a_{i1}$):

$$\sum_{i=1}^{m}(a_{i1}a_{i1})X_1 + \sum_{i=1}^{m}(a_{i1}a_{i2})X_2 = \sum_{i=1}^{m}(a_{i1}L_i)$$

$$\sum_{i=1}^{m}(a_{i2}a_{i1})X_1 + \sum_{i=1}^{m}(a_{i2}a_{i2})X_2 = \sum_{i=1}^{m}(a_{i2}L_i)$$

Substituting the appropriate values from the above table yields the required normal equations as follows:

$$2x + y = 4.5$$

$$x + 2y = 4.4$$

B-9 MATRIX METHODS IN LEAST SQUARES ADJUSTMENT

It has been previously mentioned that least squares computations are quite lengthy and are therefore most economically performed on a computer. The algebraic approach—Eqs. (B-8) and (B-9)—for forming normal equations, and for obtaining their simultaneous solution, can be programmed for computer solution or set up in a computer spreadsheet. The procedure is much more easily adapted to matrix methods, however. Readers may wish to consult references listed at the end of this appendix for a review of matrix terminology and operations.

In developing matrix equations for least squares computations, analogy will be made to the algebraic approach presented in Sec. B-8. First, observation Eqs. (B-7) may be represented in matrix form as

$$_mA^n {}_nX^1 = {}_mL^1 + {}_mV^1 \tag{B-10}$$

where

$$
{}_m A^n = \begin{bmatrix}
a_{11} & a_{12} & a_{13} & \cdots & a_{1n} \\
a_{21} & a_{22} & a_{23} & \cdots & a_{2n} \\
a_{31} & a_{32} & a_{33} & \cdots & a_{3n} \\
a_{41} & a_{42} & a_{43} & \cdots & a_{4n} \\
\cdots & \cdots & \cdots & \cdots & \cdots \\
a_{m1} & a_{m2} & a_{m3} & \cdots & a_{mn}
\end{bmatrix}
\qquad
{}_n X^1 = \begin{bmatrix}
X_1 \\ X_2 \\ X_3 \\ \vdots \\ X_n
\end{bmatrix}
$$

$$
{}_m L^1 = \begin{bmatrix}
L_1 \\ L_2 \\ L_3 \\ L_4 \\ \vdots \\ L_m
\end{bmatrix}
\qquad
{}_m V^1 = \begin{bmatrix}
V_1 \\ V_2 \\ V_3 \\ V_4 \\ \vdots \\ V_m
\end{bmatrix}
$$

Upon studying the following matrix representation, it will be noticed that normal Eqs. (B-8) are obtained as follows:

$$(A^T A)X = A^T L \tag{B-11}$$

In the above equation, $A^T A$ is the matrix of normal equation coefficients of the unknowns. Premultiplying both sides of Eq. (B-11) by $(A^T A)^{-1}$ and reducing, results in:

$$(A^T A)^{-1}(A^T A)X = (A^T A)^{-1}(A^T L)$$

$$IX = (A^T A)^{-1}(A^T L)$$

$$X = (A^T A)^{-1}(A^T L) \tag{B-12}$$

In the above reduction, I is the identity matrix. Equation (B-12) is the basic least squares matrix equation for *equally weighted* observations. The matrix X consists of most probable values for unknowns $X_1, X_2, X_3, \ldots, X_n$. For a system of *weighted* observations, the following matrix equation provides the X matrix of most probable values for the unknowns:

$$X = (A^T WA)^{-1}(A^T WL) \tag{B-13}$$

In Eq. (B-13) the matrices are identical to those of the equally weighted equations, with the inclusion of the W matrix, which is a diagonal matrix of weights and is defined as follows:

$$
W = \begin{bmatrix}
w_1 & 0 & 0 & 0 & \cdots & 0 \\
0 & w_2 & 0 & 0 & \cdots & 0 \\
0 & 0 & w_3 & 0 & \cdots & 0 \\
0 & 0 & 0 & w_4 & \cdots & 0 \\
\cdots & \cdots & \cdots & \cdots & \cdots & \cdots \\
0 & 0 & 0 & 0 & \cdots & w_m
\end{bmatrix}
$$

In the preceding W matrix, all off-diagonal elements are shown as zeros. This is proper when the individual observations are independent and uncorrelated; i.e., they are not dependent upon each other. This is often the case in photogrammetric applications.

Example B-4. Solve Example B-2 by using matrix methods.

The observation equations of Example B-2 may be expressed in matrix form as follows:

$$_3A^2 \, _3X^1 = \, _3L^1 + \, _3V^1$$

where

$$_3A^2 = \begin{bmatrix} 1 & 1 \\ 1 & 0 \\ 0 & 1 \end{bmatrix} \qquad _3X^1 = \begin{bmatrix} x \\ y \end{bmatrix} \qquad _3L^1 = \begin{bmatrix} 3.0 \\ 1.5 \\ 1.4 \end{bmatrix} \qquad _3V^1 = \begin{bmatrix} v_1 \\ v_2 \\ v_3 \end{bmatrix}$$

Solving matrix Eq. (B-12) gives

$$A^T A = \begin{bmatrix} 1 & 1 & 0 \\ 1 & 0 & 1 \end{bmatrix} \begin{bmatrix} 1 & 1 \\ 1 & 0 \\ 0 & 1 \end{bmatrix} = \begin{bmatrix} 2 & 1 \\ 1 & 2 \end{bmatrix}$$

$$(A^T A)^{-1} = \begin{bmatrix} 0.6667 & -0.3333 \\ -0.3333 & 0.6667 \end{bmatrix} \qquad A^T L = \begin{bmatrix} 4.5 \\ 4.4 \end{bmatrix}$$

$$X = (A^T A)^{-1} (A^T L) = \begin{bmatrix} 0.6667 & -0.3333 \\ -0.3333 & 0.6667 \end{bmatrix} \begin{bmatrix} 4.5 \\ 4.4 \end{bmatrix} = \begin{bmatrix} 1.533 \\ 1.433 \end{bmatrix}$$

Note that this solution yields exactly the same values for x and y as were obtained through the algebraic approach of Example B-2.

B-10   MATRIX EQUATIONS FOR PRECISIONS OF ADJUSTED QUANTITIES

The matrix equation for calculating residuals after adjustment, whether the adjustment is weighted or not, is

$$V = AX - L \tag{B-14}$$

The standard deviation of unit weight for an unweighted adjustment is

$$S_0 = \sqrt{\frac{V^T V}{r}} \tag{B-15}$$

The standard deviation of unit weight for a weighted adjustment is

$$S_0 = \sqrt{\frac{V^T W V}{r}} \tag{B-16}$$

In Eqs. (B-15) and (B-16), r is the number of degrees of freedom and equals the number of observation equations minus the number of unknowns, or $r = m - n$.

Standard deviations of the adjusted quantities are

$$S_{X_i} = S_0 \sqrt{Q_{X_i X_i}} \qquad (B\text{-}17)$$

In Eq. (B-17), S_{X_i} is the standard deviation of the ith adjusted quantity, i.e., the quantity in the ith row of the X matrix; S_0 is the standard deviation of unit weight as calculated by Eq. (B-15) or (B-16); and $Q_{X_i X_i}$ is the element in the ith row and the ith column of the matrix $(A^T A)^{-1}$ in the unweighted case, or the matrix $(A^T W A)^{-1}$ in the weighted case. The matrices $S_0^2 (A^T A)^{-1}$ and $S_0^2 (A^T W A)^{-1}$ are the *covariance* matrices.

Example B-5. Calculate the standard deviation of unit weight and the standard deviations of the adjusted quantities x and y for the unweighted problem of Example B-4.

(*a*) By Eq. (B-14) the residuals are

$$V = \begin{bmatrix} 1 & 1 \\ 1 & 0 \\ 0 & 1 \end{bmatrix} \begin{bmatrix} 1.533 \\ 1.433 \end{bmatrix} - \begin{bmatrix} 3.0 \\ 1.5 \\ 1.4 \end{bmatrix} = \begin{bmatrix} -0.033 \\ 0.033 \\ 0.033 \end{bmatrix}$$

(*b*) By Eq. (B-15) the standard deviation of unit weight is

$$V^T V = \begin{bmatrix} -0.033 & 0.033 & 0.033 \end{bmatrix} \begin{bmatrix} -0.033 \\ 0.033 \\ 0.033 \end{bmatrix} = 0.0033$$

$$S_0 = \sqrt{\frac{0.0033}{3-2}} = \pm 0.057$$

(*c*) Using Eq. (B-17), the standard deviations of the adjusted values for x and y are calculated as

$$S_x = \pm 0.057 \sqrt{0.6667} = \pm 0.047$$
$$S_y = \pm 0.057 \sqrt{0.6667} = \pm 0.047$$

In part (*c*) above, the numbers 0.6667 within the radicals are the $(1,1)$ and $(2,2)$ elements of the $(A^T A)^{-1}$ matrix of Example B-4. The interpretation of the standard deviations computed under part (*c*) is that there is a 68 percent probability that the true values for x and y are within ± 0.047 of their adjusted values. Note that for this simple example the magnitudes of the three residuals calculated in part (*a*) were equal, and that the standard deviations of x and y were equal in part (*c*). This is due to the symmetric nature of this particular problem (illustrated in Fig. B-3), but this is seldom, if ever, the case with more complex problems.

B-11   PRACTICAL EXAMPLE

The following example is presented to illustrate a practical application of least squares in photogrammetry. The example also shows the method of calculating the coefficients of a polynomial which approximates the symmetric radial-lens distortion curve for an aerial camera (see Sec. 4-13).

Example B-6. From the calibration data of an aerial camera given in the table below, calculate the coefficients of a polynomial which models the symmetric radial-lens distortion curve.

Radial distance r, mm	Radial-lens distortion Δr, mm
20.170	0.004
41.051	0.007
63.460	0.007
88.454	0.001
107.276	−0.003
128.555	−0.004

Solution. As presented in Sec. 4-13, a polynomial of the following form is the appropriate model for symmetric radial-lens distortion:

$$\Delta r = k_1 r + k_2 r^3 + k_3 r^5 + k_4 r^7 \tag{B-18}$$

In Eq. (B-18), Δr is the symmetric radial-lens distortion at a radial distance r from the principal point. The k's are coefficients which define the shape of the distortion curve. One equation of the form of Eq. (B-18) can be written for each radial distance at which the distortion is known from calibration. Since there are four k's, four equations are required to obtain a unique solution for them. From calibration, distortions are determined for six radial distances; hence six equations can be written, and the k's may be computed by least squares. In polynomial equations of this type, there is commonly a problem with ill-conditioning of the normal equations due to having numbers raised to large powers. To reduce this problem, it is useful to convert the radial distances to meters so that when raised to the seventh power, the result is less than 1. This is not a foolproof approach, although it is effective in many cases.

     Based on the calibration data, the following observation equations, in the matrix form of Eq. (B-10), may be written (note that the radial distances have been converted to meters):

$$\begin{bmatrix} (0.020170) & (0.020170)^3 & (0.020170)^5 & (0.020170)^7 \\ (0.041051) & (0.041051)^3 & (0.041051)^5 & (0.041051)^7 \\ (0.063460) & (0.063460)^3 & (0.063460)^5 & (0.063460)^7 \\ (0.088454) & (0.088454)^3 & (0.088454)^5 & (0.088454)^7 \\ (0.107276) & (0.107276)^3 & (0.107276)^5 & (0.107276)^7 \\ (0.128555) & (0.128555)^3 & (0.128555)^5 & (0.128555)^7 \end{bmatrix} \begin{bmatrix} k_1 \\ k_2 \\ k_3 \\ k_4 \end{bmatrix} = \begin{bmatrix} 0.004 \\ 0.007 \\ 0.007 \\ 0.001 \\ -0.003 \\ -0.004 \end{bmatrix} + \begin{bmatrix} v_1 \\ v_2 \\ v_3 \\ v_4 \\ v_5 \\ v_6 \end{bmatrix}$$

The least squares solution in the form of Eq. (B-11) gives

$$
\begin{bmatrix}
4.19778 \times 10^{-2} & 4.85999 \times 10^{-4} & 6.58695 \times 10^{-6} & 9.61536 \times 10^{-8} \\
4.85999 \times 10^{-4} & 6.58695 \times 10^{-6} & 9.61536 \times 10^{-8} & 1.46503 \times 10^{-9} \\
6.58695 \times 10^{-6} & 9.61536 \times 10^{-8} & 1.46503 \times 10^{-9} & 2.29302 \times 10^{-11} \\
9.61536 \times 10^{-8} & 1.46503 \times 10^{-9} & 2.29302 \times 10^{-11} & 3.65246 \times 10^{-13}
\end{bmatrix} X =
\begin{bmatrix}
6.46630 \times 10^{-5} \\
-9.20374 \times 10^{-6} \\
-1.69618 \times 10^{-7} \\
-2.73878 \times 10^{-9}
\end{bmatrix}
$$

Solving this equation for the unknowns gives the following:

$$
X = \begin{bmatrix} k_1 \\ k_2 \\ k_3 \\ k_4 \end{bmatrix} = \begin{bmatrix} 0.229581 \\ -35.8926 \\ 1{,}018.26 \\ 12{,}105.0 \end{bmatrix}
$$

By using these k's in Eq. (B-18), estimates for radial-lens distortions (in millimeters) for any value of r (in meters) may be readily calculated.

REFERENCES

American Society of Photogrammetry: *Manual of Photogrammetry,* 4th ed., Bethesda, MD, 1980, chap. 2.

Benjamin, J. R., and C. A. Cornell: *Probability, Statistics and Decision for Civil Engineers,* McGraw-Hill, New York, 1970.

Bhattacharyya, G. K., and R. A. Johnson: *Statistical Concepts and Methods,* Wiley, New York, 1977.

Crandall, K. C., and R. W. Seabloom: *Engineering Fundamentals in Measurements, Probability and Dimensions,* McGraw-Hill, New York, 1970.

Hardy, R. L.: "Least Squares Prediction," *Photogrammetric Engineering and Remote Sensing,* vol. 43, no. 4, 1977, p. 475.

Hirvonen, R. A.: *Adjustment by Least Squares in Photogrammetry and Geodesy,* Frederick Ungar Publishing, New York, 1971.

Mikhail, E. M.: "Parameter Constraints in Least Squares," *Photogrammetric Engineering,* vol. 36, no. 12, 1970, p. 1277.

————: *Observations and Least Squares,* Harper & Row, New York, 1976.

Rampal, K. K.: "Least Squares Collocation in Photogrammetry," *Photogrammetric Engineering and Remote Sensing,* vol. 42, no. 5, 1976, p. 659.

Wolf, P. R., and C. Ghilani: *Adjustment Computations: Statistics and Least Squares in Surveying and GIS,* Wiley, New York, 1997.

Wong, K. W.: "Propagation of Variance and Covariance," *Photogrammetric Engineering and Remote Sensing,* vol. 41, no. 1, 1975, p. 75.

Zimmerman, D. S.: "Least Squares by Diagonal Partitioning," *Canadian Surveyor,* vol. 28, no. 5, 1974, p. 677.

PROBLEMS

B-1. A photogrammetric distance was measured 10 times using the same equipment and procedures with the following results: 95.76, 95.68, 95.70, 95.72, 95.69, 95.75, 95.72, 95.77, 95.70, and 95.71 mm. Calculate the most probable value for the photo distance and the standard deviation of the group of measurements.

B-2. Repeat Prob. B-1, except that the following 15 measurements were obtained: 64.29, 64.26, 64.31, 64.29, 64.34, 64.28, 64.30, 64.30, 64.33, 64.29, 64.35, 64.31, 64.28, 64.32, and 64.33 mm.

B-3. Compute the most probable values of unknowns x_1, x_2, and x_3 for the following observation equations, using a computer spreadsheet, and calculate the standard deviations of the adjusted quantities.

$$2x_1 + 3x_2 + x_3 = 16$$
$$x_1 - 2x_2 + 3x_3 = -9$$
$$7x_1 + x_2 - 2x_3 = 21$$
$$-x_1 - x_2 - x_3 = -5$$

B-4. Suppose the constant terms 16, -9, 21, and -5 of the four equations of Prob. B-3 represent measurements having relative weights of 2, 3, 1, and 4, respectively. Using weighted least squares, calculate most probable values for x_1, x_2, and x_3 and determine the standard deviations of these values.

B-5. Repeat Prob. B-3, except that the four equations are as follows:

$$5x_1 + 3x_2 - 2x_3 = -15$$
$$2x_1 - x_2 + 6x_3 = 25$$
$$x_1 + 3x_2 + x_3 = 13$$
$$-4x_1 + 3x_2 - 3x_3 = 1$$

B-6. If the constant terms -15, 25, 13, and 1 of Prob. B-5 represent measurements having relative weights of 4, 2, 1, and 2, respectively, calculate the least squares solution for the unknowns and determine their standard deviations.

APPENDIX

C

COORDINATE TRANSFORMATIONS

C-1   INTRODUCTION

A problem frequently encountered in photogrammetric work is conversion from one rectangular coordinate system to another. This is because photogrammetrists commonly determine coordinates of unknown points in convenient *arbitrary* rectangular coordinate systems. These arbitrary coordinates may be read from comparators or stereoscopic plotters, or they may result from analytic computation. The arbitrary coordinates must then be converted to a *final* system, such as the camera photo coordinate system in the case of comparator measurements, or to a ground coordinate system, such as the state plane coordinate system in the case of stereoplotter or analytically derived arbitrary model coordinates. The procedure for converting from one coordinate system to another is known as *coordinate transformation*. The procedure requires that some points have their coordinates known (or measured) in both the arbitrary and the final coordinate systems. Such points are called *control points*.

C-2   TWO-DIMENSIONAL CONFORMAL COORDINATE TRANSFORMATION

The term *two-dimensional* means that the coordinate systems lie on plane surfaces. A *conformal* transformation is one in which true shape is preserved after transformation. To perform a two-dimensional conformal coordinate transformation, it is necessary that

518

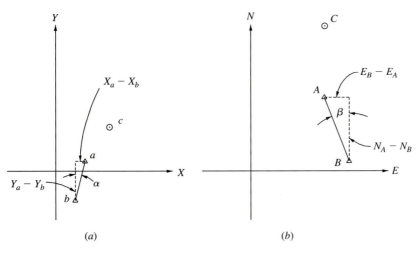

FIGURE C-1
(*a*) Arbitrary *XY* two-dimensional coordinate system. (*b*) Ground *EN* two-dimensional system.

coordinates of at least two points be known in both the arbitrary and final coordinate systems. Accuracy in the transformation is improved by choosing the two points as far apart as possible. If more than two control points are available, an improved solution may be obtained by applying the method of least squares.

A two-dimensional conformal coordinate transformation consists of three basic steps: (1) scale change, (2) rotation, and (3) translation. The example illustrated in Fig. C-1 is used to demonstrate the procedure. This example uses the minimum of two control points. Section C-4 describes the procedure when more than two control points are available. Figure C-1*a* shows the positions of points *a* through *c*, whose coordinates are known in an arbitrary *XY* system. Figure C-1*b* illustrates the positions of the same points, labeled *A* through *C* in a (ground) *EN* system. The coordinates of *A* and *B* are known in the ground system, and it is required to determine the coordinates of *C* in the ground system.

Step 1: Scale Change

By comparing Figs. C-1*a* and *b*, it is evident that the lengths of lines *ab* and *AB* are unequal, hence the scales of the two coordinate systems are unequal. The scale of the *XY* system is made equal to that of the *EN* system by multiplying each *X* and *Y* coordinate by a scale factor *s*. The scaled coordinates are designated as *X'* and *Y'*. By use of the two control points, the scale factor is calculated in relation to the two lengths *AB* and *ab* as

$$s = \frac{AB}{ab} = \frac{\sqrt{(E_B - E_A)^2 + (N_B - N_A)^2}}{\sqrt{(X_b - X_a)^2 + (Y_b - Y_a)^2}} \qquad (C\text{-}1)$$

Step 2: Rotation

If the scaled $X'Y'$ coordinate system is superimposed over the EN system of Fig. C-1b so that line AB in both systems coincides, the result is as shown in Fig. C-2. An auxiliary axis system $E'N'$ is constructed through the origin of the $X'Y'$ axis system parallel to the EN axes. It is necessary to rotate from the $X'Y'$ system to the $E'N'$ system, or in other words, to calculate $E'N'$ coordinates for the unknown points from their $X'Y'$ coordinates. The $E'N'$ coordinates of point C may be calculated in terms of the clockwise angle θ by using the following equations:

$$E'_C = X'_C \cos \theta - Y'_C \sin \theta$$
$$N'_C = X'_C \sin \theta + Y'_C \cos \theta \tag{C-2}$$

Rotation angle θ, shown in Fig. C-2, is the sum of angles α and β which are indicated on Figs. C-1a and b. From coordinates of the two control points, these angles are calculated as

$$\alpha = \tan^{-1}\left(\frac{X_a - X_b}{Y_a - Y_b}\right)$$

$$\beta = \tan^{-1}\left(\frac{E_B - E_A}{N_A - N_B}\right) \tag{C-3}$$

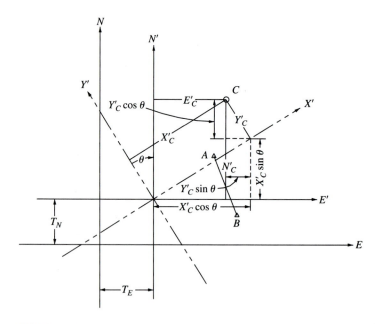

FIGURE C-2
Scaled $X'Y'$ coordinate system superimposed onto the EN ground coordinate system.

Step 3: Translation

The final step in the coordinate transformation is a translation of the origin of the $E'N'$ system to the origin of the EN system. The translation factors required are T_E and T_N, which are illustrated in Fig. C-2. Final E and N ground coordinates for points C then are

$$E_C = E'_C + T_E$$
$$N_C = N'_C + T_N \tag{C-4}$$

where the translation factors T_E and T_N are calculated as

$$T_E = E_A - E'_A = E_B - E'_B$$
$$T_N = N_A - N'_A = N_B - N'_B \tag{C-5}$$

Note from Eqs. (C-5) that these translation factors may be calculated in two different ways by using either control point A or B. It is advisable to calculate them by using both points, to obtain a computation check.

Working sketches are recommended in computing coordinate transformations to aid in reducing the likelihood of mistakes. Caution should be exercised to ensure that correct algebraic signs are applied to the coordinates used in the transformation equations.

Example C-1. Assume that in Figs. C-1a and b the arbitrary and ground coordinates of points A through C are as follows:

Point	Arbitrary coordinates		Ground coordinates	
	X	Y	E	N
A	632.17	121.45	1100.64	1431.09
B	355.20	−642.07	1678.39	254.15
C	1304.81	596.37		

It is required to compute the coordinates of point C in the ground EN system.

Solution.

(a) The scale factor is calculated from Eq. (C-1) as

$$s = \frac{\sqrt{(1678.39 - 1100.64)^2 + (254.15 - 1431.09)^2}}{\sqrt{(355.20 - 632.17)^2 + (-642.07 - 121.45)^2}}$$

$$= \frac{1311.10}{812.20} = 1.61425$$

The arbitrary coordinates are then expanded to the $X'Y'$ system, which is equal in scale to the ground coordinate system, by multiplying each of the arbitrary coordinates by the scale factor. After multiplication, the $X'Y'$ coordinates are as follows:

Point	Scaled coordinates X'	Y'
A	1020.48	196.05
B	573.38	-1036.46
C	2106.29	962.69

(b) The rotation angle, calculated by Eqs. (C-3), is

$$\tan \alpha = \frac{632.17 - 355.20}{121.45 + 642.07} = 0.362754 \qquad \alpha = 19.9384°$$

$$\tan \beta = \frac{1678.39 - 1100.64}{1431.09 + 254.15} = 0.490892 \qquad \beta = 26.1460°$$

$$\theta = 19.9384° + 26.1460° = 46.0845°$$

Rotation Eqs. (C-2) are then solved to obtain E' and N' coordinates. The solution in tabular form is as follows (with $\sin \theta = 0.720363$, and $\cos \theta = 0.693597$):

Point	$X' \cos \theta$	$Y' \sin \theta$	$X' \sin \theta$	$Y' \cos \theta$	E'	N'
A	707.80	141.23	735.12	135.98	566.57	871.10
B	397.70	-746.63	413.04	-718.89	1144.32	-305.84
C	1460.92	693.49	1517.29	667.72	767.43	2185.01

(c) The translation factors T_E and T_N are calculated next, using Eqs. (C-5) as follows:

$$T_E = E_A - E_A' = 1100.64 - 566.58 = 534.07$$

also $\quad T_E = E_B - E_B' = 1678.39 - 1144.32 = 534.07 \qquad$ *Check!*

$$T_N = N_A - N_A' = 1431.09 - 871.10 = 559.99$$

and $\quad T_N = N_B - N_B' = 254.15 + 305.84 = 559.99 \qquad$ *Check!*

By adding the translation factors to E_C' and N_C', the following final transformed E and N coordinates are obtained for point C:

$$E_C = 767.43 + 534.07 = 1301.49$$

$$N_C = 2185.01 + 559.99 = 2745.01$$

In the above example, although only one unknown point (point C) was transformed, any number of points could have been transformed by using just the two control points.

C-3   ALTERNATIVE METHOD OF TWO-DIMENSIONAL CONFORMAL COORDINATE TRANSFORMATION

When a computer is used, it is advantageous to compute coordinate transformations by an alternative method. In this method, equations involving four transformation coefficients are formulated in terms of the coordinates of the two or more points whose positions are known in both coordinate systems. The formation of the equations follows the same three steps discussed in Sec. C-2. The procedure consists of first multiplying each of the original coordinates of points a and b by a scale factor. The following four equations result:

$$X'_a = sX_a$$
$$Y'_a = sY_a$$
$$X'_b = sX_b \tag{C-6}$$
$$Y'_b = sY_b$$

Equations (C-6) are now substituted into Eqs. (C-2), except that the subscripts of Eqs. (C-2) are changed to be applicable for points A and B. This substitution yields

$$E'_A = sX_a \cos \theta - sY_a \sin \theta$$
$$N'_A = sX_a \sin \theta + sY_a \cos \theta$$
$$E'_B = sX_b \cos \theta - sY_b \sin \theta \tag{C-7}$$
$$N'_B = sX_b \sin \theta + sY_b \cos \theta$$

Finally, translation factors T_E and T_N, as described previously, are added to Eqs. (C-7) to yield the following equations:

$$E_A = sX_a \cos \theta - sY_a \sin \theta + T_E$$
$$N_A = sX_a \sin \theta + sY_a \cos \theta + T_N$$
$$E_B = sX_b \cos \theta - sY_b \sin \theta + T_E \tag{C-8}$$
$$N_B = sX_b \sin \theta + sY_b \cos \theta + T_N$$

Let $a = s \cos \theta$ and $b = s \sin \theta$. Notice that two new variables are being introduced, which are independent functions of two existing variables. This is essential so that the total number of unknown coefficients will remain the same. By substitution, Eqs. (C-8) become

$$E_A = aX_a - bY_a + T_E$$
$$N_A = aY_a + bX_a + T_N$$
$$E_B = aX_b - bY_b + T_E \tag{C-9}$$
$$N_B = aY_b + bX_b + T_N$$

Because both the *XY* and *EN* coordinates for points *A* and *B* are known, Eqs. (C-9) contain only four unknowns, the transformation parameters a, b, T_E, and T_N. The four equations may be solved simultaneously to obtain values for the unknowns. When the four transformation factors have been computed, an *E* and an *N* equation of the form of Eqs. (C-9) may be solved to obtain the final coordinates of each point whose coordinates were known only in the *XY* system.

By this method, the transformation can be performed without ever determining the scale and rotation parameters directly. If, for some reason, it is necessary to determine the scale and rotation parameters, these values can be derived from the values of a and b as follows:

$$s = \sqrt{a^2 + b^2} \tag{C-10}$$

$$\theta = \tan^{-1}\left(\frac{b}{a}\right) \tag{C-11}$$

In Eq. (C-11), it is necessary to use the full circle inverse tangent function (typically called 'atan2' in computer languages and spreadsheets) since the value of θ can cover the full range from $-180°$ to $+180°$. With a scientific calculator, this full range can generally be achieved by using the rectangular-to-polar conversion capability.

Example C-2. Solve Example C-1 by using the alternate method.

(*a*) Formulate Eqs. (C-9) for the points whose coordinates are known in both systems.

$$1100.64 = 632.17a - 121.45b + T_E$$

$$1431.09 = 121.45a + 632.17b + T_N$$

$$1678.39 = 355.20a + 642.07b + T_E$$

$$254.15 = -642.07a + 355.20b + T_N$$

(*b*) The simultaneous solution of the above four equations yields the following:

$$a = 1.11964$$

$$b = 1.16285$$

$$T_E = 534.07$$

$$T_N = 559.99$$

(*c*) Using the four transformation parameters, the final *EN* ground coordinates of point *C* are calculated as follows:

$$E_C = 1.11964(1304.81) - 1.16285(596.37) + 534.07 = 1301.49$$

$$N_C = 1.11964(596.37) - 1.16285(1304.81) + 559.99 = 2745.01$$

(*d*) (Optional step) Compute the values for *s* and θ, using Eqs. (C-10) and (C-11), respectively.

$$s = \sqrt{(1.11964)^2 + (1.16285)^2} = 1.61425$$

$$\theta = \tan^{-1}\left(\frac{1.16285}{1.11964}\right) = 46.0845°$$

C-4   COORDINATE TRANSFORMATIONS WITH REDUNDANCY

In some instances, more than two control points are available with coordinates known in both the arbitrary and final systems. In that case, redundancy exists and the transformation can be computed by using a least squares solution. In this method, as discussed in App. B, the sum of the squares of the residuals in the measurements is minimized, which, according to the theory of probability, produces the most probable solution. The least squares method has the additional advantages that mistakes in the coordinates may be detected and that the precision of the transformed coordinates may be obtained. For these reasons, it is strongly advised to use redundancy in coordinate transformations whenever possible.

In the least squares procedure, it is convenient to use the alternate method discussed in Sec. C-3. Two observation equations similar to those of Eqs. (C-9) are formed for each point whose coordinates are known in both systems. Residuals *v* are included in the equations to make them consistent, as follows:

$$aX - bY + T_E = E + v_E$$
$$aY + bX + T_N = N + v_N \tag{C-12}$$

If *n* points are available whose coordinates are known in both systems, 2*n* equations may be formed containing the four unknown transformation parameters. The equations are solved by the method of least squares to obtain the most probable transformation parameters. Transformed coordinates of all required points may then be found by using the transformation factors as illustrated in step (*c*) of Example C-2.

It is theoretically correct in least squares to associate residuals with actual observations. In Eqs. (C-12), however, the *X* and *Y* coordinates are observed, yet residuals are only associated with the *E* and *N* control coordinates. Although there is a more rigorous least squares technique available to handle this situation, the easier approach shown above is commonly used and has been found to yield entirely satisfactory results.

C-5   MATRIX METHODS IN COORDINATE TRANSFORMATIONS

Coordinate transformations involve rather lengthy calculations and are therefore best handled on a computer. Matrix algebra is ideal for computer calculations and is therefore convenient for performing transformations.

To illustrate the application of matrix algebra in two-dimensional conformal coordinate transformation by least squares, assume that coordinates of three control points A, B, and C are known in both the XY system and the EN system. Let their coordinates be of equal reliability so that their weights are equal. Assume also that transformation into the EN system is required for points D through N, whose coordinates are known only in the XY system.

First, six observation equations of the form of Eqs. (C-12) are developed, two for each control point A, B, and C, as follows:

$$X_A a - Y_A b + T_E = E_A + v_{E_A}$$
$$Y_A a + X_A b + T_N = N_A + v_{N_A}$$
$$X_B a - Y_B b + T_E = E_B + v_{E_B}$$
$$Y_B a + X_B b + T_N = N_B + v_{N_B} \qquad \text{(C-13)}$$
$$X_C a - Y_C b + T_E = E_C + v_{E_C}$$
$$Y_C a + X_C b + T_N = N_C + v_{N_C}$$

In matrix representation, the above six equations are

$$_6A^4 \,_4X^1 = \,_6L^1 + \,_6V^1 \qquad \text{(C-14)}$$

In matrix Eq. (C-14), A is the matrix of coefficients of the unknown transformation parameters, X is the matrix of unknown transformation parameters, L is the matrix of constant terms which is made up of control point coordinates, and V is the matrix of residuals in those coordinates brought about by measurement errors. More specifically, these matrices are

$$_6A^4 = \begin{bmatrix} X_A & -Y_A & 1 & 0 \\ Y_A & X_A & 0 & 1 \\ X_B & -Y_B & 1 & 0 \\ Y_B & X_B & 0 & 1 \\ X_C & -Y_C & 1 & 0 \\ Y_C & X_C & 0 & 1 \end{bmatrix} \qquad _4X^1 = \begin{bmatrix} a \\ b \\ T_E \\ T_N \end{bmatrix}$$

$$_6L^1 = \begin{bmatrix} E_A \\ N_A \\ E_B \\ N_B \\ E_C \\ N_C \end{bmatrix} \qquad _6V^1 = \begin{bmatrix} v_{E_A} \\ v_{N_A} \\ v_{E_B} \\ v_{N_B} \\ v_{E_C} \\ v_{N_C} \end{bmatrix}$$

As discussed in App. B, matrix Eq. (B-12) is used to solve this equally weighted system for the transformation parameters. The final transformation of all points D through N into the EN system is performed as discussed in step (c) of Example C-2. This phase of the computation is also readily adapted to matrix methods.

C-6   TWO-DIMENSIONAL AFFINE COORDINATE TRANSFORMATION

The two-dimensional *affine* coordinate transformation is only a slight modification of the two-dimensional conformal transformation, to include different scale factors in the x and y directions and to compensate for nonorthogonality (nonperpendicularity) of the axis system. The affine transformation achieves these additional features by including two additional unknown parameters for a total of six. As will be shown, the derivation of the transformation equations depends on the measurement characteristics of the arbitrary coordinate system.

A two-dimensional affine transformation consists of four basic steps: (1) scale change in x and y, (2) correction for nonorthogonality, (3) rotation, and (4) translation. Figure C-3 illustrates the geometric relationship between the arbitrary coordinate system xy and the final coordinate system XY. In this figure, the nonorthogonality of x and y is indicated by the angle ε. The rotation angle necessary to make the two systems parallel is θ, and translations T_X and T_Y account for the offset of the origin. The four steps of the derivation are as follows.

Step 1: Scale Change in x and y

To make the scale of the arbitrary system xy equal to that of the final system XY, each coordinate is multiplied by its associated scale factor, s_x and s_y. This results in a correctly scaled coordinate system $x'y'$ as given in Eqs. (C-15).

$$x' = s_x x$$
$$y' = s_y y$$

(C-15)

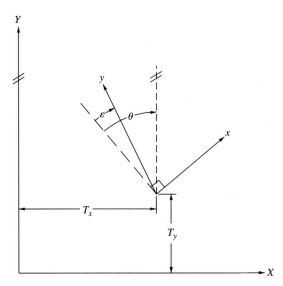

FIGURE C-3
General two-dimensional affine coordinate transformation relationship.

Step 2: Correction for Nonorthogonality

When x and y coordinates are measured from axes that intersect at a right angle, the x distance (coordinate) is measured perpendicularly from the y axis, which means the distance is at the same time parallel to the x axis. Similarly, the y distance is measured perpendicularly from the x axis and is therefore parallel to the x axis. When the xy axes are nonorthogonal, measuring an x distance perpendicularly from the y axis does not result in a distance that is parallel to the x axis. The analogous situation applies with the y distance. To derive the relationship, the specific measurement characteristics of the system must be identified.

    Several geometric possibilities exist; however, two configurations are most common and are illustrated in Figs. C-4a and b. In both figures the $x'y'$ measurement systems have already been scaled in accordance with step 1. The first configuration, illustrated in (a), is appropriate for most comparators where separate x and y carriages provide independent movement in both directions. The x' coordinate is measured parallel to the x' axis from the origin to the point, and the y' coordinate is measured parallel to the y' axis from the origin to the point. The second configuration, shown in (b), is appropriate when one is using satellite imagery that is acquired in a scanning fashion while the earth rotates beneath. The resulting image has a distinct parallelogram shape. In this configuration the x' coordinate is measured parallel to the x' axis from the y' axis to the point, and the y' coordinate is measured perpendicular to the x' axis. For the configuration of Fig. C-4a, the correction for nonorthogonality is given by Eqs. (C-16).

$$x'' = x'$$

$$y'' = \frac{y'}{\cos \varepsilon} - x' \tan \varepsilon \tag{C-16}$$

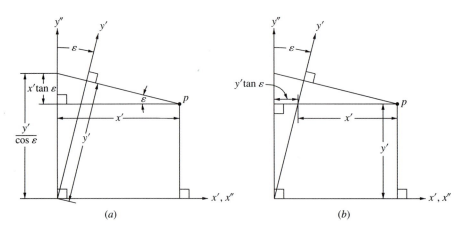

FIGURE C-4
(a) Two-dimensional affine relationship for typical comparator. (b) Two-dimensional affine relationship for typical scanning-type satellite image.

Equations (C-17) express the relationship for the configuration of Fig. C-4*b*.

$$x'' = x' + y' \tan \varepsilon$$
$$y'' = y'$$

(C-17)

Step 3: Rotation

Rotation by the angle θ is accomplished in the same fashion as in the two-dimensional conformal coordinate transformation presented in Sec. C-2. Equations (C-18) give the relationship between the $x''y''$ system and the $X'Y'$ system which is parallel to the final XY system after rotation by angle θ.

$$X' = x'' \cos \theta - y'' \sin \theta$$
$$Y' = x'' \sin \theta + y'' \cos \theta$$

(C-18)

Step 4: Translation

The final step is to translate the origin by T_X and T_Y to make it coincide with the origin of the final system, as shown in Eqs. (C-19).

$$X = X' + T_X$$
$$Y = Y' + T_Y$$

(C-19)

Combining the four steps for configuration (*a*) gives Eqs. (C-20).

$$X = s_x x \cos \theta - \left(\frac{s_y y}{\cos \varepsilon} - s_x x \tan \varepsilon \right) \sin \theta + T_X$$

$$Y = s_x x \sin \theta + \left(\frac{s_y y}{\cos \varepsilon} - s_x x \tan \varepsilon \right) \cos \theta + T_Y$$

(C-20)

Equations (C-20) can then be simplified as shown in the following steps, yielding Eqs. (C-21).

Step a

$$X = s_x x (\cos \theta + \tan \varepsilon \sin \theta) - s_y y \frac{\sin \theta}{\cos \varepsilon} + T_X$$

$$Y = s_x x (\sin \theta + \tan \varepsilon \cos \theta) + s_y y \frac{\cos \theta}{\cos \varepsilon} + T_Y$$

Step b

$$X = T_X + s_x x \frac{\cos \varepsilon \cos \theta + \sin \varepsilon \sin \theta}{\cos \varepsilon} - s_y y \frac{\sin \theta}{\cos \varepsilon}$$

$$Y = T_Y + s_x x \frac{\cos \varepsilon \sin \theta + \sin \varepsilon \cos \theta}{\cos \varepsilon} + s_y y \frac{\cos \theta}{\cos \varepsilon}$$

Step c

$$X = T_X + s_x x \frac{\cos(\varepsilon - \theta)}{\cos \varepsilon} - s_y y \frac{\sin \theta}{\cos \varepsilon}$$

$$Y = T_Y - s_x x \frac{\sin(\varepsilon - \theta)}{\cos \varepsilon} + s_y y \frac{\cos \theta}{\cos \varepsilon}$$

(C-21)

To simplify solutions involving Eqs. (C-21), the following substitutions are made.

Let $\qquad a_0 = T_X \qquad\qquad\qquad b_0 = T_Y$

$$a_1 = s_x \frac{\cos(\varepsilon - \theta)}{\cos \varepsilon} \qquad b_1 = -s_x \frac{\sin(\varepsilon - \theta)}{\cos \varepsilon}$$

$$a_2 = -s_y \frac{\sin \theta}{\cos \varepsilon} \qquad b_2 = s_y \frac{\cos \theta}{\cos \varepsilon}$$

Making these substitutions into Eqs. (C-21) gives Eqs. (C-22), which are the final form of the affine transformation. In Eqs. (C-22), the six unknown parameters s_x, s_y, ε, θ, T_X, and T_Y, which appeared in Eqs. (C-21) in a nonlinear form, have been replaced by six independent parameters a_0, a_1, a_2, b_0, b_1, and b_2, resulting in a linear form.

$$X = a_0 + a_1 x + a_2 y$$

$$Y = b_0 + b_1 x + b_2 y$$

(C-22)

After solving an affine transformation using Eqs. (C-22), if it is necessary to obtain values for the original six parameters for configuration (a), they may be obtained as follows. Note that in the first two expressions, the full circle inverse tangent function (e.g., atan2) must be used.

$$\theta = \tan^{-1}\left(\frac{-a_2}{b_2}\right)$$

$$\varepsilon - \theta = \tan^{-1}\left(\frac{-b_1}{a_1}\right)$$

$$\varepsilon = (\varepsilon - \theta) + \theta$$

$$s_x = a_1 \frac{\cos \varepsilon}{\cos(\varepsilon - \theta)}$$

$$s_y = b_2 \frac{\cos \varepsilon}{\cos \theta}$$

$$T_X = a_0$$

$$T_Y = b_0$$

Combining the four steps for configuration (b) gives Eqs. (C-23).

$$X = (s_x x + s_y y \tan \varepsilon)\cos \theta - s_y y \sin \theta + T_X$$

$$Y = (s_x x + s_y y \tan \varepsilon)\sin \theta - s_y y \cos \theta + T_Y$$

(C-23)

Equations (C-23) can then be simplified as shown in the following steps, yielding Eqs. (C-24).

Step a

$$X = s_x x \cos \theta + s_y y (\tan \varepsilon \cos \theta - \sin \theta) + T_X$$

$$Y = s_x x \sin \theta + s_y y (\tan \varepsilon \sin \theta + \cos \theta) + T_Y$$

Step b

$$X = T_X + s_x x \cos \theta + s_y y \frac{\sin \varepsilon \cos \theta - \cos \varepsilon \sin \theta}{\cos \varepsilon}$$

$$Y = T_Y + s_x x \sin \theta + s_y y \frac{\sin \varepsilon \sin \theta + \cos \varepsilon \cos \theta}{\cos \varepsilon}$$

Step c

$$X = T_X + s_x x \cos \theta + s_y y \frac{\sin(\varepsilon - \theta)}{\cos \varepsilon}$$

$$Y = T_Y + s_x x \sin \theta + s_y y \frac{\cos(\varepsilon - \theta)}{\cos \varepsilon}$$

(C-24)

To simplify solutions involving Eqs. (C-24), the following substitutions are made.

Let

$$a_0 = T_X \qquad\qquad b_0 = T_Y$$

$$a_1 = s_x \cos \theta \qquad\qquad b_1 = s_x \sin \theta$$

$$a_2 = s_y \frac{\sin(\varepsilon - \theta)}{\cos \varepsilon} \qquad b_2 = s_y \frac{\cos(\varepsilon - \theta)}{\cos \varepsilon}$$

After making these substitutions into Eqs. (C-24), the same linear form of the affine transformation is obtained as before, i.e., Eqs. (C-22). After solving an affine transformation using Eqs. (C-22), if it is necessary to obtain values for the original six parameters for configuration (b), they may be obtained as follows. Note that in the first two expressions, the full circle inverse tangent function (e.g., atan2) must be used.

$$\theta = \tan^{-1}\left(\frac{b_1}{a_1}\right)$$

$$\varepsilon - \theta = \tan^{-1}\left(\frac{a_2}{b_2}\right)$$

$$\varepsilon = (\varepsilon - \theta) + \theta$$

$$s_x = \frac{a_1}{\cos \theta}$$

$$s_y = b_2 \frac{\cos \varepsilon}{\cos(\varepsilon - \theta)}$$

$$T_X = a_0$$

$$T_Y = b_0$$

As noted above, the transformation equations [Eqs. (C-22)] are identical for either configuration (*a*) or (*b*), and thus the method of solution is the same for both. The two derivations have been presented, however, to enable computation of the transformation parameters after solution, a factor that can be important in certain situations, for example in evaluating or calibrating equipment.

As with the two-dimensional conformal transformation, the application of the affine transformation is a two-step procedure of (1) determining the *a* and *b* coefficients, using points whose coordinates are known in both the *XY* and *xy* systems, and (2) applying these coefficients to calculate transformed *XY* coordinates for all other points from their *xy* coordinates. In correcting photo coordinates, the fiducial marks are used to perform step 1, since their calibrated *XY* coordinates are known from camera calibration, and their *xy* coordinates are available from comparator measurements. For a given photograph, a pair of equations of the form of Eqs. (C-22) can be written for each fiducial mark. If there are four fiducials, four *X* and four *Y* equations are obtained. Any three of the fiducial marks could be used to obtain a solution for the unknown *a*'s and *b*'s. An improved solution may be obtained, however, if all four fiducial marks are used and the system is solved by least squares. In other applications where the geometric arrangement of common points (control) is not predetermined, it is important that the control points do not approximate a straight line, since this would lead to a very weak or even an indeterminate solution.

Example C-3. Calibrated coordinates and comparator-measured coordinates of the four fiducial marks for a certain photograph are given in the following table. The comparator-measured coordinates of other points 1, 2, and 3 are also given. It is required to compute the corrected coordinates of points 1, 2, and 3 by using the affine transformation.

	Comparator coordinates		Calibrated coordinates	
Point	*x*, mm	*y*, mm	*X*, mm	*Y*, mm
Fiducial *A*	228.170	129.730	112.995	0.034
Fiducial *B*	2.100	129.520	−113.006	0.005
Fiducial *C*	115.005	242.625	0.003	112.993
Fiducial *D*	115.274	16.574	−0.012	−113.000
1	206.674	123.794		
2	198.365	132.856		
3	91.505	18.956		

Solution. Equations of the form of Eqs. (C-22), with residuals added for consistency, are formulated for the four fiducial marks as follows:

$$112.995 + v_{X_A} = a_0 + 228.170a_1 + 129.730a_2$$

$$0.034 + v_{Y_A} = b_0 + 228.170b_1 + 129.730b_2$$

$$-113.006 + v_{X_B} = a_0 + 2.100a_1 + 129.520a_2$$

$$0.005 + v_{Y_B} = b_0 + 2.100b_1 + 129.520b_2$$

$$0.003 + v_{X_C} = a_0 + 115.005a_1 + 242.625a_2$$

$$112.993 + v_{Y_C} = b_0 + 115.005b_1 + 242.625b_2$$

$$-0.012 + v_{X_D} = a_0 + 115.274a_1 + 16.574a_2$$

$$-113.000 + v_{Y_D} = b_0 + 115.274b_1 + 16.574b_2$$

In matrix representation, after switching of the terms to the opposite side, the above eight equations are

$$_8A^6\,_6X^1 = {}_8L^1 + {}_8V^1 \tag{C-25}$$

In matrix Eq. (C-25), A is the matrix of coefficients of the unknown transformation parameters, X is the matrix of unknown transformation parameters, L is the matrix of constant terms which consists of calibrated fiducial coordinates, and V is the matrix of residuals in those coordinates brought about by measurement errors. More specifically, these matrices are

$$_8A^6 = \begin{bmatrix} 1 & 228.170 & 129.730 & 0 & 0 & 0 \\ 0 & 0 & 0 & 1 & 228.170 & 129.730 \\ 1 & 2.100 & 129.520 & 0 & 0 & 0 \\ 0 & 0 & 0 & 1 & 2.100 & 129.520 \\ 1 & 115.005 & 242.625 & 0 & 0 & 0 \\ 0 & 0 & 0 & 1 & 115.005 & 242.625 \\ 1 & 115.274 & 16.574 & 0 & 0 & 0 \\ 0 & 0 & 0 & 1 & 115.274 & 16.574 \end{bmatrix} \qquad _6X^1 = \begin{bmatrix} a_0 \\ a_1 \\ a_2 \\ b_0 \\ b_1 \\ b_2 \end{bmatrix}$$

$$_8L^1 = \begin{bmatrix} 112.995 \\ 0.034 \\ -113.006 \\ 0.005 \\ 0.003 \\ 112.993 \\ -0.012 \\ -113.000 \end{bmatrix} \qquad _8V^1 = \begin{bmatrix} v_{X_A} \\ v_{Y_A} \\ v_{X_B} \\ v_{Y_B} \\ v_{X_C} \\ v_{Y_C} \\ v_{X_D} \\ v_{Y_D} \end{bmatrix}$$

Solving this system of equations using the method of least squares with equal weights as presented in Sec. B-9, the following solution is obtained:

$$X = \begin{bmatrix} a_0 \\ a_1 \\ a_2 \\ b_0 \\ b_1 \\ b_2 \end{bmatrix} = \begin{bmatrix} -115.270 \\ 0.999694 \\ 0.001256 \\ -129.479 \\ -0.000800 \\ 0.999742 \end{bmatrix}$$

Calculation of residual matrix V gives the following:

$$AX - L = V = \begin{bmatrix} v_{X_A} \\ v_{Y_A} \\ v_{X_B} \\ v_{Y_B} \\ v_{X_C} \\ v_{Y_C} \\ v_{X_D} \\ v_{Y_D} \end{bmatrix} = \begin{bmatrix} -0.0017 \\ 0.0012 \\ -0.0017 \\ 0.0012 \\ 0.0017 \\ -0.0012 \\ 0.0017 \\ -0.0012 \end{bmatrix}$$

Since all the residuals have small magnitudes, the solution is deemed acceptable and the transformed coordinates of points 1, 2, and 3 are then computed as follows:

$X_1 = -115.270 + (0.999694)(206.674) + (0.001256)(123.794) = 91.496$

$Y_1 = -129.479 + (-0.000800)(206.674) + (0.999742)(123.794) = -5.882$

$X_2 = -115.270 + (0.999694)(198.365) + (0.001256)(132.856) = 83.201$

$Y_2 = -129.479 + (-0.000800)(198.365) + (0.999742)(132.856) = 3.184$

$X_3 = -115.270 + (0.999694)(91.505) + (0.001256)(18.956) = -23.769$

$Y_3 = -129.479 + (-0.000800)(91.505) + (0.999742)(18.956) = -110.601$

In computing affine transformations, it is strongly recommended that more than the minimum number of common points be used, and that the solution be obtained by least squares. In addition to other benefits, least squares can reveal the presence of measurement mistakes through calculation of residuals.

C-7   THREE-DIMENSIONAL CONFORMAL COORDINATE TRANSFORMATION

As implied by its name, a three-dimensional conformal coordinate transformation involves converting from one three-dimensional system to another. In the transformation, true shape is retained. This type of coordinate transformation is essential in analytical or computational photogrammetry for two basic problems: (1) to transform arbitrary stereomodel coordinates to a ground or object space system and (2) to form continuous three-dimensional "strip models" from independent stereomodels. Three-dimensional conformal coordinate transformation equations are developed here in general, while their application to specific photogrammetry problems is described elsewhere in the text where appropriate.

In Fig. C-5 it is required to transform coordinates of points from an xyz system to an XYZ system. As illustrated in the figure, the two coordinate systems are at different scales, they are not parallel, and their origins do not coincide. The necessary

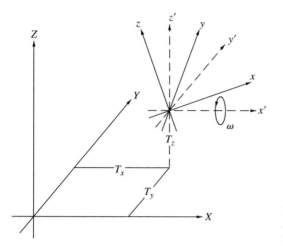

FIGURE C-5
XYZ and xyz right-handed three-dimensional coordinate systems.

transformation equations can be expressed in terms of seven independent parameters: the three rotation angles omega (ω), phi (ϕ), and kappa (κ); a scale factor s; and three translation parameters T_X, T_Y, and T_Z. Before proceeding with the development of the transformation equations, it is important to define sign conventions. All coordinates shall be defined as right-handed, i.e., systems in which positive X, Y, and Z are as shown in Fig. C-5. Rotation angles ω, ϕ, and κ are positive if they are counterclockwise when viewed from the positive end of their respective axes. For example, positive ω rotation about the x' axis, is shown in Fig. C-5.

The transformation equations will be developed in the following two basic steps: (1) rotation and (2) scaling and translation.

Step 1: Rotation

In Fig. C-5 an $x'y'z'$ coordinate system parallel to the XYZ object system is constructed with its origin at the origin of the xyz system. In the development of rotation formulas, it is customary to consider the three rotations as taking place so as to convert from the $x'y'z'$ system to the xyz system. The rotation equations are developed in a sequence of three independent two-dimensional rotations. These rotations, illustrated in Fig. C-6, are, first, ω rotation about the x' axis which converts coordinates from the $x'y'z'$ system to an $x_1y_1z_1$ system; second, ϕ rotation about the once-rotated y_1 axis which converts coordinates from the $x_1y_1z_1$ system to an $x_2y_2z_2$ system; and third, κ rotation about the twice-rotated z_2 axis which converts coordinates from the $x_2y_2z_2$ system to the xyz system of Fig. C-5. The exact amount and direction of the rotations for any three-dimensional coordinate transformation will depend upon the orientation relationship between the xyz and XYZ coordinate systems.

The development of the rotation formulas is as follows:

1. Rotation through the angle ω about the x' axis is illustrated in Fig. C-7. The coordinates of any point A in the once-rotated $x_1y_1z_1$ system, as shown graphically in

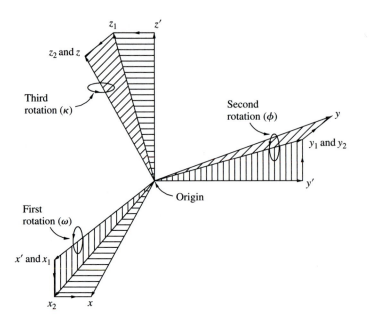

FIGURE C-6
The three sequential angular rotations.

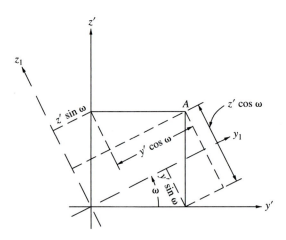

FIGURE C-7
Omega rotation about the x' axis.

Fig. C-7, are

$$x_1 = x'$$

$$y_1 = y' \cos \omega + z' \sin \omega \qquad \text{(C-26)}$$

$$z_1 = -y' \sin \omega + z' \cos \omega$$

Since this rotation was about x', the x' and x_1 axes are coincident and therefore the x coordinate of A is unchanged.

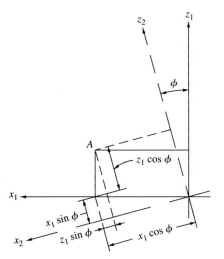

FIGURE C-8
Phi rotation about the y_1 axis.

2. Rotation through ϕ about the y_1 axis is illustrated in Fig. C-8. The coordinates of A in the twice-rotated $x_2y_2z_2$ coordinate system, as shown graphically in Fig. C-8, are

$$x_2 = -z_1 \sin \phi + x_1 \cos \phi$$
$$y_2 = y_1 \qquad\qquad\qquad (C\text{-}27)$$
$$z_2 = z_1 \cos \phi + x_1 \sin \phi$$

In this rotation about y_1, the y_1 and y_2 axes are coincident, and therefore the y coordinate of A is unchanged. Substituting Eqs. (C-26) into (C-27) gives

$$x_2 = -(-y' \sin \omega + z' \cos \omega) \sin \phi + x' \cos \phi$$
$$y_2 = y' \cos \omega + z' \sin \omega \qquad\qquad (C\text{-}28)$$
$$z_2 = (-y' \sin \omega + z' \cos \omega) \cos \phi + x_1 \sin \phi$$

3. Rotation through κ about the z_2 axis is illustrated in Fig. C-9. The coordinates of A in the three-times-rotated coordinate system, which has now become the xyz system as shown graphically in Fig. C-9, are

$$x = x_2 \cos \kappa + y_2 \sin \kappa$$
$$y = -x_2 \sin \kappa + y_2 \cos \kappa \qquad\qquad (C\text{-}29)$$
$$z = z_2$$

In this rotation about z_2, the z_2 and z axes are coincident, and therefore the z coordinate of A is unchanged. Substituting Eqs. (C-28) into (C-29) gives

$$x = [(y' \sin \omega - z' \cos \omega)\sin \phi + x' \cos \phi] \cos \kappa + (y' \cos \omega + z' \sin \omega) \sin \kappa$$
$$y = [(-y' \sin \omega + z' \cos \omega)\sin \phi - x' \cos \phi] \sin \kappa + (y' \cos \omega + z' \sin \omega) \cos \kappa$$
$$z = (-y' \sin \omega + z' \cos \omega) \cos \phi + x' \sin \phi \qquad\qquad (C\text{-}30)$$

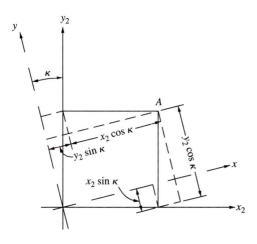

FIGURE C-9
Kappa rotation about the z_2 axis.

Factoring Eqs. (C-30) gives

$$x = x' (\cos \phi \cos \kappa) + y' (\sin \omega \sin \phi \cos \kappa + \cos \omega \sin \kappa)$$
$$+ z'(-\cos \omega \sin \phi \cos \kappa + \sin \omega \sin \kappa)$$

$$y = x'(-\cos \phi \sin \kappa) + y'(-\sin \omega \sin \phi \sin \kappa + \cos \omega \cos \kappa) \qquad \text{(C-31)}$$
$$+ z'(\cos \omega \sin \phi \sin \kappa + \sin \omega \cos \kappa)$$

$$z = x' (\sin \phi) + y'(-\sin \omega \cos \phi) + z' (\cos \omega \cos \phi)$$

Substituting m's for the coefficients of x', y', and z' in Eqs. (C-31) gives

$$x = m_{11}x' + m_{12}y' + m_{13}z'$$
$$y = m_{21}x' + m_{22}y' + m_{23}z' \qquad \text{(C-32)}$$
$$z = m_{31}x' + m_{32}y' + m_{33}z'$$

where
$$m_{11} = \cos \phi \cos \kappa$$
$$m_{12} = \sin \omega \sin \phi \cos \kappa + \cos \omega \sin \kappa$$
$$m_{13} = -\cos \omega \sin \phi \cos \kappa + \sin \omega \sin \kappa$$
$$m_{21} = -\cos \phi \sin \kappa$$
$$m_{22} = -\sin \omega \sin \phi \sin \kappa + \cos \omega \cos \kappa \qquad \text{(C-33)}$$
$$m_{23} = \cos \omega \sin \phi \sin \kappa + \sin \omega \cos \kappa$$
$$m_{31} = \sin \phi$$
$$m_{32} = -\sin \omega \cos \phi$$
$$m_{33} = \cos \omega \cos \phi$$

Equations (C-32) may be expressed in matrix form as

$$X = MX' \qquad \text{(C-34)}$$

where

$$X = \begin{bmatrix} x \\ y \\ z \end{bmatrix} \qquad M = \begin{bmatrix} m_{11} & m_{12} & m_{13} \\ m_{21} & m_{22} & m_{23} \\ m_{31} & m_{32} & m_{33} \end{bmatrix} \qquad \text{and} \qquad X' = \begin{bmatrix} x' \\ y' \\ z' \end{bmatrix}$$

The matrix M is commonly called the *rotation matrix*. The individual elements of the rotation matrix are *direction cosines* which relate the two axis systems. These matrix elements, expressed in terms of direction cosines, are

$$M = \begin{bmatrix} \cos xx' & \cos xy' & \cos xz' \\ \cos yx' & \cos yy' & \cos yz' \\ \cos zx' & \cos zy' & \cos zz' \end{bmatrix} \qquad (\text{C-35})$$

In the above matrix, cos xx' is the direction cosine relating the x and x' axes, cos xy' relates the x and y' axes, etc. Direction cosines are simply the cosines of the angles in space between the respective axes, the angles being taken between 0° and 180°. It is an important property that *the sum of the squares of the three direction cosines in any row or in any column is unity*. This property may be used to check the computed elements of the rotation matrix for correctness.

The rotation matrix is an *orthogonal* matrix, which has the property that *its inverse is equal to its transpose,* or

$$M^{-1} = M^T \qquad (\text{C-36})$$

By using this property, Eq. (C-34) may rewritten, expressing $x'y'z'$ coordinates in terms of xyz coordinates as follows:

$$X' = M^T X \qquad (\text{C-37})$$

In expanded form this equation is:

$$x' = m_{11}x + m_{21}y + m_{31}z$$
$$y' = m_{12}x + m_{22}y + m_{32}z \qquad (\text{C-38})$$
$$z' = m_{13}x + m_{23}y + m_{33}z$$

Step 2: Scaling and Translation

To arrive at the final three-dimensional coordinate transformation equations, i.e., equations that yield coordinates in the XYZ system of Fig. C-5, it is necessary to multiply each of Eqs. (C-38) by a scale factor s and to add the translation factors T_X, T_Y, and T_Z. [Recall that the $x'y'z'$ coordinates given by Eqs. (C-38) are in a system that is parallel the XYZ system.] This step makes the lengths of any lines equal in both coordinate systems, and it translates from the origin of $x'y'z'$ to the origin of the XYZ system. Performing this step yields

$$X = sx' + T_X = s(m_{11}x + m_{21}y + m_{31}z) + T_X$$
$$Y = sy' + T_Y = s(m_{12}x + m_{22}y + m_{32}z) + T_Y \qquad (\text{C-39})$$
$$Z = sz' + T_Z = s(m_{13}x + m_{23}y + m_{33}z) + T_Z$$

In matrix form, Eqs. (C-39) are

$$\overline{X} = sM^T X + T \tag{C-40}$$

In Eq. (C-40), matrices M and X are as previously defined, s is the scale factor, and

$$\overline{X} = \begin{bmatrix} X \\ Y \\ Z \end{bmatrix} \quad \text{and} \quad T = \begin{bmatrix} T_X \\ T_Y \\ T_Z \end{bmatrix}$$

In Eqs. (C-39), the nine m's are not independent of one another, but rather, as seen in Eqs. (C-33), they are functions of the three rotation angles ω, ϕ, and κ. In addition to these three unknown angles, there are three unknown translations and one scale factor in Eqs. (C-39), for a total of seven unknowns. A unique solution is obtained for the unknowns if the x and y coordinates of two horizontal points and the z coordinates of three vertical points are known in both coordinate systems. If more than the minimum of seven coordinates are known in both systems, redundant equations may be written, which makes possible an improved solution through least squares techniques.

Solution of the three-dimensional conformal coordinate transformation is more complex than that of the two-dimensional transformations presented earlier in this appendix. This added complexity is due to the fact that Eqs. (C-39) are nonlinear in terms of the unknowns s, ω, ϕ, and κ. While it is possible to directly compute values for the translation factors since their terms exist in a linear form, it is more convenient to treat them as if they appeared in a nonlinear form. To solve these equations, they are linearized by using a Taylor series expansion including only the first-order terms. Application of Taylor's series requires that initial approximations be obtained for each of the seven unknowns. Each point P contributes as many as three equations (two if the point is horizontal only and one if the point has just vertical control) which are linearized as shown in Eqs. (C-41).

$$X_P = (X_P)_0 + \left(\frac{\partial X_P}{\partial s}\right)_0 ds + \left(\frac{\partial X_P}{\partial \omega}\right)_0 d\omega + \left(\frac{\partial X_P}{\partial \phi}\right)_0 d\phi + \left(\frac{\partial X_P}{\partial \kappa}\right)_0 d\kappa$$

$$+ \left(\frac{\partial X_P}{\partial T_X}\right)_0 dT_X + \left(\frac{\partial X_P}{\partial T_Y}\right)_0 dT_Y + \left(\frac{\partial X_P}{\partial T_Z}\right)_0 dT_Z$$

$$Y_P = (Y_P)_0 + \left(\frac{\partial Y_P}{\partial s}\right)_0 ds + \left(\frac{\partial Y_P}{\partial \omega}\right)_0 d\omega + \left(\frac{\partial Y_P}{\partial \phi}\right)_0 d\phi + \left(\frac{\partial Y_P}{\partial \kappa}\right)_0 d\kappa$$

$$+ \left(\frac{\partial Y_P}{\partial T_X}\right)_0 dT_X + \left(\frac{\partial Y_P}{\partial T_Y}\right)_0 dT_Y + \left(\frac{\partial Y_P}{\partial T_Z}\right)_0 dT_Z \tag{C-41}$$

$$Z_P = (Z_P)_0 + \left(\frac{\partial Z_P}{\partial s}\right)_0 ds + \left(\frac{\partial Z_P}{\partial \omega}\right)_0 d\omega + \left(\frac{\partial Z_P}{\partial \phi}\right)_0 d\phi + \left(\frac{\partial Z_P}{\partial \kappa}\right)_0 d\kappa$$

$$+ \left(\frac{\partial Z_P}{\partial T_X}\right)_0 dT_X + \left(\frac{\partial Z_P}{\partial T_Y}\right)_0 dT_Y + \left(\frac{\partial Z_P}{\partial T_Z}\right)_0 dT_Z$$

In Eqs. (C-41), $(X_P)_0$, $(Y_P)_0$, and $(Z_P)_0$ are the right-hand sides of Eqs. (C-39) evaluated at the initial approximations; $(\partial X_P/\partial s)_0$, $(\partial X_P/\partial \omega)_0$, etc., are the partial derivatives with respect to the indicated unknowns evaluated at the initial approximations; and ds, $d\omega$, $d\phi$, $d\kappa$, dT_X, dT_Y, and dT_Z are corrections to the initial approximations which will be computed during the solution. The units of $d\omega$, $d\phi$, and $d\kappa$ are radians.

Substituting letters for partial derivative coefficients, adding residuals to make the equations suitable for a least squares solution, and rearranging terms, the following equations result.

$$a_{11}\,ds + a_{12}\,d\omega + a_{13}\,d\phi + a_{14}\,d\kappa + a_{15}\,dT_X + a_{16}\,dT_Y + a_{17}\,dT_Z = [X_P - (X_P)_0] + v_{X_P}$$

$$a_{21}\,ds + a_{22}\,d\omega + a_{23}\,d\phi + a_{24}\,d\kappa + a_{25}\,dT_X + a_{26}\,dT_Y + a_{27}\,dT_Z = [Y_P - (Y_P)_0] + v_{Y_P}$$

$$a_{31}\,ds + a_{32}\,d\omega + a_{33}\,d\phi + a_{34}\,d\kappa + a_{35}\,dT_X + a_{36}\,dT_Y + a_{37}\,dT_Z = [Z_P - (Z_P)_0] + v_{Z_P}$$

$$\text{(C-42)}$$

To clarify the coefficients of Eqs. (C-42), the partial derivative terms, which must be evaluated at the initial approximations, are as follows:

$$a_{11} = m_{11}x_P + m_{21}y_P + m_{31}z_P$$

$$a_{12} = 0$$

$$a_{13} = [(-\sin\phi\cos\kappa)\,x_P + \sin\phi\sin\kappa\,(y_P) + \cos\phi\,(z_P)]s$$

$$a_{14} = (m_{21}x_P - m_{11}y_P)s$$

$$a_{15} = a_{26} = a_{37} = 1$$

$$a_{16} = a_{17} = a_{25} = a_{27} = a_{35} = a_{36} = 0$$

$$a_{21} = m_{12}x_P + m_{22}y_P + m_{32}z_P$$

$$a_{22} = (-m_{13}x_P - m_{23}y_P - m_{33}z_P)s$$

$$a_{23} = [(\sin\omega\cos\phi\cos\kappa)\,x_P + (-\sin\omega\cos\phi\sin\kappa)\,y_P + (\sin\omega\sin\phi)\,z_P]s$$

$$a_{24} = (m_{22}x_P - m_{12}y_P)s$$

$$a_{31} = m_{13}x_P + m_{23}y_P + m_{33}z_P$$

$$a_{32} = (m_{12}x_P + m_{22}y_P + m_{32}z_P)s$$

$$a_{33} = [(-\cos\omega\cos\phi\cos\kappa)x_P + (\cos\omega\cos\phi\sin\kappa)y_P + (-\cos\omega\sin\phi)z_P]s$$

$$a_{34} = (m_{23}x_P - m_{13}y_P)s$$

In a least squares solution, each point contributes up to three rows of coefficients to the A matrix, as well as terms in the L and V matrices. In general, assuming points $1, 2, \ldots, n$ are three-dimensional control points, the following matrix equation results:

$$_{3n}A^7\,_7X^1 = \,_{3n}L^1 + \,_{3n}V^1 \tag{C-43}$$

where

$$
_{3n}A^7 = \begin{bmatrix}
a_{1_{11}} & a_{1_{12}} & a_{1_{13}} & a_{1_{14}} & 1 & 0 & 0 \\
a_{1_{21}} & a_{1_{22}} & a_{1_{23}} & a_{1_{24}} & 0 & 1 & 0 \\
a_{1_{31}} & a_{1_{32}} & a_{1_{33}} & a_{1_{34}} & 0 & 0 & 1 \\
a_{2_{11}} & a_{2_{12}} & a_{2_{13}} & a_{2_{14}} & 1 & 0 & 0 \\
a_{2_{21}} & a_{2_{22}} & a_{2_{23}} & a_{2_{24}} & 0 & 1 & 0 \\
a_{2_{31}} & a_{2_{32}} & a_{2_{33}} & a_{2_{34}} & 0 & 0 & 1 \\
\cdots & & & & & & \\
a_{n_{11}} & a_{n_{12}} & a_{n_{13}} & a_{n_{14}} & 1 & 0 & 0 \\
a_{n_{21}} & a_{n_{22}} & a_{n_{23}} & a_{n_{24}} & 0 & 1 & 0 \\
a_{n_{31}} & a_{n_{32}} & a_{n_{33}} & a_{n_{34}} & 0 & 0 & 1
\end{bmatrix}
$$

$$
_7X^1 = \begin{bmatrix}
ds \\
d\omega \\
d\phi \\
d\kappa \\
dT_X \\
dT_Y \\
dT_Z
\end{bmatrix}
\qquad
_{3n}L^1 = \begin{bmatrix}
X_1 - (X_1)_0 \\
Y_1 - (Y_1)_0 \\
Z_1 - (Z_1)_0 \\
X_2 - (X_2)_0 \\
Y_2 - (Y_2)_0 \\
Z_2 - (Z_2)_0 \\
\vdots \\
X_n - (X_n)_0 \\
Y_n - (Y_n)_0 \\
Z_n - (Z_n)_0
\end{bmatrix}
\qquad
_{3n}V^1 = \begin{bmatrix}
v_{X_1} \\
v_{Y_1} \\
v_{Z_1} \\
v_{X_2} \\
v_{Y_2} \\
v_{Z_2} \\
\vdots \\
v_{X_n} \\
v_{Y_n} \\
v_{Z_n}
\end{bmatrix}
$$

Equation (C-43) may be solved by using least squares Eq. (B-11), giving corrections which are then added to the initial approximations for the unknowns, resulting in a better set of approximations. The solution must be iterated (since only the first-order terms of Taylor's series were used) until negligibly small values are obtained in matrix X. Other techniques of testing for the convergence of an iterative solution, such as convergence of the computed estimate of the standard deviation of unit weight [see Eq. (B-15) for an unweighted adjustment or Eq. (B-16) for a weighted adjustment], may also be used. The reader may consult references listed at the end of this appendix for a discussion of these convergence techniques.

Once the solution has reached satisfactory convergence, the latest approximations for the unknowns are the values for the transformation parameters. Then the transformed coordinates for each point whose coordinates are known only in the original system are obtained by applying Eqs. (C-39).

In solving nonlinear equations using Taylor's theorem, it is important that good initial approximations be obtained. If approximations are chosen that are not reasonably close to the final computed values, the solution will generally diverge (i.e., the corrections get larger with each subsequent iteration instead of smaller), or the solution may converge to an erroneous set of transformation parameters. In cases involving stereomodels formed from near-vertical photography, ω and ϕ may be assumed to be zero. An approximation for κ may be determined from the difference in azimuths of a common line in both systems, and an approximation for s may be determined from the ratio of the lengths of a common line. Approximations for the translation factors may then be found by rearranging Eqs. (C-39) so that T_X, T_Y, and T_Z are isolated on the left side. By using previously determined initial approximations for s, ω, ϕ, and κ, and the coordinates of a common point on the right-hand side, initial approximations for the translations may thus be determined. When the assumption of near-vertical photography is invalid, such as when one is dealing with terrestrial photography, an alternative method must be used. The reader may consult Dewitt (1996) in the references for a description of this technique.

C-8   TWO-DIMENSIONAL PROJECTIVE COORDINATE TRANSFORMATION

Two-dimensional projective transformation equations enable the analytical computation of the XY coordinates of points after they have been projected onto a plane from another nonparallel plane. The most common use of these equations is in analytical rectification, i.e., calculating coordinates of points in a rectified-ratioed photo plane based upon their coordinates in a tilted photograph. This situation is illustrated in Fig. C-10. In the figure, a tilted photo with its xy fiducial coordinate axis system (shown in the plane of the photo) is illustrated. The fiducial coordinates of the principal point o are $x_o y_o$. The projection center (the origin of the $x'y'z'$ system) is at L, which is on the end of a perpendicular to the photo plane from point o, such as $Lo = f$. The $x'y'$ axes are parallel to the fiducial xy axes. The projection of points a, b, c, and d from the tilted photo onto the plane of the rectified-ratioed photo occurs as A, B, C, and D, respectively. Positions of projected points in the rectified-ratioed photo plane are expressed in the XYZ coordinate system shown in the figure.

In a simplified development of the equations for the two-dimensional projective transformation, an $X'Y'Z'$ coordinate system is adopted which is parallel to the XYZ system and has its origin at L. Using Eqs. (C-32), which were developed in the preceding section, the $x'y'z'$ coordinates of any point, such as p of Fig. C-10, can be expressed in terms of $X'Y'Z'$ coordinates as follows:

$$x'_p = m_{11}X'_p + m_{12}Y'_p + m_{13}Z'_p$$
$$y'_p = m_{21}X'_p + m_{22}Y'_p + m_{23}Z'_p \qquad \text{(C-44)}$$
$$z'_p = m_{31}X'_p + m_{32}Y'_p + m_{33}Z'_p$$

In Eqs. (C-44), the m's are functions of the rotation angles omega, phi, and kappa which define the tilt relationships between the two plane coordinate systems xy and XY. These functions are described in the preceding section. The other terms in Eqs. (C-44) are coordinates as previously described. Consider now Fig. C-11, which shows the parallel

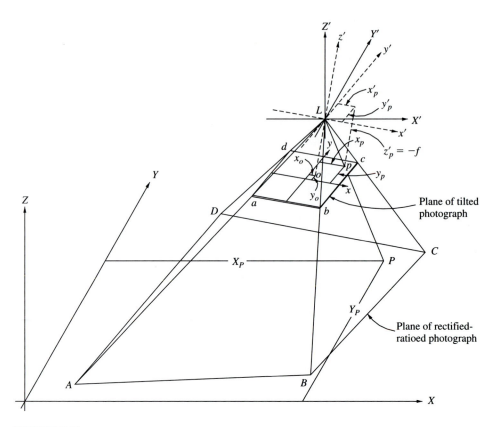

FIGURE C-10
Geometry of two-dimensional projective transformation.

relationships between the $X'Y'Z'$ and XYZ planes after rotation. From similar triangles of Fig. C-11,

$$\frac{X'_p}{X_P - X_L} = \frac{-Z'_p}{Z_L}$$

from which

$$X'_p = \frac{-Z'_p}{Z_L}(X_P - X_L) \tag{a}$$

Again from similar triangles of Fig. C-11,

$$\frac{Y'_p}{Y_P - Y_L} = \frac{-Z'_p}{Z_L}$$

from which

$$Y'_p = \frac{-Z'_p}{Z_L}(Y_P - Y_L) \tag{b}$$

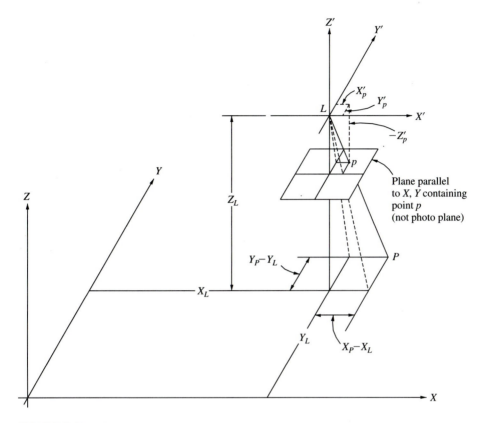

FIGURE C-11
Parallel relationships that exist after rotation between $X'Y'Z'$ and XYZ planes in two-dimensional projective transformation.

Also, intuitively the following equation may be written:

$$Z'_p = \frac{-Z'_p}{Z_L}(-Z_L) \tag{c}$$

Substituting Eqs. (a), (b), and (c) into Eq. (C-44) gives

$$x'_p = m_{11}\frac{-Z'_p}{Z_L}(X_P - X_L) + m_{12}\frac{-Z'_p}{Z_L}(Y_P - Y_L) + m_{13}\frac{-Z'_p}{Z_L}(-Z_L)$$

$$y'_p = m_{21}\frac{-Z'_p}{Z_L}(X_P - X_L) + m_{22}\frac{-Z'_p}{Z_L}(Y_P - Y_L) + m_{23}\frac{-Z'_p}{Z_L}(-Z_L) \tag{C-45}$$

$$z'_p = m_{31}\frac{-Z'_p}{Z_L}(X_P - X_L) + m_{32}\frac{-Z'_p}{Z_L}(Y_P - Y_L) + m_{33}\frac{-Z'_p}{Z_L}(-Z_L)$$

Factoring Eqs. (C-45) gives

$$x'_p = \frac{-Z'_p}{Z_L}[m_{11}(X_P - X_L) + m_{12}(Y_P - Y_L) + m_{13}(-Z_L)] \qquad (d)$$

$$y'_p = \frac{-Z'_p}{Z_L}[m_{21}(X_P - X_L) + m_{22}(Y_P - Y_L) + m_{23}(-Z_L)] \qquad (e)$$

$$z'_p = \frac{-Z'_p}{Z_L}[m_{31}(X_P - X_L) + m_{32}(Y_P - Y_L) + m_{33}(-Z_L)] \qquad (f)$$

Dividing Eqs. (d) and (e) by Eq. (f) yields

$$\frac{x'_p}{z'_p} = \frac{m_{11}(X_P - X_L) + m_{12}(Y_P - Y_L) + m_{13}(-Z_L)}{m_{31}(X_P - X_L) + m_{32}(Y_P - Y_L) + m_{33}(-Z_L)} \qquad (g)$$

$$\frac{y'_p}{z'_p} = \frac{m_{21}(X_P - X_L) + m_{22}(Y_P - Y_L) + m_{23}(-Z_L)}{m_{31}(X_P - X_L) + m_{32}(Y_P - Y_L) + m_{33}(-Z_L)} \qquad (h)$$

Referring to Fig. C-10, it can be seen that the $x'y'$ coordinates are offset from the fiducial coordinates xy by x_o and y_o. Furthermore, for a photograph, the z coordinates are equal to $-f$. The following equations provide for these relationships:

$$x_p = x'_p + x_o \qquad (i)$$

$$y_p = y'_p + y_o \qquad (j)$$

$$z'_p = -f \qquad (k)$$

Substituting Eqs. (i), (j), and (k) into Eqs. (g) and (h) and rearranging gives

$$x_p = x_o - f\left[\frac{m_{11}(X_P - X_L) + m_{12}(Y_P - Y_L) + m_{13}(-Z_L)}{m_{31}(X_P - X_L) + m_{32}(Y_P - Y_L) + m_{33}(-Z_L)}\right] \qquad (l)$$

$$y_p = y_o - f\left[\frac{m_{21}(X_P - X_L) + m_{22}(Y_P - Y_L) + m_{23}(-Z_L)}{m_{31}(X_P - X_L) + m_{32}(Y_P - Y_L) + m_{33}(-Z_L)}\right] \qquad (m)$$

Equations (l) and (m) contain a total of nine unknown parameters (x_o, y_o, f, ω, ϕ, κ, X_L, Y_L, and Z_L) as well as the measured coordinates x_p, y_p, X_P, and Y_P in the two coordinate systems. Two of the unknown parameters, f and Z_L, however, are not independent. By using the ratio of these two dependent parameters as a constant, a single parameter H can be used in their place. The relationship is shown in Eq. (n).

$$\frac{f}{Z_L} = \frac{1}{H} \qquad (n)$$

Given Eq. (n), substitutions can be made for f and Z_L in Eqs. (l) and (m) such that the ratio of the two parameters is maintained. A value of 1 (unity) will be used in place of f, and H will be used in place of Z_L, giving the following equations:

$$x_p = x_o - \frac{m_{11}(X_P - X_L) + m_{12}(Y_P - Y_L) + m_{13}(-H)}{m_{31}(X_P - X_L) + m_{32}(Y_P - Y_L) + m_{33}(-H)}$$

$$y_p = y_o - \frac{m_{21}(X_P - X_L) + m_{22}(Y_P - Y_L) + m_{23}(-H)}{m_{31}(X_P - X_L) + m_{32}(Y_P - Y_L) + m_{33}(-H)} \tag{C-46}$$

While Eqs. (C-46) can be used in their given form for a two-dimensional projective transformation, additional simplification and rearrangement facilitate their use. The subscript, P's will be dropped from the xy and XY coordinates at the start of this process. First, bring terms x_o and y_o into the fractions on the right side.

$$x = \frac{x_o m_{31}(X - X_L) + x_o m_{32}(Y - Y_L) + x_o m_{33}(-H) - m_{11}(X - X_L) - m_{12}(Y - Y_L) - m_{13}(-H)}{m_{31}(X - X_L) + m_{32}(Y - Y_L) + m_{33}(-H)}$$

$$y = \frac{y_o m_{31}(X - X_L) + y_o m_{32}(Y - Y_L) + y_o m_{33}(-H) - m_{21}(X - X_L) - m_{22}(Y - Y_L) - m_{23}(-H)}{m_{31}(X - X_L) + m_{32}(Y - Y_L) + m_{33}(-H)}$$

Next, collect terms in the numerator.

$$x = \frac{(x_o m_{31} - m_{11})(X - X_L) + (x_o m_{32} - m_{12})(Y - Y_L) + (x_o m_{33} - m_{13})(-H)}{m_{31}(X - X_L) + m_{32}(Y - Y_L) + m_{33}(-H)}$$

$$y = \frac{(y_o m_{31} - m_{21})(X - X_L) + (y_o m_{32} - m_{22})(Y - Y_L) + (y_o m_{33} - m_{23})(-H)}{m_{31}(X - X_L) + m_{32}(Y - Y_L) + m_{33}(-H)}$$

Then, separate the terms containing X and Y in both the numerator and denominator.

$$x = \frac{(x_o m_{31} - m_{11})X + (x_o m_{32} - m_{12})Y + (x_o m_{31} - m_{11})(-X_L) + (x_o m_{32} - m_{12})(-Y_L) + (x_o m_{33} - m_{13})(-H)}{m_{31}X + m_{32}Y + m_{31}(-X_L) + m_{32}(-Y_L) + m_{33}(-H)}$$

$$y = \frac{(y_o m_{31} - m_{21})X + (y_o m_{32} - m_{22})Y + (y_o m_{31} - m_{21})(-X_L) + (y_o m_{32} - m_{22})(-Y_L) + (y_o m_{33} - m_{23})(-H)}{m_{31}X + m_{32}Y + m_{31}(-X_L) + m_{32}(-Y_L) + m_{33}(-H)}$$

Finally, divide both the numerator and denominator in each equation by $m_{31}(-X_L) + m_{32}(-Y_L) + m_{33}(-H)$ to give the final form of the two-dimensional projective transformation equations.

$$x = \frac{a_1 X + b_1 Y + c_1}{a_3 X + b_3 Y + 1}$$

$$y = \frac{a_2 X + b_2 Y + c_2}{a_3 X + b_3 Y + 1} \tag{C-47}$$

In Eqs. (C-47), a_1, b_1, c_1, a_2, b_2, c_2, a_3, and b_3 are a new set of eight independent transformation parameters which are functions of the original eight unknowns x_o, y_o, ω, ϕ, κ, X_L, Y_L, and H from Eqs. (C-46).

As developed, Eqs. (C-47) yield x and y tilted photo coordinates from X and Y rectified-ratioed coordinates. It is customary, however, to perform rectification in the opposite sense, i.e., to compute X and Y rectified-ratioed coordinates from x and y coordinates measured on a tilted photo. Since Eqs. (C-47) are general and simply express the projective relationship between any two nonparallel planes, they can be written in the following form to enable computation of X and Y rectified-ratioed coordinates in terms of x and y tilted photo coordinates.

$$X = \frac{a_1 x + b_1 y + c_1}{a_3 x + b_3 y + 1}$$

$$Y = \frac{a_2 x + b_2 y + c_2}{a_3 x + b_3 y + 1}$$

(C-48)

In using Eqs. (C-48) for rectification, X and Y are ground coordinates of control points, and x and y are coordinates of the same points in the fiducial system of the tilted photograph. As listed, these equations are nonlinear due to their rational form. To facilitate their use with least squares, it is common for Eqs. (C-48) to be rearranged as follows:

$$a_1 x + b_1 y + c_1 - a_3 xX - b_3 yX = X + v_X$$

$$a_2 x + b_2 y + c_2 - a_3 xY - b_3 yY = Y + v_Y$$

(C-49)

In Eqs. (C-49), measured xy and XY coordinates appear on the left-hand side. Technically, equations of this form should be treated by general least squares techniques instead of the simpler method of observation equations. If, however, the measured values of x, y, X, and Y on the left side of the equations are treated as constants and if X and Y on the right side are treated as measured values, then satisfactory results will generally be obtained.

A pair of Eqs. (C-49) can be written for each control point, and since there are eight unknown parameters, four control points are needed for a unique solution. It is strongly recommended that more than four control points be used in the least squares solution, however. When equations of the type of Eqs. (C-49) have been written for all control points, a solution can be obtained for the unknown parameters. These parameters are then used in Eqs. (C-48) to compute the rectified and ratioed coordinates of all other points in the tilted photo whose xy coordinates have been measured. Besides using these equations for rectification, they can be used to transform comparator coordinates into the photo coordinate system defined by the fiducial marks, although this should be done only if more than four fiducials are available. In this instance, X and Y are calibrated coordinates of fiducial marks, and x and y are their comparator coordinates.

REFERENCES

American Society of Photogrammetry: *Manual of Photogrammetry*, 4th ed., Bethesda, MD, 1980, chaps. 2 and 14.

Blais, J. A. R.: "Three-Dimensional Similarity," *Canadian Surveyor,* vol. 26, no. 1, 1972, p. 71.

Dewitt, B. A.: "Initial Approximations for the Three-Dimensional Conformal Coordinate Transformation," *Photogrammetric Engineering and Remote Sensing,* vol. 62, no. 1, 1996, p. 79.

Erio, G.: "Three-Dimensional Transformations for Independent Models," *Photogrammetric Engineering and Remote Sensing,* vol. 41, no. 9, 1975, p. 1117.

Mikhail, E. M.: "Simultaneous Three-Dimensional Transformation of Higher Degree," *Photogrammetric Engineering,* vol. 30, no. 4, 1964, p. 588.

———: "Discussion Paper: Simultaneous Three-Dimensional Transformation," *Photogrammetric Engineering,* vol. 32, no. 2, 1966, p. 180.

Schut, G. H.: "Conformal Transformations and Polynomials," *Photogrammetric Engineering*, vol. 32, no. 5, 1966, p. 826.

Wolf, P. R., and C. Ghilani: *Adjustment Computations: Statistics and Least Squares in Surveying and GIS,* Wiley, New York, 1997.

Yassa, G.: "Orthogonal Transformations," *Photogrammetric Engineering,* vol. 40, no. 8, 1974, p. 961.

PROBLEMS

C-1. The following table contains arbitrary *x* and *y* coordinates of a group of points which were measured from a map by using a tablet digitizer. The table also includes UTM coordinates for three of the points. Using a *two-dimensional conformal* coordinate transformation, and least squares, calculate UTM coordinates for the other points.

Point	Arbitrary coordinates from tablet digitizer		UTM coordinates	
	x, mm	*y*, mm	*X*, m	*Y*, m
A	59.31	102.95	618,526.23	3,280,543.63
B	119.07	498.33	619,366.77	3,290,104.25
C	430.72	451.10	626,902.24	3,289,437.27
1	82.27	307.98		
2	186.56	147.01		
3	253.47	293.64		
4	414.68	260.15		
5	328.90	519.32		

C-2. The following table contains *x* and *y* comparator coordinates for eight fiducial marks and three additional photo image points. The table also contains calibrated coordinates for the eight fiducials. Using a *two-dimensional affine* transformation, and least squares, calculate coordinates for the three image points in the fiducial axis system.

Point	Comparator coordinates		Calibrated coordinates	
	x, mm	*y*, mm	*X*, mm	*Y*, mm
1	229.096	127.310	112.995	0.034
2	3.029	126.910	−113.006	0.005
3	115.832	240.108	0.003	112.993
4	116.291	14.070	−0.012	−113.000
5	228.864	240.287	112.996	112.990
6	3.280	13.893	−113.002	−112.999
7	2.799	239.932	−112.995	113.001
8	229.331	14.237	112.998	−112.999
A	103.426	111.531		
B	143.769	68.984		
C	7.508	212.623		

C-3. For the data of Prob. C-2, use the *two-dimensional projective* transformation and least squares to calculate coordinates for the three image points in the fiducial axis system.

C-4. Coordinates X_1, Y_1, and Z_1 for model I and X_2, Y_2, and Z_2 for model II of an independent model aerotriangulation are contained in the table below. Transform the model II coordinates into the model I coordinate system, using a *three-dimensional conformal* coordinate transformation and least squares. (Refer to Ex. Prob. 17-1 for information on available computer program.)

Point	Model II coordinates (arbitrary)			Model I coordinates (control)		
	X_2, mm	Y_2, mm	Z_2, mm	X_1, mm	Y_1, mm	Z_1, mm
100	0.000	0.000	0.000	89.919	2.841	−2.000
1	−17.407	78.746	−148.327	69.817	76.041	−157.144
2	0.619	64.828	−149.365	88.425	62.016	−157.713
3	−8.787	16.352	−152.516	79.426	12.202	−158.529
4	−1.030	14.574	−152.773	87.383	10.472	−158.792
5	−6.714	−87.446	−158.704	82.867	−94.117	−159.940
6	−6.813	−96.562	−159.262	82.869	−103.492	−160.106
200	72.357	1.230	−0.487			
11	70.827	76.422	−150.108			
12	74.001	76.411	−150.151			
13	67.763	−12.802	−154.747			
14	73.600	−8.354	−154.631			
15	55.065	−79.707	−158.200			
16	70.481	−73.410	−158.135			

D

DEVELOPMENT OF COLLINEARITY CONDITION EQUATIONS

D-1 INTRODUCTION

Collinearity, as illustrated in Fig. D-1, is the condition in which the exposure station of any photograph, an object point, and its photo image all lie on a straight line. The equations expressing this condition are called the *collinearity condition equations*. They are perhaps the most useful of all equations to the photogrammetrist.

In Fig. D-2 exposure station L of an aerial photo has coordinates X_L, Y_L, and Z_L with respect to the object (ground) coordinate system XYZ. Image a of object point A, shown in a *rotated* image plane, has image space coordinates x'_a, y'_a, and z'_a, where the rotated image space coordinate system $x'y'z'$ is parallel to object space coordinate system XYZ. Initially, it is assumed that the principal point o is located at the origin of the xy photo coordinate system. A correction that compensates for this assumption is introduced at the end of the development.

D-2 ROTATION IN TERMS OF OMEGA, PHI, AND KAPPA

By means of the three-dimensional rotation formulas developed in App. C, an image point a having coordinates x_a, y_a, and z_a in a tilted photo such as that of Fig. D-1 may have its coordinates rotated into the $x'y'z'$ coordinate system (parallel to XYZ), as shown in Fig. D-3. The rotated image coordinates x'_a, y'_a, and z'_a are related to the measured

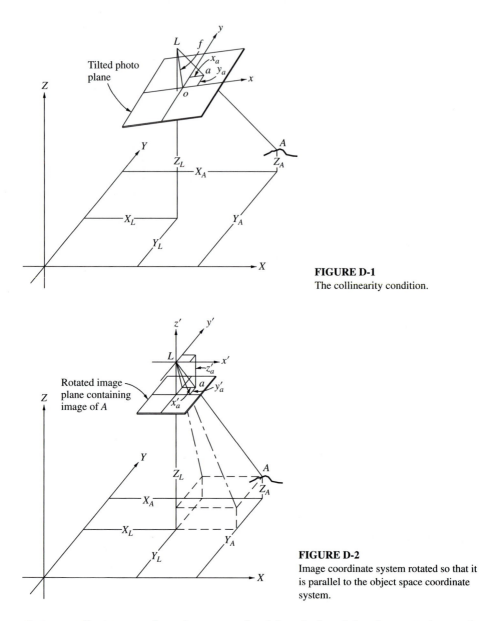

FIGURE D-1
The collinearity condition.

FIGURE D-2
Image coordinate system rotated so that it is parallel to the object space coordinate system.

photo coordinates x_a and y_a, the camera focal length f, and the three rotation angles omega, phi, and kappa. The rotation formulas are Eqs. (C-32). For convenience they are repeated here:

$$
\begin{aligned}
x_a &= m_{11}x_a' + m_{12}y_a' + m_{13}z_a' \\
y_a &= m_{21}x_a' + m_{22}y_a' + m_{23}z_a' \\
z_a &= m_{31}x_a' + m_{32}y_a' + m_{33}z_a'
\end{aligned}
\tag{D-1}
$$

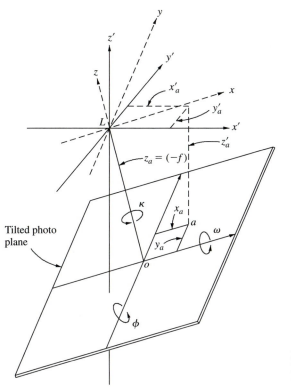

FIGURE D-3
Measurement xyz and rotated $x'y'z'$
image coordinate systems.

In Eqs. (D-1) the m's are functions of the rotation angles omega, phi, and kappa. These functions are given in Eqs. (C-33). Also note in Fig. D-3 that the value for z_a is equal to $-f$.

D-3   DEVELOPMENT OF THE COLLINEARITY CONDITION EQUATIONS

The collinearity condition equations are developed from similar triangles of Fig. D-2 as follows:

$$\frac{x'_a}{X_A - X_L} = \frac{y'_a}{Y_A - Y_L} = \frac{-z'_a}{Z_L - Z_A}$$

Reducing gives

$$x'_a = \left(\frac{X_A - X_L}{Z_A - Z_L}\right)z'_a \tag{a}$$

$$y'_a = \left(\frac{Y_A - Y_L}{Z_A - Z_L}\right)z'_a \tag{b}$$

Also, by identity,

$$z_a' = \left(\frac{Z_A - Z_L}{Z_A - Z_L}\right) z_a' \tag{c}$$

Substituting (a), (b), and (c) into Eqs. (D-1) gives

$$x_a = m_{11}\left(\frac{X_A - X_L}{Z_A - Z_L}\right) z_a' + m_{12}\left(\frac{Y_A - Y_L}{Z_A - Z_L}\right) z_a' + m_{13}\left(\frac{Z_A - Z_L}{Z_A - Z_L}\right) z_a' \tag{D-2}$$

$$y_a = m_{21}\left(\frac{X_A - X_L}{Z_A - Z_L}\right) z_a' + m_{22}\left(\frac{Y_A - Y_L}{Z_A - Z_L}\right) z_a' + m_{23}\left(\frac{Z_A - Z_L}{Z_A - Z_L}\right) z_a' \tag{D-3}$$

$$z_a = m_{31}\left(\frac{X_A - X_L}{Z_A - Z_L}\right) z_a' + m_{32}\left(\frac{Y_A - Y_L}{Z_A - Z_L}\right) z_a' + m_{33}\left(\frac{Z_A - Z_L}{Z_A - Z_L}\right) z_a' \tag{D-4}$$

Factoring the term $z_a'/(Z_A - Z_L)$ from Eqs. (D-2) through (D-4), dividing (D-2) and (D-3) by (D-4), substituting $-f$ for z_a, and adding corrections for offset of the principal point (x_o, y_o), the following collinearity equations result:

$$x_a = x_o - f\left[\frac{m_{11}(X_A - X_L) + m_{12}(Y_A - Y_L) + m_{13}(Z_A - Z_L)}{m_{31}(X_A - X_L) + m_{32}(Y_A - Y_L) + m_{33}(Z_A - Z_L)}\right] \tag{D-5}$$

$$y_a = y_o - f\left[\frac{m_{21}(X_A - X_L) + m_{22}(Y_A - Y_L) + m_{23}(Z_A - Z_L)}{m_{31}(X_A - X_L) + m_{32}(Y_A - Y_L) + m_{33}(Z_A - Z_L)}\right] \tag{D-6}$$

D-4   LINEARIZATION OF THE COLLINEARITY EQUATIONS

Equations (D-5) and (D-6) are nonlinear and involve nine unknowns: the three rotation angles omega, phi, and kappa which are inherent in the m's; the three exposure station coordinates X_L, Y_L, and Z_L; and the three object point coordinates X_A, Y_A, and Z_A. The photo coordinate measurements x_a and y_a are constant terms, as well as the calibration parameters x_o, y_o, and f which are considered to be constants in most applications of collinearity. The nonlinear collinearity equations are linearized by using Taylor's theorem. In linearizing them, Eqs. (D-5) and (D-6) are rewritten as follows:

$$F = x_o - f\frac{r}{q} = x_a \tag{D-7}$$

$$G = y_o - f\frac{s}{q} = y_a \tag{D-8}$$

where
$$q = m_{31}(X_A - X_L) + m_{32}(Y_A - Y_L) + m_{33}(Z_A - Z_L)$$
$$r = m_{11}(X_A - X_L) + m_{12}(Y_A - Y_L) + m_{13}(Z_A - Z_L)$$
$$s = m_{21}(X_A - X_L) + m_{22}(Y_A - Y_L) + m_{23}(Z_A - Z_L)$$

According to Taylor's theorem, Eqs. (D-7) and (D-8) may be expressed in linearized form by taking partial derivatives with respect to the unknowns:

$$F_0 + \left(\frac{\partial F}{\partial \omega}\right)_0 d\omega + \left(\frac{\partial F}{\partial \phi}\right)_0 d\phi + \left(\frac{\partial F}{\partial \kappa}\right)_0 d\kappa + \left(\frac{\partial F}{\partial X_L}\right)_0 dX_L + \left(\frac{\partial F}{\partial Y_L}\right)_0 dY_L$$

$$+ \left(\frac{\partial F}{\partial Z_L}\right)_0 dZ_L + \left(\frac{\partial F}{\partial X_A}\right)_0 dX_A + \left(\frac{\partial F}{\partial Y_A}\right)_0 dY_A + \left(\frac{\partial F}{\partial Z_A}\right)_0 dZ_A = x_a \tag{D-9}$$

$$G_0 + \left(\frac{\partial G}{\partial \omega}\right)_0 d\omega + \left(\frac{\partial G}{\partial \phi}\right)_0 d\phi + \left(\frac{\partial G}{\partial \kappa}\right)_0 d\kappa + \left(\frac{\partial G}{\partial X_L}\right)_0 dX_L + \left(\frac{\partial G}{\partial Y_L}\right)_0 dY_L$$

$$+ \left(\frac{\partial G}{\partial Z_L}\right)_0 dZ_L + \left(\frac{\partial G}{\partial X_A}\right)_0 dX_A + \left(\frac{\partial G}{\partial Y_A}\right)_0 dY_A + \left(\frac{\partial G}{\partial Z_A}\right)_0 dZ_A = y_a \tag{D-10}$$

In Eqs. (D-9) and (D-10), F_0 and G_0 are functions F and G of Eqs. (D-7) and (D-8) evaluated at the initial approximations for the nine unknowns; the terms $(\partial F/\partial \omega)_0$, $(\partial G/\partial \omega)_0$, $(\partial F/\partial \phi)_0$, $(\partial G/\partial \phi)_0$, etc., are partial derivatives of functions F and G with respect to the indicated unknowns evaluated at the initial approximations; and $d\omega$, $d\phi$, $d\kappa$, etc., are unknown corrections to be applied to the initial approximations. The units of $d\omega$, $d\phi$, and $d\kappa$ are radians. Since the photo coordinates x_a and y_a are measured values, if the equations are to be used in a least squares solution, residual terms must be included to make the equations consistent. The following simplified forms of the linearized collinearity equations include these residuals.

$$b_{11}\, d\omega + b_{12}\, d\phi + b_{13}\, d\kappa - b_{14}\, dX_L - b_{15}\, dY_L - b_{16}\, dZ_L$$
$$+ b_{14}\, dX_A + b_{15}\, dY_A + b_{16}\, dZ_A = J + v_{x_a} \tag{D-11}$$

$$b_{21}\, d\omega + b_{22}\, d\phi + b_{23}\, d\kappa - b_{24}\, dX_L - b_{25}\, dY_L - b_{26}\, dZ_L$$
$$+ b_{24}\, dX_A + b_{25}\, dY_A + b_{26}\, dZ_A = K + v_{y_a} \tag{D-12}$$

In Eqs. (D-11) and (D-12), J and K are equal to $x_a - F_0$ and $y_a - G_0$, respectively. The b's are coefficients equal to the partial derivatives. For convenience these coefficients are given below and on the next page. In these coefficients ΔX, ΔY, and ΔZ are equal to $X_A - X_L$, $Y_A - Y_L$, and $Z_A - Z_L$, respectively. Numerical values for these coefficient terms are obtained by using initial approximations for the unknowns.

$$b_{11} = \frac{f}{q^2}\left[r(-m_{33}\,\Delta Y + m_{32}\,\Delta Z) - q(-m_{13}\,\Delta Y + m_{12}\,\Delta Z)\right]$$

$$b_{12} = \frac{f}{q^2}\left[r(\cos\phi\,\Delta X + \sin\omega\sin\phi\,\Delta Y - \cos\omega\sin\phi\,\Delta Z)\right.$$
$$\left. - q(-\sin\phi\cos\kappa\,\Delta X + \sin\omega\cos\phi\cos\kappa\,\Delta Y - \cos\omega\cos\phi\cos\kappa\,\Delta Z)\right]$$

$$b_{13} = \frac{-f}{q}(m_{21}\,\Delta X + m_{22}\,\Delta Y + m_{23}\,\Delta Z)$$

$$b_{14} = \frac{f}{q^2}(rm_{31} - qm_{11})$$

$$b_{15} = \frac{f}{q^2}(rm_{32} - qm_{12})$$

$$b_{16} = \frac{f}{q^2}(rm_{33} - qm_{13})$$

$$J = x_a - x_o + f\frac{r}{q}$$

$$b_{21} = \frac{f}{q^2}[s(-m_{33}\,\Delta Y + m_{32}\,\Delta Z) - q(-m_{23}\,\Delta Y + m_{22}\,\Delta Z)]$$

$$b_{22} = \frac{f}{q^2}[s(\cos\phi\,\Delta X + \sin\omega\sin\phi\,\Delta Y - \cos\omega\sin\phi\,\Delta Z)$$
$$- q(\sin\phi\sin\kappa\,\Delta X - \sin\omega\cos\phi\sin\kappa\,\Delta Y + \cos\omega\cos\phi\sin\kappa\,\Delta Z)]$$

$$b_{23} = \frac{f}{q}(m_{11}\,\Delta X + m_{12}\,\Delta Y + m_{13}\,\Delta Z)$$

$$b_{24} = \frac{f}{q^2}(sm_{31} - qm_{21})$$

$$b_{25} = \frac{f}{q^2}(sm_{32} - qm_{22})$$

$$b_{26} = \frac{f}{q^2}(sm_{33} - qm_{23})$$

$$K = y_a - y_o + f\frac{s}{q}$$

D-5 APPLICATIONS OF COLLINEARITY

The collinearity equations are applicable to the analytical solution of almost every photogrammetry problem. As examples, Sec. 11-6 describes their use in *space resection,* in which the six elements of exterior orientation of a tilted photograph are computed; and Sec. 11-10 explains how collinearity is applied in analytic *relative orientation,* which is necessary in analytically extending control photogrammetrically. Other applications are described elsewhere in this book. Regardless of the particular problem, an x equation [Eq. (D-11)] and y equation [Eq. (D-12)] are written for each point whose image or images appear in the photo or photos involved in the problem. The equations will contain unknowns, the number of which will vary with the particular problem. If the number of equations is equal to or greater than the number of unknowns, a solution is generally possible.

Initial approximations are needed for all unknowns, and these are usually easily obtained by making certain assumptions, such as vertical photography. The initial approximations do not have to be extremely close, but the closer they are to the unknowns, the faster a satisfactory solution will be reached; and the result is a savings in computer time.

In solving a system of collinearity equations of the form of Eqs. (D-11) and (D-12) for any problem, the quantities that are determined are corrections to the initial approximations. After the first solution, the computed corrections are added to the initial approximations to obtain revised approximations. The solution is then repeated to find new corrections. This procedure is continued (iterated) until the magnitudes of the corrections become insignificant. A system of collinearity equations of the form of Eqs. (D-11) and (D-12) may be expressed in matrix form as

$$_mA^n{}_nX^1 = {}_mL^1 + {}_mV^1 \tag{D-13}$$

In Eq. (D-13), m is the number of equations; n is the number of unknowns; $_mV^1$ is the matrix of residual errors in the measured x and y photo coordinates; $_mA^n$ is the matrix of b's, the coefficients of the unknowns; $_nX^1$ is the matrix of unknown corrections to the initial approximations; and $_mL^1$ is the matrix of constant terms J and K. If the number of equations exceeds the number of unknowns, a least squares solution may be obtained for the most probable values for the unknowns by using matrix Eq. (B-12) or (B-13). Precision of the unknowns may be computed by applying matrix Eqs. (B-14) through (B-17).

D-6 ROTATION IN TERMS OF AZIMUTH, TILT, AND SWING

Instead of using omega, phi, and kappa as the rotation angles for transforming the tilted photo coordinates into an $x'y'z'$ coordinate system parallel to the ground system, the rotation angles of azimuth, tilt, and swing may be used. A tilted photo illustrating the azimuth α, tilt t, and swing s angles is shown in Fig. D-4. In the figure the photo principal

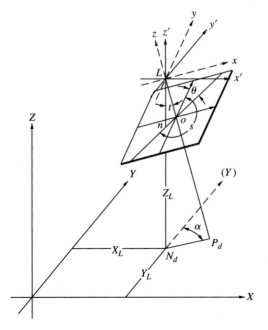

FIGURE D-4
Azimuth, tilt, and swing rotation angles.

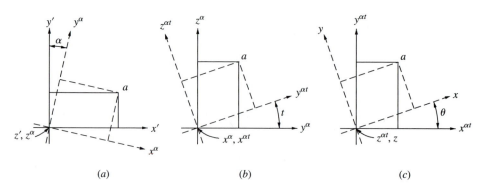

FIGURE D-5
Rotation in azimuth, tilt, and swing. (a) First rotation; (b) second rotation; (c) third rotation.

plane intersects the datum plane along line $N_d P_d$. The rotation formulas are developed by initially assuming an $x'y'z'$ coordinate system parallel to XYZ and then, by means of rotations, converting it to the xyz photo measurement system. The origins of both image coordinate systems are taken at exposure station L.

The rotation equations are developed in a sequence of three separate two-dimensional rotations. The $x'y'z'$ coordinate system is first rotated about the z' axis through a clockwise angle α to create an $x^\alpha y^\alpha z^\alpha$ coordinate system. After rotation the y^α axis will be in the principal plane of the photo. With reference to Fig. D-5a, the coordinates of any point in the $x^\alpha y^\alpha z^\alpha$ system are

$$x^\alpha = x' \cos \alpha - y' \sin \alpha$$
$$y^\alpha = x' \sin \alpha + y' \cos \alpha \qquad \text{(D-14)}$$
$$z^\alpha = z'$$

The second rotation is a counterclockwise rotation t about the x^α axis to create an $x^{\alpha t} y^{\alpha t} z^{\alpha t}$ coordinate system. After rotation, the $x^{\alpha t}$ and $y^{\alpha t}$ axes are in the plane of the tilted photograph. With reference to Fig. D-5b, the coordinates of any point in the $x^{\alpha t} y^{\alpha t} z^{\alpha t}$ system are

$$x^{\alpha t} = x^\alpha$$
$$y^{\alpha t} = y^\alpha \cos t + z^\alpha \sin t \qquad \text{(D-15)}$$
$$z^{\alpha t} = -y^\alpha \sin t + z^\alpha \cos t$$

The third rotation is about the $z^{\alpha t}$ axis through the counterclockwise angle θ. Angle θ is defined as

$$\theta = s - 180°$$

This third rotation creates an $x^{\alpha t \theta} y^{\alpha t \theta} z^{\alpha t \theta}$ coordinate system which coincides with the xyz tilted photo system. With reference to Fig. D-5c, the coordinates of any point in the *xyz* system are

$$x = x^{\alpha t \theta} = x^{\alpha t} \cos \theta + y^{\alpha t} \sin \theta$$

$$y = y^{\alpha t \theta} = -x^{\alpha t} \sin \theta + y^{\alpha t} \cos \theta \qquad \text{(D-16)}$$

$$z = z^{\alpha t \theta} = z^{\alpha t}$$

Because $\sin \theta$ equals $-\sin s$, and $\cos \theta$ equals $-\cos s$, these substitutions may be made into Eqs. (D-16), from which

$$x = -x^{\alpha t} \cos s - y^{\alpha t} \sin s$$

$$y = x^{\alpha t} \sin s - y^{\alpha t} \cos s \qquad \text{(D-17)}$$

$$z = z^{\alpha t}$$

Substituting Eq. (D-14) into Eq. (D-15), in turn substituting into Eq. (D-17), and factoring, the following expressions for the x, y, and z coordinates of any point are obtained:

$$x = m_{11}x' + m_{12}y' + m_{13}z'$$

$$y = m_{21}x' + m_{22}y' + m_{23}z' \qquad \text{(D-18)}$$

$$z = m_{31}x' + m_{32}y' + m_{33}z'$$

In Eqs. (D-18) the m's are

$$m_{11} = -\cos \alpha \cos s - \sin \alpha \cos t \sin s$$

$$m_{12} = \sin \alpha \cos s - \cos \alpha \cos t \sin s$$

$$m_{13} = -\sin t \sin s$$

$$m_{21} = \cos \alpha \sin s - \sin \alpha \cos t \cos s$$

$$m_{22} = -\sin \alpha \sin s - \cos \alpha \cos t \cos s \qquad \text{(D-19)}$$

$$m_{23} = -\sin t \cos s$$

$$m_{31} = -\sin \alpha \sin t$$

$$m_{32} = -\cos \alpha \sin t$$

$$m_{33} = \cos t$$

D-7   COLLINEARITY EQUATIONS USING AZIMUTH-TILT-SWING ROTATION

By simply substituting Eqs. (D-19) for the m's into Eqs. (D-5) and (D-6), collinearity equations are obtained which include azimuth, tilt, and swing as unknowns instead of omega, phi, and kappa. By applying Taylor's theorem, these azimuth, tilt, and swing equations can be linearized and used to solve photogrammetry problems analytically. More frequently, however, the omega-phi-kappa equations are used, and if the azimuth, tilt, and swing angles are desired, they are determined from omega, phi, and kappa as described in the next section.

D-8   CONVERTING FROM ONE ROTATION SYSTEM TO THE OTHER

Although the azimuth-tilt-swing expressions [Eqs. (D-19)] for the m's differ from their corresponding omega-phi-kappa expressions [Eqs. (C-33)], for any given tilted photo their numerical values are equal. This is true because the m's are actually direction cosines relating image and object coordinate systems as described in Sec. C-7. Because of their equality, corresponding m's may be set equal to each other; for example, $m_{11} = \cos \phi \cos \kappa = -\cos \alpha \cos s - \sin \alpha \cos t \sin s$. This equality of m's enables conversions to be made between the omega-phi-kappa and azimuth-tilt-swing systems.

Before discussing the conversion method, it is necessary to establish limits on the ranges of values for the two sets of angles. This is required since if each angle is allowed to have a full 360° range, any particular rotation configuration will have two equally valid sets of angles in either the azimuth-tilt-swing or the omega-phi-kappa system. For example, azimuth = 0°, tilt = 5°, and swing = 0° will generate the same rotation matrix as azimuth = 180°, tilt = −5°, and swing = 180°. Ranges for azimuth, tilt, and swing which avoid this multiple-definition problem are

$$-180° < \text{azimuth} \le 180°$$

$$0° \le \text{tilt} \le 180°$$

$$-180° < \text{swing} \le 180°$$

and ranges for omega, phi, and kappa are

$$-180° < \text{omega} \le 180°$$

$$-90° \le \text{phi} \le 90°$$

$$-180° < \text{kappa} \le 180°$$

Although not necessary, any of the ranges for azimuth, swing, omega, or kappa could be chosen as 0° to 360°, if desired.

If omega, phi, and kappa for a particular photograph are known, numerical values for the m's can be calculated by Eqs. (C-33) and the tilt, swing, and azimuth determined from the following:

$$t = \cos^{-1}(m_{33}) \tag{D-20}$$

$$s = \tan^{-1}\left(\frac{-m_{13}}{-m_{23}}\right) \tag{D-21}$$

$$\alpha = \tan^{-1}\left(\frac{-m_{31}}{-m_{32}}\right) \tag{D-22}$$

In Eqs. (D-21) and (D-22) it is essential that a full-circle inverse tangent function (such as atan2) be used so that the full ranges for s and α can be determined. In the rare case where tilt is exactly 0° or 180°, both the numerator and denominator in each of Eqs. (D-21) and (D-22) will equal zero, resulting in invalid results from a full-circle inverse tangent function. (Note that if the denominator is zero but the numerator is nonzero, a properly implemented full-circle inverse tangent function will return valid results.) In this situation where tilt is exactly zero, no principal line exists, and swing and azimuth are undefined. However, by arbitrarily defining azimuth to be equal to zero, a value for swing can be obtained from the rotation matrix. The original definitions of swing and azimuth will no longer apply (see Sec. 10-2); however, the resulting values can still be properly used in photogrammetric equations. When the tilt is exactly 0° (or 180°), the values for swing and azimuth can be obtained by

$$ s = \tan^{-1} \left(\frac{-m_{12}}{-m_{22}} \right) \tag{D-23} $$

$$ \alpha = 0° \tag{D-24} $$

If the azimuth, tilt, and swing are known for a particular photo, conversion to omega, phi, and kappa is also readily made as follows:

$$ \phi = \sin^{-1}(m_{31}) \tag{D-25} $$

$$ \omega = \tan^{-1} \left(\frac{-m_{32}}{m_{33}} \right) \tag{D-26} $$

$$ \kappa = \tan^{-1} \left(\frac{-m_{21}}{m_{11}} \right) \tag{D-27} $$

Once again, it is essential that a full-circle inverse tangent function be used with Eqs. (D-26) and (D-27) so that values of omega and kappa in the proper ranges are computed. In the rare case where phi is exactly ±90°, both the numerator and denominator in each of Eqs. (D-26) and (D-27) will be zero, and the values for omega and kappa will be undefined. By giving an arbitrary definition of zero for omega as indicated by Eq. (D-28), Eq. (D-29) may then be used to compute kappa.

$$ \omega = 0° \tag{D-28} $$

$$ \kappa = \tan^{-1} \left(\frac{m_{12}}{m_{22}} \right) \tag{D-29} $$

The foregoing conversion method is completely general and will work for any possible rotation configuration. In conventional aerial photography, the value for tilt should never be greater than 90°; however, in terrestrial photography or when using an earth-centered coordinate system with aerial photography, any rotation is possible.

REFERENCE

American Society of Photogrammetry: *Manual of Photogrammetry,* 4th ed., Bethesda, MD, 1980, chap. 2.

PROBLEMS

D-1. State the condition of collinearity in photogrammetry.

D-2. Explain why linearized collinearity equations must be iterated a number of times before a satisfactory solution is achieved.

D-3. Read the portion of sec. 2.6.5 of the *Manual of Photogrammetry* (1980) that discusses three convergence criteria for an iterative solution. Outline a method for convergence testing that utilizes the estimate for the standard error of unit weight.

D-4. Given the following values, compute the photo coordinates x_a and y_a, using Eqs. (D-5) and (D-6). Express your answers to the nearest 0.001 mm.

$$f = 152.916 \text{ mm} \qquad x_o = 0.010 \text{ mm} \qquad y_o = -0.004 \text{ mm}$$
$$\omega = -0.4052° \qquad \phi = 1.2095° \qquad \kappa = 102.8006°$$
$$X_L = 1027.863 \text{ m} \qquad Y_L = 1043.998 \text{ m} \qquad Z_L = 611.032 \text{ m}$$
$$X_A = 974.435 \text{ m} \qquad Y_A = 956.592 \text{ m} \qquad Z_A = 14.619 \text{ m}$$

D-5. Convert the following values of azimuth, tilt, and swing to omega, phi, and kappa. Express your answers in decimal degrees to the nearest thousandth of a degree.
(a) $\alpha = 10.000°$, $t = 1.500°$, $s = -165.000°$
(b) $\alpha = -135.000°$, $t = 97.000°$, $s = 20.000°$
(c) $\alpha = 90.000°$, $t = 90.000°$, $s = -85.000°$

D-6. Convert the following values of omega, phi, and kappa to azimuth, tilt, and swing. Express your answers in decimal degrees to the nearest thousandth of a degree.
(a) $\omega = 2.800°$, $\phi = -1.500°$, $\kappa = 0.000°$
(b) $\omega = -105.000°$, $\phi = 35.000°$, $\kappa = -10.000°$
(c) $\omega = 0.000°$, $\phi = 0.000°$, $\kappa = 0.000°$

APPENDIX
E

DIGITAL
RESAMPLING

E-1 INTRODUCTION

The acquisition of a digital image involves discrete sampling of a continuous analog signal, for example, the reflected energy from the ground. These digital samples are made in a (distorted) grid pattern, with each grid cell or pixel containing a digital number (DN) representing the lightness or darkness at its corresponding ground location. When a digital image is acquired, no attempt is made to have the pixels line up with any particular map projection coordinates. It is therefore necessary to perform resampling to obtain a digital sample at an intermediate (i.e., fractional) row, column location. Resampling involves interpolation between existing pixels (DNs) to synthesize pixels that correspond to fractional locations. Determination of the appropriate fractional locations is often the result of a coordinate transformation (see App. C).

There are several techniques available for resampling digital images, although three particular ones are by far, most prevalent. They are known as *nearest-neighbor* interpolation, *bilinear* interpolation, and *bicubic* interpolation. Other, more computationally intensive techniques are generally not employed since they tend to be sensitive to sensor noise which exists in digital imagery.

E-2 NEAREST NEIGHBOR

The nearest-neighbor interpolation is simplest of the three. As its name implies, the DN chosen will be that of the image pixel whose center is closest to the center of the grid cell. From a computational standpoint, all that is required is to round off the fractional

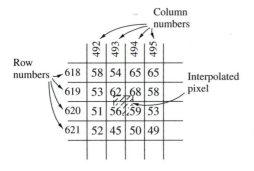

FIGURE E-1
A 4 × 4 subarea of image pixels with superimposed grid cell at a fractional location.

row and column values to the nearest integral value. Figure E-1 shows the DNs for a 4 × 4 subarea from a digital image. A pixel is superimposed at a fractional location ($R = 619.71$, $C = 493.39$). Rounding these values to the nearest integer yields 620 and 493 for the row and column indices, respectively. Thus, the resampled value is 56.

E-3   BILINEAR INTERPOLATION

A second resampling method, which involves greater computational complexity, is bilinear interpolation. In this approach, the four surrounding pixels are selected, and linear interpolation is performed in two directions. This is illustrated by example, using the values from Fig. E-1. First, linearly interpolated values DN_1 and DN_2 are computed along rows 619 and 620, respectively. Equations (E-1) and (E-2) illustrate the calculation.

$$DN_1 = 0.39(68 - 62) + 62 = 64.34 \tag{E-1}$$

$$DN_2 = 0.39(59 - 56) + 56 = 57.17 \tag{E-2}$$

Next, a linear interpolation is performed in the column direction, yielding the result (DN) given in Eq. (E-3).

$$DN = 0.71(57.17 - 64.34) + 64.34 = 59.25 \tag{E-3}$$

Finally, since DNs are generally integers, the value from Eq. (E-3) is rounded to 59. Note that this differs from the DN value of 56 obtained by using the nearest-neighbor method.

E-4   BICUBIC INTERPOLATION

Bicubic interpolation, also known as *cubic convolution,* is a third resampling technique which is commonly used. Explanation of this technique requires a little background in sampling theory. First, an assumption is made that the original signal has been sampled above the *Nyquist rate,* which is generally satisfied for imaging sensors. The Nyquist rate is, in essence, the sampling frequency required to faithfully record the highest (spatial) frequency content of the scene. Given this assumption, the "sinc" function allows

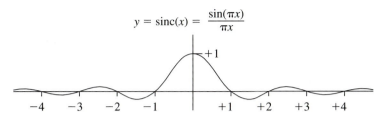

$$y = \text{sinc}(x) = \frac{\sin(\pi x)}{\pi x}$$

FIGURE E-2
Form of the sinc function.

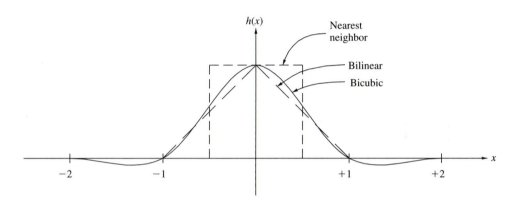

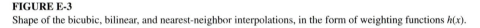

FIGURE E-3
Shape of the bicubic, bilinear, and nearest-neighbor interpolations, in the form of weighting functions $h(x)$.

an (almost) exact reconstruction of the original scene. The form of the sinc function is shown in Fig. E-2. (Note that in this figure, the argument for the sine function is in radians.) If the images had an infinite number of rows and columns, and all pixels were used for the interpolation, the sinc function would yield a perfect reconstruction. Practicality, however, dictates that interpolations be carried out using only small neighborhoods surrounding the interpolated pixel. A cubic spline approximation to the sinc function is the form generally used for bicubic interpolation. The shape of the spline is given in Fig. E-3 while Eqs. (E-4) through (E-6) express the functional relationship. For comparison, Fig. E-3 also shows the shape of nearest-neighbor and bilinear interpolations expressed in the form of similar weighting functions. Note that the cubic spline most nearly approximates the sinc function of Fig. E-2, whereas bilinear and nearest-neighbor interpolations are less consistent approximations.

$$f_1(x) = (a + 2)x^3 - (a + 3)x^2 + 1 \qquad \text{for } 0 \leq x \leq 1 \qquad \text{(E-4)}$$

$$f_2(x) = ax^3 - 5ax^2 + 8ax - 4a \qquad \text{for } 1 \leq x \leq 2 \qquad \text{(E-5)}$$

$$f_3(x) = 0 \qquad \text{for } x \geq 2 \qquad \text{(E-6)}$$

where a = free parameter equal to slope of weighting function at $x = 1$ (generally $a = -0.5$ yields best results)

x = absolute value of difference between whole-number row or column and fractional row or column

The computational process is analogous to that of bilinear interpolation in that it is performed first along the rows and then down the single, fractional column. (The procedure is demonstrated with a numerical example below.) The computations are conveniently expressed in matrix form, as shown in Eq. (E-7). In this equation, the R and C matrices consist of coefficients derived from Eqs. (E-4) and (E-5), and the D matrix contains the digital numbers from the 4×4 neighborhood surrounding the interpolated pixel.

$$RDC = [r_1 \, r_2 \, r_3 \, r_4] \begin{bmatrix} d_{11} & d_{12} & d_{13} & d_{14} \\ d_{21} & d_{22} & d_{23} & d_{24} \\ d_{31} & d_{32} & d_{33} & d_{34} \\ d_{41} & d_{42} & d_{43} & d_{44} \end{bmatrix} \begin{bmatrix} c_1 \\ c_2 \\ c_3 \\ c_4 \end{bmatrix} \quad \text{(E-7)}$$

Interpolating across the rows (based on the fractional column position) is done by forming the product DC. Subsequently, R is multiplied by the product to obtain the final interpolated value. In fact, the sequence of multiplications does not matter; they can be performed from left to right in Eq. (E-7).

Referring to the example of Fig. E-1, the bicubic interpolation computation procedure begins by computing the elements of the R matrix. Given the fractional row location (619.71), an interpolation weighting function will be computed for the two rows above (618 and 619) and the two rows below (620 and 621). The distance x from the fractional row is determined for each of the four surrounding rows, and the corresponding function [Eq. (E-4) or (E-5)] is selected. The result of this computation is listed in Table E-1. In a similar fashion, the elements of matrix C are computed, as shown in Table E-2.

Following these preliminary calculations, the matrix product from Eq. (E-7) is formed. This product, using the example values, is shown in Eq. (E-8). The resultant value of 60.66 is then rounded to 61, the nearest integer. Note that again this DN

TABLE E-1
Row interpolation weight matrix R computed for example of Fig. E-1. Slope value of $a = -0.5$ was used.

Element	Row	Distance x	Weighting function	Value
r_1	618	1.71	$f_2(x)$	−0.02986
r_2	619	0.71	$f_1(x)$	0.27662
r_3	620	0.29	$f_1(x)$	0.82633
r_4	621	1.29	$f_2(x)$	−0.07309

TABLE E-2
Column interpolation weight matrix C computed for example of Fig. E-1. Slope value of $a = -0.5$ was used.

Element	Column	Distance x	Weighting function	Value
c_1	492	1.39	$f_2(x)$	-0.07256
c_2	493	0.39	$f_1(x)$	0.70873
c_3	494	0.61	$f_1(x)$	0.41022
c_4	495	1.61	$f_2(x)$	-0.04639

value differs from both the 56 of nearest-neighbor interpolation and the 59 of bilinear interpolation.

$$[-0.02986 \quad 0.27662 \quad 0.82633 \quad -0.07309] \begin{bmatrix} 58 & 54 & 65 & 65 \\ 53 & 62 & 68 & 58 \\ 51 & 56 & 59 & 53 \\ 52 & 45 & 50 & 49 \end{bmatrix} \begin{bmatrix} -0.07256 \\ 0.70873 \\ 0.41022 \\ -0.04639 \end{bmatrix} = 60.66$$

(E-8)

E-5   DISCUSSION OF THE THREE METHODS

The primary advantage of the nearest-neighbor method is that it is the fastest of the three techniques in terms of computational time. It also has the advantage of not modifying the original image data, which is important if remote-sensing image classification will be performed. However, since a continuous interpolation is not being performed, the resultant appearance can be somewhat jagged or blocky.

The primary advantage of bilinear interpolation is the smoother appearance of the result. This appearance is slightly compromised by the fact that some high-frequency detail is filtered out. In other words, edges in the scene are slightly less distinct. In terms of computational time, bilinear interpolation is slower than nearest-neighbor interpolation but faster than bicubic interpolation.

Bicubic interpolation is the most rigorous resampling technique of the three on the basis of signal processing theory. It achieves the smooth appearance without sacrificing as much high-frequency (edge) detail. However, this enhanced appearance comes at a penalty in terms of computational time.

REFERENCES

American Society of Photogrammetry: *Manual of Remote Sensing,* 2d ed., Bethesda, MD, 1983, chap. 17.
Billingsley, F. C.: "Review of Image Processing Fundamentals," *Digital Image Processing, SPIE,* vol. 528, 1985.
Moik, J. G.: *Digital Processing of Remotely Sensed Images,* NASA SP-431, Government Printing Office, Washington, 1980.

PROBLEMS

E-1. Using the data from the example presented in this appendix, calculate a bicubic interpolation using a slope parameter $a = -1$.

E-2. Using the 4×4 array of digital numbers from Fig. E-1, compute a resampled pixel value by each of the three given methods corresponding to row = 619.20, column = 493.55. Use a value of $a = -0.5$ for the slope parameter of the bicubic interpolation method.

APPENDIX
F

CONVERSIONS BETWEEN OBJECT SPACE COORDINATE SYSTEMS

F-1 INTRODUCTION

In Chap. 5, descriptions of four fundamental object space coordinate systems were given. In this appendix, mathematical formulas which can be used to convert coordinates between the various systems are presented. *Conversions* differ from *transformations* in that conversions are exact mathematical processes which simply change the mathematical representation of positions from one form to another. Changing two-dimensional rectangular coordinates to polar coordinates is an example of a conversion. A transformation, on the other hand, is not an exact mathematical process per se, but is based on measurements and therefore contains errors. This is the key distinguishing characteristic between conversions and transformations. If a different set of measurements is used, a transformation may yield different results; but since a conversion is not based on measurements, it will always be consistent.

Figure F-1 illustrates the sequence in which the conversions are accomplished. At the top of this figure are map projections which assign plane coordinates XY based on developable surfaces. These coordinates can be converted to geodetic coordinates of latitude ϕ and longitude λ. The two-way arrow between the two coordinate systems in the figure indicates that the conversion can also be made in reverse, i.e., conversion from geodetic to map projection coordinates. Following downward in Fig. F-1, the next conversion represented is from geodetic latitude ϕ, longitude λ, and ellipsoid height h to geocentric coordinates X, Y, and Z. Again, this conversion can be made in reverse. At the

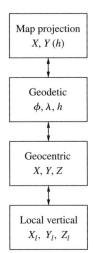

FIGURE F-1
Object space coordinate conversion sequence.

bottom of the figure, the conversion of geocentric coordinates to local vertical coordinates X_l, Y_l, and Z_l is illustrated; and as before, the conversion can be reversed. The conversions will always follow the path indicated by Fig. F-1. If conversion is to be made from map projection coordinates to local vertical, for instance, calculation of geodetic and geocentric coordinates will be performed as intermediate steps.

In many cases, values of coordinates in the aforementioned systems may have as many as 10 or 11 significant figures. When performing coordinate conversions using values this precise, it is essential to carry a large number of significant figures in the computations. A hand calculator having 10 significant figures may give marginal results. It is suggested that solutions be programmed on a computer using double-precision variables or solved with a spreadsheet having at least 12 significant figures. Off-the-shelf computer programs that perform these coordinate conversions generally use suitable precision for internal calculations, but this should be verified to ensure accurate results. *Note:* The examples in this appendix were solved using 16 significant figures.

F-2 STANDARD ELLIPSOID CONSTANTS FOR GRS80

To define the size and shape of the ellipsoid, at least two constants are required. Generally, the semimajor axis of the ellipsoid a and the flattening f are used. From these two defining constants, other ellipsoid parameters can be derived.

The examples in this appendix are based on the GRS80 ellipsoid. Since many of the conversions presented in this chapter require values for standard ellipsoid constants, they are presented here. From Table 5-1 the values for a and $1/f$ are

$$a = 6{,}378{,}137 \text{ m (exact by definition)}$$

$$f = \frac{1}{298.25722210088}$$

From these quantities, the values of b, e, e^2, and e'^2 can be computed as follows:
From Eq. (5-2):

$$b = (6{,}378{,}137 \text{ m}) \left(1 - \frac{1}{298.25722210088} \right) = 6{,}356{,}752.314 \text{ m}$$

From Eq. (5-4):

$$e^2 = \left(\frac{1}{298.25722210088} \right) \left(2 - \frac{1}{298.25722210088} \right) = 0.006694380023$$

$$e = \sqrt{0.006694380023} = 0.08181919104$$

From Eq. (5-6):

$$e'^2 = \frac{0.006694380023}{(1 - 1/298.25722210088)^2} = 0.006739496775$$

F-3 CONVERSION BETWEEN GEODETIC AND GEOCENTRIC COORDINATES

Conversion between geodetic $\phi\lambda h$ and geocentric XYZ is a three-dimensional process. Figure 5-5 illustrates the relationship between the two systems. Since geodetic coordinates are related to the reference surface used, certain ellipsoid constants (a, e^2, and e'^2) are required in order to achieve the conversion. To convert a position expressed in geodetic coordinates to geocentric coordinates, the following equations, which can be found in standard geodesy texts, are used:

$$X = (N + h) \cos \phi \cos \lambda \tag{F-1}$$

$$Y = (N + h) \cos \phi \sin \lambda \tag{F-2}$$

$$Z = [N(1 - e^2) + h] \sin \phi \tag{F-3}$$

In Eqs. (F-1) through (F-3), the value N is the length of the normal to the ellipsoid at latitude ϕ and is computed by

$$N = \frac{a}{\sqrt{1 - e^2 \sin^2 \phi}} \tag{F-4}$$

Conversion from geocentric coordinates to geodetic coordinates may be accomplished by the following approach. First, longitude is calculated by dividing Eq. (F-2) by (F-1) and taking the inverse tangent as follows:

$$\lambda = \tan^{-1} \left(\frac{Y}{X} \right) \tag{F-5}$$

In Eq. (F-5) it is essential that the full-circle inverse tangent be used since longitude can range from $-180°$ to $+180°$.

Latitude and height are not as easily isolated for direct solution. A number of methods exist for this conversion, some of which are exact and others iterative. A particularly elegant method attributed to Bowring (1976) is essentially an iterative approach, although accuracy to within about 1 μm can be obtained for terrestrial applications without iteration. For applications far beyond the earth's surface (e.g., satellite applications), a single iteration will yield submillimeter accuracy for points anywhere within the solar system. The applicable equations follow.

$$\beta = \tan^{-1}\left(\frac{Z}{(1 - f)\sqrt{X^2 + Y^2}}\right) \tag{F-6}$$

$$\phi = \tan^{-1}\left(\frac{Z + e'^2 b \sin^3 \beta}{\sqrt{X^2 + Y^2} - ae^2 \cos^3 \beta}\right) \tag{F-7}$$

To iterate, using the value of ϕ computed by Eq. (F-7), compute an updated value for β by

$$\beta = \tan^{-1}[(1 - f) \tan \phi] \tag{F-8}$$

and use the updated β value in Eq. (F-7) to obtain an updated value for ϕ.

Once the value for latitude has been computed, the ellipsoid height can be computed by

$$h = \frac{\sqrt{X^2 + Y^2}}{\cos \phi} - N \tag{F-9}$$

In Eq. (F-9), N must be computed from Eq. (F-4) while using the final value of ϕ. As the latitude approaches the pole, the cosine term in Eq. (F-9) approaches zero, causing the solution for h to be unstable. In that case, Eq. (F-3) can be rearranged to solve for h.

Example F-1. Given the following values of latitude, longitude, and ellipsoid height for a point, and the parameters for the GRS80 ellipsoid, compute the XYZ geocentric coordinates of the point. Check by performing the conversion in reverse.

$$\phi = 29°47'12.31265'' \text{ north}$$

$$\lambda = 81°39'08.75520'' \text{ west}$$

$$h = 35.478 \text{ m}$$

Solution. Compute value of normal N by Eq. (F-4).

$$N = \frac{6,378,137}{\sqrt{1 - 0.006694380023 \sin^2 (29°47'12.31265'')}} = 6,383,412.083 \text{ m}$$

Compute XYZ coordinates by Eqs. (F-1) through (F-3).

$$X = (6,383,412.083 + 35.478) \cos (29°47'12.31265'') \cos (-81°39'08.75520'')$$

$$= 804,294.265 \text{ m}$$

$$Y = (6,383,412.083 + 35.478) \cos (29°47'12.31265'') \sin (-81°39'08.75520'')$$

$$= -5,481,374.865 \text{ m}$$

$$Z = [(6,383,412.083) (1 - 0.006694380023) + 35.478] \sin (29°47'12.31265'')$$

$$= 3,149,897.861 \text{ m}$$

Check. Compute the value of longitude by Eq. (F-5).

$$\lambda = \tan^{-1}\left(\frac{-5,481,374.865}{804,294.265}\right) = -81°39'08.75520'' \qquad Check!$$

Compute an initial value for β, using Eq. (F-6).

$$\beta = \tan^{-1}\left[\frac{3,149,897.861}{(1 - 1/298.25722210088)(5,540,068.562)}\right] = 29.703863135°$$

where

$$\sqrt{X^2 + Y^2} = \sqrt{(804,294.265)^2 + (-5,481,374.865)^2} = 5,540,068.562$$

Compute the initial value for ϕ, using Eq. (F-7)

$$\phi = \tan^{-1}\left[\frac{3,149,897.861 + 0.006739496775\,(6,356,752.314)\,\sin^3(29.703863135°)}{5,540,068.562 - 6,378,137\,(0.006694380023)\,\cos^3(29.703863135°)}\right]$$

$$= 29.786753514°$$

Since this example was based on a point at the earth's surface, the resulting latitude value should be sufficiently accurate; however, for demonstration purposes an additional iteration will be performed.

Compute an updated value for β, using Eq. (F-8).

$$\beta = \tan^{-1}[(1 - 1/298.25722210088)\tan(29.786753514°)] = 29.703862211°$$

Compute an updated value for ϕ, using Eq. (F-7).

$$\phi = \tan^{-1}\left[\frac{3,149,897.861 + 0.006739496775\,(6,356,752.314)\sin^3(29.703862211°)}{5,540,068.562 - 6,378,137\,(0.006694380023)\cos^3(29.703862211°)}\right]$$

$$= 29.786753514° = 29°47'12.31265'' \qquad Check!$$

The final step is to compute the ellipsoid height. Using the value for ϕ computed above, compute the length of the normal N.

$$N = \frac{6,378,137}{\sqrt{1 - 0.006694380023\,\sin^2(29.786753514°)}} = 6,383,412.083 \text{ m}$$

Compute the value for h by Eq. (F-9).

$$h = \frac{5,540,068.562}{\cos(29.786753514°)} - 6,383,412.083 = 35.478 \text{ m} \qquad Check!$$

F-4   CONVERSION BETWEEN GEOCENTRIC AND LOCAL VERTICAL COORDINATES

Geocentric and local vertical coordinates are both three-dimensional cartesian coordinate systems. The steps required to convert between these systems involve only rotations and translations. Figure 5-6 shows the relationship between the geocentric XYZ and local vertical $X_l Y_l Z_l$ systems. Initially, two parameters ϕ_o and λ_o are selected which specify the geodetic coordinates of the origin for the local vertical system.

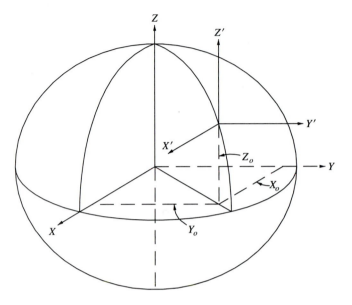

FIGURE F-2
The $X'Y'Z'$ coordinate system translated to the local vertical origin.

Step 1: Translation to Local Vertical Origin

Beginning with the geocentric XYZ axes with the origin at the center of the earth, a new system $X'Y'Z'$ with its axes parallel to the geocentric axes has been translated by X_o, Y_o, and Z_o such that its origin is at the local vertical system origin. This new system is illustrated in Fig. F-2. Equations (F-10) express the translated coordinates of a point P in the $X'Y'Z'$ system in terms of its $X_P Y_P Z_P$ geocentric coordinates after the translation.

$$X'_P = X_P - X_o$$

$$Y'_P = Y_P - Y_o \hspace{3cm} \text{(F-10)}$$

$$Z'_P = Z_P - Z_o$$

In Eqs. (F-10), X_o, Y_o, and Z_o are the coordinates of the local vertical origin computed from Eqs. (F-1) through (F-3) using the values of ϕ_o and λ_o, and with $h_o = 0$.

Step 2: Rotation of $90° + \lambda_o$ about the Z' Axis

Figure F-3 illustrates the geocentric XYZ axes and the $X'Y'Z'$ axes as viewed from positive Z looking toward the earth's equatorial plane. A rotation of $90° + \lambda_o$ about the Z' axis is applied to the $X'Y'Z'$ system to create an $X''Y''Z''$ system with its Y'' axis in the direction of local north (i.e., pointing toward the pole in the meridian through the origin). Equations (F-11) express the coordinates of point P in the $X''Y''Z''$ system in terms

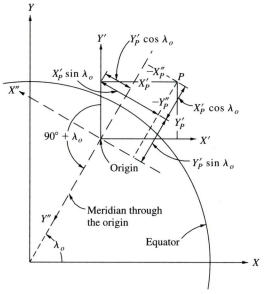

FIGURE F-3
Rotation of $90° + \lambda_o$ about the Z' axis.

of X'_P, Y'_P, Z'_P, and the longitude of the local origin λ_o.

$$X''_P = -X'_P \sin \lambda_o + Y'_P \cos \lambda_o$$

$$Y''_P = -X'_P \cos \lambda_o - Y'_P \sin \lambda_o \qquad \text{(F-11)}$$

$$Z''_P = Z'_P$$

Step 3: Rotation of $90° - \phi_o$ about the X'' Axis

Figure F-4 illustrates the $X''Y''Z''$ system as viewed from the positive X'' axis toward the local vertical origin. A rotation of $90° - \phi_o$ will be applied to align the Z'' axis with the normal to the local vertical origin. This gives the final $X_l Y_l Z_l$ system. Equations (F-12) express the coordinates of point P in the $X_l Y_l Z_l$ system in terms of X''_P, Y''_P, Z''_P, and the latitude of the local vertical origin ϕ_o.

$$X_{l_P} = X''_P$$

$$Y_{l_P} = Y''_P \sin \phi_o + Z''_P \cos \phi_o \qquad \text{(F-12)}$$

$$Z_{l_P} = -Y''_P \cos \phi_o + Z''_P \sin \phi_o$$

Substituting Eqs. (F-10) into Eqs. (F-11) and in turn substituting the result into Eqs. (F-12), and expressing the result in matrix form, gives

$$
\begin{bmatrix} X_{l_P} \\ Y_{l_P} \\ Z_{l_P} \end{bmatrix}
=
\begin{bmatrix} 1 & 0 & 0 \\ 0 & \sin \phi_o & \cos \phi_o \\ 0 & -\cos \phi_o & \sin \phi_o \end{bmatrix}
\begin{bmatrix} -\sin \lambda_o & \cos \lambda_o & 0 \\ -\cos \lambda_o & -\sin \lambda_o & 0 \\ 0 & 0 & 1 \end{bmatrix}
\begin{bmatrix} X_P - X_o \\ Y_P - Y_o \\ Z_P - Z_o \end{bmatrix}
\qquad \text{(F-13)}
$$

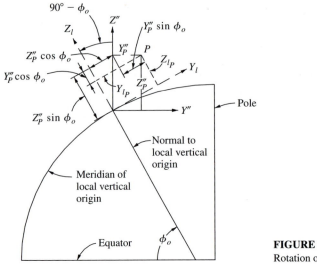

FIGURE F-4
Rotation of $90° - \phi_o$ about the X'' axis.

Combining the two rotation matrices from Eq. (F-13) into a single rotation matrix M and dropping the P subscripts gives

$$\begin{bmatrix} X_l \\ Y_l \\ Z_l \end{bmatrix} = \begin{bmatrix} m_{11} & m_{12} & m_{13} \\ m_{21} & m_{22} & m_{23} \\ m_{31} & m_{32} & m_{33} \end{bmatrix} \begin{bmatrix} X - X_o \\ Y - Y_o \\ Z - Z_o \end{bmatrix} = M \begin{bmatrix} X - X_o \\ Y - Y_o \\ Z - Z_o \end{bmatrix} \tag{F-14}$$

where

$$m_{11} = -\sin \lambda_o$$

$$m_{12} = \cos \lambda_o$$

$$m_{13} = 0$$

$$m_{21} = -\sin \phi_o \cos \lambda_o$$

$$m_{22} = -\sin \phi_o \sin \lambda_o$$

$$m_{23} = \cos \phi_o$$

$$m_{31} = \cos \phi_o \cos \lambda_o$$

$$m_{32} = \cos \phi_o \sin \lambda_o$$

$$m_{33} = \sin \phi_o$$

Equation (F-14) gives the final form of the equation for converting coordinates from geocentric XYZ to local vertical $X_l Y_l Z_l$. In an optional step, false XYZ offsets can be added as a convenience to cause all coordinates within the project area to be positive.

To convert coordinates of points from local vertical $X_l Y_l Z_l$ to geocentric XYZ, Eq. (F-14) is rearranged as follows:

$$\begin{bmatrix} X \\ Y \\ Z \end{bmatrix} = M^{-1} \begin{bmatrix} X_l \\ Y_l \\ Z_l \end{bmatrix} + \begin{bmatrix} X_o \\ Y_o \\ Z_o \end{bmatrix} \qquad\text{(F-15)}$$

Matrix M is an orthogonal matrix which has the property that its inverse is equal to its transpose, and therefore Eq. (F-15) can be rewritten as

$$\begin{bmatrix} X \\ Y \\ Z \end{bmatrix} = M^{T} \begin{bmatrix} X_l \\ Y_l \\ Z_l \end{bmatrix} + \begin{bmatrix} X_o \\ Y_o \\ Z_o \end{bmatrix} \qquad\text{(F-16)}$$

Equation (F-16) gives the final form of the equation for converting coordinates from local vertical $X_l Y_l Z_l$ to geocentric XYZ. If the optional false XYZ offsets were previously added, they must be subtracted before Eq. (F-16) is used.

Example F-2. Convert the following XYZ geocentric coordinates to local vertical coordinates $X_l Y_l Z_l$ having the parameters listed below. Use ellipsoid parameters for GRS80. Check by performing the conversion in reverse.

$$X = 815,519.912 \text{ m}$$

$$Y = -5,482,723.662 \text{ m}$$

$$Z = 3,144,707.279 \text{ m}$$

$$\phi_o = 29°45'00'' \text{ north}$$

$$\lambda_o = 81°35'00'' \text{ west}$$

Solution. Compute the length of the normal to the local vertical origin by Eq. (F-4).

$$N_o = \frac{6{,}378{,}137}{\sqrt{1 - 0.006694380023 \sin^2(29°45')}} = 6{,}383{,}400.249 \text{ m}$$

Compute geocentric coordinates of the local vertical origin by Eqs. (F-1) through (F-3).

$$X_o = 6{,}383{,}400.249 \cos(29°45') \cos(-81°35') = 811{,}195.778 \text{ m}$$

$$Y_o = 6{,}383{,}400.249 \cos(29°45') \sin(-81°35') = -5{,}482{,}371.413 \text{ m}$$

$$Z_o = 6{,}383{,}400.249(1 - 0.006694380023) \sin(29°45') = 3{,}146{,}343.779 \text{ m}$$

Compute rotation matrix terms from Eq. (F-14).

$$m_{11} = -\sin(-81°35') = 0.98922979728$$

$$m_{12} = \cos(-81°35') = 0.14637079003$$

$$m_{13} = 0$$

$$m_{21} = -\sin(29°45')\cos(-81°35') = -0.072631601671$$

$$m_{22} = -\sin(29°45')\sin(-81°35') = 0.49087215134$$

$$m_{23} = \cos(29°45') = 0.86819881449$$

$$m_{31} = \cos(29°45')\cos(-81°35') = 0.12707894638$$

$$m_{32} = \cos(29°45')\sin(-81°35') = -0.85884813725$$

$$m_{33} = \sin(29°45') = 0.49621650368$$

$$M = \begin{bmatrix} 0.98922979728 & 0.14637079003 & 0 \\ -0.072631601671 & 0.49087215134 & 0.86819881449 \\ 0.12707894638 & -0.85884813725 & 0.49621650368 \end{bmatrix}$$

Now solve Eq. (F-14), using the above rotation matrix M, for the local vertical coordinates.

$$\begin{bmatrix} X_l \\ Y_l \\ Z_l \end{bmatrix} = M \begin{bmatrix} 815{,}519.912 - 811{,}195.778 \\ -5{,}482{,}723.662 - (-5{,}482{,}371.413) \\ 3{,}144{,}707.279 - 3{,}146{,}343.779 \end{bmatrix} = \begin{bmatrix} 4226.003 \text{ m} \\ -1907.785 \text{ m} \\ 39.976 \text{ m} \end{bmatrix}$$

Check. Convert back to the original geocentric coordinates, using Eq. (F-16).

$$\begin{bmatrix} X \\ Y \\ Z \end{bmatrix} = M^T \begin{bmatrix} 4226.003 \\ -1907.785 \\ 39.976 \end{bmatrix} + \begin{bmatrix} 811{,}195.778 \\ -5{,}482{,}371.413 \\ 3{,}146{,}343.779 \end{bmatrix} = \begin{bmatrix} 815{,}519.912 \text{ m} \\ -5{,}482{,}723.662 \text{ m} \\ 3{,}144{,}707.279 \text{ m} \end{bmatrix} \quad \textit{Check!}$$

F-5 CONVERSION BETWEEN GEODETIC AND LAMBERT CONFORMAL CONIC COORDINATES

Conversion of coordinates between the geodetic system of latitude and longitude, and the Lambert conformal conic map projection, involves complex mathematics which are beyond the scope of this text. The reader may consult references at the end of this appendix for details. The equations presented in this section are from Snyder (1987).

As mentioned in Sec. 5-6, a Lambert conformal conic coordinate zone has a number of defining constants in addition to the required standard ellipsoid parameters. These defining constants are, respectively, ϕ_1 and ϕ_2, the latitudes of the standard parallels; ϕ_o

and λ_o, the latitude and longitude of the grid origin; and E_o and N_o, the false easting and false northing.

A forward procedure that can convert latitude and longitude ϕ and λ of a point to X and Y Lambert conformal conic coordinates starts by computing a number of additional parameters which will be constant for a given set of zone constants and ellipsoid parameters. They are

$$n = \frac{\ln m_{\phi_1} - \ln m_{\phi_2}}{\ln t_{\phi_1} - \ln t_{\phi_2}} \tag{F-17}$$

$$F = \frac{m_{\phi_1}}{n t_{\phi_1}^n} \tag{F-18}$$

$$\rho_o = aFt_{\phi_o}^n \tag{F-19}$$

where

$$m_\phi = \frac{\cos \phi}{1 - \sqrt{e^2 \sin^2 \phi}} \tag{F-20}$$

$$t_\phi = \sqrt{\frac{1 - \sin \phi}{1 + \sin \phi} \left(\frac{1 + e \sin \phi}{1 - e \sin \phi} \right)^e} \tag{F-21}$$

and

a = length of semimajor axis

e = first eccentricity

$\ln$ = natural log function

After these initial parameters have been computed, the remainder of the forward procedure is as follows. Beginning with the latitude ϕ and longitude λ of a point, compute the following:

$$\rho = aFt_\phi^n \tag{F-22}$$

$$\theta = n(\lambda - \lambda_o) \tag{F-23}$$

$$X = \rho \sin \theta + E_o \tag{F-24}$$

$$Y = \rho_o - \rho \cos \theta + N_o \tag{F-25}$$

An inverse procedure that can convert XY Lambert conformal conic coordinates of a point to latitude ϕ and longitude λ requires the same additional parameters that were computed in Eqs. (F-17) through (F-19). Given these parameters, and the XY coordinates, the latitude and longitude ϕ and λ can be computed as follows:

$$\rho = \text{sign}(n) \sqrt{(X - E_o)^2 + (\rho_o - Y + N_o)^2} \tag{F-26}$$

$$t = \left(\frac{\rho}{aF} \right)^{1/n} \tag{F-27}$$

$$\theta = \tan^{-1} \left[\frac{\text{sign}(n)(X - E_o)}{\text{sign}(n)(\rho_o - Y + N_o)} \right] \tag{F-28}$$

$$\lambda = \frac{\theta}{n} + \lambda_o \qquad (F\text{-}29)$$

$$\phi \approx 90° - 2\tan^{-1}(t) \qquad (F\text{-}30)$$

$$\phi = 90° - 2\tan^{-1}\left[t\left(\frac{1 - e\sin\phi}{1 + e\sin\phi}\right)^{e/2}\right] \qquad (F\text{-}31)$$

where
a = length of semimajor axis

e = first eccentricity

In Eqs. (F-26) and (F-28) the function sign(n) is +1 if n [obtained from Eq. (F-17)] is positive, and it is −1 if n is negative. The inverse tangent function in Eq. (F-28) should be of full-circle range since θ can theoretically range from −180° to +180°, although in practical cases the range will be far less. Equation (F-30) provides an initial approximation for ϕ. This value is then used on the right-hand side of Eq. (F-31) to compute a better approximation for ϕ. This and other successively better approximations can be substituted on the right side of Eq. (F-31) until the value for ϕ stabilizes.

Example F-3. Given the following values ϕ and λ for latitude and longitude of a point, and the specified zone constants for a Lambert conformal conic projection, compute the X and Y coordinates for the point. Use the parameters for the GRS80 ellipsoid. Check by performing the inverse conversion.

$$\phi = 30°50'44.78012'' \text{ north}$$

$$\lambda = 82°02'39.90825'' \text{ west}$$

$$\phi_1 = 29°35'00'' \text{ north}$$

$$\phi_2 = 30°45'00'' \text{ north}$$

$$\phi_o = 29°00'00'' \text{ north}$$

$$\lambda_o = 84°30'00'' \text{ west}$$

$$E_o = 600,000 \text{ m}$$

$$N_o = 0$$

Solution. Compute required m_ϕ values by Eq. (F-20).

$$m_{\phi_1} = \frac{\cos(29°35')}{\sqrt{1 - (0.006694380023)^2 \sin^2(29°35')}} = 0.8703489007$$

$$m_{\phi_2} = \frac{\cos(30°45')}{\sqrt{1 - (0.006694380023)^2 \sin^2(30°45')}} = 0.8601594016$$

Compute required t_ϕ values by Eq. (F-21).

$$t_{\phi_o} = \sqrt{\left(\frac{1 - \sin 29°}{1 + \sin 29°}\right)\left(\frac{1 + 0.08181919104 \sin 29°}{1 - 0.08181919104 \sin 29°}\right)^{0.08181919104}} = 0.5909608745$$

$$t_{\phi_1} = \sqrt{\left(\frac{1 - \sin 29°35'}{1 + \sin 29°35'}\right)\left(\frac{1 + 0.08181919104 \sin 29°35'}{1 - 0.08181919104 \sin 29°35'}\right)^{0.08181919104}} = 0.5841370260$$

$$t_{\phi_2} = \sqrt{\left(\frac{1 - \sin 30°45'}{1 + \sin 30°45'}\right)\left(\frac{1 + 0.08181919104 \sin 30°45'}{1 - 0.08181919104 \sin 30°45'}\right)^{0.08181919104}} = 0.5706072177$$

Compute the additional parameters by Eqs. (F-17) through (F-19).

$$n = \frac{\ln 0.8703489007 - \ln 0.8601594016}{\ln 0.5841370260 - \ln 0.5706072177} = 0.5025259027$$

$$F = \frac{0.8703489007}{0.5025259027 \, (0.5841370260)^{0.5025259027}} = 2.269171531$$

$$\rho_o = 6{,}378{,}137 \, (2.269171531)(0.5909608745)^{0.5025259027} = 11{,}111{,}265.207 \text{ m}$$

Compute the t_ϕ value required in Eq. (F-22).

$$t_\phi = \sqrt{\left(\frac{1 - \sin 30°50'44.78012''}{1 + \sin 30°50'44.78012''}\right)\left(\frac{1 + 0.08181919104 \sin 30°50'44.78012''}{1 - 0.08181919104 \sin 30°50'44.78012''}\right)^{0.08181919104}}$$

$$= 0.5695034001$$

Compute values for Eqs. (F-22) through (F-25).

$$\rho = 6{,}378{,}137 \, (2.269171531) \, (0.5695034001)^{0.5025259027} = 10{,}906{,}659.071 \text{ m}$$

$$\theta = 0.5025259027 \, [-82°02'39.90825'' - (-84°30'00'')] = 1.2339930795°$$

$$X = 10{,}906{,}659.071 \sin 1.2339930795° + 600{,}000 = 834{,}881.198 \text{ m}$$

$$Y = 11{,}111{,}265.207 - (10{,}906{,}659.071 \text{ m}) \cos (1.2339930795°) + 0$$

$$= 207{,}135.581 \text{ m}$$

Check. The check will be performed by inverse conversion, back to ϕ and λ. Compute values for ρ, t, and θ by Eqs. (F-26) through (F-28).

$$\rho = +1 \sqrt{(834{,}881.198 - 600{,}000)^2 + (11{,}111{,}265.207 - 207{,}135.581 + 0)^2}$$

$$= 10{,}906{,}659.071 \text{ m}$$

$$t = \left(\frac{10{,}906{,}659.071}{6{,}378{,}137 \times 2.269171531}\right)^{1/0.5025259027} = 0.5695034001$$

$$\theta = \tan^{-1}\left[\frac{+1(834{,}881.198 - 600{,}000)}{+1(11{,}111{,}265.207 - 207{,}135.581 + 0)}\right] = 1.2339930795°$$

Compute longitude by Eq. (F-29).

$$\lambda = \frac{1.2339930795°}{0.5025259027} + (-84°30') = -82°02'39.90825'' \qquad Check!$$

Now for the iterative part. First compute an initial approximation using Eq. (F-30).

$$\phi \approx 90° - 2 \tan^{-1} (0.5695034001) = 30.676680106°$$

Apply Eq. (F-31) to compute updated values for ϕ until it converges.

$$\phi = 90° - 2 \tan^{-1} \left[0.5695034001 \left(\frac{1 - 0.08181919104 \sin 30.676680106°}{1 + 0.08181919104 \sin 30.676680106°} \right)^{0.08181919104/2} \right]$$

$$= 30.844935670°$$

$$\phi = 90° - 2 \tan^{-1} \left[0.5695034001 \left(\frac{1 - 0.08181919104 \sin 30.844935670°}{1 + 0.08181919104 \sin 30.844935670°} \right)^{0.08181919104/2} \right]$$

$$= 30.845768120°$$

$$\phi = 90° - 2 \tan^{-1} \left[0.5695034001 \left(\frac{1 - 0.08181919104 \sin 30.845768120°}{1 + 0.08181919104 \sin 30.845768120°} \right)^{0.08181919104/2} \right]$$

$$= 30.845772235°$$

$$\phi = 90° - 2 \tan^{-1} \left[0.5695034001 \left(\frac{1 - 0.08181919104 \sin 30.845772235°}{1 + 0.08181919104 \sin 30.845772235°} \right)^{0.08181919104/2} \right]$$

$$= 30.845772255°$$

$$\phi = 90° - 2 \tan^{-1} \left[0.5695034001 \left(\frac{1 - 0.08181919104 \sin 30.845772255°}{1 + 0.08181919104 \sin 30.845772255°} \right)^{0.08181919104/2} \right]$$

$$= 30.845772256°$$

The next iteration produces a negligible change. Therefore,

$$\phi = 30.845772256° = 30°50'44.78012'' \qquad Check!$$

F-6 CONVERSION BETWEEN GEODETIC AND TRANSVERSE MERCATOR COORDINATES

Conversion of coordinates between the geodetic system of latitude and longitude and the transverse Mercator map projection also involves complex mathematics which are beyond the scope of this text. Again, the reader may consult references at the end of this appendix for details. The equations presented in this section are also from Snyder (1987).

As mentioned in Sec. 5-7, a transverse Mercator coordinate zone has a number of defining constants in addition to the required standard ellipsoid parameters. These defining constants are k_o, the scale factor along the central meridian; ϕ_o and λ_o, the latitude and longitude, respectively, of the grid origin; and E_o and N_o, the false easting and false northing. The longitude of the grid origin λ_o is conventionally referred to as the longitude of the central meridian.

Unlike the Lambert conformal conic conversion equations, those of the transverse Mercator involve series expansions, truncated to a limited number of terms. As such, the accuracy is not limited strictly to the number of significant figures in the computations; instead, it is limited by the truncation of the series. When the following procedures are used, the accuracy should be satisfactory (i.e., subcentimeter) as long as the points are limited to the defined region of the particular transverse Mercator zone. Beyond that, the accuracy deteriorates, particularly in the direction of longitude. The reader is therefore cautioned to use these conversion procedures only within the defined limits of the zone.

A key parameter involved in transverse Mercator conversions is the meridional distance M from the equator to a specific latitude ϕ. Calculation of M can be performed by a truncated series expansion which is given in the following equation.

$$
\begin{aligned}
M = a\bigg[& \bigg(1 - \frac{e^2}{4} - \frac{3e^4}{64} - \frac{5e^6}{256}\bigg)\phi - \bigg(\frac{3e^2}{8} + \frac{3e^4}{32} + \frac{45e^6}{1024}\bigg)\sin 2\phi \\
& + \bigg(\frac{15e^4}{256} + \frac{45e^6}{1024}\bigg)\sin 4\phi - \frac{35e^6}{3072}\sin 6\phi\bigg]
\end{aligned}
$$
(F-32)

In Eq. (F-32), a is the semimajor axis, and e the eccentricity. The value of ϕ (latitude) in the first term must be in radians. This equation is accurate to within 1 mm for any latitude.

A forward procedure that can convert latitude ϕ and longitude λ of a point to X and Y transverse Mercator coordinates begins by computing the following preliminary quantities T, C, and A.

$$T = \tan^2 \phi \tag{F-33}$$

$$C = e'^2 \cos^2 \phi \tag{F-34}$$

$$A = (\lambda - \lambda_o)\cos \phi \qquad (\lambda \text{ and } \lambda_o \text{ in radians}) \tag{F-35}$$

where e' = second eccentricity and λ_o = longitude of the grid origin (central meridian). Next, compute the values

$$N = \text{length of normal at latitude } \phi \qquad \text{by Eq. (F-4)}$$

$$M = \text{meridional distance at latitude } \phi \qquad \text{by Eq. (F-32)}$$

$$M_o = \text{meridional distance at latitude } \phi_o \qquad \text{by Eq. (F-32)}$$

The following equations complete the forward conversion to X and Y.

$$X = k_o N\bigg[A + (1 - T + C)\frac{A^3}{6} + (5 - 18T + T^2 + 72C - 58e'^2)\frac{A^5}{120}\bigg] + E_o \tag{F-36}$$

$$
\begin{aligned}
Y = k_o\bigg\{ & M - M_o + N\tan\phi\bigg[\frac{A^2}{2} + (5 - T + 9C + 4C^2)\frac{A^4}{24} \\
& + (61 - 58T + T^2 + 600C - 330e'^2)\frac{A^6}{720}\bigg]\bigg\} + N_o
\end{aligned}
\tag{F-37}
$$

where $\quad k_o$ = scale factor at central meridian

$\qquad e'$ = second eccentricity

$\qquad E_o, N_o$ = false easting and false northing, respectively

An inverse procedure that can convert transverse Mercator coordinates X and Y of a point to latitude ϕ and longitude λ begins by computing the *footprint latitude ϕ_1*. The footprint latitude is the latitude at the central meridian that has the same Y coordinate as the point being converted. The footprint latitude is computed by the following equations:

$$M = M_o + \frac{Y - N_o}{k_o} \tag{F-38}$$

$$\mu = \frac{M}{a(1 - e^2/4 - 3e^4/64 - 5e^6/256)} \tag{F-39}$$

$$e_1 = \frac{1 - \sqrt{1 - e^2}}{1 + \sqrt{1 - e^2}} \tag{F-40}$$

$$\phi_1 = \mu + \left(\frac{3e_1}{2} - \frac{27e_1^3}{32}\right) \sin(2\mu) + \left(\frac{21e_1^2}{16} - \frac{55e_1^4}{32}\right) \sin(4\mu)$$

$$+ \frac{151e_1^3}{96} \sin(6\mu) + \frac{1097e_1^4}{512} \sin(8\mu) \tag{F-41}$$

In Eq. (F-38), M_o is the meridional distance from the equator to the latitude of the grid origin ϕ_o, as computed by Eq. (F-32); N_o is the false northing; and k_o is the scale factor at the central meridian. The variable μ, in Eqs. (F-39) and (F-41), is the *rectifying latitude* and has units of radians. The units of ϕ_1 in Eqs. (F-41) are radians.

Next, compute R_1, the radius of the meridional arc at the footprint latitude.

$$R_1 = \frac{a(1 - e^2)}{(1 - e^2 \sin^2 \phi_1)^{3/2}} \tag{F-42}$$

Then compute the parameters C_1, T_1, and N_1 corresponding to the footprint latitude ϕ_1 by Eqs. (F-34), (F-33), and (F-4), respectively.

$$C_1 = e'^2 \cos^2 \phi_1$$

$$T_1 = \tan^2 \phi_1$$

$$N_1 = \frac{a}{\sqrt{1 - e^2 \sin^2 \phi_1}}$$

Now compute the parameter D.

$$D = \frac{X - E_o}{N_1 k_o} \tag{F-43}$$

The inverse conversion is then completed by solving the following equations:

$$\phi = \phi_1 - \frac{N_1 \tan \phi_1}{R_1}\left[\frac{D^2}{2} - (5 + 3T_1 + 10C_1 - 4C_1^2 - 9e'^2)\frac{D^4}{24}\right.$$

$$\left. + (61 + 90T_1 + 298C_1 + 45T_1^2 - 252e'^2 - 3C_1^2)\frac{D^6}{720}\right] \qquad \text{(F-44)}$$

$$\lambda = \lambda_o + \frac{1}{\cos \phi_1}\left[D - (1 + 2T_1 + C_1)\frac{D^3}{6}\right.$$

$$\left. + (5 - 2C_1 + 28T_1 - 3C_1^2 + 8e'^2 + 24T_1^2)\frac{D^5}{120}\right] \qquad \text{(F-45)}$$

Example F-4. Given the following values for latitude ϕ and longitude λ of a point, and the specified zone constants for the universal transverse Mercator zone 17 projection, compute the X and Y coordinates for the point. Use the parameters for the GRS80 ellipsoid. Check by performing the inverse conversion.

$$\phi = 27°59'07.15355'' \text{ north}$$

$$\lambda = 82°31'57.04696'' \text{ west}$$

$$\phi_o = 0°00'00'' \text{ (equator)}$$

$$\lambda_o = 81°00'00'' \text{ west}$$

$$E_o = 500,000 \text{ m}$$

$$N_o = 0$$

$$k_o = 0.9996$$

Solution. Convert ϕ and λ to radians.

$$\phi = 0.48843598374 \text{ rad}$$

$$\lambda = -1.4404640926 \text{ rad}$$

Compute values T, C, and A by Eqs. (F-33) through (F-35).

$$T = \tan^2(27°59'07.15355'') = 0.28236559298$$

$$C = 0.006739496775 \cos^2(27°59'07.15355'') = 0.005255519029$$

$$A = [-1.4404640926 - (-1.4137166941)] \cos(27°59'07.15355'')$$
$$= -0.023619767532$$

Compute the value of N by Eq. (F-4) and M and M_o by Eq. (F-32).

$$N = \frac{6,378,137}{\sqrt{1 - 0.006694380023 \sin^2(27°59'07.15355'')}} = 6,382,843.029 \text{ m}$$

$$M = 6{,}378{,}137\{[1 - 0.006694380023/4 - 3(0.006694380023)^2/64$$
$$- 5(0.006694380023)^3/256]0.48843598374$$
$$- [3(0.006694380023)/8 - 3(0.006694380023)^2/32$$
$$- 45(0.006694380023)^3/1024] \sin[2(27°59'07.15355'')]$$
$$+ [15(0.006694380023)^2/256$$
$$+ 45(0.006694380023)^3/1024] \sin[4(27°59'07.15355'')]$$
$$- [35(0.006694380023)^3/3072] \sin[6(27°59'07.15355'')]\}$$
$$= 3{,}096{,}814.967 \text{ m}$$

$$M_o = 0$$

Now solve Eqs. (F-36) and (F-37) for X and Y.

$$X = 0.9996(6{,}382{,}843.029)\{-0.023619767532$$
$$+ (1 - 0.28236559298 + 0.005255519029)$$
$$\times (-0.023619767532)^3/6 + [5 - 18(0.28236559298)$$
$$+ (0.28236559298)^2 + 72(0.005255519029)$$
$$- 58(0.006739496775)] (-0.023619767532)^5/120\} + 500{,}000$$
$$= 349{,}288.906 \text{ m}$$

$$Y = 0.9996\{3{,}096{,}814.967 - 0 + 6{,}382{,}843.029 \tan(27°59'07.15355'')$$
$$\times [(-0.023619767532)^2/2 + [5 - 0.28236559298$$
$$+ 9(0.005255519029) + 4(0.005255519029)^2]$$
$$\times (-0.023619767532)^4/24 + [61 - 58(0.28236559298)$$
$$+ (0.28236559298)^2 + 600(0.005255519029)$$
$$- 330(0.006739496775)] (-0.023619767532)^6/720]\} + 0$$
$$= 3{,}096{,}522.182 \text{ m}$$

Check. First, compute the footprint latitude by solving Eqs. (F-38) through (F-41).

$$M = 0 + \frac{3096522.182 - 0}{0.9996} = 3{,}097{,}761.286$$

$$\mu = \frac{3{,}097{,}761.286}{6{,}378{,}137\left[1 - \dfrac{0.006694380023}{4} - \dfrac{3(0.006694380023)^2}{64} - \dfrac{5(0.006694380023)^3}{256}\right]}$$

$$= 0.48649957230 \text{ rad} = 27.874372228°$$

$$e_1 = \frac{1 - \sqrt{1 - 0.006694380023}}{1 + \sqrt{1 - 0.006694380023}} = 0.001679220395$$

$$\phi_1 = 0.48649957230$$

$$+ \left[\frac{3(0.001679220395)}{2} - \frac{27(0.001679220395)^3}{32} \right] \sin[2(27.874372228°)]$$

$$+ \left[\frac{21(0.001679220395)^2}{16} - \frac{55(0.001679220395)^4}{32} \right] \sin[4(27.874372228°)]$$

$$+ \frac{151(0.001679220395)^3}{96} \sin[6(27.874372228°)]$$

$$+ \frac{1097(0.001679220395)^4}{512} \sin[8(27.874372228°)]$$

$$= 0.48858502260 \text{ rad} = 27.993859728°$$

Compute R_1 by Eq. (F-42).

$$R_1 = \frac{6,378,137(1 - 0.006694380023)}{[1 - 0.006694380023 \sin^2(27.993859728°)]^{3/2}} = 6,349,481.139 \text{ m}$$

Compute C_1, T_1, and N_1 by Eqs. (F-34), (F-33), and (F-4), respectively.

$$C_1 = 0.006739496775 \cos^2(27.993859728°) = 0.005254686508$$

$$T_1 = \tan^2(27.993859728°) = 0.28256876307$$

$$N_1 = \frac{6,378,137}{\sqrt{1 - 0.006694380023 \sin^2(27.993859728°)}} = 6,382,845.672 \text{ m}$$

Compute parameter D by Eq. (F-43).

$$D = \frac{349,288.906 - 500,000}{6,382,845.672(0.9996)} = -0.02362134537$$

Compute ϕ and λ by Eqs. (F-44) and (F-45).

$$\phi = 27.993859728° - \frac{6,382,845.672 \tan 27.993859728°}{6,349,481.139} \left\{ \frac{(-0.02362134537)^2}{2} \right.$$

$$- [5 + 3(0.28256876307) + 10(0.005254686508) - 4(0.005254686508)^2$$

$$- 9(0.006739496775)] \frac{(-0.02362134537)^4}{24} + [61 + 90(0.28256876307)$$

$$+ 298(0.005254686508) + 45(0.28256876307)^2 - 252(0.006739496775)$$

$$\left. - 3(0.005254686508)^2] \frac{(-0.02362134537)^6}{720} \right\}$$

$$= 27.985320430° = 27°59'07.15355''   \textit{Check!}$$

$$
\lambda = -81° + \frac{1}{\cos 27.993859728°}\left\{ -0.02362134537 \right.
$$

$$
- [1 + 2(0.28256876307) + 0.005254686508]
$$

$$
\times \frac{(-0.02362134537)^3}{6} + [5 - 2(0.005254686508)
$$

$$
+ 28(0.28256876307) - 3(0.005254686508)^2
$$

$$
+ 8(0.006739496775) + 24(0.28256876307)^2]
$$

$$
\left. \times \frac{(-0.02362134537)^5}{120} \right\} = -82.532513044°
$$

$$
= -82°31'57.04696" \quad \textit{Check!}
$$

REFERENCES

American Society of Photogrammetry: *Manual of Photogrammetry,* 4th ed., Bethesda, MD, 1980, chaps. 8 and 9.

Anderson, J. M., and E. M. Mikhail: *Surveying: Theory and Practice,* WCB/McGraw-Hill, New York, 1998.

Bomford, G.: *Geodesy,* 4th ed., Clarendon Press, Oxford, 1980.

Ewing, C. E., and M. M. Mitchell: *Introduction to Geodesy,* Elsevier, New York, 1970.

Snyder, J. P.: *Map Projections—A Working Manual,* U.S. Geological Survey Professional Paper 1395, U.S. Geological Survey, Washington, 1987.

Stern, J. E.: *State Plane Coordinate System of 1983,* NOAA Manual NOS NGS 5, National Oceanic and Atmospheric Administration, Rockville, MD, 1989.

Wolf, P. R., and R. C. Brinker: *Elementary Surveying,* 9th ed., HarperCollins, New York, 1993.

PROBLEMS

F-1. Prepare a computer spreadsheet that will convert geodetic latitude, longitude, and ellipsoid height to geocentric *XYZ* coordinates. Using the spreadsheet, convert the following geodetic coordinates to geocentric. Use the GRS80 ellipsoid.
(*a*) $\phi = 42°06'52.3245"$ north, $\lambda = 93°33'40.2165"$ west, $h = 276.409$ m
(*b*) $\phi = 37°02'41.2227"$ north, $\lambda = 122°31'56.1093"$ west, $h = 153.131$ m
(*c*) $\phi = 27°59'06.9389"$ north, $\lambda = 86°56'11.0057"$ east, $h = 8847.734$ m
(*d*) $\phi = 64°25'30.1990"$ north, $\lambda = 152°01'43.6737"$ west, $h = 6193.536$ m

F-2. Prepare a computer spreadsheet that will convert geocentric *XYZ* coordinates to geodetic latitude, longitude, and ellipsoid height. Using the spreadsheet, convert the following geocentric coordinates to geodetic. Use the GRS80 ellipsoid.
(*a*) $X = -1,777,470.052$ m, $Y = 2,921,801.730$ m, $Z = 5,370,764.340$ m
(*b*) $X = 855,515.030$ m, $Y = -5,487,911.094$ m, $Z = 3,125,090.292$ m
(*c*) $X = -4,592,691.234$ m, $Y = 2,537,269.009$ m, $Z = -3,617,528.719$ m
(*d*) $X = 4,054,206.962$ m, $Y = 2996.527$ m, $Z = 4,907,448.862$ m

F-3. Prepare a computer spreadsheet that will convert geocentric *XYZ* coordinates to local vertical coordinates $X_l Y_l Z_l$. Using the spreadsheet, convert the following geocentric coordinates to local vertical.

Use the GRS80 ellipsoid and a local vertical origin latitude $\phi_o = 29°30'00''$ north and longitude $\lambda_o = 82°12'00''$ west.

(a) $X = 780,666.787$ m, $Y = -5,501,666.889$ m, $Z = 3,120,529.196$ m
(b) $X = 778,950.096$ m, $Y = -5,453,563.214$ m, $Z = 3,203,727.478$ m
(c) $X = 720,410.933$ m, $Y = -5,481,868.690$ m, $Z = 3,169,180.707$ m
(d) $X = 734,246.533$ m, $Y = -5,527,194.279$ m, $Z = 3,086,712.792$ m

F-4. Prepare a computer spreadsheet that will convert local vertical coordinates $X_l Y_l Z_l$ to geocentric XYZ coordinates. Using the spreadsheet, convert the following local vertical coordinates to geocentric. Use the GRS80 ellipsoid and a local vertical origin latitude $\phi_o = 29°24'00''$ north and longitude $\lambda_o = 81°50'00''$ west.

(a) $X_l = -18,728.640$ m, $Y_l = 14,779.927$ m, $Z_l = 1.981$ m
(b) $X_l = 7770.644$ m, $Y_l = 2278.638$ m, $Z_l = 45.884$ m
(c) $X_l = 5191.700$ m, $Y_l = -6019.912$ m, $Z_l = 43.388$ m
(d) $X_l = -25,078.884$ m, $Y_l = -20,805.035$ m, $Z_l = -30.919$ m

F-5. Prepare a computer spreadsheet that will convert geodetic latitude and longitude to Lambert conformal conic XY coordinates. Using the spreadsheet, convert the following geodetic coordinates to Lambert conformal conic. Use the GRS80 ellipsoid and the following Lambert zone constants: $\phi_1 = 42°44'$ north, $\phi_2 = 44°04'$ north, $\phi_o = 42°00'$ north, $\lambda_o = 90°00'$ west, $E_o = 600,000$ m, and $N_o = 0$.

(a) $\phi = 42°46'23.5233''$ north, $\lambda = 91°15'59.0517''$ west
(b) $\phi = 44°23'21.3480''$ north, $\lambda = 90°09'40.2099''$ west
(c) $\phi = 43°43'52.6206''$ north, $\lambda = 89°24'08.0503''$ west
(d) $\phi = 44°27'25.0788''$ north, $\lambda = 91°21'06.7792''$ west

F-6. Prepare a computer spreadsheet that will convert Lambert conformal conic XY coordinates to geodetic latitude and longitude. Using the spreadsheet, convert the following Lambert conformal conic coordinates to geodetic. Use the GRS80 ellipsoid and the following Lambert zone constants: $\phi_1 = 45°34'$ north, $\phi_2 = 46°46'$ north, $\phi_o = 45°10'$ north, $\lambda_o = 90°00'$ west, $E_o = 600,000$ m, and $N_o = 0$.

(a) $X = 498,109.633$ m, $Y = 142,262.591$ m
(b) $X = 565,889.445$ m, $Y = 85,870.635$ m
(c) $X = 616,933.581$ m, $Y = 40,649.790$ m
(d) $X = 534,986.172$ m, $Y = 184,266.842$ m

F-7. Prepare a computer spreadsheet that will convert geodetic latitude and longitude to transverse Mercator XY coordinates. Using the spreadsheet, convert the following geodetic coordinates to transverse Mercator. Use the GRS80 ellipsoid and the following transverse Mercator zone constants: $k_o = 0.9999411765$, $\phi_o = 24°20'$ north, $\lambda_o = 81°00'$ west, $E_o = 200,000$ m, and $N_o = 0$.

(a) $\phi = 28°37'49.5323''$ north, $\lambda = 82°02'45.0368''$ west
(b) $\phi = 26°38'39.5273''$ north, $\lambda = 81°49'47.0588''$ west
(c) $\phi = 27°25'47.3755''$ north, $\lambda = 81°46'50.8965''$ west
(d) $\phi = 25°18'30.9991''$ north, $\lambda = 80°44'19.2206''$ west

F-8. Prepare a computer spreadsheet that will convert universal transverse Mercator (UTM) XY coordinates to geodetic latitude and longitude. Using the spreadsheet, convert the following UTM coordinates to geodetic. Use the GRS80 ellipsoid and the following UTM zone constants: zone 17 $k_o = 0.9996$, $\phi_o = 0°00'$, $\lambda_o = 81°00'$ west, $E_o = 500,000$ m, and $N_o = 0$.

(a) $X = 502,831.680$ m, $Y = 3,588,953.376$ m
(b) $X = 394,100.551$ m, $Y = 3,257,118.082$ m
(c) $X = 465,252.738$ m, $Y = 2,801,516.948$ m
(d) $X = 367,122.167$ m, $Y = 3,016,206.144$ m

INDEX